Universitext

AF068933

Springer-Verlag Berlin Heidelberg GmbH

J. J. Duistermaat
J. A. C. Kolk

Lie Groups

Springer

J. J. Duistermaat
J. A. C. Kolk
Department of Mathematics
Utrecht University
3508 TA Utrecht
The Netherlands
e-mail: duis@math.uu.nl
kolk@math.uu.nl

Mathematics Subject Classification (1991):
22-01, 22E10, 22E15, 22E30, 22E45, 22E46, 22E60, 22E67, 43A80, 53C30, 57S25

Library of Congress Cataloging-in-Publication Data

Duistermaat, J. J. (Johannes Jisse), 1942-
Lie groups / Johannes J. Duistermaat, Johan A.C. Kolk.
p. cm. -- (Universitext)
Includes bibliographical references and index.
ISBN 978-3-540-15293-4 ISBN 978-3-642-56936-4 (eBook)
DOI 10.1007/978-3-642-56936-4
1. Lie groups. I. Kolk, Johan A. C., 1947- II. Title.

QA387 .D85 1999
512'.55--dc21
99-050197

This work is subject to copyright. All rights are reserved, whether the whole or part of the material is concerned, specifically the rights of translation, reprinting, reuse of illustrations, recitation, broad-casting, reproduction on microfilms or in any other way, and storage in data banks. Duplication of this publication or parts thereof is permitted only under the provisions of the German Copyright Law of September 9, 1965, in its current version, and permission for use must always be obtained from Springer-Verlag. Violations are liable for prosecution under the German Copyright Law.

© Springer-Verlag Berlin Heidelberg 2000
Originally published by Springer-Verlag Berlin Heidelberg New York in 2000

Typesetting: By the authors using a Springer TEX macro package
Cover design: *design & production* GmbH, Heidelberg

Printed on acid-free paper SPIN: 12076666 41/3180 5 4 3 2

Preface

This book is devoted to an exposition of the theory of finite-dimensional Lie groups and Lie algebras, which is a beautiful and central topic in modern mathematics. At the end of the nineteenth century this theory came to life in the works of Sophus Lie. It had its origins in Lie's idea of applying Galois theory to differential equations and in Klein's "Erlanger Programm" of treating symmetry groups as the fundamental objects in geometry. Lie's approach to many problems of analysis and geometry was mainly local, that is, valid in local coordinate systems only. At the beginning of the twentieth century É. Cartan and Weyl began a systematic treatment of the global aspects of Lie's theory. Since then this theory has ramified tremendously and now, as the twentieth century is coming to a close, its concepts and methods pervade mathematics and theoretical physics.

Despite the plethora of books devoted to Lie groups and Lie algebras we feel there is justification for a text that puts emphasis on Lie's principal idea, namely, geometry treated by a blend of algebra and analysis. Lie groups are geometrical objects whose structure can be described conveniently in terms of group actions and fiber bundles. Therefore our point of view is mainly differential geometrical. We have made no attempt to discuss systematically the theory of infinite-dimensional Lie groups and Lie algebras, which is currently an active area of research.

We now give a short description of the contents of each chapter.

Chapter 1 treats the fundamental properties of Lie groups and Lie algebras and their remarkable mathematical relations. It ends with a demonstration of Lie's third fundamental theorem, on the existence of a simply connected Lie group with a prescribed finite-dimensional Lie algebra. Our proof of this is not a standard one, in that we use the path space of the Lie algebra, which is a Banach Lie group of infinite dimension, and that we obtain the simply connected Lie group as a quotient of the path space by a suitable normal subgroup.

Proper actions of groups, in particular actions of compact groups, on manifolds are the main topic of Chapter 2. We introduce the stratification of the manifold into orbit types and into local action types and then show how blowing up simplifies the action.

In Chapter 3 the theory developed in Chapter 2 is used to study the structure of compact Lie groups. In particular, our proof of the conjugacy of maximal tori in a compact Lie group depends on results from Chapter 2. The complete classification of compact Lie groups is not given in this volume as it relies on special algebraic calculations. However, Chapter 3 does contain a discussion of nonconnected compact Lie groups.

The first part of Chapter 4 treats the main facts about representations of a general compact group, culminating in the Peter-Weyl theorem. In the second part, we classify all representations of a compact Lie group from both the infinitesimal and the global points of view. We give Weyl's formula for the character of a representation and Cartan's highest weight theorem. Also we realize every representation in the space of sections of a line bundle over an associated flag manifold. Again we end with a discussion of the case of nonconnected groups.

Apart from a certain amount of mathematical maturity, the formal prerequisites for reading this book are a familiarity with the rudiments of group theory, the basic aspects of analysis in several real variables, and the elementary concepts of differential geometry. The appendices contain a discussion of some more advanced notions from differential geometry and a formulation of some basic facts on ordinary differential equations on manifolds.

Each chapter ends with notes and references. Given the vastness of the subject we are certain to have missed many relevant contributions. Despite of this shortcoming, we hope that these notes provide the reader an accurate historical perspective.

The text contains references to chapters belonging to a future volume. Nevertheless the main body of the text can be studied independently of these references.

It is a pleasant duty to thank our friends and colleagues, in particular Tonny Springer, for many discussions concerning this text. André de Meijer was essential for solving numerous problems concerning computers and software. The second author is most grateful to his cardiologist, dr. H.W.M. Plokker, for contributing in his own ways to the completion of this book. Finally we want to record our indebtedness to the staff of Springer-Verlag at Heidelberg, and in particular to dr. C. Byrne, for patience and understanding throughout this long project.

J.J. Duistermaat
J.A.C. Kolk
Utrecht, October 29, 1999

Table of Contents

Chapter 1

Lie Groups and Lie Algebras

1.1 Lie Groups and their Lie Algebras

(1.1.1) Definition. *A Lie group is a group G that at the same time is a finite-dimensional manifold of differentiability class* C^2*, in such a way that the two group operations of G:*

$$\mu\colon (x, y) \mapsto xy\colon G \times G \to G \qquad \textit{(multiplication)}, \tag{1.1.1}$$

$$\iota\colon x \mapsto x^{-1} \quad : G \to G \qquad \textit{(inversion)}, \tag{1.1.2}$$

are C^2 *mappings: $G \times G \to G$, and $G \to G$, respectively.*

Later on we shall also discuss group satisfying weaker conditions, such as topological groups or infinite-dimensional transformation groups. (cf. Sections 11.5 or 11.1). On the other hand there are natural smoothness assumptions stronger than C^2, like real-analytic or real affine algebraic, or the complex-analytic and complex affine algebraic versions (Section 1.6 and Chapter 14). Using the implicit function theorem, one can show that condition (1.1.2) by itself is already sufficient.

The most basic example of a Lie group is the general linear group $\mathbf{GL}(V)$ of a finite-dimensional vector space V over $\mathbf{R}$. Write $\mathbf{L}(V, V)$ for the vector space of all linear mappings: $V \to V$. The group

$$\mathbf{GL}(V) = \{\, A \in \mathbf{L}(V, V) \mid \det A \neq 0 \,\} \tag{1.1.3}$$

of invertible linear transformations of V is an open subset of $\mathbf{L}(V, V)$, so it can be considered as a (real-analytic) manifold of dimension equal to $(\dim V)^2$, with only one coordinate chart. In this case the multiplication μ is the restriction to $\mathbf{GL}(V) \times \mathbf{GL}(V)$ of the mapping:

$$\mu\colon (A, B) \mapsto A \circ B\colon\ \mathbf{L}(V, V) \times \mathbf{L}(V, V) \to \mathbf{L}(V, V), \tag{1.1.4}$$

which is a quadratic polynomial in the coordinates, Furthermore, Cramer's formula exhibits the inversion ι as the restriction to $\mathbf{GL}(V)$ of a rational mapping; this is regular precisely on $\mathbf{GL}(V)$, because only the polynomial $A \mapsto \det A$ appears in the denominator. So this mapping certainly is C^2, making $\mathbf{GL}(V)$ into a Lie group according to Definition 1.1.1. In the meantime the reader will have guessed that the definition of real linear algebraic group will be such that $\mathbf{GL}(V)$ actually belongs to that much more special category, cf. Section 14.1.

The *identity element* of any group G will be denoted by 1, or 1_G if confusion might be possible. We denote the tangent space of the Lie group G at the identity element by:

$$\mathfrak{g} = \mathrm{T}_1 G. \tag{1.1.5}$$

Using that $x \mapsto xy$, and $y \mapsto xy$, is the identity if $y = 1$, and $x = 1$, respectively, and combining this with:

$$\begin{pmatrix} X \\ Y \end{pmatrix} = \begin{pmatrix} X \\ 0 \end{pmatrix} + \begin{pmatrix} 0 \\ Y \end{pmatrix},$$

one obtains:

$$\mathrm{T}_{1,1}\,\mu\colon (X, Y) \mapsto X + Y : \mathfrak{g} \times \mathfrak{g} \to \mathfrak{g}. \tag{1.1.6}$$

Applying this to $x \mapsto x\iota(x) \equiv 1$ and using the sum rule for differentiation, one also gets:

$$\mathrm{T}_1\,\iota\colon X \mapsto -X : \mathfrak{g} \to \mathfrak{g}. \tag{1.1.7}$$

These two mappings just define the *additive group structure* of $\mathfrak{g}$, making $\mathfrak{g}$ into a commutative Lie group, sometimes denoted by $(\mathfrak{g}, +)$, of the same dimension as G.

In order to detect a possible noncommutativity of G on the infinitesimal level at 1 (that is, in terms of finitely many derivatives at 1), we therefore have to turn to second-order derivatives at 1. This can be done as follows. Write, for each $x \in G$,

$$\mathbf{Ad}\,x\colon y \mapsto xyx^{-1}, \tag{1.1.8}$$

for the *conjugation* by x in the group G. Noncommutativity of G means that $\mathbf{Ad}\,x$ is not equal to the identity: $G \to G$, for each $x \in G$. Because $(\mathbf{Ad}\,x)(1) = 1$, the tangent mapping of $\mathbf{Ad}\,x$ at 1 is a linear mapping:

$$\mathrm{Ad}\,x := \mathrm{T}_1(\mathbf{Ad}\,x)\colon \mathfrak{g} \to \mathfrak{g}, \tag{1.1.9}$$

called the *adjoint mapping* of x, or the *infinitesimal conjugation* by x in $\mathfrak{g}$. Because:

$$\mathbf{Ad}(ab) = (\mathbf{Ad}\,a) \circ (\mathbf{Ad}\,b) \qquad (a, b \in G), \tag{1.1.10}$$

an application of the chain rule for tangent mappings shows that:

$$\text{(1.1.11)} \qquad \operatorname{Ad}(ab) = (\operatorname{Ad} a) \circ (\operatorname{Ad} b) \qquad (a, b \in G),$$

as well. That is, the mapping

$$\text{(1.1.12)} \qquad \operatorname{Ad}\colon x \mapsto \operatorname{Ad} x\colon G \to \mathbf{GL}(\mathfrak{g})$$

is a homomorphism of groups; it is called the *adjoint representation* of G in $\mathfrak{g} = \mathrm{T}_1\, G$.

The next step is to observe that $\operatorname{Ad}\colon G \to \mathbf{GL}(\mathfrak{g})$ is a C^1 mapping. So again we can take the tangent mapping at 1, and we obtain a linear mapping:

$$\text{(1.1.13)} \qquad \operatorname{ad} := \mathrm{T}_1 \operatorname{Ad}\colon \mathfrak{g} \to \mathbf{L}(\mathfrak{g}, \mathfrak{g}),$$

if we identify the tangent space at I of the open subset $\mathbf{GL}(\mathfrak{g})$ of $\mathbf{L}(\mathfrak{g}, \mathfrak{g})$ with $\mathbf{L}(\mathfrak{g}, \mathfrak{g})$, as is usual. For each $X, Y \in \mathfrak{g}$:

$$\text{(1.1.14)} \qquad [X, Y] := (\operatorname{ad} X)(Y) \in \mathfrak{g}$$

is called the *Lie bracket* of X and Y. The condition that ad is a linear mapping from $\mathfrak{g}$ into $\mathbf{L}(\mathfrak{g}, \mathfrak{g})$ just means that:

$$\text{(1.1.15)} \qquad (X, Y) \mapsto [X, Y]\colon \mathfrak{g} \times \mathfrak{g} \to \mathfrak{g}$$

is a bilinear mapping; hence it can be viewed as a product structure turning $\mathfrak{g}$ into an algebra over $\mathbf{R}$.

(1.1.2) Definition. *The tangent space $\mathfrak{g} = \mathrm{T}_1\, G$, provided with the Lie bracket (1.1.14) as the product structure, is called the* Lie algebra *of the Lie group G.*

In the example $G = \mathbf{GL}(V)$ of the general linear group, we get:

$$\text{(1.1.16)} \qquad \mathfrak{g} = \mathrm{T}_{\mathrm{I}}\, \mathbf{GL}(V) = \mathbf{L}(V, V),$$

$$\text{(1.1.17)} \qquad \operatorname{Ad} x\colon Y \mapsto x \circ Y \circ x^{-1} \qquad (x \in \mathbf{GL}(V),\ Y \in \mathbf{L}(V, V)).$$

So in this case $\mathbf{Ad}\, x$ is just the restriction of $\operatorname{Ad} x$ to the open subset $\mathbf{GL}(V)$ of $\mathbf{L}(V, V)$. Anyhow, differentiating the right hand side in (1.1.17) with respect to x, at $x = \mathrm{I}$ and in the direction of $X \in \mathbf{L}(V, V)$, we get:

$$\text{(1.1.18)} \qquad [X, Y] = (\operatorname{ad} X)(Y) = X \circ Y - Y \circ X,$$

the *commutator* of X and Y in $\mathbf{L}(V, V)$.

Having determined the Lie bracket in $\mathbf{L}(V, V)$, considered as the Lie algebra of $\mathbf{GL}(V)$, let us now collect some elementary facts about the Lie bracket for general Lie groups.

(1.1.3) Proposition. *Let G, and G', be Lie groups with Lie algebras $\mathfrak{g}$, and $\mathfrak{g}'$, respectively, and suppose that Φ is a group homomorphism: $G \to G'$ that is differentiable at $1 \in G$. Write $\phi := \mathrm{T}_1 \Phi$. Then:*

(a) $\phi \circ \mathrm{Ad}\, x = \mathrm{Ad}\, \Phi(x) \circ \phi$, *for every $x \in G$.*

(b) ϕ *is a homomorphism of Lie algebras: $\mathfrak{g} \to \mathfrak{g}'$. That is, it is a linear mapping: $\mathfrak{g} \to \mathfrak{g}'$ such that in addition:*

$$\mathrm{T}_1 \Phi([X, Y]_{\mathfrak{g}}) = [\mathrm{T}_1 \Phi(X), \mathrm{T}_1 \Phi(Y)]_{\mathfrak{g}'} \qquad (X, Y \in \mathfrak{g}). \tag{1.1.19}$$

Proof. Differentiating:

$$\Phi\big((\mathbf{Ad}\, x)(y)\big) = \Phi(xyx^{-1}) = \Phi(x)\Phi(y)\Phi(x)^{-1} = \big(\mathbf{Ad}\, \Phi(x)\big)(\Phi(y))$$

with respect to y at $y = 1$, in the direction of $Y \in \mathfrak{g}$, we get:

$$\mathrm{T}_1 \Phi(\mathrm{Ad}\, x(Y)) = \big(\mathrm{Ad}\, \Phi(x)\big)(\mathrm{T}_1 \Phi(Y)).$$

Differentiating this with respect to x at $x = 1$, in the direction of $X \in \mathfrak{g}$ (and again using the chain rule), we obtain (1.1.19). □

(1.1.4) Theorem. *Let $(X, Y) \mapsto [X, Y]$ be the Lie bracket in the Lie algebra $\mathfrak{g}$ of a Lie group G. Then:*

$$[X, Y] = -[Y, X] \qquad (X, Y \in \mathfrak{g}), \tag{1.1.20}$$

$$[[X, Y], Z] = [X, [Y, Z]] - [Y, [X, Z]] \qquad (X, Y, Z \in \mathfrak{g}). \tag{1.1.21}$$

Proof. Because of (1.1.12) Ad is a C^1 homomorphism: $G \to \mathbf{GL}(\mathfrak{g})$, hence Proposition 1.1.3 gives $\mathrm{ad} = \mathrm{T}_1 \mathrm{Ad}$ is a homomorphism of Lie algebras: $\mathfrak{g} \to \mathbf{L}(\mathfrak{g}, \mathfrak{g})$. Recall now that the Lie algebra structure of $\mathbf{L}(\mathfrak{g}, \mathfrak{g})$ is given by the commutator (1.1.18), therefore:

$$\mathrm{ad}[X, Y] = \mathrm{ad}\, X \circ \mathrm{ad}\, Y - \mathrm{ad}\, Y \circ \mathrm{ad}\, X, \tag{1.1.22}$$

for all $X, Y \in \mathfrak{g}$. Applying this to $Z \in \mathfrak{g}$ we get (1.1.21). For (1.1.20) we observe that:

$$(x, y) \mapsto xyx^{-1}y^{-1}, \quad \text{the } \textit{group commutator}, \tag{1.1.23}$$

defines a C^2 mapping: $G \times G \to G$, with tangent mapping at $(1, 1)$ equal to zero. (This expresses the commutativity of $(\mathfrak{g}, +)$.) As a consequence, its second-order derivative at (1,1) is an intrinsically defined symmetric bilinear mapping from $(\mathfrak{g} \times \mathfrak{g}) \times (\mathfrak{g} \times \mathfrak{g})$ into $\mathfrak{g}$. The derivative of (1.1.23) with respect to y at $y = 1$, in the direction of $Y \in \mathfrak{g}$, is equal to $(\mathrm{Ad}\, x - \mathrm{I})(Y)$. Then the derivative of this with respect to x at $x = 1$, in the direction of $X \in \mathfrak{g}$, is equal to $[X, Y]$. Applying these differentiations in the opposite order, we first get $(\mathrm{I} - \mathrm{Ad}\, y)(X)$, and then $-[Y, X]$. Because the order of differentiations can be interchanged for C^2 mappings (which is equivalent to the symmetry of the second-order derivative as a bilinear mapping), the conclusion (1.1.20) follows. □

Using (1.1.20), we can write Formula (1.1.21) also in the form:

$$[X,[Y,Z]]+[Y,[Z,X]]+[Z,[X,Y]]=0 \qquad (X,Y,Z\in\mathfrak{g}). \tag{1.1.24}$$

Note that the second and third term in the left hand side of (1.1.24) can be obtained from the first one by performing all cyclic permutations of X, Y and Z; this makes (1.1.24) easier to memorize than (1.1.21).

(1.1.5) Definition. *A real Lie algebra is a vector space* $\mathfrak{g}$ *over* **R**, *together with a bilinear mapping* $(X,Y)\mapsto[X,Y]\colon \mathfrak{g}\times\mathfrak{g}\to\mathfrak{g}$, *called the Lie bracket of* $\mathfrak{g}$, *satisfying (1.1.20) (which is called the anti-symmetry), and (1.1.24) (the Jacobi identity).*

(1.1.6) Definition. *For later use we also mention here the complex version. A complex Lie algebra is a vector space* $\mathfrak{g}$ *over* **C** *together with a Lie bracket (anti-symmetric and satisfying the Jacobi identity) that is a* **complex** *bilinear mapping:* $\mathfrak{g}\times\mathfrak{g}\to\mathfrak{g}$.

In terms of Definition 1.1.5, Theorem 1.1.4 says that the Lie algebra of a Lie group is a (real, finite-dimensional) Lie algebra. However, the surprising fact is that, at least locally near the identity element, the group structure of a Lie group G can be recovered completely from the algebraic information contained in the Lie bracket on $\mathfrak{g}$, which a priori only reflects some information about second-order derivatives at 1 of the group structure. How this comes about will be explained in the next sections. This stimulates the study of Lie groups by studying their Lie algebras as a problem in (multi-)linear algebra, because in principle no local information is lost this way.

Another motivation for the name "Lie algebra" in Definition 1.1.5 is that every finite-dimensional real Lie algebra actually is the Lie algebra of a Lie group, see Theorem 1.14.3.

In the next section we will consider a few simple, basic examples of noncommutative Lie groups G in quite some detail. In particular, we shall pay attention to the *conjugacy classes*:

$$C(y)=\{\,xyx^{-1}\mid x\in G\,\} \tag{1.1.25}$$

of the elements $y\in G$; their size expresses the noncommutativity of G. (Note that $C(y)=\{y\}$ if and only if $xy=yx$ for all $x\in G$, that is, if and only if y commutes with every element of G.) Already in these simple examples, manifolds occur with an interesting differential geometric structure, showing that it is not just a pedantic mathematical desire for generality to study groups in terms of "global differential geometric" concepts like manifolds. Of course, if one wants to proceed faster with the general theory, this section with explicit calculations may be skipped, since it is meant mainly as an illustration.

1.2 Examples

1.2.A The Real Orthogonal Groups and Rotation Groups

The group:

$$\mathbf{O}(n,\mathbf{R}) = \{\, A \in \mathbf{L}(\mathbf{R}^n,\mathbf{R}^n) \mid {}^tAA = \mathrm{I}\,\} \tag{1.2.1}$$

of orthogonal linear transformations A of $\mathbf{R}^n$ (here ${}^tA$ denotes the adjoint of A) is a smooth real-analytic submanifold of $\mathbf{L}(\mathbf{R}^n,\mathbf{R}^n)$. It is a Lie group, with Lie algebra equal to the space:

$$\mathfrak{o}(n,\mathbf{R}) = \{\, X \in \mathbf{L}(\mathbf{R}^n,\mathbf{R}^n) \mid {}^tX + X = 0\,\} \tag{1.2.2}$$

of anti-symmetric matrices. (So $\dim \mathbf{O}(n,\mathbf{R}) = \frac{1}{2}n(n-1)$.) The easiest proof is by showing that the total derivative of $A \mapsto {}^tAA$ maps $\mathbf{L}(\mathbf{R}^n,\mathbf{R}^n)$ onto the space of symmetric matrices, at each $A \in \mathbf{O}(n,\mathbf{R})$, and by then applying the implicit function theorem. For each $A \in \mathbf{O}(n,\mathbf{R})$, we have $\det A = \pm 1$; and $\mathbf{O}(n,\mathbf{R})$ does have elements A with $\det A = -1$, for instance: A a diagonal matrix with coefficients $-1, 1, \dots, 1$. The map $A \mapsto \det A$ is a homomorphism from $\mathbf{O}(n,\mathbf{R})$ to the multiplicative group $\{1,-1\}$; its kernel:

$$\mathbf{SO}(n,\mathbf{R}) = \{\, A \in \mathbf{O}(n,\mathbf{R}) \mid \det A = 1\,\} \tag{1.2.3}$$

is called the *special orthogonal group* or the *rotation group* in $\mathbf{R}^n$.

For $n = 2$,

$$\alpha \mapsto \begin{pmatrix} \cos\alpha & -\sin\alpha \\ \sin\alpha & \cos\alpha \end{pmatrix} \ (= \text{the rotation through the angle } \alpha) \tag{1.2.4}$$

is a homomorphism from $(\mathbf{R},+)$ onto $\mathbf{SO}(2,\mathbf{R})$, with kernel equal $2\pi\mathbf{Z}$. This leads to an identification of $\mathbf{SO}(2,\mathbf{R})$ with the circle $\mathbf{R}/2\pi\mathbf{Z}$, a commutative one-dimensional connected compact Lie group. For general n and each $A \in \mathbf{SO}(n,\mathbf{R})$, the Jordan normal form theory (in rudimentary form) leads to an orthogonal sum decomposition:

$$\mathbf{R}^n = \bigoplus_{i=1}^{k} P_i, \quad \text{if } n = 2k, \text{ and } \bigoplus_{i=1}^{k} P_i \oplus L, \quad \text{if } n = 2k+1, \tag{1.2.5}$$

where P_i, and L, are a two-dimensional, and a one-dimensional, linear subspace, respectively, such that $A|_{P_i}$ is a rotation: $P_i \to P_i$, say through an angle α_i, and A leaves L pointwise fixed. This decomposition into planar rotations explains the name rotation group; if n is odd, then L is called an *axis of rotation*, and it is uniquely determined if all $\alpha_i \notin 2\pi\mathbf{Z}$. Conversely, all A described in (1.2.5) clearly belong to $\mathbf{SO}(n, \mathbf{R})$. Replacing each α_i by $t\alpha_i$ and letting t run from 0 to 1, one sees that $\mathbf{SO}(n, \mathbf{R})$ is connected. Because both $\mathbf{SO}(n, \mathbf{R})$ and its complement in $\mathbf{O}(n, \mathbf{R})$ are closed, it follows that $\mathbf{SO}(n, \mathbf{R})$ **is the connected component of** $\mathbf{O}(n, \mathbf{R})$ containing the identity element.

As a final general remark, note that the operator norm of each $A \in \mathbf{O}(n, \mathbf{R})$ is equal to 1, so $\mathbf{O}(n, \mathbf{R})$ is a closed and bounded subset of $\mathbf{L}(\mathbf{R}^n, \mathbf{R}^n)$ and therefore **compact**. In Corollary 4.6.5 and Corollary 14.6.2 we shall see that actually every compact Lie group can be obtained as a closed analytic (even algebraic) subgroup of $\mathbf{O}(n, \mathbf{R})$, for some n; an analogue of the Whitney embedding theorem, which asserts that every smooth compact manifold can be smoothly embedded into the unit sphere of a Euclidean space of sufficiently high dimension.

Next to the planar case ($n = 2$), the best known example is the group $\mathbf{SO}(3, \mathbf{R})$ of rotations in $\mathbf{R}^3$ which, according to the results above, is a three-dimensional compact connected Lie group. Coordinates on $\mathbf{SO}(3, \mathbf{R})$, for instance near the identity element I, could be obtained by substituting $A = \mathrm{I} + B + C$, with B an arbitrary anti-symmetric matrix near 0 and C a symmetric one near 0, and solving the matrix $C = C(B)$ from the equation ${}^tAA = \mathrm{I}$. Such a parametrization however is not very well adapted to the geometric interpretation of the A as rotations in $\mathbf{R}^3$. For this purpose it is a better choice to consider, for every $x \in \mathbf{R}^3$ (see Section 1.4, (1.4.16-18), for a more elaborate explanation of these facts):

(1.2.6)
$$R_x := \mathrm{I} + \frac{\sin |x|}{|x|} A_x + \frac{1 - \cos |x|}{|x|^2} A_x^2,$$

where $|x| = (x_1^2 + x_2^2 + x_3^2)^{1/2}$ is the Euclidean length of x, and
(1.2.7)
$$A_x = \begin{pmatrix} 0 & -x_3 & x_2 \\ x_3 & 0 & -x_1 \\ -x_2 & x_1 & 0 \end{pmatrix}, \quad A_x^2 = \begin{pmatrix} -x_3^2 - x_2^2 & x_2 x_1 & x_3 x_1 \\ x_1 x_2 & -x_3^2 - x_1^2 & x_3 x_2 \\ x_1 x_3 & x_2 x_3 & -x_2^2 - x_1^2 \end{pmatrix}.$$

The point is that, if $|x| \notin 2\pi\mathbf{Z}$, then R_x is a rotation around the axis through x, through the angle $|x|$. Furthermore, if $0 < |x| < \pi$, then R_x is the unique rotation such that, for each nonzero vector y in the plane of rotation, the triple $y, R_x(y), x$ has positive orientation, that is, $\det(y, R_x(y), x) > 0$.

The mapping $x \mapsto R_x$ is surjective, from the closed ball:

(1.2.8)
$$\mathcal{B}(\pi) = \{\, x \in \mathbf{R}^3 \mid |x| \leq \pi \,\}$$

of radius π in $\mathbf{R}^3$, to $\mathbf{SO}(3, \mathbf{R})$. It is a diffeomorphism from the interior:

(1.2.9)
$$\mathcal{B}(\pi)^{\mathrm{int}} = \{\, x \in \mathbf{R}^3 \mid |x| < \pi \,\}$$

onto the open dense subset of $\mathbf{SO}(3, \mathbf{R})$ of rotations through an angle $< \pi$. In order to see what happens near the boundary sphere:

$$\mathcal{S}(\pi) = \{ x \in \mathbf{R}^3 \mid |x| = \pi \}, \tag{1.2.10}$$

one checks that $x \mapsto R_x$ is a smooth two-fold covering (see Definition A.3 in Appendix A) from $\mathcal{B}(2\pi)^{\text{int}} \setminus \{0\}$, mapping x and $(|x| - 2\pi)\frac{x}{|x|}$ to the same element of $\mathbf{SO}(3, \mathbf{R})$. In particular, $x \mapsto R_x$ is a smooth two-fold covering from the sphere $\mathcal{S}(\pi)$ onto the subset of rotations through the angle π, mapping the antipodal points x and $-x$ to the same element of $\mathbf{SO}(3, \mathbf{R})$. Identifying $\mathcal{B}(\pi)^{\text{int}}$ with $\mathbf{R}^3$, and $\mathcal{S}(\pi)$, with antipodal points identified, with the projective plane, attached to $\mathbf{R}^3$ at infinity, we see that $\mathbf{SO}(3, \mathbf{R})$ is diffeomorphic to the three-dimensional *real projective space* $\mathbf{RP}(3)$. Below, in the discussion of $\mathbf{SU}(2)$, we shall get another identification of $\mathbf{SO}(3, \mathbf{R})$ with $\mathbf{RP}(3)$ where $\mathbf{RP}(3)$ is viewed as the space of one-dimensional linear subspaces of $\mathbf{R}^4$, that is, the unit sphere in $\mathbf{R}^4$ with antipodal points identified.

The conjugacy classes in $\mathbf{SO}(3, \mathbf{R})$ are the sets:

$$\mathcal{C}_c = \{ R_x \mid x \in \mathbf{R}^3, |x| = c \} \qquad (0 \leq c \leq \pi), \tag{1.2.11}$$

since the eigenvalues determine the conjugacy classes for diagonalizable matrices. Observe that $\mathcal{C}_0 = \{1\}$, $\mathcal{C}_c$ is diffeomorphic (via $x \mapsto R_x$) to a sphere for $0 < c < \pi$, whereas $\mathcal{C}_\pi$, the conjugacy class of:

$$\begin{pmatrix} -1 & 0 & 0 \\ 0 & -1 & 0 \\ 0 & 0 & 1 \end{pmatrix},$$

is diffeomorphic to the real projective plane $\mathbf{RP}(2)$, that is, the sphere of dimension 2 with antipodal points identified. The dimension of the conjugacy classes is quite big, expressing the strong noncommutativity of $\mathbf{SO}(3, \mathbf{R})$.

If e_i, $(i = 1, 2, 3)$ is the standard basis in $\mathbf{R}^3$, the R_{te_3}, the rotations in the e_1, e_2-plane through the angle t, form a subgroup T of $\mathbf{SO}(3, \mathbf{R})$, isomorphic to $\mathbf{SO}(2, \mathbf{R}) \cong \mathbf{R}/2\pi\mathbf{Z}$, which intersects each conjugacy class of $\mathbf{SO}(3, \mathbf{R})$. In fact it intersects $\mathcal{C}_0$ and $\mathcal{C}_\pi$ in one point, and each $\mathcal{C}_c$, for $0 < c < \pi$, in two antipodal points. The subgroup T is commutative, and in some sense complementary to the conjugacy classes.

The curve $\gamma \colon t \mapsto R_{te_3}$, with t running from 0 to 2π, is a closed differentiable curve starting and ending at 1, intersecting the compact smooth submanifold $\mathcal{C}_\pi$ once and transversely. It follows that each continuous loop γ', starting and ending at 1, that is homotopic to γ, must intersect $\mathcal{C}_\pi$ as well; and this shows that the homotopy class of γ is not trivial.

On the other hand $2\gamma \colon t \mapsto R_{te_3}$, with t running from 0 to 4π, is homotopic to the constant curve $\gamma'(t) \equiv 1$. This can be seen by introducing a continuous curve $\delta(\tau)$ on the unit sphere in $\mathbf{R}^3$ such that $\delta(0) = e_3$, and $\delta(1) = -e_3$. Now let $\gamma_\tau(t) = \gamma(t)$ for $t \in [0, 2\pi]$, followed by $\gamma_\tau(t) = R_{t\delta(\tau)}$ for $t \in [0, 2\pi]$. This defines a homotopy from 2γ to the curve γ_1 which is

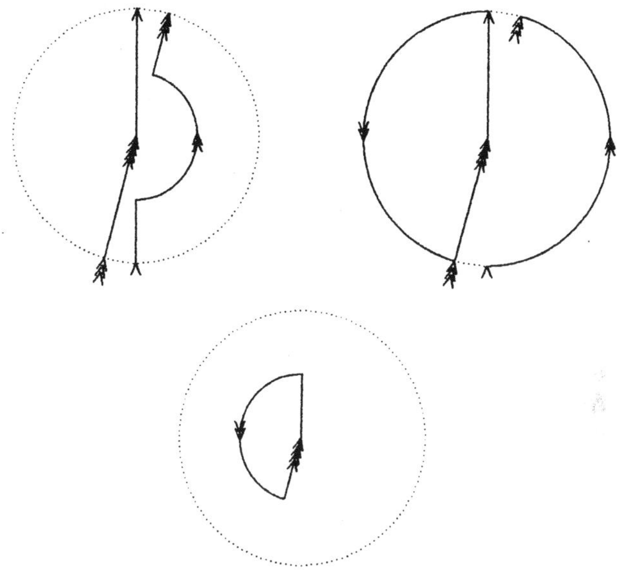

Fig. 1.2.1.

equal to γ followed by $-\gamma$, that is, γ traversed in the opposite direction. It is easy to see that γ_1 is homotopic to the constant curve. Combined with the information that $\mathbf{SO}(3,\mathbf{R}) \setminus \mathcal{C}_\pi$ is diffeomorphic to $\mathcal{B}(\pi)^{\text{int}}$, and hence contractible, the conclusion is that $\pi_1(\mathbf{SO}(3,\mathbf{R}))$, the fundamental group of $\mathbf{SO}(3,\mathbf{R})$, is isomorphic to $\mathbf{Z}/2\mathbf{Z}$, the group with 2 elements. (See Guillemin and Pollack [1974], p.35, or Greenberg [1967] for the elementary facts about intersection theory, fundamental groups and coverings which have been used in the discussion above). For a more systematic discussion of coverings of Lie groups, see Section 1.13, where the assertion $\pi_1(\mathbf{SO}(3,\mathbf{R})) \cong \mathbf{Z}/2\mathbf{Z}$ is confirmed in another way.

The strong noncommutativity of $\mathbf{SO}(3,\mathbf{R})$ is also reflected by the nontrivial structure of its Lie algebra $\mathfrak{so}(3,\mathbf{R}) = \mathfrak{o}(3,\mathbf{R})$ of antisymmetric 3×3-matrices. In fact, the mapping $x \mapsto A_x$ sets up an isomorphism between the Lie algebra $\mathbf{R}^3$ (with the exterior product as Lie bracket) and $\mathfrak{so}(3,\mathbf{R})$. In the notation of (1.2.7):

$$[A_{e_1}, A_{e_2}] = A_{e_3}, \quad [A_{e_2}, A_{e_3}] = A_{e_1}, \quad [A_{e_3}, A_{e_1}] = A_{e_2}. \tag{1.2.12}$$

Another indication of the strong noncommutativity of $\mathbf{SO}(3,\mathbf{R})$, even on the infinitesimal level at 1, is that the adjoint representation has kernel equal to 1, that is, $\mathrm{Ad}\colon x \mapsto \mathrm{Ad}\, x$ is a group isomorphism: $\mathbf{SO}(3,\mathbf{R}) \to \mathrm{Ad}\,\mathbf{SO}(3,\mathbf{R})$.

1.2.B The Unitary Groups and the Covering $\mathbf{SU}(2) \to \mathbf{SO}(3, \mathbf{R})$

If E and F are complex linear spaces, then we denote by $\mathbf{L}_{\mathbf{C}}(E, F)$ the space of complex-linear mappings: $E \to F$. Another classical group is the group of unitary transformations in $\mathbf{C}^n$:

$$\mathbf{U}(n) = \{ A \in \mathbf{L}_{\mathbf{C}}(\mathbf{C}^n, \mathbf{C}^n) \mid A^* A = \mathrm{I} \}, \tag{1.2.13}$$

where A^* denotes the adjoint of A with respect to the standard Hermitian inner product in $\mathbf{C}^n$. With analogous proofs as for $\mathbf{O}(n, \mathbf{R})$, we see that $\mathbf{U}(n)$ is a compact connected real-analytic submanifold of $\mathbf{L}_{\mathbf{C}}(\mathbf{C}^n, \mathbf{C}^n)$, hence a Lie group, called the *unitary group*, with Lie algebra equal to the space:

$$\mathfrak{u}(n) = \{ A \in \mathbf{L}_{\mathbf{C}}(\mathbf{C}^n, \mathbf{C}^n) \mid A^* + A = 0 \} \tag{1.2.14}$$

of anti-selfadjoint elements of $\mathbf{L}_{\mathbf{C}}(\mathbf{C}^n, \mathbf{C}^n)$. The real dimension of $\mathbf{U}(n)$ is equal to n^2. (Note that $\mathbf{U}(n)$ is not at all a complex-analytic submanifold of $\mathbf{L}_{\mathbf{C}}(\mathbf{C}^n, \mathbf{C}^n)$!). This can be seen by observing that each $A \in \mathbf{L}_{\mathbf{C}}(\mathbf{C}^n, \mathbf{C}^n)$ can be uniquely written as the sum of an anti-selfadjoint and a selfadjoint element of $\mathbf{L}_{\mathbf{C}}(\mathbf{C}^n, \mathbf{C}^n)$, and that $A \mapsto iA$ is a linear isomorphism from the space of self-adjoint elements to the space of anti-selfadjoint elements of $\mathbf{L}_{\mathbf{C}}(\mathbf{C}^n, \mathbf{C}^n)$. This shows that:

$$\dim_{\mathbf{R}} \mathbf{U}(n) = \dim_{\mathbf{R}} \mathfrak{u}(n) = \frac{1}{2} \dim_{\mathbf{R}} \mathbf{L}_{\mathbf{C}}(\mathbf{C}^n, \mathbf{C}^n) = \dim_{\mathbf{C}} \mathbf{L}_{\mathbf{C}}(\mathbf{C}^n, \mathbf{C}^n) = n^2.$$

In this case, for $n = 1$, the group:

$$\mathbf{U}(1) = \{ \text{multiplications in } \mathbf{C} \text{ by } z \in \mathbf{C} \text{ such that } |z| = 1 \} \tag{1.2.15}$$

is one-dimensional, and isomorphic to the unit circle in $\mathbf{C}$, viewed as a multiplicative group. Furthermore, for each n, the group $\mathbf{U}(n)$ contains the subgroup:

$$\mathrm{Z}(\mathbf{U}(n)) = \{ z\mathrm{I} \bigm| z \in \mathbf{C}, |z| = 1 \}, \tag{1.2.16}$$

isomorphic to $\mathbf{U}(1)$, as the subgroup of elements of $\mathbf{U}(n)$ that commute with all other elements of $\mathbf{U}(n)$. This one-dimensional "center" can be largely eliminated by passing to the *special unitary group*:

$$\mathbf{SU}(n) = \{ A \in \mathbf{U}(n) \mid \det A = 1 \}, \tag{1.2.17}$$

which is a compact connected Lie group of real dimension equal to $n^2 - 1$, with Lie algebra equal to the space:

$$\mathfrak{su}(n) = \{ A \in \mathbf{L}_{\mathbf{C}}(\mathbf{C}^n, \mathbf{C}^n) \mid A^* + A = 0 \text{ and } \operatorname{tr} A = 0 \} \tag{1.2.18}$$

of anti-selfadjoint complex $n \times n$-matrices with trace equal to zero. Note that:

$$\mathbf{SU}(n) \cap \mathrm{Z}(\mathbf{U}(n)) = \{ z\mathrm{I} \mid z \in \mathbf{C} \text{ and } z^n = 1 \} \cong \mathbf{Z}/n\mathbf{Z}.$$

We have $\mathbf{SU}(1) = \{\mathrm{I}\}$, but for $n = 2$ we already get the interesting group:

$$\mathbf{SU}(2) = \left\{ \begin{pmatrix} a & b \\ -\bar{b} & \bar{a} \end{pmatrix} \,\middle|\, a, b \in \mathbf{C},\ |a|^2 + |b|^2 = 1 \right\}; \tag{1.2.19}$$

it is diffeomorphic to the three-dimensional unit sphere in $\mathbf{C}^2 \cong \mathbf{R}^4$. One remarkable property is that:

$$\mathbf{R} \cdot \mathbf{SU}(2) = \left\{ \begin{pmatrix} a & b \\ -\bar{b} & \bar{a} \end{pmatrix} \,\middle|\, a, b \in \mathbf{C} \right\} \tag{1.2.20}$$

is a 4-dimensional real-linear subspace of the (real 8-dimensional) space of complex 2×2-matrices. This subspace is also closed under matrix multiplication and in this way forms a noncommutative division algebra. It has the basis:

$$1 = \begin{pmatrix} 1 & 0 \\ 0 & 1 \end{pmatrix}, \quad i = \begin{pmatrix} i & 0 \\ 0 & -i \end{pmatrix}, \quad j = \begin{pmatrix} 0 & 1 \\ -1 & 0 \end{pmatrix}, \quad k = \begin{pmatrix} 0 & i \\ i & 0 \end{pmatrix},$$

with the relations $i^2 = j^2 = k^2 = ijk = -1$. So $\mathbf{R} \cdot \mathbf{SU}(2)$ can be identified with Hamilton's system $\mathbf{H}$ of *quaternions*, and $\mathbf{SU}(2)$ with the multiplicative group of quaternions of Euclidean length equal to 1. The relations $ij = -ji = k$, $jk = -kj = i$, and $ki = -ik = j$ express the noncommutativity of $\mathbf{H}$ in a strong way.

The conjugacy classes in $\mathbf{SU}(2)$ are the sets:

$$C_c^\sim = \left\{ \begin{pmatrix} a & b \\ -\bar{b} & \bar{a} \end{pmatrix} \in \mathbf{SU}(2) \,\middle|\, \mathrm{Re}\, a = c \right\} \qquad (-1 \leq c \leq 1), \tag{1.2.21}$$

since the eigenvalues of $\begin{pmatrix} a & b \\ -\bar{b} & \bar{a} \end{pmatrix}$ are completely determined by $\mathrm{Re}\, a$. In particular, $C_1^\sim = \{\mathrm{I}\}$, $C_{-1}^\sim = \{-\mathrm{I}\}$, and for each $-1 < c < 1$, the set $C_c^\sim$ is a two-dimensional sphere. The one-dimensional circle group:

$$T^\sim = \left\{ \begin{pmatrix} a & 0 \\ 0 & \bar{a} \end{pmatrix} \,\middle|\, a \in \mathbf{C}, |a| = 1 \right\} \cong \mathbf{U}(1) \tag{1.2.22}$$

intersects each conjugacy class $C_c^\sim$ with $-1 < c < 1$ in two points, corresponding to a that are complex conjugate on $|a| = 1$.

The Lie algebra of $\mathbf{SU}(2)$ is equal to the space:

$$\mathfrak{su}(2) = \left\{ \begin{pmatrix} i\alpha & \beta \\ -\bar{\beta} & -i\alpha \end{pmatrix} \,\middle|\, \alpha \in \mathbf{R}, \beta \in \mathbf{C} \right\} \tag{1.2.23}$$

of anti-selfadjoint complex 2×2-matrices, with trace equal to zero, a real three-dimensional linear subspace of $\mathbf{L}_{\mathbf{C}}(\mathbf{C}^2, \mathbf{C}^2)$. Using the notation of (1.2.19), we get (compare Euler [1770/71], p.309):

$$\text{(1.2.24)}\quad \operatorname{Ad}\begin{pmatrix} a & b \\ -\bar b & \bar a \end{pmatrix} = \begin{pmatrix} |a|^2-|b|^2 & 2\operatorname{Im}(a\bar b) & 2\operatorname{Re}(a\bar b) \\ 2\operatorname{Im}(ab) & \operatorname{Re}(a^2+b^2) & -\operatorname{Im}(a^2-b^2) \\ -2\operatorname{Re}(ab) & \operatorname{Im}(a^2+b^2) & \operatorname{Re}(a^2-b^2) \end{pmatrix}.$$

Comparing this matrix with (1.2.6), (1.2.7), we see it can be identified with the rotation R_x, for:

$$\text{(1.2.25)}\quad x = \operatorname{sgn}(\operatorname{Re} a)\frac{\arccos\big(2(\operatorname{Re} a)^2-1\big)}{\sqrt{1-(\operatorname{Re} a)^2}}\begin{pmatrix} \operatorname{Im} a \\ \operatorname{Re} b \\ \operatorname{Im} b \end{pmatrix}.$$

It follows that the adjoint representation of $\mathbf{SU}(2)$ defines a group homomorphism from $\mathbf{SU}(2)$ onto $\mathbf{SO}(3,\mathbf{R})$ with kernel equal to $\{-\mathrm{I},\mathrm{I}\}$, so this is a two-fold covering of $\mathbf{SO}(3,\mathbf{R})$ with the three-dimensional sphere, where antipodal points of the sphere are mapped to the same element of $\mathbf{SO}(3,\mathbf{R})$. This leads to the identification of $\mathbf{SO}(3,\mathbf{R})$ with $\mathbf{RP}(3)$ announced following (1.2.10).

Note that the adjoint representation of $\mathbf{SU}(2)$ maps the two conjugacy classes (two-dimensional spheres) $\mathcal{C}_c^{\sim}$, $\mathcal{C}_{-c}^{\sim}$, for $0<c<1$, onto the conjugacy class $\mathcal{C}_\alpha$, with $\alpha = \arccos(2c^2-1)$ in $\mathbf{SO}(3,\mathbf{R})$, which is a sphere as well because $0<\alpha<\pi$. On the other hand it maps the two-dimensional sphere $\mathcal{C}_0^{\sim}$ as a two-fold cover onto $\mathcal{C}_\pi \cong \mathbf{RP}(2)$.

It seems that the main motivation for Hamilton for inventing his quaternions was his desire to introduce a multiplication of quantities $p = a+bi+cj$ with $i^2=j^2=-1$, such that $|pq| = |p|.|q|$. Formulae to compactly represent rotations appeared soon as a byproduct. For us, one point of the story is that, although $\mathbf{SU}(2)$ is geometrically simpler than $\mathbf{SO}(3,\mathbf{R})$ both as a manifold and with respect to the structure of its conjugacy classes, the group $\mathbf{SO}(3,\mathbf{R})$ nevertheless appears in the study of $\mathbf{SU}(2)$ as it adjoint group.

1.2.C SL(2, R)

We conclude this section with a basic example of a noncompact noncommutative group:

$$\text{(1.2.26)}\quad \mathbf{SL}(2,\mathbf{R}) = \left\{ \begin{pmatrix} a & b \\ c & d \end{pmatrix} \in \mathbf{L}(\mathbf{R}^2,\mathbf{R}^2) \mid ad-bc=1 \right\},$$

the simplest nontrivial one among the *special linear groups*:

$$\text{(1.2.27)}\quad \mathbf{SL}(n,\mathbf{R}) = \{\, A \in \mathbf{L}(\mathbf{R}^n,\mathbf{R}^n) \mid \det A = 1 \,\} \qquad (n \in \mathbf{Z}_{>0}).$$

Clearly $\mathbf{SL}(2,\mathbf{R})$ is a three-dimensional analytic submanifold of $\mathbf{L}(\mathbf{R}^2,\mathbf{R}^2)$, actually a hyperboloid; and therefore it is a three-dimensional Lie group with Lie algebra equal to:

$$\mathfrak{sl}(2, \mathbf{R}) = \left\{ \begin{pmatrix} \alpha & \beta \\ \gamma & -\alpha \end{pmatrix} \middle| \alpha, \beta, \gamma \in \mathbf{R} \right\}, \tag{1.2.28}$$

consisting of real 2×2-matrices with trace equal zero. (One may verify that in general $\mathbf{SL}(n, \mathbf{R})$ is a Lie group with Lie algebra equal to the space of $n \times n$-matrices with trace equal to zero).

To see the topological structure of $\mathbf{SL}(2, \mathbf{R})$, it is convenient to make the substitutions:

$$a = p + q, \quad d = p - q; \quad b = r + s, \quad c = r - s, \tag{1.2.29}$$

so that the equation $ad - bc = 1$ reads:

$$p^2 + s^2 = q^2 + r^2 + 1. \tag{1.2.30}$$

For each $(q, r) \in \mathbf{R}^2$, the point (p, s) runs over the circle in $\mathbf{R}^2$ around the origin with radius equal to $(q^2 + r^2 + 1)^{1/2}$. This makes $\mathbf{SL}(2, \mathbf{R})$ diffeomorphic to the Cartesian product of the circle and the plane. In order to make a picture we instead use the diffeomorphism:

$$(\theta, (u, v)) \mapsto (1 - u^2 - v^2)^{-1/2} \begin{pmatrix} \cos\theta + u & -\sin\theta + v \\ \sin\theta + v & \cos\theta - u \end{pmatrix} \tag{1.2.31}$$

between the Cartesian product of the θ-circle $\mathbf{R}/2\pi\mathbf{Z}$ and the unit disc:

$$D = \{ (u, v) \in \mathbf{R}^2 \mid u^2 + v^2 < 1 \}, \tag{1.2.32}$$

and $\mathbf{SL}(2, \mathbf{R})$.

The eigenvalues of $A \in \mathbf{SL}(2, \mathbf{R})$ are given by:

$$\lambda_{1,2} = \frac{\operatorname{tr} A \pm \sqrt{(\operatorname{tr} A)^2 - 4}}{2}; \tag{1.2.33}$$

so they are determined by prescribing:

$$\operatorname{tr} A = \frac{2\cos\theta}{\sqrt{1 - u^2 - v^2}} = c. \tag{1.2.34}$$

See Fig.1.2.2 below, where the varieties $\operatorname{tr} A = 2$, and $\operatorname{tr} A = -2$, corresponding to the case that $\lambda_1 = \lambda_2 = 1$, and $\lambda_1 = \lambda_2 = -1$, respectively, are drawn, because these are the only singular ones. In fact, they have a cone-like singularity at $A = \mathrm{I}$, and $A = -\mathrm{I}$, respectively. Clearly the elements A on $\operatorname{tr} A = 2$ with $A \neq \mathrm{I}$ are precisely the *unipotent* elements of $\mathbf{SL}(2, \mathbf{R})$. The "boundary at infinity" $u^2 + v^2 = 1$ does not belong to the group.

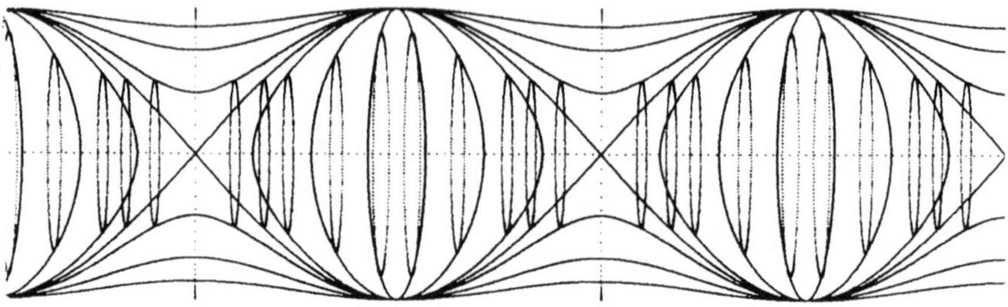

Fig. 1.2.2.

For $-2 < c < 2$, the variety $\operatorname{tr} A = c$ has two components, each of which is the conjugacy class of a rotation, but over opposite angles.

These two components look, for c close to 2 near $A = \mathrm{I}$, like the two sheets of the two-sheeted hyperboloid. For $c > 2$, and $c < -2$, the variety $\operatorname{tr} A = c$ has one component (for c near 2 looking near $A = \mathrm{I}$ like the one-sheeted hyperboloid), which is the conjugacy class of $\begin{pmatrix} \lambda & 0 \\ 0 & \lambda^{-1} \end{pmatrix}$ with $\lambda > 1$, and $\lambda < -1$, respectively.

Bending Fig.1.2.2 around in $\mathbf{R}^3$ in order to attach $\theta = -\pi$ to $\theta = \pi$, we get Fig.1.2.3 below. This figure exhibits $\mathbf{SL}(2,\mathbf{R})$ as the interior of a solid torus. We have also drawn the group T of rotations:

$$\left\{ \begin{pmatrix} \cos\theta & -\sin\theta \\ \sin\theta & \cos\theta \end{pmatrix} \,\middle|\, \theta \in \mathbf{R}/2\pi\mathbf{Z} \right\}$$

by $\cdots$; this group meets each conjugacy class in the region $|\operatorname{tr} A| < 2$ exactly once. These are called the *elliptic* elements of $\mathbf{SL}(2,\mathbf{R})$. Similarly, the group:

$$A = \left\{ \begin{pmatrix} \lambda & 0 \\ 0 & \lambda^{-1} \end{pmatrix} \,\middle|\, \lambda > 0 \right\}$$

and the set:

$$-A = \left\{ \begin{pmatrix} \lambda & 0 \\ 0 & \lambda^{-1} \end{pmatrix} \,\middle|\, \lambda < 0 \right\}$$

are drawn by $\cdots$; they intersect each conjugacy class in the region $|\operatorname{tr} A| > 2$ exactly twice. These are called the *hyperbolic* elements. T intersects A, and $-A$, in I, and in $-$I, respectively. Note that T, A and $-A$ miss precisely the two conjugacy classes of $\begin{pmatrix} 1 & 1 \\ 0 & 1 \end{pmatrix}$, $\begin{pmatrix} 1 & -1 \\ 0 & 1 \end{pmatrix}$ (the *unipotent* elements), and the two conjugacy classes of $\begin{pmatrix} -1 & 1 \\ 0 & -1 \end{pmatrix}$, $\begin{pmatrix} -1 & -1 \\ 0 & -1 \end{pmatrix}$.

The adjoint representation has $\{\mathrm{I}, -\mathrm{I}\}$ as its kernel, so the image $\operatorname{Ad}\mathbf{SL}(2,\mathbf{R})$ of $\mathbf{SL}(2,\mathbf{R})$ under the adjoint representation is isomorphic to $\mathbf{SL}(2,\mathbf{R})/\{\mathrm{I}, -\mathrm{I}\}$. A picture is given in Fig.1.2.4.

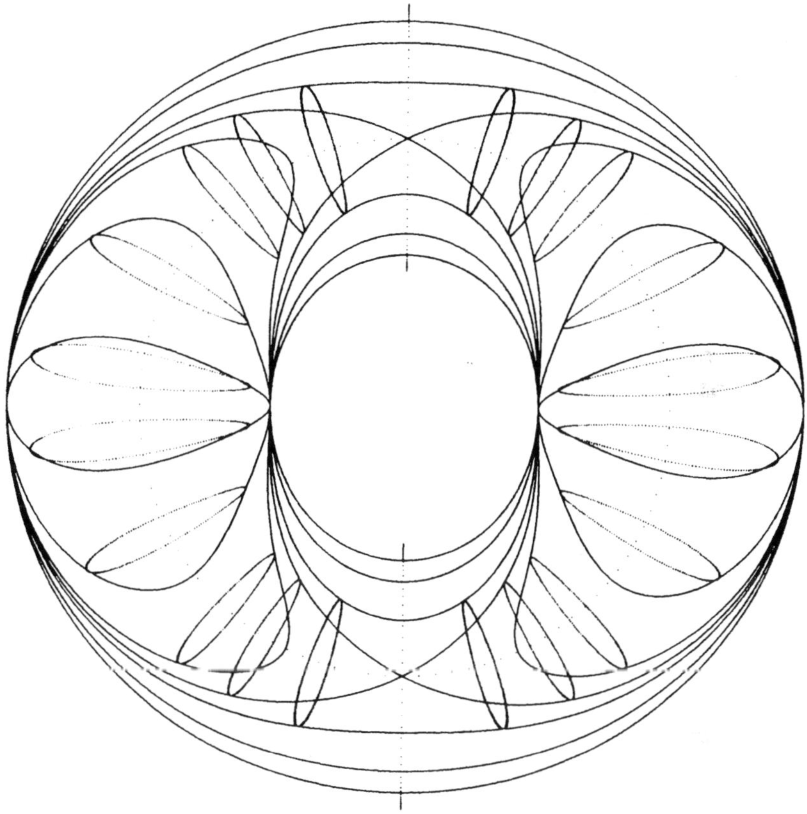

Fig. 1.2.3.

We conclude this section with an example of a two-dimensional noncommutative Lie group: the group of upper triangular elements with positive eigenvalues:

$$U = \left\{ \begin{pmatrix} a & b \\ 0 & a^{-1} \end{pmatrix} \,\middle|\, a > 0, b \in \mathbf{R} \right\}.$$

The conjugacy classes are here:

$$\left\{ \begin{pmatrix} 1 & 0 \\ 0 & 1 \end{pmatrix} \right\}, \quad \left\{ \begin{pmatrix} 1 & b \\ 0 & 1 \end{pmatrix} \,\middle|\, b > 0 \right\},$$
$$\left\{ \begin{pmatrix} 1 & b \\ 0 & 1 \end{pmatrix} \,\middle|\, b < 0 \right\}, \quad \left\{ \begin{pmatrix} a & b \\ 0 & a^{-1} \end{pmatrix} \,\middle|\, b \in \mathbf{R} \right\},$$

where a runs over the set $a > 0$ and $a \neq 1$. Of these, the second and third are the nonclosed ones.

Fig. 1.2.4.

1.3 The Exponential Map

We return to the general Lie group G of Section 1.1. For each $x \in G$, write:

(1.3.1) $\mathrm{L}(x)\colon y \mapsto xy\colon G \to G$ (left multiplication by x),

(1.3.2) $\mathrm{R}(x)\colon y \mapsto yx\colon G \to G$ (right multiplication by x).

The associative law $(xy)z = x(yz)$ is equivalent with each of the following rules:

(1.3.3) $$\mathrm{R}(z) \circ \mathrm{L}(x) = \mathrm{L}(x) \circ \mathrm{R}(z) \qquad (x, z \in G),$$

(1.3.4) $$\mathrm{L}(xy) = \mathrm{L}(x) \circ \mathrm{L}(y) \qquad (x, y \in G),$$

(1.3.5) $$\mathrm{R}(yz) = \mathrm{R}(z) \circ \mathrm{R}(y) \qquad (y, z \in G).$$

That is, all left multiplications commute with all right multiplications; and $\mathrm{L}\colon x \mapsto \mathrm{L}(x)$ is a homomorphism, and $\mathrm{R}\colon x \mapsto \mathrm{R}(x)$, an anti-homomorphism, respectively, from G to the group of C^2 diffeomorphisms: $G \to G$.

As a consequence, we recover that:

$$\mathbf{Ad}\colon x \mapsto \mathbf{Ad}(x) = \mathrm{L}(x) \circ \mathrm{R}(x)^{-1} \tag{1.3.6}$$

is a homomorphism from G to the group of C^2 diffeomorphisms of G, as we already observed in (1.1.10).

In the sequel we shall freely use results on vector fields and on the flows associated with them; more details can be found in Appendix B.

A vector field v on G is called *left invariant*, and *right invariant*, if:

$$v_{\mathrm{L}(x)y} = \mathrm{T}_y\,\mathrm{L}(x)v_y, \tag{1.3.7}$$

and:

$$v_{\mathrm{R}(x)y} = \mathrm{T}_y\,\mathrm{R}(x)v_y, \tag{1.3.8}$$

respectively, for all $y, x \in G$. Taking $y = 1$ in (1.3.7) and (1.3.8) we see that any left, and right, invariant vector field is determined by its value $v_1 = X \in \mathfrak{g}$ at the identity element, by the formula:

$$v_x = \mathrm{T}_1\,\mathrm{L}(x)X, \tag{1.3.9}$$

and:

$$v_x = \mathrm{T}_1\,\mathrm{R}(x)X, \tag{1.3.10}$$

respectively, for all $x \in G$. Conversely, in view of (1.3.4,5) and the chain rule for tangent mappings, (1.3.9,10) define a left, and right, invariant vector field on G, which we will denote by X^{L}, and X^{R}, respectively. Note that the mappings:

$$\mathcal{L}\colon (x, X) \mapsto X^{\mathrm{L}}_x, \tag{1.3.11}$$

and

$$\mathcal{R}\colon (x, X) \mapsto X^{\mathrm{R}}_x, \tag{1.3.12}$$

respectively, are C^1 isomorphisms: $G \times \mathfrak{g} \to \mathrm{T}\,G$ of vector bundles over G. This shows that the tangent bundle $\mathrm{T}\,G$, as a C^1 vector bundle over G, is **trivial**, with two natural trivializations $\mathcal{L}^{-1}$, and $\mathcal{R}^{-1}$, respectively. In particular, all left or right invariant vector fields on G are C^1.

(1.3.1) Lemma. *Let Φ^t be the flow of a left or right invariant vector field v on G. This is a C^2 mapping $G \to G$ satisfying:*

$$\Phi^t = \mathrm{R}(\Phi^t(1)), \quad \text{if } v \text{ is left invariant,} \tag{1.3.13}$$

$$\Phi^t = \mathrm{L}(\Phi^t(1)), \quad \text{if } v \text{ is right invariant.} \tag{1.3.14}$$

Proof. Let v be left invariant. Write:

$$x(t) = \mathrm{R}(\Phi^t(1))(x) = x\Phi^t(1) = \mathrm{L}(x)(\Phi^t(1)).$$

Then $x(0) = x$ and:

$$\frac{dx}{dt}(t) = \mathrm{T}_{\Phi^t(1)}\,\mathrm{L}(x)\big(v(\Phi^t(1))\big) = v\big(\mathrm{L}(x)(\Phi^t(1))\big) = v(x(t)),$$

for all t, hence $x(t) = \Phi^t(x)$. This proves (1.3.13). If v is right invariant, interchange L and R. □

(1.3.2) Theorem. *For every $X \in \mathfrak{g}$, there is a unique homomorphism $h = h_X : (\mathbf{R}, +) \to (G, \cdot)$ that is differentiable at $t = 0$ and satisfies $\frac{dh}{dt}(0) = X$. It is equal to the solution curve of both X^{L} and of X^{R} starting at the identity element of G. The flows of X^{L} and X^{R} are globally defined for all $t \in \mathbf{R}$.*

Proof. Let Φ^t denote the flow of X^{L}. As far as defined,

$$\Phi^{t+s}(1) = \Phi^s(\Phi^t(1)) = \Phi^t(1) \cdot \Phi^s(1)$$

because of (1.3.13). But this shows that there is an $\epsilon > 0$ such that, if $\Phi^t(1)$ is defined, then $\Phi^{t'}(1)$ is defined for all $t' \in \,]\,t - \epsilon, t + \epsilon\,[$. Therefore $t \mapsto \Phi^t(1)$ is defined for all $t \in \mathbf{R}$, and defines a C^1 homomorphism $h \colon (\mathbf{R}, +) \to (G, \cdot)$, while $\frac{dh}{dt}(0) = X$. Again applying (1.3.13) we also get that $\Phi^t : G \to G$ exists for all $t \in \mathbf{R}$. Replacing X^{L} by X^{R} we obtain a similar conclusion for the flow of X^{R}.

Conversely, suppose h is a homomorphism: $(\mathbf{R}, +) \to (G, \cdot)$, differentiable at $t = 0$ and such that $\frac{dh}{dt}(0) = X$. Differentiating:

$$h(t + s) = h(t)h(s) = h(s)h(t)$$

with respect to s at $s = 0$, we then see that h is differentiable at t, and:

$$\frac{dh}{dt}(t) = \mathrm{T}_1\,\mathrm{L}(h(t))(X) = X^{\mathrm{L}}(h(t)),$$

and:

$$\frac{dh}{dt}(t) = \mathrm{T}_1\,\mathrm{R}(h(t))(X) = X^{\mathrm{R}}(h(t));$$

so $h(t)$ is a solution curve for X^{L}, and X^{R}, respectively, with $h(0) = 1$. Therefore it is uniquely determined by $X \in \mathfrak{g}$. □

(1.3.3) Definition. *For each* $X \in \mathfrak{g}$, *the element* $\exp X \in G$ *is defined as* $h(1)$, *where* h *is the differentiable homomorphism:* $(\mathbf{R}, +) \to (G, \cdot)$ *such that* $\frac{dh}{dt}(0) = X$. *The mapping:*

$$\exp\colon X \mapsto \exp X\colon \mathfrak{g} \to G$$

is called the exponential mapping from $\mathfrak{g}$ *into* G.

Now the mapping $\mathcal{L}$ in (1.3.11) is C^1 and $\exp X$ is equal to the solution curve of the vector field X^{L} at time $t = 1$, starting at time $t = 0$ at the identity element. Hence it follows from the theorem about smooth dependence of solutions of ordinary differential equations on parameters that the exponential mapping is C^1 from $\mathfrak{g}$ into G. Also note that:

$$h\colon s \mapsto h_X(st)$$

is a differentiable homomorphism: $(\mathbf{R}, +) \to (G, \cdot)$ such that $\frac{dh}{dt}(0) = tX$, and it follows that:

$$h_X(st) = h_{tX}(s).$$

Taking $s = 1$, we find that:

(1.3.15) $$h_X(t) = \exp(tX) \qquad (X \in \mathfrak{g},\ t \in \mathbf{R});$$

and this implies also that $t \mapsto \exp(tX)$ is a homomorphism: $(\mathbf{R}, +) \to (G, \cdot)$. Differentiating (1.3.15) with respect to t at $t = 0$, we obtain:

() $$X = \mathrm{T}_0(\exp)(X) \qquad (X \in \mathfrak{g}),$$

that is, $\mathrm{T}_0(\exp)$ is equal to the identity: $\mathfrak{g} \to \mathfrak{g}$. Applying the open mapping theorem, we get:

(1.3.4) Proposition. *There is an open neighborhood* U *of* 0 *in* $\mathfrak{g}$, *and* V *of* 1 *in* G, *respectively, such that the exponential mapping is a* C^1 *diffeomorphism from* U *onto* V.

(1.3.5) Definition. *If* U, V *are as in Proposition 1.3.4, then the inverse of the exponential mapping:* $U \to V$ *is called a logarithmic chart (for* G*) and is denoted by* $\log\colon V \to U$.

In the next sections we shall show that the translates of the logarithmic charts make G into a real-analytic Lie group, and also that the group structure, at least in a neighborhood of the identity element, is completely determined by the Lie algebra structure. However, before turning to these topics, let us take a look at the exponential mapping for $G = \mathbf{GL}(V)$.

1.4 The Exponential Map for a Vector Space

If V is a finite-dimensional real or complex vector space and $X \in \mathbf{L}(V, V)$, then the power series:

$$\mathrm{e}^X := \sum_{k=0}^{\infty} \frac{1}{k!} X^k \tag{1.4.1}$$

converges in $\mathbf{L}(V, V)$. In fact, if $\|\,.\,\|$ denotes the operator norm, the k-th term has norm $\frac{1}{k!}\|X^k\| \leq \frac{1}{k!}\|X\|^k$; so the sum of the norms of all terms can be estimated by $\mathrm{e}^{\|X\|} < \infty$. The mapping $X \mapsto \mathrm{e}^X$ is real-analytic: $\mathbf{L}(V, V) \to \mathbf{L}(V, V)$, and complex-analytic: $\mathbf{L_C}(V, V) \to \mathbf{L_C}(V, V)$, respectively. One easily verifies that $A(t) = \mathrm{e}^{tX}$ satisfies:

$$\frac{dA}{dt}(t) = X A(t) = \big(\mathrm{T_I}\, \mathrm{R}(A(t))\big)(X) = X^{\mathrm{R}}(A(t)), \quad A(0) = \mathrm{I}; \tag{1.4.2}$$

hence $t \mapsto \mathrm{e}^{tX}$ is equal to the homomorphism: $(\mathbf{R}, +) \to (\mathbf{GL}(V), \cdot)$ of Theorem 1.3.2. That is, $X \mapsto \mathrm{e}^X$ is equal to the exponential mapping from the Lie algebra $\mathbf{L}(V, V)$ of $\mathbf{GL}(V)$, into $\mathbf{GL}(V)$. Note that in this case (1.3.14) says that every solution $A(t)$ of $\frac{dA}{dt}(t) = XA(t)$ in $\mathbf{GL}(V)$ is of the form $\mathrm{e}^{tX} C$ for some constant $C \in \mathbf{GL}(V)$, a familiar statement about fundamental solutions of linear differential equations. Also, $\delta \colon t \mapsto \det A(t)$ is a differentiable homomorphism:

$$(\mathbf{R}, +) \to (\mathbf{R} \setminus \{0\}, \cdot) \quad \text{with} \quad \frac{d\delta}{dt}(0) = (\mathrm{T_I} \det) X = \operatorname{tr} X,$$

leading to the well-known formula:

$$\det(\mathrm{e}^X) = \mathrm{e}^{\operatorname{tr} X}\,. \tag{1.4.3}$$

This confirms that the matrices with trace equal to zero form the Lie algebra of the group $\mathbf{SL}(n, \mathbf{R})$ mentioned in (1.2.27).

Because of (1.4.3), or just using the continuity of the exponential mapping, one sees that $\exp \mathbf{L}(V, V)$ is contained in the connected component:

$$\mathbf{GL}(V)^\circ = \{\, A \in \mathbf{GL}(V) \mid \det A > 0 \,\} \tag{1.4.4}$$

of the identity element in $\mathbf{GL}(V)$. As we shall see below, already for a two-dimensional vector space V over $\mathbf{R}$, the exponential mapping $\mathbf{L}(V, V) \to \mathbf{GL}(V)^\circ$ is not surjective. On the other hand, using Jordan normal forms, one may verify that for a **complex** vector space V we have:

$$\exp \mathbf{L_C}(V, V) = \mathbf{GL_C}(V). \tag{1.4.5}$$

Examples. If $G = \mathbf{SL}(2, \mathbf{C})$, then:

$$X = \begin{pmatrix} a & b \\ c & -a \end{pmatrix} \in \mathfrak{g} \qquad (a, b, c \in \mathbf{C}) \tag{1.4.6}$$

satisfies the equation:

$$X^2 = \lambda^2 \mathrm{I}, \quad \text{where } \lambda = \pm\sqrt{a^2 + bc} \tag{1.4.7}$$

are the eigenvalues of X. Using the formulae:

$$\sum_{n=0}^{\infty} \frac{\lambda^{2n}}{(2n)!} = \frac{1}{2}(\mathrm{e}^{\lambda} + \mathrm{e}^{-\lambda}) = \cosh\lambda, \tag{1.4.8}$$

$$\sum_{n=0}^{\infty} \frac{\lambda^{2n}}{(2n+1)!} = \frac{1}{2\lambda}(\mathrm{e}^{\lambda} - \mathrm{e}^{-\lambda}) = \frac{\sinh\lambda}{\lambda}, \tag{1.4.9}$$

we get the explicit formula:

$$\mathrm{e}^X = \cosh\lambda\, \mathrm{I} + \frac{\sinh\lambda}{\lambda} X. \tag{1.4.10}$$

Observe that both $\cosh\lambda$ and $\frac{\sinh\lambda}{\lambda}$ are entire analytic functions of $\lambda^2 = a^2 + bc$, hence they are entire analytic functions of X.

If $\lambda = 0$ we recover the result:

$$\mathrm{e}^X = \mathrm{I} + X \quad \text{if } X \text{ is nilpotent } (\iff X^2 = 0). \tag{1.4.11}$$

If $\lambda = it$, **with** $t \in \mathbf{R}$, one usually rewrites (1.4.10) as:

$$\mathrm{e}^X = \cos t\, \mathrm{I} + \frac{\sin t}{t} X. \tag{1.4.12}$$

This occurs if $a, b, c \in \mathbf{R}$ ($\iff X \in \mathfrak{sl}(2, \mathbf{R})$) and $a^2 + bc < 0$, that is, "X is in the elliptic domain". The case $\lambda = it$, with $t \in \mathbf{R}$, occurs also for all:

$$X = \begin{pmatrix} i\alpha & \beta \\ -\bar{\beta} & -i\alpha \end{pmatrix} \in \mathfrak{su}(2);$$

in this case:

$$t = \sqrt{\alpha^2 + |\beta|^2} = \text{ the Euclidean length of } (\alpha, \beta) \in \mathbf{R} \times \mathbf{C} \cong \mathbf{R}^3. \tag{1.4.13}$$

For $X = t\begin{pmatrix} 0 & -1 \\ 1 & 0 \end{pmatrix}$ we recover the rotation:

$$R(t) = \mathrm{e}^{t\begin{pmatrix} 0 & -1 \\ 1 & 0 \end{pmatrix}} = \begin{pmatrix} \cos t & -\sin t \\ \sin t & \cos t \end{pmatrix} \in \mathbf{SO}(2, \mathbf{R}) \tag{1.4.14}$$

in the plane, through the angle t. In fact, it makes perfect sense to **define** $R(t)$ as the solution of the differential equation with initial condition:

$$\frac{dR}{dt}(t) = \begin{pmatrix} 0 & -1 \\ 1 & 0 \end{pmatrix} \circ R(t), \quad R(0) = \mathrm{I}, \tag{1.4.15}$$

and to **define** the trigonometric functions $\cos t$ and $\sin t$, as the first and second coordinate of the uniform motion $t \mapsto R(t)\begin{pmatrix} 1 \\ 0 \end{pmatrix}$ on the circle.

If $X \in \mathfrak{sl}(2, \mathbf{R})$, then e^X has eigenvalues equal to -1 if and only if there is $k \in \mathbf{Z}$ such that X has eigenvalues equal to $\pm(i\pi + 2\pi ik)$ (with both signs occurring). This implies that X is diagonalizable over $\mathbf{C}$, hence e^X is diagonalizable over $\mathbf{C}$, so $e^X = -\mathrm{I}$. It follows that the elements $A \in \mathbf{SL}(2, \mathbf{R})$ that have eigenvalues equal to -1 without themselves being equal to $-\mathrm{I}$, that is, the $A \in \mathbf{SL}(2, \mathbf{R})$ which are conjugate to $\begin{pmatrix} -1 & \pm 1 \\ 0 & -1 \end{pmatrix}$, **are not in the image of the exponential mapping**: $X \mapsto e^X : \mathfrak{sl}(2, \mathbf{R}) \to \mathbf{SL}(2, \mathbf{R})$. (And they are the only ones).

We conclude this section by computing the exponential mapping:

$$\mathfrak{so}(3, \mathbf{R}) \to \mathbf{SO}(3, \mathbf{R}).$$

If $A_x \in \mathfrak{so}(3, \mathbf{R})$ is given as in (1.2.7), then it has eigenvalues 0 (with the eigenvector $x = \begin{pmatrix} x_1 \\ x_2 \\ x_3 \end{pmatrix}$) and $\pm i|x|$. So A_x satisfies the equation:

$$A_x^3 = -|x|^2 A_x. \tag{1.4.16}$$

Inserting this in (1.4.1) leads to:

$$e^{A_x} = \mathrm{I} + \frac{\sin|x|}{|x|} A_x + \frac{1 - \cos|x|}{|x|^2} A_x^2. \tag{1.4.17}$$

Because $A_x(x) = 0$, we get $e^{tA_x}(x) = x$; so $t \mapsto e^{tA_x}$ is a rotation around the axis through x. Furthermore, because the eigenvalues of the restriction of A_x to the orthogonal complement of x are equal to $\pm i|x|$, this rotation has uniform speed equal to $|x|$. This shows:

$$e^{A_x} = \text{ the rotation } R_x \text{ discussed after (1.2.6), (1.2.7).} \tag{1.4.18}$$

1.5 The Tangent Map of Exp

In this short section we collect three basic identities about the exponential mapping: $\mathfrak{g} \to G$, for an arbitrary Lie group G.

(1.5.1) Lemma. *Let G, and H, be Lie groups, with Lie algebra equal to $\mathfrak{g}$, and $\mathfrak{h}$, respectively. Suppose that Φ is a homomorphism: $G \to H$ that is differentiable at $1 \in G$. Then $\Phi(\exp X) = \exp(\mathrm{T}_1 \Phi(X))$, for each $X \in \mathfrak{g}$.*

Proof. See Theorem 1.3.2 and Definition 1.3.3. The mapping h given by $t \mapsto \Phi(\exp tX)$ is a homomorphism: $(\mathbf{R}, +) \to (G, \cdot)$. By the chain rule, it is differentiable at $t = 0$, and $\frac{dh}{dt}(0) = \mathrm{T}_1 \Phi(X)$. □

(1.5.2) Theorem. *We have the following formulae:*

(a) $\mathrm{Ad}(\exp X) = \mathrm{e}^{\mathrm{ad}\, X}$, *for each* $X \in \mathfrak{g}$.
(b) $x \exp X x^{-1} = \exp\big((\mathrm{Ad}\, x)(X)\big)$, *for each* $x \in G$, $X \in \mathfrak{g}$.

Proof. For (a), apply Lemma 1.5.1 to the homomorphism $\mathrm{Ad}\colon G \to \mathbf{L}(\mathfrak{g}, \mathfrak{g})$, and use (1.1.13). For (b), consider the homomorphism $\mathbf{Ad}\, x\colon G \to G$, and use (1.1.9). □

(1.5.3) Theorem. *For any $X \in \mathfrak{g}$, the linear mapping $\mathrm{T}_X \exp\colon \mathfrak{g} \to \mathrm{T}_{\exp X} G$ is given by:*

$$
\begin{aligned}
\mathrm{T}_X \exp &= \mathrm{T}_1 \mathrm{R}(\exp X) \circ \int_0^1 \mathrm{e}^{s\, \mathrm{ad}\, X}\, ds \\
&= \mathrm{T}_1 \mathrm{L}(\exp X) \circ \int_0^1 \mathrm{e}^{-s\, \mathrm{ad}\, X}\, ds.
\end{aligned}
\tag{1.5.1}
$$

Proof. If $x \mapsto v_\epsilon(x)$ is a C^1 vector field on a C^2 manifold M, depending in a C^1 fashion on a real parameter ϵ, and if Φ_ϵ^t denotes its flow after time t, then $\epsilon \mapsto \Phi_\epsilon^t(x)$ is differentiable, and satisfies:

$$
\frac{\partial}{\partial \epsilon} \Phi_\epsilon^t(x) = \int_0^t \mathrm{T}_{\Phi_\epsilon^s(x)} \big(\Phi_\epsilon^{t-s}\big) \frac{\partial v_\epsilon}{\partial \epsilon} \big(\Phi_\epsilon^s(x)\big)\, ds \in \mathrm{T}_{\Phi_\epsilon^t(x)} M.
\tag{1.5.2}
$$

See Appendix B (Formula (B.10)). Now (1.5.1) follows from (1.5.2). Indeed, according to (1.3.13) the mapping: $s \mapsto \mathrm{R}\big(\exp s(X + \epsilon Y)\big)$ is the flow Φ_ϵ^s of the ϵ-dependent left invariant vector field $(X + \epsilon Y)^{\mathrm{L}}$ on G. Hence:

$$
\begin{aligned}
(\mathrm{T}_X \exp)(Y) &= \frac{d}{d\epsilon}\bigg|_{\epsilon=0} \exp(X+\epsilon Y) = \frac{d}{d\epsilon}\bigg|_{\epsilon=0} \mathrm{R}\big(\exp(X+\epsilon Y)\big)(1) \\
&= \int_0^1 \mathrm{T}_{\exp sX}\, \mathrm{R}(\exp(1-s)X)\big(Y^{\mathrm{L}}(\exp sX)\big)\, ds \\
&= \int_0^1 \mathrm{T}_1\, \mathrm{R}(\exp X) \circ \mathrm{T}_{\exp sX}\, \mathrm{R}(\exp -sX) \circ \mathrm{T}_1\, \mathrm{L}(\exp sX)(Y)\, ds \\
&= \mathrm{T}_1\, \mathrm{R}(\exp X) \circ \int_0^1 \mathrm{T}_1\big(\mathrm{R}(\exp -sX) \circ \mathrm{L}(\exp sX)\big)\, ds\, (Y) \\
&= \mathrm{T}_1\, \mathrm{R}(\exp X) \circ \int_0^1 \mathrm{Ad}(\exp sX)\, ds\, (Y) \\
&= \mathrm{T}_1\, \mathrm{R}(\exp X) \circ \int_0^1 \mathrm{e}^{s\,\mathrm{ad}\,X}\, ds\, (Y).
\end{aligned}
$$

In last equality we have used Theorem 1.5.2.(a). This proves the first identity in (1.5.1). The second one follows from:

$$
\begin{aligned}
&\mathrm{T}_1\, \mathrm{R}(\exp X)^{-1} \circ \mathrm{T}_1\, \mathrm{L}(\exp X) \circ \int_0^1 \mathrm{e}^{-s\,\mathrm{ad}\,X}\, ds \\
&= \mathrm{Ad}(\exp X) \circ \int_0^1 \mathrm{e}^{-s\,\mathrm{ad}\,X}\, ds = \int_0^1 \mathrm{e}^{(1-s)\,\mathrm{ad}\,X}\, ds = \int_0^1 \mathrm{e}^{u\,\mathrm{ad}\,X}\, du,
\end{aligned}
$$

by means of the substitution $1 - s = u$. □

Remark. As our proof expresses, we like to consider (1.5.1) as a direct consequence of the variational equation (1.5.2) for solution curves of vector fields with respect to parameters. Conversely, one can view (1.5.2) as being equal to the Formula (1.5.1) for G equal to the pseudogroup of local C^1 diffeomorphisms of M and $\mathfrak{g}$ equal to "its Lie algebra" of C^1 vector fields on M, see Section 1.11.

If $A \in \mathbf{L}(V,V)$, with V a finite-dimensional vector space, then:

$$
(1.5.3) \qquad f(A) := \int_0^1 \mathrm{e}^{sA}\, ds = \sum_{k=0}^{\infty} \frac{1}{(k+1)!} A^k = A^{-1}(\mathrm{e}^A - \mathrm{I}),
$$

where the second identity holds if A is invertible. If A is not invertible, we will use (1.5.3) as the **definition** of $A^{-1}(\mathrm{e}^A - \mathrm{I})$. In particular, this happens if $A = \mathrm{ad}\,X \in \mathbf{L}(\mathfrak{g}, \mathfrak{g})$: if $X = 0$ then $A = 0$, and if $X \neq 0$ then $A(X) = [X, X] = 0$; so $\mathrm{ad}\,X$ is never invertible. With these conventions, one may also write (1.5.1) as:

$$
(1.5.4) \qquad \mathrm{T}_X \exp = \mathrm{T}_1\, \mathrm{R}(\exp X) \circ \frac{\mathrm{e}^{\mathrm{ad}\,X} - \mathrm{I}}{\mathrm{ad}\,X} = \mathrm{T}_1\, \mathrm{L}(\exp X) \circ \frac{\mathrm{I} - \mathrm{e}^{-\,\mathrm{ad}\,X}}{\mathrm{ad}\,X}.
$$

Finally, using complex Jordan normal forms (regarding A as an element of $\mathbf{L}_{\mathbf{C}}(V_{\mathbf{C}}, V_{\mathbf{C}})$, with $V_{\mathbf{C}} = V \oplus iV$, the complexification of V), we see that $\int_0^1 e^{sA}\, ds$ has the eigenvalues:

$$\int_0^1 e^{s\lambda}\, ds = \frac{e^{\lambda} - 1}{\lambda} \quad (= 1 \text{ if } \lambda = 0),$$

if λ ranges over the eigenvalues of A.

(1.5.4) Corollary. *The singular points of the exponential mapping:* $\mathfrak{g} \to G$, *that is, the* $X \in \mathfrak{g}$ *such that* $\mathrm{T}_X \exp$ *is not invertible, are precisely the* $X \in \mathfrak{g}$ *such that* $\operatorname{ad} X \in \mathbf{L}(\mathfrak{g}, \mathfrak{g})$ *has an eigenvalue of the form* $2\pi i k$, *with* $k \in \mathbf{Z} \backslash \{0\}$.

In other words, the singular set of the exponential mapping is equal to the disjoint union:

$$\Sigma = \bigcup_{k \in \mathbf{Z} \backslash \{0\}} k\Sigma_1, \tag{1.5.5}$$

where Σ_1 is the algebraic variety in $\mathfrak{g}$:

$$\Sigma_1 = \{\, X \in \mathfrak{g} \mid \det\big((\operatorname{ad} X)_{\mathbf{C}} - 2\pi i\, \mathrm{I}\big) = 0 \,\}. \tag{1.5.6}$$

Examples. If $G = \mathbf{SO}(3, \mathbf{R})$ then, in the notation of (1.2.7),

$$\Sigma_1 = \big\{\, A_x \in \mathfrak{so}(3, \mathbf{R}) \;\big|\; |x| = 2\pi \,\big\}$$

(see before (1.4.16)). The exponential mapping is a local diffeomorphism from the ball:

$$\{\, A_x \in \mathfrak{so}(3, \mathbf{R}) \;\big|\; |x| < 2\pi \,\}$$

onto $\mathbf{SO}(3, \mathbf{R})$, which restricts to a two-fold covering from the punctured ball:

$$\{\, A_x \in \mathfrak{so}(3, \mathbf{R}) \;\big|\; 0 < |x| < 2\pi \,\}$$

onto $\mathbf{SO}(3, \mathbf{R}) \setminus \{\mathrm{I}\}$. The set:

$$\{\, A_x \in \mathfrak{so}(3, \mathbf{R}) \;\big|\; |x| < \pi \,\}$$

is a maximal open subset of $\mathfrak{so}(3, \mathbf{R})$ on which the exponential mapping is injective; it is mapped diffeomorphically by the exponential mapping onto the complement in $\mathbf{SO}(3, \mathbf{R})$ of the conjugacy class of:

$$\begin{pmatrix} 1 & 0 & 0 \\ 0 & -1 & 0 \\ 0 & 0 & -1 \end{pmatrix}.$$

If $G = \mathbf{SL}(2, \mathbf{C})$, then:

$$\Sigma_1 = \left\{ \begin{pmatrix} a & b \\ c & -a \end{pmatrix} \in \mathfrak{sl}(2, \mathbf{C}) \mid a^2 + bc = -\pi^2 \right\}.$$

So, for $G = \mathbf{SU}(2)$,

$$\Sigma_1 = \left\{ \begin{pmatrix} i\alpha & \beta \\ -\bar{\beta} & -i\alpha \end{pmatrix} \in \mathfrak{su}(2) \mid \alpha^2 + |\beta|^2 = \pi^2 \right\},$$

that is, the sphere in $\mathfrak{su}(2)$ of Euclidean radius equal to π; and exp maps the ball:

$$\left\{ \begin{pmatrix} i\alpha & \beta \\ -\bar{\beta} & -i\alpha \end{pmatrix} \in \mathfrak{su}(2) \mid \alpha^2 + |\beta|^2 < \pi^2 \right\}$$

diffeomorphically onto $\mathbf{SU}(2) \setminus \{-\mathrm{I}\}$. On the other hand, for $G = \mathbf{SL}(2, \mathbf{R})$, the set Σ_1 is the two-sheeted hyperboloid:

$$\left\{ \begin{pmatrix} a & b \\ c & -a \end{pmatrix} \in \mathfrak{sl}(2, \mathbf{R}) \mid a^2 + bc = -\pi^2 \right\},$$

mapped to $\{-\mathrm{I}\}$ by exp. Here exp is a diffeomorphism from:

$$\left\{ \begin{pmatrix} a & b \\ c & -a \end{pmatrix} \in \mathfrak{sl}(2, \mathbf{R}) \mid a^2 + bc > -\pi^2 \right\}$$

onto the complement in $\mathbf{SL}(2, \mathbf{R})$ of $\{-\mathrm{I}\}$ and the conjugacy classes of $\begin{pmatrix} -1 & \pm 1 \\ 0 & -1 \end{pmatrix}$, the latter being the elements that are not in the image of $\exp\colon \mathfrak{sl}(2, \mathbf{R}) \to \mathbf{SL}(2, \mathbf{R})$ at all, as we have seen after (1.4.15). See Fig.1.2.3, where these conjugacy classes are the two components of the smooth part of the bent cone through $-\mathrm{I}$. Note that $\exp \mathfrak{sl}(2, \mathbf{R})$ is neither open in $\mathbf{SL}(2, \mathbf{R})$ ($-\mathrm{I}$ belongs to it, but is not an interior point), nor closed.

1.6 The Product in Logarithmic Coordinates

For any finite-dimensional Lie algebra $\mathfrak{g}$ over $\mathbf{R}$, let $\mathfrak{g}_e := \mathfrak{g} \setminus \Sigma$ be the set of $X \in \mathfrak{g}$ such that $f(\operatorname{ad} X) = \frac{e^{\operatorname{ad} X} - \mathrm{I}}{\operatorname{ad} X}$ (cf. (1.5.3)) is invertible. See also (1.5.5,6) for another description of the complementary set Σ. The set $\mathfrak{g}_e$ is an open neighborhood of 0 in $\mathfrak{g}$ and, in view of the chain rule,

$$X \mapsto (f(\operatorname{ad} X))^{-1} = \frac{\operatorname{ad} X}{e^{\operatorname{ad} X} - \mathrm{I}} \tag{1.6.1}$$

is an analytic mapping: $\mathfrak{g}_e \to \mathbf{L}(\mathfrak{g}, \mathfrak{g})$. (Notice that the coefficients B_k in $\frac{x}{e^x - 1} = \sum_{k=0}^{\infty} \frac{B_k}{k!} x^k$ are the Bernoulli numbers, cf. Walter [1985], p.160). Let $\mathfrak{g}_e^2$ be the set of $(X, Y) \in \mathfrak{g} \times \mathfrak{g}_e$ such that the solution $t \mapsto Z(t)$ of the equation:

(1.6.2) $$\frac{dZ}{dt}(t) = \frac{\operatorname{ad} Z(t)}{\mathrm{e}^{\operatorname{ad} Z(t)} - \mathrm{I}}(X), \quad Z(0) = Y$$

(in $\mathfrak{g}_e$) is defined for all $t \in [0,1]$. Set:

(1.6.3) $$\mu(X,Y) = Z(1) \qquad ((X,Y) \in \mathfrak{g}_e^2).$$

(1.6.1) Theorem. *The set $\mathfrak{g}_e^2$ is an open neighborhood of $(0,0)$ in $\mathfrak{g} \times \mathfrak{g}$ and μ is a real-analytic mapping: $\mathfrak{g}_e^2 \to \mathfrak{g}$. If $\mathfrak{g}$ is the Lie algebra of a Lie group G, with exponential mapping* $\exp \colon \mathfrak{g} \to G$, *then:*

(1.6.4) $$\exp X \exp Y = \exp \mu(X,Y) \qquad ((X,Y) \in \mathfrak{g}_e^2).$$

If $\mathfrak{g}$ is a complex vector space and $\operatorname{ad} \colon \mathfrak{g} \times \mathfrak{g} \to \mathfrak{g}$ *is complex bilinear, then μ is complex-analytic.*

Proof. That $\mathfrak{g}_e^2$ is open and μ is real- (and complex-)analytic, on $\mathfrak{g}_e^2$ follows from the fact that $(X,Z) \mapsto \frac{\operatorname{ad} Z}{\mathrm{e}^{\operatorname{ad} Z} - I}(X)$ is real-, and complex-, analytic: $\mathfrak{g} \times \mathfrak{g}_e \to \mathfrak{g}$, respectively, combined with the analytic dependence on initial values and parameters of solutions of analytic vector fields depending analytically on the parameters (see Appendix B). Now (1.5.4) combined with (1.6.2) gives:

(1.6.5) $$\begin{aligned}\frac{d}{dt}(\exp Z(t)) &= (\mathrm{T}_{Z(t)} \exp)\frac{dZ}{dt}(t) = \mathrm{T}_1 \,\mathrm{R}(\exp Z(t))(X)\\ &= X^{\mathrm{R}}(\exp Z(t)).\end{aligned}$$

In view of Lemma 1.3.1. this shows that:

(1.6.6) $$\exp Z(t) = \exp tX \exp Z(0) = \exp tX \exp Y,$$

for all t in the interval of definition of $t \mapsto Z(t)$. Taking $t = 1$ gives (1.6.4). □

In Theorem 1.14.3 we shall see that every finite-dimensional real Lie algebra is the Lie algebra of a Lie group; this then, together with (1.6.4), justifies the terminology of calling the mapping μ the *product in logarithmic coordinates.* As a first application of Theorem 1.6.1 we shall show that every C^2 Lie group G in the sense of Definition 1.1.1 can be provided with the structure of a real-analytic manifold for which it becomes a real-analytic Lie group in the sense of the following:

(1.6.2) Definition. *A real-, and complex-analytic, Lie group G is a group G that at the same time is a real-, and complex-analytic, manifold in such a way that the group operations $(x,y) \mapsto xy \colon G \times G \to G$ and $x \mapsto x^{-1} \colon G \to G$ are real-, and complex-analytic mappings, respectively.*

To show the analyticity of G, take open neighborhoods U and U_0 of 0 in $\mathfrak{g}$, and V of 1 in G, respectively, such that:

(1.6.7) $$\exp \text{ is a diffeomorphism from } U \text{ to } V$$

and, for for all X, Y, $Z \in U_0$:

$$(1.6.8) \qquad (X, -Y) \in \mathfrak{g}_e^2, \quad (\mu(X, -Y), Z) \in \mathfrak{g}_e^2, \qquad \mu(\mu(X, -Y), Z) \in U.$$

The existence of such U, U_0, V follows from the invertibility of $\mathrm{T}_0 \exp = \mathrm{I}\colon \mathfrak{g} \to \mathfrak{g}$, if we apply the inverse function theorem to exp, use $\mu(0, 0) = 0$, and the continuity of μ at $(0, 0)$. For each $x \in G$, write:

$$(1.6.9) \qquad V_0^x = \mathrm{L}(x)(\exp U_0), \quad \kappa^x(y) = \log(x^{-1}y) \qquad (y \in V_0^x).$$

(1.6.3) Theorem. *The $\kappa^x : V_0^x \to U_0$, for $x \in G$, form a real-analytic atlas for G, making G into a real-analytic Lie group G_{an}, such that the identity in G is a C^2 diffeomorphism between G and G_{an}. Furthermore, if $\mathfrak{g}$ is a complex Lie algebra as in Definition 1.1.6, then this atlas is complex-analytic. It makes G into a complex-analytic group if in addition $\mathrm{Ad}\, x$ is complex-linear: $\mathfrak{g} \to \mathfrak{g}$, for every $x \in G$.*

Proof. First note that (1.5.1) shows that $X \mapsto \mathrm{T}_X \exp$ is C^1, hence exp is a C^2 mapping: $\mathfrak{g} \to G$. It follows that κ^x is a C^2 diffeomorphism: $V_0^x \to \mathfrak{g}$, for each $x \in G$. Now suppose that $V_0^x \cap V_0^y \neq \emptyset$, that is:

$$x \exp X_0 = y \exp Y_0, \qquad \text{for some } X_0, Y_0 \in U_0.$$

Then $Y = \kappa^y \circ (\kappa^x)^{-1}(X)$ means that $x \exp X = y \exp Y$, or $\exp Y = \exp Y_0 \exp -X_0 \exp X$, which is equivalent to:

$$Y = \mu(\mu(Y_0, -X_0), X).$$

This proves the desired real-, and complex-, analyticity, respectively, of the atlas. (Here we actually only need the analyticity of $Y \mapsto \mu(X, Y)$, that is the analytic dependence of the solutions of (1.6.2) on the initial value.)

For the analyticity of the group operations, write:

$$\begin{aligned} x \exp X (y \exp Y)^{-1} &= x \exp X \exp -Y y^{-1} = (xy^{-1}) y \exp \mu(X, -Y) y^{-1} \\ &= xy^{-1} \exp\big(\mathrm{Ad}\, y(\mu(X, -Y))\big). \end{aligned}$$

This shows that:

$$(X, Y) \mapsto \kappa^{xy^{-1}}\left((\kappa^x)^{-1}(X)\big((\kappa^y)^{-1}(Y)\big)^{-1}\right) = \mathrm{Ad}\, y(\mu(X, -Y))$$

is real-, and complex-, analytic, respectively. The proof is complete because the real-, and complex-, analyticity, respectively, of $(x, y) \mapsto xy^{-1}$ implies that of $y \mapsto y^{-1} = 1y^{-1}$ and of $(x, y) \mapsto xy = x(y^{-1})^{-1}$. □

Remark. If G is a complex-analytic Lie group, then $\mathfrak{g} = \mathrm{T}_1 G$ is a complex vector space, $\mathrm{Ad}\, x$ is complex-linear: $\mathfrak{g} \times \mathfrak{g}$ for each $x \in G$, $x \mapsto \mathrm{Ad}\, x$ is complex-analytic: $G \to \mathbf{GL}_{\mathbb{C}}(\mathfrak{g})$, so finally $\mathrm{ad} = \mathrm{T}_1 \mathrm{Ad}$ is complex bilinear: $\mathfrak{g} \times \mathfrak{g} \to \mathfrak{g}$. The second part of the theorem provides a converse to these observations. In Prop. 1.9.4 we shall see that, if $\mathfrak{g}$ is a complex Lie algebra, then $\mathrm{Ad}\, x$ is automatically complex-linear: $\mathfrak{g} \to \mathfrak{g}$ for all x in the connected component G° of 1 in G. So the last condition in the theorem is really a condition on the action of G/G°, the discrete group of connected components of G, on the space of complex structures in $\mathfrak{g}$.

The following Proposition expresses the **uniqueness** of the real or complex structures of a Lie group G with a given Lie algebra $\mathfrak{g}$.

(1.6.4) Proposition. *If G is a real-, and complex-, analytic Lie group, then the exponential mapping is real-, and complex-, analytic: $\mathfrak{g} \to G$, respectively. If G is provided with another structure of a real-, and complex-, analytic Lie group $\widetilde{G}$ such that the identity is differentiable: $G \to \widetilde{G}$ and differentiable: $\widetilde{G} \to G$ (at 1), then the identity is a real-, and complex-, analytic diffeomorphism, respectively: $G \to \widetilde{G}$.*

Proof. X^{R} is a real-, or complex-analytic vector field on G, because in logarithmic coordinates it is given by (1.6.2), cf. (1.6.5). It depends in a linear, or complex-linear way on X. Further $\exp X$ is the solution after time 1 of X^{R}, starting at $1 \in G$. In view of the theorem about analytic dependence on parameters of solutions of analytic vector fields depending analytically on the parameters, the analyticity of the exponential mapping follows. For the second statement we observe that the assumption insures that G and $\widetilde{G}$ have the same Lie algebra. □

(1.6.5) *Warning.* One needs an assumption to the effect that the Lie algebra $\mathfrak{g}$ of G is specified, because if $\widetilde{G} = G$ with the discrete topology, then $\widetilde{G}$ is a 0-dimensional Lie group (with Lie algebra $\tilde{\mathfrak{g}} = 0$). The identity: $\widetilde{G} \to G$ is analytic but the identity: $G \to \widetilde{G}$ is not even continuous if $\dim G > 0$.

From now on we shall assume all Lie groups to be real-, and complex-, analytic, with the unique real-, and complex-, analytic structure determined by its given real, and complex, Lie algebra, respectively.

1.7 Dynkin's Formula

As an intermezzo, we derive an explicit power series expansion for the product in logarithmic coordinates. From (1.6.6) and Theorem 1.5.2.(a) we read off that:

$$\mathrm{e}^{\mathrm{ad}\, Z(t)} = \mathrm{e}^{t\, \mathrm{ad}\, X} \circ \mathrm{e}^{\mathrm{ad}\, Y} . \tag{1.7.1}$$

We can also verify this using only the Lie algebra structure of $\mathfrak{g}$ (that is, without using that $\mathfrak{g}$ is the Lie algebra of a Lie group) by applying the Lie algebra homomorphism $\mathrm{ad}\colon \mathfrak{g} \mapsto \mathbf{L}(\mathfrak{g}, \mathfrak{g})$ to (1.6.2) and then reading (1.6.6) with $Z(t)$, X, and Y replaced by $\mathrm{ad}\, Z(t)$, $\mathrm{ad}\, X$, and $\mathrm{ad}\, Y$, respectively. Now:

$$A = \log(\mathrm{I} + (\mathrm{e}^A - \mathrm{I})) = \sum_{k=0}^{\infty} \frac{(-1)^k}{k+1} (\mathrm{e}^A - \mathrm{I})^{k+1},$$

so:

$$\frac{A}{\mathrm{e}^A - \mathrm{I}} = \sum_{k=0}^{\infty} \frac{(-1)^k}{k+1} (\mathrm{e}^A - \mathrm{I})^k. \tag{1.7.2}$$

Inserting this into (1.6.2) leads to:

$$\begin{aligned}
\frac{dZ}{dt}(t) &= \sum_{k=0}^{\infty} \frac{(-1)^k}{k+1} (\mathrm{e}^{\mathrm{ad}\, Z(t)} - \mathrm{I})^k (X) = \sum_{k=0}^{\infty} \frac{(-1)^k}{k+1} \left(\mathrm{e}^{t\, \mathrm{ad}\, X} \circ \mathrm{e}^{\mathrm{ad}\, Y} - \mathrm{I}\right)^k (X) \\
&= \sum_{k=0}^{\infty} \frac{(-1)^k}{k+1} \Bigg(\sum_{\{\, l, m \geq 0,\, l+m>0\,\}} t^l \, \frac{(\mathrm{ad}\, X)^l}{l!} \circ \frac{(\mathrm{ad}\, Y)^m}{m!} \Bigg)^k (X) \\
&= \sum_{k=0}^{\infty} \frac{(-1)^k}{k+1} \sum t^{l_1 + \ldots + l_k} \\
&\qquad \times \left(\frac{(\mathrm{ad}\, X)^{l_1}}{l_1!} \circ \frac{(\mathrm{ad}\, Y)^{m_1}}{m_1!} \circ \cdots \circ \frac{(\mathrm{ad}\, X)^{l_k}}{l_k!} \circ \frac{(\mathrm{ad}\, Y)^{m_k}}{m_k!} \right)(X),
\end{aligned}$$

where the sum is taken over all $l_1, \ldots, l_k, m_1, \ldots, m_k \geq 0$ such that $l_j + m_j > 0$, for all j. Integrating this over t from 0 to 1 and using that $Z(0) = Y$, we find:

Dynkin's formula:

$$\begin{aligned}
\mu(X, Y) = Y + X + \sum_{k=1}^{\infty} \frac{(-1)^k}{k+1} &\sum_{\substack{l_1, \ldots, l_k \geq 0, \\ m_1, \ldots, m_k \geq 0, \\ l_j + m_j > 0}} \frac{1}{l_1 + \ldots + l_k + 1} \\
&\times \left(\frac{(\mathrm{ad}\, X)^{l_1}}{l_1!} \circ \frac{(\mathrm{ad}\, Y)^{m_1}}{m_1!} \circ \cdots \circ \frac{(\mathrm{ad}\, X)^{l_k}}{l_k!} \circ \frac{(\mathrm{ad}\, Y)^{m_k}}{m_k!} \right)(X).
\end{aligned} \tag{1.7.3}$$

This is the desired power series expansion. The Taylor expansion of μ at $(0, 0)$ up to the order 2 reads:

$$\mu(X, Y) = X + Y + \frac{1}{2}[X, Y] + \mathcal{O}\big(|(X, Y)|^3\big), \quad \text{as } (X, Y) \to (0, 0), \tag{1.7.4}$$

and for many purposes this is already sufficient. In view of the definition of the Lie algebra structure in terms of the second-order derivatives at $1 \in G$ of the group structure (in (1.1.13) and the remark preceding (1.1.8)), it is not surprising that the Lie bracket can be recovered from the Taylor expansion of

μ at $(0,0)$ up to the order 2. Moreover, any two norms on a finite-dimensional linear space are comparable; therefore we don't need to specify which norm $|\cdot|$ we use here.

As a curiosity, we also give the expansion up to the order 4:

$$\begin{aligned}\mu(X,Y) &= X + Y + \frac{1}{2}[X,Y] + \frac{1}{12}[X,[X,Y]] + \frac{1}{12}[Y,[Y,X]] \\ &+ \frac{1}{24}[Y,[X,[Y,X]]] + \mathcal{O}(|(X,Y)|^5), \quad \text{as } (X,Y) \to (0,0).\end{aligned} \tag{1.7.5}$$

Note that, using the Jacobi identity, we can rewrite the term of order 4 as:

$$[Y,[X,[Y,X]]] = [X,[Y,[Y,X]]].$$

This is only a first example of the phenomenon that one can get quite differently looking expressions for the terms in the Taylor expansion; moreover, the number of possibilities increases rapidly with the order of the terms.

1.8 Lie's Fundamental Theorems

The definition of μ in (1.6.3) applies to any finite-dimensional Lie algebra $\mathfrak{g}$. This suggests the construction of a Lie group G with Lie algebra $\mathfrak{g}$ by simply taking for G an open neighborhood U of 0 in $\mathfrak{g}$ with μ as the product structure. The problem is that in general there is no open neighborhood U of 0 in $\mathfrak{g}$ such that $\mu(X,-Y) \in U$ whenever $X, Y \in U$, so that U cannot become a group with μ as multiplication and $X \mapsto -X$ as inversion. This is one motivation behind the introduction of a *local Lie group* as an open neighborhood U of 0 in a finite-dimensional vector space (written as $\mathfrak{g}$), together with real-analytic mappings $\mu\colon U \times U \to \mathfrak{g}$ (multiplication) and $\iota\colon U \to \mathfrak{g}$ (inversion), such that, for X, Y, Z sufficiently close to 0 in $\mathfrak{g}$:

$$\begin{aligned}\mu(X,0) &= X = \mu(0,X), \\ \mu(X,\iota(X)) &= \mu(\iota(X),X) = 0, \\ \mu(\mu(X,Y),Z) &= \mu(X,\mu(Y,Z)).\end{aligned} \tag{1.8.1}$$

That is, the group laws, with 0 in the role of the identity element, hold locally near this identity element. Local Lie groups (U,μ,ι) and (U',μ',ι') are considered to be isomorphic if there are open neighborhoods U_{00}, U_0 of 0 in U, with $U_{00} \subset U_0 \subset U$, and a diffeomorphism ϕ from U_0 onto an open neighborhood U_0' of 0 in U', such that $\mu(X,Y) \in U_0$, $\phi(\mu(X,Y)) = \mu'(\phi(X),\phi(Y))$ and $\iota(X) \in U_0$, $\phi(\iota(X)) = \iota'(\phi(X))$, whenever $X, Y \in U_{00}$. An isomorphism class of triples (U,μ,ι) is called a *Lie group germ*.

It is almost obvious that the definition of the Lie algebra and the exponential mapping in Sections 1.1 and 1.3, (together with all the theorems about them) have local versions which are valid for any local Lie group. The following is a converse to this.

(1.8.1) Theorem. *For any finite-dimensional Lie algebra $\mathfrak{g}$, an open neighborhood of 0 in $\mathfrak{g}$ becomes a local Lie group, provided with μ, defined as in (1.6.3), as the local multiplication and $X \mapsto -X$ as the local inversion.*

Proof. The identities in (1.8.1) are easily verified, except for the last one, expressing the local associativity of the product. Trying to prove this, for instance, by using (1.7.3) becomes quite a messy calculation. Instead, we first prove the following lemma. □

(1.8.2) Lemma. *Let, for each $X \in \mathfrak{g}$, the vector field X^{R} on $\mathfrak{g}_e$ be defined by:*

$$X^{\mathrm{R}}(Z) = \frac{\operatorname{ad} Z}{\mathrm{e}^{\operatorname{ad} Z} - \mathrm{I}}(X) \qquad (Z \in \mathfrak{g}_e). \tag{1.8.2}$$

Then for each $X, Y \in \mathfrak{g}$, the Lie bracket (commutator) $[X^{\mathrm{R}}, Y^{\mathrm{R}}]$ of the vector fields X^{R} and Y^{R} is equal to $([X, Y])^{\mathrm{R}}$. Moreover, the mapping $X \mapsto X^{\mathrm{R}}$ is injective.

Proof. Let v and w be smooth vector fields on an open subset U of a finite-dimensional vector space. In coordinates, v and w are given by differentiable mappings from an open subset of $\mathbf{R}^n$ to $\mathbf{R}^n$, and we denote the total derivative at x of such a mapping v by $\mathrm{D}\, v(x) \in \mathbf{L}(\mathbf{R}^n, \mathbf{R}^n)$. Then the Lie bracket of these is given by:

$$[v, w](x) = \mathrm{D}\, v(x)\, w(x) - \mathrm{D}\, w(x)\, v(x), \tag{1.8.3}$$

cf. Formula (11.2.7.). So, in order to compute $[X^{\mathrm{R}}, Y^{\mathrm{R}}](Z)$, we start by differentiating (see Formula (1.5.3)):

$$\int_0^1 \mathrm{e}^{t \operatorname{ad} Z}\, dt (X^{\mathrm{R}}(Z)) = X \qquad (Z \in \mathfrak{g}_e), \tag{1.8.4}$$

with respect to Z, in the direction of $Y^{\mathrm{R}}(Z)$. We get, using Formula (1.5.1):

$$\begin{aligned} &\int_0^1 \left(\int_0^1 \mathrm{e}^{st \operatorname{ad}(\operatorname{ad} Z)}\, ds \circ t \operatorname{ad} Y^{\mathrm{R}}(Z) \right) \circ \mathrm{e}^{t \operatorname{ad} Z}\, dt \left(X^{\mathrm{R}}(Z) \right) \\ &\qquad + \int_0^1 \mathrm{e}^{t \operatorname{ad} Z}\, dt \circ \mathrm{D}\, X^{\mathrm{R}}(Z) \left(Y^{\mathrm{R}}(Z) \right) = 0. \end{aligned} \tag{1.8.5}$$

Because ad is a Lie algebra homomorphism: $\mathfrak{g} \to \mathfrak{gl}(\mathfrak{g})$ (this is equivalent to the Jacobi identity, see (1.1.21)), one has:

$$\mathrm{e}^{\operatorname{ad}(\operatorname{ad} X)} \circ \operatorname{ad} Y = \operatorname{ad}(\mathrm{e}^{\operatorname{ad} X}\, Y) \qquad (X, Y \in \mathfrak{g}). \tag{1.8.6}$$

Reading this with X, and Y, replaced by stZ, and $Y^{\mathrm{R}}(Z)$, respectively, and substituting $st = u$, we can recognize the first term in the left hand side of (1.8.5) as:

$$\iint\limits_{0\le u\le t\le 1} \left[\mathrm{e}^{u \operatorname{ad} Z}(Y^{\mathrm{R}}(Z)), \mathrm{e}^{t \operatorname{ad} Z}(X^{\mathrm{R}}(Z)) \right] du\, dt. \tag{1.8.7}$$

Interchanging X and Y in (1.8.7) and subtracting the resulting expression from (1.8.7) we get, also using the antisymmetry of the Lie bracket, the quantity:

$$\begin{aligned}
&- \iint\limits_{0\le u\le t\le 1} \left[\mathrm{e}^{t \operatorname{ad} Z}(X^{\mathrm{R}}(Z)), \mathrm{e}^{u \operatorname{ad} Z}(Y^{\mathrm{R}}(Z)) \right] du\, dt \\
&\qquad - \iint\limits_{0\le u\le t\le 1} \left[\mathrm{e}^{u \operatorname{ad} Z}(X^{\mathrm{R}}(Z)), \mathrm{e}^{t \operatorname{ad} Z}(Y^{\mathrm{R}}(Z)) \right] du\, dt \\
&= - \int_0^1 \int_0^1 \left[\mathrm{e}^{u \operatorname{ad} Z}(X^{\mathrm{R}}(Z)), \mathrm{e}^{t \operatorname{ad} Z}(Y^{\mathrm{R}}(Z)) \right] du\, dt \\
&= - \left[\int_0^1 \mathrm{e}^{u \operatorname{ad} Z}\, du\, (X^{\mathrm{R}}(Z)), \int_0^1 \mathrm{e}^{t \operatorname{ad} Z}\, dt\, (Y^{\mathrm{R}}(Z)) \right] \\
&= -[X, Y].
\end{aligned}$$

(We have used the bilinearity of the Lie bracket in order to bring the integrations inside.)

So, if we interchange X and Y in (1.8.5) and subtract, we arrive, using (1.8.3) with $v = X^{\mathrm{R}}$, $w = Y^{\mathrm{R}}$, and $x = Z$, at:

$$[X, Y] = \int_0^1 \mathrm{e}^{t \operatorname{ad} Z}\, dt\, ([X^{\mathrm{R}}, Y^{\mathrm{R}}](Z)),$$

which shows that $[X^{\mathrm{R}}, Y^{\mathrm{R}}](Z) = [X, Y]^{\mathrm{R}}(Z)$ in view of (1.8.2) (with X replaced by $[X, Y]$). The linearity of the mapping $X \mapsto X^{\mathrm{R}}$ is obvious from (1.8.2), and its injectivity follows from:

$$X^{\mathrm{R}}(0) = X \qquad (X \in \mathfrak{g}). \tag{1.8.8}$$

□

We now complete the proof of Theorem 1.8.1 using the following theorem about general Lie algebras of vector fields, which is of independent interest. It is phrased in a context that is described in more detail in Section 11.2.

(1.8.3) Theorem. *Let M be a C^2 manifold and let $\mathfrak{g}$ be a finite-dimensional vector space of C^1 vector fields on M such that $[X, Y] \in \mathfrak{g}$, whenever $X, Y \in \mathfrak{g}$. Then, for every relatively compact open subset V of M, there is an open neighborhood U of 0 in $\mathfrak{g}$ such that:*

(a) For every $X \in U$, the set V_X, the domain of definition of the flow Φ^X of X after time $t = 1$, contains V.

(b) If $X, Y \in U$, then $\Phi^Y(V) \subset V_X$ and $(\Phi^X \circ \Phi^Y)|_V = \Phi^{\mu(X,Y)}|_V$.
(c) $X \mapsto \Phi^X|_V$ is injective on U.
In particular, the Φ^X define a local Lie group with Lie algebra equal to $\mathfrak{g}$.

Proof. Assertion (a) follows from the local existence theorem for solutions of ordinary differential equations, combined with the locally uniform estimates for the domains of existence with respect to the parameters on which the vector fields depend continuously. Similarly we obtain $\Phi^Y(V) \subset V_X$, for all $X, Y \in U$, by sufficiently shrinking U.

In order to prove the identity in (b), consider, for $X, Y \in \mathfrak{g}$, the vector field on M:

$$Y(t) = (\Phi^{tX})_*^{-1}(\mathrm{e}^{t\,\mathrm{ad}\,X}(Y)); \tag{1.8.9}$$

here $\mathrm{e}^{t\,\mathrm{ad}\,X}(Y) \in \mathfrak{g}$ is computed in terms of the finite-dimensional Lie algebra structure of $\mathfrak{g}$ and Φ^{tX} is the flow of the vector field X after time t. For any local diffeomorphism Φ and vector field v, the vector field $\Phi_* v$ is defined by:

$$(\Phi_* v)(\Phi(x)) = \mathrm{T}_x\,\Phi(v(x)), \quad \text{for } x \text{ in the domain of definition of } \Phi. \tag{1.8.10}$$

$\Phi_* v$ can be interpreted as $(\mathrm{Ad}\,\Phi)(v)$ if the space of vector fields on M is viewed as the Lie algebra of the group of transformations, and then Φ^X can be interpreted as $\exp X$. In order to avoid confusion we don't yet make these identifications here, but the following arguments will provide further support for this point of view.

Differentiating (1.8.9) with respect to t, or using Formula (11.2.6), we get:

$$\frac{dY}{dt}(t) = -[X, Y(t)] + (\Phi^{tX})_*^{-1}[X, \mathrm{e}^{t\,\mathrm{ad}\,X}(Y)] = -[X, Y(t)] + [X, Y(t)] = 0,$$

where we have used that $\Phi_*^{-1}[X, Y] = [\Phi_*^{-1}X, \Phi_*^{-1}Y]$ and $(\Phi^{tX})_*^{-1}(X) = X$. Because $Y(0) = Y$, the conclusion is that $Y(t) = Y$, or:

$$(\Phi^{tX})_*(Y) = \mathrm{e}^{t\,\mathrm{ad}\,X}(Y) \quad \text{on } V_{tX}. \tag{1.8.11}$$

(This confirms the interpretation above, see Theorem 1.5.2.(a).) Now consider:

$$Z(t) = \mu(tX, Y). \tag{1.8.12}$$

Then, using (1.5.2) for the computation of $\frac{d}{dt}\Phi^{Z(t)}(x)$, replacing ϵ, and t, by t, and 1, respectively, we obtain:

$$\begin{aligned} \frac{d}{dt}(\Phi^{-tX} \circ \Phi^{Z(t)}(x)) &= -X(\Phi^{-tX} \circ \Phi^{Z(t)}(x)) \\ &+ \mathrm{T}_{\Phi^{Z(t)}(x)}(\Phi^{-tX}) \circ \mathrm{T}_x(\Phi^{Z(t)}) \circ \int_0^1 (\Phi^{sZ(t)})_*^{-1}\Big(\frac{dZ}{dt}(t)\Big)\, ds\, (x). \end{aligned} \tag{1.8.13}$$

Now, combining (1.8.11) with (1.5.3) and (1.6.2), we get:

$$\int_0^1 \left(\Phi^{sZ(t)}\right)_*^{-1}\left(\frac{dZ}{dt}(t)\right) ds = \int_0^1 \mathrm{e}^{-s\,\mathrm{ad}\,Z(t)}\, ds\left(\frac{dZ}{dt}(t)\right)$$
$$= \mathrm{e}^{-\,\mathrm{ad}\,Z(t)}(X) = \left(\Phi^{Z(t)}\right)_*^{-1}(X).$$

If we apply (1.8.10) with Φ, and v, replaced by $\Phi^{Z(t)}$, and X, respectively, we see that the right hand side in (1.8.13) is equal to 0. Hence $\Phi^{-tX} \circ \Phi^{Z(t)} = \Phi^{Z(0)} = \Phi^Y$, or:

(1.8.14) $$\Phi^{Z(t)} = \Phi^{tX} \circ \Phi^Y,$$

which for $t = 1$ becomes the required identity in (b).

The last statement follows because otherwise we would get sequences X_j, Y_j in $\mathfrak{g}$ converging to 0, and sets $V_j \subset M$ such that $X_j \neq Y_j$, for all j, but $\Phi^{X_j}|_{V_j} = \Phi^{Y_j}|_{V_j}$, and such that, for each relatively compact subset V of M, there is a j_0 with $V_j \supset V$, for all $j \geq j_0$. Passing to $Z_j = \mu(X_j, -Y_j)$, we get a sequence in $\mathfrak{g}$, converging to 0, such that $Z_j \neq 0$ and $\Phi^{Z_j}|_{V_j}$ is the identity on V_j, for a sequence V_j of subsets of M as above. Passing to a suitable subsequence, we may assume that $\frac{1}{|Z_j|} Z_j$ converges in $\mathfrak{g}$ to, say Z. Here $X \mapsto |X|$ denotes a norm in $\mathfrak{g}$; clearly $|Z| = 1$, so in particular $Z \neq 0$. Let $t > 0$ and $m_j = \left[\frac{t}{|Z_j|}\right]$, the integral part of $\frac{t}{|Z_j|}$. Because $|Z_j| \to 0$ we have that $m_j|Z_j| \to t$ as $j \to \infty$, so $m_j Z_j = (m_j|Z_j|)(\frac{1}{|Z_j|} Z_j) \to tZ$.

On the other hand,

$$\Phi^{m_j Z_j} = (\Phi^{Z_j})^{m_j} = \text{ identity on } V_j,$$

and therefore we conclude that $\Phi^{tZ} =$ identity on M, for all $t \geq 0$. Differentiation with respect to t at $t = 0$, now yields $Z = 0$, a contradiction. □

The required associativity for μ now follows from the associativity for transformations, in combination with (b) and (c).

(1.8.4) *Remarks.* If (and only if) the vector fields $X \in \mathfrak{g}$ are *complete*, that is, if Φ^X is defined on all of M (and then is a diffeomorphism: $M \to M$) for each $X \in \mathfrak{g}$, we can form the group G of C^1 diffeomorphisms: $M \to M$ generated by the Φ^X, for $X \in \mathfrak{g}$. Using the charts (1.6.9) with "log" replaced by the inverse of $X \mapsto \Phi^X$ (with X in a suitable neighborhood of 0 in $\mathfrak{g}$), we can turn G into a Lie group with Lie algebra equal to $\mathfrak{g}$.

Unfortunately the vector fields X^{R} on $\mathfrak{g}_e$ defined in (1.8.2) are not always complete, so this does not lead yet to the conclusion that every finite-dimensional Lie algebra is the Lie algebra of a Lie group.

(1.8.5) *Remark.* The whole theory of Sections 1.6 - 1.8, possibly except (c) in Theorem 1.8.3 (because we used local compactness) goes through if we relax the condition that $\mathfrak{g}$ is finite-dimensional to the assumption that $\mathfrak{g}$ is a *Banach Lie algebra*, that is, a Lie algebra that is provided with a norm $X \mapsto |X|$ such that: (i) $\mathfrak{g}$ is complete and (ii) $X \mapsto \operatorname{ad} X$ is a continuous linear mapping from $\mathfrak{g}$ to the space of continuous linear mappings: $\mathfrak{g} \to \mathfrak{g}$. This second requirement is equivalent to:

$$\exists C \geq 0 \text{ such that } |[X, Y]| \leq C|X||Y| \qquad (X, Y \in \mathfrak{g}). \tag{1.8.15}$$

Actually, the smallest constant C in (1.8.15) is just the operator norm $|\operatorname{ad}|$ of $\operatorname{ad}: \mathfrak{g} \to \mathbf{L}(\mathfrak{g}, \mathfrak{g})$, where $\mathbf{L}(\mathfrak{g}, \mathfrak{g})$ is provided with the operator norm.

Warning. The condition above is quite restrictive, for instance, the space of all C^1 vector fields on a manifold M is not a Banach Lie algebra. For an infinite-dimensional example of a Banach Lie algebra, see Section 1.14.

(1.8.6) *Remark.* If X^{R} and Y^{R} are represented by convergent power series (at $0 \in \mathfrak{g}$) then the integrals in the proof of Lemma 1.8.2 can be expressed by recursive algebraic expressions for the coefficients. In this fashion, the whole theory of Sections 1.6 - 1.8 can be given a much more algebraic flavor (if desired).

1.9 The Component of the Identity

Let us recall that a topological space X is said to be *connected* if it cannot be written as the union of two disjoint nonvoid open subsets. It follows that X is connected if it is the union of a collection of connected subsets of X with a nonvoid common intersection. This makes the relation, for $x, y \in X$, that x and y belong to a connected subset of X, into an equivalence relation. The equivalence classes are called the *connected components* of X, they are the maximal connected subsets of X. If (and only if) X is *locally connected*, that is each $x \in X$ has a connected neighborhood in X, then we may conclude that the connected components of X are open subsets of X. But then they are also closed subsets of X, being the complement of the other connected components. An example of a topological space that is not locally connected, is $\mathbf{Q}$ provided with the usual interval topology.

Also, the image of a connected topological space under a continuous mapping is connected. Because an interval in $\mathbf{R}$ is connected, one obtains that X is connected if it is *pathwise connected*, that is, if for any two points $x, y \in X$ there is a continuous mapping $\gamma: [a, b] \to X$ such that $\gamma(a) = x$, $\gamma(b) = y$. Obviously $\mathbf{R}^n$ is pathwise connected, and therefore a topological manifold X is locally pathwise connected: each $x \in X$ has a pathwise connected neighborhood in X. For any topological space X the relation that $x, y \in X$ can be connected by a continuous curve γ as above, is an equivalence relation.

The equivalence classes are called the *pathwise connected components* of X, they are the maximal pathwise connected subsets of X. In general each connected component of X is a union of pathwise connected components. If (and only if) X is *locally pathwise connected*, that is, each $x \in X$ has a pathwise connected neighborhood in X, then the pathwise connected components of X are open (and closed) subsets of X, and coincide with the connected components of X. In particular this is true if X is a topological manifold.

(1.9.1) Theorem. *If G is a Lie group, then G°, the connected component of the identity element, is a pathwise connected, open and closed subset of G, and also a normal subgroup of G. The connected components of G are precisely the cosets $xG^\circ = G^\circ x$, with $x \in G$. The connected component G° is contained in each open subgroup of G. For each neighborhood V of 1 in G°, we have that G° is equal to the subgroup of G generated by V. In particular, G° is generated by $\exp U$ for every neighborhood U of 0 in $\mathfrak{g}$. The set G° is equal to the union of countably many compact subsets, and G is paracompact.*

Proof. Since G is locally pathwise connected, it follows that G° is a pathwise connected open and closed subset of G. Because multiplication and inversion are continuous, we obtain that $\{ xy^{-1} \mid x, y \in G^\circ \}$ is a connected set containing 1, hence this set is contained in G°; and it follows that G° is a subgroup of G. Further, $xG^\circ x^{-1}$ is connected and contains 1, for any $x \in G$; thus it is contained in G°, and this implies that G° is a normal subgroup of G. The cosets xG° are connected and open, hence they are connected components of G. Because each $x \in G$ is contained in xG°, every connected component of G is of this form. Now let V be a neighborhood of 1 in G°, then it contains a symmetric, that is, invariant under $x \mapsto x^{-1}$, neighborhood W of 1. Therefore $H = \cup_{n=1}^{\infty} W^n$ is a subgroup and open, but this implies that $G \setminus H = \cup_{x \notin H} xH$ is open too. We obtain: $G^\circ \subset H \subset \cup_{n=1}^{\infty} V^n \subset G^\circ$, i.e.:

$$G^\circ = \bigcup_{n=1}^{\infty} V^n. \tag{1.9.1}$$

If, in addition, we take V to be compact, then V^n is compact. Accordingly $U_n = V^{\text{int}} V^n$ is an open neighborhood of V^n whose closure $\overline{U}_n$ is compact, being contained in V^{n+1}. If now $\mathcal{A}$ is an open covering of G°, then we can find a finite subset $\mathcal{B}_n$ of the set $\{ A \setminus \overline{U}_{n-2} \mid A \in \mathcal{A} \}$ such that $\mathcal{B}_n$ is a covering of $\overline{U}_n \setminus U_{n-1}$. Then $\cup_{n=1}^{\infty} \mathcal{B}_n$ is a locally finite refinement of $\mathcal{A}$, and thus G° satisfies the definition of paracompactness; but this implies that G is paracompact. □

Remark. In the course of the proof above we have established the fact that an open subgroup of a topological group is closed.

The quotient group $G/G^\circ = \{ xG^\circ = G^\circ x \mid x \in G \}$ is called the *group of connected components* of G or the *component group* of G. Each open subgroup H of G, because it contains G°, is the union of cosets $xG^\circ = G^\circ x$, for $x \in H$; and the map $H \mapsto H/G^\circ \subset G/G^\circ$ is a bijection between the collection of open subgroups of G and the collection of subgroups of G/G°. Any open subgroup H of G (in particular also G°), provided with the manifold structure of G, is a Lie group with the same Lie algebra as G. Note that if we change the manifold structure of G to the 0-dimensional one, then the connected component of G is equal to $\{1\}$ and the component group is equal to G.

(1.9.2) Proposition. *Let G be a Lie group with Lie algebra $\mathfrak{g}$. If $X, Y \in \mathfrak{g}$, and $[X, Y] = 0$, then $\exp X \exp Y = \exp Y \exp X$. Moreover, $[X, Y] = 0$ for all $X, Y \in \mathfrak{g}$ if and only if G° is commutative (Abelian).*

Proof. Using Theorem 1.5.1, we have:

$$(\mathbf{Ad}(\exp X))(\exp Y) = \exp(e^{\operatorname{ad} X}(Y)),$$

and $e^{\operatorname{ad} X}(Y) = Y$ if $(\operatorname{ad} X)(Y) = 0$. This proves the first assertion. It now follows, if $\operatorname{ad} = 0$, that the group generated by $\exp \mathfrak{g}$, which is equal to G°, is commutative. Conversely, if G° is commutative, then, for each $x \in G^\circ$, the mapping $\mathbf{Ad}\, x|_{G^\circ}$ is equal to the identity: $G^\circ \to G^\circ$; and hence $\operatorname{Ad} x = \mathrm{I}$ on $\mathfrak{g}$ because G° is open in G. Differentiating this equality with respect to x at $x = 1$, we get that $\operatorname{ad} = 0$, again using that G° is open in G. □

(1.9.3) Definition. *A Lie algebra $\mathfrak{g}$ is said to be Abelian if $\operatorname{ad} = 0$, that is, if $[X, Y] = 0$ for all $X, Y \in \mathfrak{g}$.*

Proposition 1.9.2 now says that the Lie algebra of a Lie group G is Abelian if and only if the connected component G° of the identity element is Abelian. In particular, if G is connected, then it is Abelian if and only if its Lie algebra is Abelian.

(1.9.4) Proposition. *If the Lie algebra $\mathfrak{g}$ of the Lie group G is a complex Lie algebra (cf. Definition 1.1.6) then G°, provided with the logarithmic coordinate charts, is a complex Lie group. (Conversely it is obvious that if G° is a complex Lie group, then ad is a complex-bilinear mapping: $\mathfrak{g} \times \mathfrak{g} \to \mathfrak{g}$.)*

Proof. If $X \in \mathfrak{g}$ and $\operatorname{ad} X$ is complex-linear: $\mathfrak{g} \to \mathfrak{g}$, then $\operatorname{Ad}(\exp X) = e^{\operatorname{ad} X}$ is complex-linear: $\mathfrak{g} \to \mathfrak{g}$. Therefore, if $\operatorname{ad} X$ is complex-linear: $\mathfrak{g} \to \mathfrak{g}$ for each $X \in \mathfrak{g}$, then $\operatorname{Ad} x$ is complex-linear: $\mathfrak{g} \to \mathfrak{g}$ for each x in the subgroup of G generated by $\exp \mathfrak{g}$, which is G°. Now apply the last assertion of Theorem 1.6.3, with G replaced by G°. □

Propositions 1.9.2 and 1.9.4 are examples of the general principle that many properties of the connected Lie group G° are equivalent to properties of the Lie algebra $\mathfrak{g}$. It cannot be expected that information about the Lie algebra $\mathfrak{g}$ (infinitesimal properties of G at the identity element) extends to the other connected components of G.

Example. Illustration to Proposition 1.9.2. In $\mathbf{SU}(2)$ we have the circle group:

$$T = \left\{ \begin{pmatrix} a & 0 \\ 0 & \bar{a} \end{pmatrix} \middle| \, a \in \mathbf{C}, |a| = 1 \right\},$$

which is an Abelian, one-dimensional, compact Lie group. Its *normalizer*:

$$\mathrm{N}(T) = \{ x \in \mathbf{SU}(2) \mid xTx^{-1} = T \}$$

in $\mathbf{SU}(2)$ is also one-dimensional and compact, but has two connected components: T and the circle (not a group):

$$T' = \left\{ \begin{pmatrix} 0 & b \\ -\bar{b} & 0 \end{pmatrix} \middle| \, b \in \mathbf{C}, |b| = 1 \right\} = \begin{pmatrix} 0 & 1 \\ -1 & 0 \end{pmatrix} T.$$

The component group $\mathrm{N}(T)/T$ is isomorphic to the multiplicative group $\{\pm 1\}$ of two elements. For more details about the group structure of $\mathrm{N}(T)$, observe that:

$$\begin{pmatrix} 0 & b \\ -\bar{b} & 0 \end{pmatrix} \begin{pmatrix} 0 & b' \\ -\bar{b}' & 0 \end{pmatrix} = \begin{pmatrix} -b\bar{b}' & 0 \\ 0 & -\bar{b}b' \end{pmatrix} = \begin{pmatrix} -b(b')^{-1} & 0 \\ 0 & -b^{-1}b' \end{pmatrix},$$

so we have in $\mathrm{N}(T)$:

$$\begin{pmatrix} 0 & b \\ -\bar{b} & 0 \end{pmatrix}^{-1} = \begin{pmatrix} 0 & -b \\ \bar{b} & 0 \end{pmatrix}.$$

Furthermore:

$$\begin{pmatrix} a & 0 \\ 0 & \bar{a} \end{pmatrix} \begin{pmatrix} 0 & b \\ -\bar{b} & 0 \end{pmatrix} = \begin{pmatrix} 0 & ab \\ -\overline{ab} & 0 \end{pmatrix}, \quad \begin{pmatrix} 0 & b \\ -\bar{b} & 0 \end{pmatrix} \begin{pmatrix} a & 0 \\ 0 & \bar{a} \end{pmatrix} = \begin{pmatrix} 0 & b\bar{a} \\ -\bar{b}a & 0 \end{pmatrix},$$

so $\mathrm{L}\begin{pmatrix} a & 0 \\ 0 & \bar{a} \end{pmatrix}$ acts on T' as a rotation over the angle $\arg a$, and $\mathrm{R}\begin{pmatrix} a & 0 \\ 0 & \bar{a} \end{pmatrix}$ acts on T' as a rotation over $-\arg a$. In particular, conjugation by $\begin{pmatrix} a & 0 \\ 0 & \bar{a} \end{pmatrix} \in T$ acts on T' as multiplication by a^2, that is rotation over $2 \arg a$. This shows that T' is also a single conjugacy class, making the group $\mathrm{N}(T) = T \cup T'$ highly noncommutative. On the other hand, conjugation by $\begin{pmatrix} 0 & b \\ -\bar{b} & 0 \end{pmatrix} \in T'$ acts on $\begin{pmatrix} a & 0 \\ 0 & \bar{a} \end{pmatrix}$ by interchanging a and $\bar{a}$: the reflection of the unit circle in the complex plane about the real axis. (That this conjugation does not depend on the choice of the element in T' is explained by the fact that $T' = \begin{pmatrix} 0 & 1 \\ -1 & 0 \end{pmatrix} T$ and that T is Abelian.) So the conjugacy classes of elements in T consist of two points, except for the elements $\pm\mathrm{I}$, which are a single conjugacy class.

Clearly all these noncommutativity properties of $\mathrm{N}(T)$ cannot be read off from the Lie algebra of $\mathrm{N}(T)$, or equivalently, from T. Even knowledge of T and the component group $\mathrm{N}(T)/T \simeq \{\pm 1\}$ together does not give any hint (both being Abelian), that is, $\mathrm{N}(T)$ is certainly not isomorphic (as a group) to the Cartesian product of $\mathrm{N}(T)/T$ and T. For instance, every element in T' has its square equal to $-\mathrm{I}$, so $\mathrm{N}(T)$ does not have a subgroup which is isomorphic to $\mathrm{N}(T)/T$ under the quotient mapping: $\mathrm{N}(T) \to \mathrm{N}(T)/T$. This shows that neither $\mathrm{N}(T)$ is equal to the semidirect product of $\mathrm{N}(T)/T$ and T.

Example. Illustration to Proposition 1.9.4. Let $G = \mathbf{GL}(n, \mathbf{C})$, with $n \in \mathbf{Z}_{>0}$. Regarding each complex-linear mapping $A\colon \mathbf{C}^n \to \mathbf{C}^n$ as a real-linear one: $\mathbf{R}^{2n} \to \mathbf{R}^{2n}$, we can consider G as a subgroup of $\mathbf{GL}(2n, \mathbf{R})$. The complex conjugation:

$$C : z = x + iy \mapsto \bar{z} = x - iy \qquad (x, y \in \mathbf{R}^n) \tag{1.9.2}$$

belongs to $\mathbf{GL}(2n, \mathbf{R})$: for each $A \in \mathbf{GL}(n, \mathbf{C})$ one verifies that $\bar{A} = CAC^{-1} = CAC \in \mathbf{GL}(n, \mathbf{C})$. In fact, the matrix entries of $\bar{A}$ (as a complex $n \times n$-matrix) are the complex conjugates of the matrix entries of A. The group G' generated by G and C is equal to the union of G and CG, it is a Lie group with two components, and with Lie algebra equal to $\mathfrak{gl}(n, \mathbf{C})$. However, $\operatorname{Ad} C\colon A \mapsto \bar{A}$ is definitely not complex-linear, so G' is not a complex Lie group, although its connected component is one and its Lie algebra is complex too.

1.10 Lie Subgroups and Homomorphisms

(1.10.1) Definition. *If H and G are Lie groups, then a homomorphism of Lie groups from H to G is a mapping $\Phi\colon H \to G$ that is a homomorphism of groups, and at the same time an analytic mapping from the analytic manifold H to the analytic manifold G. If $G = \mathbf{GL}(V)$, then Φ is called a (linear) representation (of Lie groups) of H in the vector space V. A mapping Φ is called an isomorphism of Lie groups: $H \to G$ if Φ is a homomorphism of Lie groups: $H \to G$, and moreover, if Φ is bijective and Φ^{-1} is a homomorphism of Lie groups: $G \to H$. A Lie subgroup of a Lie group G is a subgroup H of G with the structure of a Lie group in such a way that the inclusion: $H \to G$ is an analytic immersion (and therefore a homomorphism of Lie groups: $H \to G$). Notice that a Lie subgroup H is not necessarily closed in G.*

If Φ is a homomorphism of Lie groups: $H \to G$, then Φ is an isomorphism of Lie groups: $H \to G$ if Φ is bijective and Φ^{-1} is analytic, because Φ^{-1} is automatically a homomorphism. In turn Φ^{-1} is analytic as soon as $\mathrm{T}_1 \Phi\colon \mathfrak{h} \to \mathfrak{g}$ is invertible. Indeed, the chain rule applied to $\Phi(xy) = \Phi(x)\Phi(y)$ gives that $\mathrm{T}_x \Phi \circ \mathrm{T}_1 \mathrm{L}(x) = \mathrm{T}_1 \mathrm{L}(\Phi(x)) \circ \mathrm{T}_1 \Phi$; and this shows that $\mathrm{T}_x \Phi$ is invertible for every $x \in H$. Then we may apply the inverse function theorem to conclude that Φ^{-1} is analytic.

Warning 1.6.5 shows that the topology of a Lie subgroup H of G may be strictly finer than the topology of G restricted to H. For a less trivial example, see the dense subgroups of the torus described in Section 1.12.

(1.10.2) Definition. *If $\mathfrak{h}$ and $\mathfrak{g}$ are Lie algebras, then a homomorphism of Lie algebras from $\mathfrak{h}$ to $\mathfrak{g}$ is a linear mapping $\phi\colon \mathfrak{h} \to \mathfrak{g}$ such that:*

$$[\phi(X), \phi(Y)]_{\mathfrak{g}} = \phi([X, Y]_{\mathfrak{h}}) \qquad (X, Y \in \mathfrak{h}). \tag{1.10.1}$$

ϕ is called an isomorphism of Lie algebras: $\mathfrak{h} \to \mathfrak{g}$ if ϕ is a bijective homomorphism of Lie algebras: $\mathfrak{h} \to \mathfrak{g}$; in that case ϕ^{-1} is automatically a homomorphism of Lie algebras: $\mathfrak{g} \to \mathfrak{h}$. A Lie subalgebra of a Lie algebra $\mathfrak{g}$ is a linear subspace $\mathfrak{h}$ of $\mathfrak{g}$ such that:

$$[X, Y] \in \mathfrak{h} \qquad (X, Y \in \mathfrak{h}). \tag{1.10.2}$$

It follows that the restriction of the Lie bracket to $\mathfrak{h} \times \mathfrak{h}$ makes $\mathfrak{h}$ into a Lie algebra, and that the identity: $\mathfrak{h} \to \mathfrak{g}$ is a homomorphism of Lie algebras from $\mathfrak{h}$ to $\mathfrak{g}$. In particular, the Lie algebra $\mathfrak{h}$ of a Lie subgroup H of a Lie group G becomes a Lie subalgebra of the Lie algebra $\mathfrak{g}$ of G.

Proposition 1.1.3 can be reformulated as follows. If Φ is a homomorphism of Lie groups: $H \to G$, with Lie algebras $\mathfrak{h} = \mathrm{T}_1 H$, and $\mathfrak{g} = \mathrm{T}_1 G$, respectively, then $\phi = \mathrm{T}_1 \Phi$ is a homomorphism of Lie algebras: $\mathfrak{h} \to \mathfrak{g}$.

We have already seen a nontrivial example in Lemma 1.8.2: for every Lie algebra $\mathfrak{g}$ the mapping $X \mapsto X^{\mathrm{R}}$ is an injective homomorphism from $\mathfrak{g}$ to the Lie algebra $\mathcal{X}^{\omega}(\mathfrak{g}_e)$ of analytic vector fields on $\mathfrak{g}_e$, making $\mathfrak{g}$ isomorphic to a Lie subalgebra of $\mathcal{X}^{\omega}(\mathfrak{g}_e)$. If $\mathfrak{g}$ is the Lie algebra of a Lie group G, then $X \mapsto X^{\mathrm{R}}$ is an isomorphism between the Lie algebra $\mathfrak{g}$ and the Lie algebra of right invariant vector fields on G. (The inverse is given by evaluating the vector field at $1 \in G$.)

Many authors **define** the Lie algebra of G as the Lie algebra of right invariant vector fields on G; the remark above shows that this definition is equivalent to the one of Section 1.1. We have preferred the definition $\mathfrak{g} = \mathrm{T}_1 G$, because if G is a group of transformations, then $\mathrm{T}_1 G$ is the corresponding system of infinitesimal transformations (that is, vector fields, in modern language), which is the way Lie thought about the subject.

Warning. $X \mapsto X^{\mathrm{L}}$ is **not** a homomorphism of Lie algebras (unless $[X,Y] = 0$, for all $X, Y \in \mathfrak{g}$), instead we have the rule:

$$[X^{\mathrm{L}}, Y^{\mathrm{L}}] = -([X,Y])^{\mathrm{L}} \qquad (X, Y \in \mathfrak{g}). \tag{1.10.3}$$

One way of seeing this is to observe that the inversion $\iota \colon x \mapsto x^{-1} \colon G \to G$ satisfies $\iota(xy) = (xy)^{-1} = y^{-1}x^{-1} = \iota(y)\iota(x)$, for all $x, y \in G$. One expresses this by saying that ι is an *anti-homomorphism of groups*: $G \to G$ rather than a homomorphism, or that ι is a homomorphism from G to the "opposite" group G', equal to G provided with the product $\cdot'$ defined by $x \cdot' y = y \cdot x$. It follows that ι_* maps left invariant vector fields to right invariant vector fields, and because $\mathrm{T}_1\, \iota = -\mathrm{I} \colon \mathfrak{g} \to \mathfrak{g}$, we see that:

$$X^{\mathrm{L}} = -\iota_*(X^{\mathrm{R}}) \qquad (X \in \mathfrak{g}). \tag{1.10.4}$$

Now (1.10.3) follows from:

$$\begin{aligned} [X^{\mathrm{L}}, Y^{\mathrm{L}}] &= [-\iota_*(X^{\mathrm{R}}), -\iota_*(Y^{\mathrm{R}})] = [\iota_*(X^{\mathrm{R}}), \iota_*(Y^{\mathrm{R}})] \\ &= \iota_*[X^{\mathrm{R}}, Y^{\mathrm{R}}] = \iota_*([X,Y]^{\mathrm{R}}) = -([X,Y])^{\mathrm{L}}. \end{aligned}$$

One expresses (1.10.3) also by saying that $X \mapsto X^{\mathrm{L}}$ is an *anti-homomorphism of Lie algebras.*

The next theorem is a converse to the last statement in Definition 1.10.2. In it we need the discussion of connected Lie groups of Section 1.9.

(1.10.3) Theorem. *Let G be a Lie group with Lie algebra $\mathfrak{g}$. For each Lie subalgebra $\mathfrak{h}$ of $\mathfrak{g}$ there is a unique connected Lie subgroup H of G with Lie algebra equal to $\mathfrak{h}$. (Such a subgroup H is often called an analytic subgroup.) In fact, as a subset of G, the group H is equal to the subgroup of G generated by* $\exp(\mathfrak{h})$.

Proof. For each $X \in \mathfrak{h}$, the vector field X^{R} on $\mathfrak{g}_e$ defined in (1.8.2) is tangent to $\mathfrak{h}$ in the sense that $X^{\mathrm{R}}(Z) \in \mathfrak{h}$ for all $Z \in \mathfrak{h} \cap \mathfrak{g}_e$. This follows because $\mathrm{ad}\, Z \in \mathbf{L}(\mathfrak{h}, \mathfrak{h})$ if $Z \in \mathfrak{h}$. Moreover $\mathfrak{h}$ is a closed subspace of $\mathfrak{g}$ as a finite-dimensional linear subspace, hence the power series in the right hand side of (1.8.2) converges to an element of $\mathfrak{h}$. (In the case of Banach Lie groups, we have to make the additional assumption that $\mathfrak{h}$ is a closed linear subspace of $\mathfrak{g}$, but then the theorem remains valid.) If we solve (1.6.2) with $Z(0) = Y \in \mathfrak{h}$, regarded as a differential equation in $\mathfrak{h} \cap \mathfrak{g}_e$, the solution will be in $\mathfrak{h} \cap \mathfrak{g}_e$. However, it is also equal to the solution of (1.6.2) as a differential equation in $\mathfrak{g}_e$, and according to (1.6.6) it is equal to $\log(\exp tX \exp Y)$. So the conclusion is that $\log(\exp X \exp Y) \in \mathfrak{h}$ if $(X,Y) \in (\mathfrak{h} \times \mathfrak{h}) \cap \mathfrak{g}_e^2$. Denote by H the subgroup generated by $\exp(\mathfrak{h})$, then the $\kappa^x|_H$, for $x \in H$, with κ defined as in (1.6.9), form an atlas for H of local coordinatizations, mapping into $\mathfrak{h}$. Replacing G by H and $\mathfrak{g}$ by $\mathfrak{h}$ in the proof of Theorem 1.6.3, we see that this atlas makes H into an analytic Lie group. The inclusion: $H \to G$ is an immersion, because in the charts κ^x it is equal to the inclusion from open subsets of $\mathfrak{h}$ into $\mathfrak{g}$. In view of Theorem 1.9.1, H is connected, and it is the only connected Lie subgroup of G having $\mathfrak{h}$ as its Lie algebra. For the uniqueness of the structure of a real-analytic group on H with prescribed Lie algebra $\mathfrak{h}$, see Proposition 1.6.4. □

(1.10.4) *Example.* Let $\mathfrak{g}$ be a finite-dimensional Lie algebra. Because ad is a homomorphism of Lie algebras: $\mathfrak{g} \to \mathbf{L}(\mathfrak{g}, \mathfrak{g})$, see (1.1.22), $\operatorname{ad} \mathfrak{g}$ is a Lie subalgebra of $\mathbf{L}(\mathfrak{g}, \mathfrak{g})$, the Lie algebra of $\mathbf{GL}(\mathfrak{g})$. According to Theorem 1.10.3, the subgroup of $\mathbf{GL}(\mathfrak{g})$ generated by the $\mathrm{e}^{\operatorname{ad} X}$, for $X \in \mathfrak{g}$, is the unique connected Lie subgroup of $\mathbf{GL}(\mathfrak{g})$ with Lie algebra equal to $\operatorname{ad} \mathfrak{g}$. It is called the *adjoint group* $\operatorname{Ad} \mathfrak{g}$ *of the Lie algebra* $\mathfrak{g}$.

If $X, Y, Z \in \mathfrak{g}$, then:

$$\begin{aligned}
&\frac{d}{dt}\, \mathrm{e}^{-t \operatorname{ad} X} [\, \mathrm{e}^{t \operatorname{ad} X}(Y), \mathrm{e}^{t \operatorname{ad} X}(Z) \,] \\
&= - \mathrm{e}^{-t \operatorname{ad} X} \circ \operatorname{ad} X \circ [\, \mathrm{e}^{t \operatorname{ad} X}(Y), \mathrm{e}^{t \operatorname{ad} X}(Z) \,] \\
&+ \mathrm{e}^{-t \operatorname{ad} X} \Big([\, \operatorname{ad} X \circ \mathrm{e}^{t \operatorname{ad} X}(Y), \mathrm{e}^{t \operatorname{ad} X}(Z)] \\
&+ [\, \mathrm{e}^{t \operatorname{ad} X}(Y), \operatorname{ad} X \circ \mathrm{e}^{t \operatorname{ad} X}(Z) \,] \Big) = 0,
\end{aligned}$$

because of the Jacobi identity. So $t \mapsto \mathrm{e}^{-t \operatorname{ad} X} [\, \mathrm{e}^{t \operatorname{ad} X}(Y), \mathrm{e}^{t \operatorname{ad} X}(Z) \,]$ is constant, equal to its value $[\, Y, Z \,]$ at $t = 0$. In other words,

$$[\, \mathrm{e}^{\operatorname{ad} X}(Y), \mathrm{e}^{\operatorname{ad} X}(Z) \,] = \mathrm{e}^{\operatorname{ad} X} [\, Y, Z \,];$$

and this shows that $\mathrm{e}^{\operatorname{ad} X}$ is an automorphism of $\mathfrak{g}$, for any $X \in \mathfrak{g}$. It follows that $\operatorname{Ad} \mathfrak{g}$ is a subgroup of the *group* $\operatorname{Aut} \mathfrak{g}$ *of all automorphisms* of $\mathfrak{g}$. The elements of $\operatorname{Ad} \mathfrak{g}$ therefore are also called the *inner automorphisms* of $\mathfrak{g}$. The reader may verify directly from an analysis of the equations:

$$\Phi([\, X_i, X_j \,]) = [\, \Phi(X_i), \Phi(X_j) \,], \quad \text{with } i, j, = 1, \ldots, n,$$

for the elements $\Phi \in \operatorname{Aut} \mathfrak{g}$, where $X_1, \ldots, X_n$ is a basis of $\mathfrak{g}$, that $\operatorname{Aut} \mathfrak{g}$ is an analytic submanifold, hence a closed Lie subgroup, of $\mathbf{GL}(\mathfrak{g})$ (a shorter proof is to use Corollary 1.10.7 below). Furthermore, the Lie algebra of $\operatorname{Aut} \mathfrak{g}$ is equal to:

$$\begin{aligned}
&\operatorname{Der} \mathfrak{g} = \\
&\{\, \phi \in \mathbf{L}(\mathfrak{g}, \mathfrak{g}) \mid \phi([\, X, Y \,]) = [\, \phi(X), Y \,] + [\, X, \phi(Y) \,], \forall\, X, Y \in \mathfrak{g} \,\},
\end{aligned} \tag{1.10.5}$$

the Lie subalgebra of $\mathbf{L}(\mathfrak{g}, \mathfrak{g})$ consisting of the *infinitesimal automorphisms* or *derivations* of $\mathfrak{g}$. Notice that $\operatorname{ad} \mathfrak{g} \subset \operatorname{Der} \mathfrak{g}$. In general $\operatorname{Ad} \mathfrak{g}$ need not be closed in $\operatorname{Aut} \mathfrak{g}$, hence in $\mathbf{GL}(\mathfrak{g})$, whereas on the other hand $\operatorname{Aut} \mathfrak{g}$ need not be connected.

Further it should be observed that **if** G is a Lie group with Lie algebra $\mathfrak{g}$, then Ad is a homomorphism of Lie groups: $G \to \mathbf{GL}(\mathfrak{g})$, with $\mathrm{T}_1 \operatorname{Ad} = \operatorname{ad}$, so it follows that Ad actually maps G° homomorphically onto $\mathrm{Ad}\,\mathfrak{g}$. In formula: $\mathrm{Ad}(G^\circ) = \mathrm{Ad}\,\mathfrak{g}$. The point is that nonisomorphic connected Lie groups with isomorphic Lie algebras do exist; the result above shows that their adjoint groups then are isomorphic and can be defined intrinsically in terms of the Lie algebras.

In Theorem 1.5.2.(b) we have seen that $x \exp X x^{-1} = \exp\big((\mathrm{Ad}\, x)(X)\big)$, for any $x \in G$ and $X \in \mathfrak{g}$. So if $x \in \ker \mathrm{Ad}$, then $xyx^{-1} = y$ for all $y \in \exp \mathfrak{g}$, hence also for all $y \in G^\circ$, the subgroup of G generated by $\exp \mathfrak{g}$. Conversely, this implies that $x \in \ker \mathrm{Ad}$. In particular, $\ker \mathrm{Ad} \cap G^\circ = \mathrm{Z}(G^\circ)$, the *center* of G°; this is defined as the set of all $x \in G^\circ$ which commute with all $y \in G^\circ$. From Proposition 1.11.8 below we get that $\mathrm{Z}(G^\circ)$ is a closed Lie subgroup of G° and that $\mathrm{Ad}\colon G^\circ \twoheadrightarrow \mathrm{Ad}\,\mathfrak{g}$ induces an isomorphism of Lie groups: $G^\circ / \mathrm{Z}(G^\circ) \to \mathrm{Ad}\,\mathfrak{g}$, and this completes our somewhat lengthy discussion of the group $\mathrm{Ad}\,\mathfrak{g}$ of interior automorphisms of $\mathfrak{g}$.

If H is a connected Lie group then a *Lie group covering* of H is a homomorphism of Lie groups $\pi'\colon H' \to H$ where H' is a connected Lie group and $\mathrm{T}_1\,\pi'$ is an isomorphism of Lie algebras. This implies that $\pi'\colon H' \to H$ is an analytic covering, that is, a fibration with a discrete fiber (see Definition A.3 in Appendix A). We now have the following application of Theorem 1.10.3, which is a converse to Proposition 1.1.3:

(1.10.5) Corollary. *Let H, and G, be Lie groups with Lie algebras $\mathfrak{h}$, and $\mathfrak{g}$, respectively. Assume that H is connected and that each Lie group covering of H is injective. Then for every homomorphism of Lie algebras $\phi\colon \mathfrak{h} \to \mathfrak{g}$, there is a unique homomorphism of Lie groups $\Phi\colon H \to G$ such that $\phi = \mathrm{T}_1\,\Phi$.*

Proof. The condition that ϕ is a homomorphism of Lie algebras just means that its graph is a Lie subalgebra of $\mathfrak{h} \times \mathfrak{g}$, which is regarded as the Lie algebra of $H \times G$. According to Theorem 1.10.3 there is a connected Lie subgroup F of $H \times G$ having $\mathrm{graph}\,\phi$ as its Lie algebra.

Let π_1 be the composition of the inclusion: $F \to H \times G$ and the projection $(x, y) \mapsto x\colon H \times G \to H$. Then π_1 is a homomorphism of Lie groups: $F \to H$, being the composition of two such ones. Its tangent map at 1_F is equal to the projection of $\mathrm{graph}\,\phi \subset \mathfrak{h} \times \mathfrak{g}$ onto the first variable, which is invertible. Using the same argument as in the observation following Definition 1.10.1, we get that $\mathrm{T}_f\,\pi_1$ is invertible for $f \in F$. So $\pi_1(F)$ is an open subgroup of H, and H is connected, so $\pi_1(F) = H$ by Theorem 1.9.1. Because also F is connected, π_1 is a covering of Lie groups: $F \to H$, and thus by assumption π_1 is an isomorphism. By the inverse function theorem, π_1^{-1} is analytic, hence a homomorphism of Lie groups: $H \to F$. Writing π_2 for the composition of the inclusion: $F \to H \times G$ with the projection $(x, y) \mapsto y\colon H \times G \to G$, we get that $\Phi = \pi_2 \circ \pi_1^{-1}$ is a homomorphism of Lie groups: $H \to G$, having F as its graph, and $\mathrm{T}_1\,\Phi = \phi$ because $\mathrm{T}_{(1,1)}\,F = \mathrm{graph}\,\phi$.

Finally the uniqueness of Φ follows from the observation that if Φ is a homomorphism as desired, then its graph is a connected Lie subgroup of $H \times G$ having graph ϕ as its tangent space at $(1,1)$, so it follows from the uniqueness in Theorem 1.10.3 that graph $\Phi = F$. □

The condition on H in Corollary 1.10.5 is clearly necessary. Indeed, let $\pi'\colon H' \to H$ be a covering of Lie groups, with Lie algebras $\mathfrak{h}'$, and $\mathfrak{h}$, respectively. If there is a homomorphism of Lie groups $\Phi\colon H \to H'$ with $\mathrm{T}_1 \Phi = (\mathrm{T}_1 \pi')^{-1}$, then the uniqueness argument at the end of the proof, combined with the connectedness of H', and H, gives that $\Phi \circ \pi'$, and $\pi' \circ \Phi$, is equal to the identity in H', and H, respectively; so π' is an isomorphism in that case.

A different proof of Corollary 1.10.5, not based on Theorem 1.10.3 or Section 1.6, will be given in Remark 1.13.5.

From Proposition 1.13.2 we shall obtain that a connected Lie group H has the property that every Lie group covering of it is an isomorphism if and only if the topological space H is *simply connected*, that is, every loop in H can be shrunk (within H) to a point.

As an example, the Lie group $\mathbf{SU}(2)$ described in Section 1.2.(b) is simply connected, because this is the case for every sphere of dimension ≥ 2. Therefore, every homomorphism $\phi\colon \mathfrak{su}(2) \to \mathfrak{g}$, with $\mathfrak{g}$ the Lie algebra of any Lie group G, is equal to $\mathrm{T}_1 \Phi$ for a unique homomorphism of Lie groups $\Phi\colon \mathbf{SU}(2) \to G$. On the other hand, $\mathrm{Ad}\colon \mathbf{SU}(2) \to \mathbf{SO}(3, \mathbf{R})$ is a two-fold covering, $\ker \mathrm{Ad} = \{-\mathrm{I}, +\mathrm{I}\}$, so $\mathbf{SO}(3, \mathbf{R})$ does not satisfy the condition for H in Corollary 1.10.5. If ψ is a homomorphism of Lie algebras: $\mathfrak{so}(3, \mathbf{R}) \to \mathfrak{g}$, then it is equal to $\mathrm{T}_1 \Psi$ for a homomorphism of Lie groups $\Psi\colon \mathbf{SO}(3, \mathbf{R}) \to G$, if and only if:

$$\exp \psi \begin{pmatrix} 0 & 0 & 0 \\ 0 & 0 & -2\pi \\ 0 & 2\pi & 0 \end{pmatrix} = 1_G.$$

To see this, apply the previous statement to $\phi = \psi \circ \mathrm{ad}$, from which Ψ is determined by $\Phi = \Psi \circ \mathrm{Ad}$; such a homomorphism of Lie groups Ψ exists if and only if $\Phi(-\mathrm{I}) = 1_G$. Now use that:

$$-\mathrm{I} = \exp \mathrm{ad}^{-1} \begin{pmatrix} 0 & 0 & 0 \\ 0 & 0 & -2\pi \\ 0 & 2\pi & 0 \end{pmatrix}.$$

According to the Remarks 1.13.6, one can specify a finite set of $X_i \in \mathfrak{h}$ such that any homomorphism of Lie algebras $\phi\colon \mathfrak{h} \to \mathfrak{g}$ is equal to $\mathrm{T}_1 \Phi$ for a homomorphism of Lie groups $\Phi\colon H \to G$, if and only if $\exp \phi(X_i) = 1_G$ for all i. This generalizes the criterion above for $H = \mathbf{SO}(3, \mathbf{R})$.

We can weaken analyticity assumptions in Lie groups not only to C^2 assumptions, but even to assumptions of a purely topological nature, like continuity, still getting the same end results. The following is a typical example of this principle.

(1.10.6) Theorem. *Let H_0 be a closed subset of G that at the same time is a "local subgroup of G" in the sense that $1 \in H_0$ and that there is a symmetric neighborhood V of 1 in G such that $x^{-1} \in H_0$ and $xy \in H_0$, for all $x, y \in V \cap H_0$. Then there is a Lie subalgebra $\mathfrak{h}$ of $\mathfrak{g}$ such that $\exp U$ is a neighborhood of 1 in H_0, for each sufficiently small neighborhood U of 0 in $\mathfrak{h}$. In particular, H_0 is a neighborhood of 1 in the connected Lie subgroup H of G with Lie algebra equal to $\mathfrak{h}$.*

Proof. Because the statements are local, we may restrict ourselves to a logarithmic chart. We will also make use of a norm $X \mapsto |X|$ in $\mathfrak{g}$.

Let $\mathfrak{h}$ be the set of $X \in \mathfrak{g}$ for which there exist sequences $h_n \in H_0$ and $t_n \in \mathbf{R}_{\geq 0}$, such that:

$$\lim_{n \to \infty} h_n = 1 \quad \text{in} \quad H_0, \tag{1.10.6}$$

and

$$\lim_{n \to \infty} t_n \log h_n = X \quad \text{in} \quad \mathfrak{g}. \tag{1.10.7}$$

We first show that there is a neighborhood U of 0 in $\mathfrak{h}$ such that $\exp U \subset H_0$. Because of the continuity of exp at 0, there is an $\eta > 0$ such that $\exp X \in V$, whenever $X \in \mathfrak{g}$, and $|X| < \eta$. Choose $0 < \epsilon < \eta$ and select $X \in U = \{ X \in \mathfrak{h} \mid |X| \leq \epsilon \}$. Let h_n, t_n be as above. Because $\lim_{n \to \infty} \log h_n = 0$, we get $(t_n - [t_n]) \log h_n \to 0$, if $[t]$ denotes the integral part of t; so $[t_n] \log h_n \to X$. If $0 \leq k \leq [t_n] - 1$, then:

$$\begin{aligned} |k \log h_n| &= \frac{k}{[t_n]} |[t_n] \log h_n| \leq \frac{k}{[t_n]} (|X| + |[t_n] \log h_n - X|) \\ &\leq \epsilon + |[t_n] \log h_n - X| < \eta, \end{aligned}$$

for sufficiently large n. So $h_n^k = \exp(k \log h_n) \in V$, for $0 \leq k \leq [t_n] - 1$; but now $h_n \in H_0 \cap V$ yields that $h_n^{[t_n]} \in H_0$ because of the assumption. Using the continuity of exp, we get that:

$$\exp X = \lim_{n \to \infty} \exp([t_n] \log h_n) = \lim_{n \to \infty} h_n^{[t_n]},$$

and the limit belongs to H_0 because H_0 is closed. This proves $\exp U \subset H_0$.

Next we show that $\mathfrak{h}$ is a linear subspace of $\mathfrak{g}$. Obviously $tX \in \mathfrak{h}$, if $X \in \mathfrak{h}$ and $t \in \mathbf{R}_{\geq 0}$. Also $-X \in \mathfrak{h}$, because $V \cap H_0$ is invariant under inversion, which in logarithmic coordinates is the mapping $X \mapsto -X$. If now $X, Y \in \mathfrak{h}$, then by the result just proved $\exp tX, \exp tY \in H_0 \cap V$, for all sufficiently small t; hence $\exp tX \exp tY \in H_0$, for all sufficiently small t. On the other hand (cf. (1.7.4)):

$$\lim_{t \downarrow 0} \frac{1}{t} \log(\exp tX \exp tY) = X + Y,$$

from which we read off that $X + Y \in \mathfrak{h}$.

Thirdly we prove that, for any neighborhood U of 0 in $\mathfrak{h}$, the set $\exp U$ contains a neighborhood of 1 in H_0. For if not, then there exists a sequence $h_n \in H_0 \setminus \exp U$ such that $h_n \to 1$. Choose a linear subspace $\mathfrak{k}$ of $\mathfrak{g}$ such that $\mathfrak{g} = \mathfrak{h} \oplus \mathfrak{k}$. Then $(X, Y) \mapsto \exp X \exp Y$ is a diffeomorphism from a neighborhood of $(0, 0)$ in $\mathfrak{h} \times \mathfrak{k}$ to a neighborhood of 1 in G. Hence there exist sequences $X_n \in \mathfrak{h}$, and $Y_n \in \mathfrak{k}$, both converging to 0, such that $h_n = \exp X_n \exp Y_n$, for sufficiently large n. Note that $Y_n \neq 0$, because $h_n \notin \exp U$. Because $h_n \in H_0$, $\exp X_n \in H_0$, it follows that $\exp Y_n \in H_0$, for sufficiently large n, according to the assumption in the first line of the theorem. On the other hand, a subsequence of $\frac{1}{|Y_n|} Y_n$ converges to some $Y \in \mathfrak{k}$ with $|Y| = 1$. By the definition of $\mathfrak{h}$ we conclude that $Y \in \mathfrak{h}$, in contradiction with $\mathfrak{h} \cap \mathfrak{k} = 0$.

Taking $U = U' \cap \mathfrak{h}$, with U' a sufficiently small open neighborhood of 0 in $\mathfrak{g}$ such that $\exp|_{U'}$ is a diffeomorphism from U' onto an open neighborhood V' of 1 in G, one obtains that $V = \exp U$ is a real-analytic submanifold of G and $\exp|_U$ is a real-analytic diffeomorphism from U onto V. Because on the other hand V is an open neighborhood of 1 in H_0, it follows that V is a real-analytic local Lie group, with Lie algebra $\mathfrak{h}$, which apparently is a Lie subalgebra of $\mathfrak{g}$. If H denotes the group generated by V, then H is the connected Lie subgroup of G with Lie algebra equal to $\mathfrak{h}$ that was introduced in Theorem 1.10.3. □

(1.10.7) Corollary. *Let H be a subgroup of a Lie group G. Then the following statements are equivalent:*

(i) For some $h \in H$ there is a closed neighborhood V of h in G such that $H \cap V$ is closed in G.

(ii) H is a closed Lie subgroup of G. The Lie group structure of H is unique if we require that its topology as a Lie group is equal to the restriction topology as a subset of G.

Proof. Suppose (i) is valid. Applying the homeomorphism $\mathrm{L}(h)^{-1}$, we see that we may assume that $h = 1$ in (i). Now let $x \in G$ be a limit of a sequence $x_n \in H$. Then:

$$\lim_{n,m\to\infty} x_n x_m^{-1} = e \Rightarrow x_n x_m^{-1} \in V, \quad \text{for } n, m \text{ sufficiently large} .$$

It follows that:

$$x x_m^{-1} = \lim_{n\to\infty} x_n x_m^{-1} \in H \cap V,$$

for m sufficiently large: this implies that $x \in H$, because $x_m \in H$ and H is a group. This shows that H is closed in G. From Theorem 1.10.6 we see that H is a Lie subgroup of G; actually the proof of Theorem 1.10.6 becomes slightly simpler if we may assume that H is a closed subgroup of G instead of only a closed local group. Note that in Theorem 1.10.6 the exponential mapping is a homeomorphism from U to a neighborhood of 1 in H with respect to the restriction topology making the Lie group topology of H equal to the restriction topology. □

Remark. It is also known that H is a Lie subgroup of G if H is a subgroup of a Lie group G such that, for every $x \in H$, there is a continuous curve $\gamma\colon [0,1] \to G$ satisfying $\gamma(0) = 1$, $\gamma(1) = x$, $\gamma(t) \in H$ for all $t \in [0,1]$. Here the Lie group structure of H is unique if we require that H is connected with respect to its Lie group topology.

The proof of Yamabe [1950] combines elements of the proofs of Theorem 1.10.6 and of Corollary 11.3.2 with a degree argument (continuous mappings that are uniformly close to diffeomorphisms are surjective).

If in the assumption the continuous curves γ are replaced by C^1 curves γ, then the result follows from Corollary 11.3.2, by looking at the action of H on G by left (or right) multiplications. The theorem that every pathwise connected subgroup of a Lie group is a Lie subgroup, is attributed in the literature to Freudenthal [1941], who proved it in the case of real-analytic curves γ, instead of continuous curves γ.

(1.10.8) Proposition. *Let G, H be Lie groups and assume that G has only countably many connected components. Then every homomorphism of groups $\Phi\colon H \to G$ with a closed graph $\subset H \times G$ is analytic, that is, Φ is a homomorphism of Lie groups: $H \to G$.*

Proof. Write $F = \operatorname{graph} \Phi$. Because Φ is a homomorphism, F is a subgroup of $H \times G$, and because it is closed, F is a Lie subgroup in view of Corollary 1.10.7. Let π_1, and π_2, be the projections: $F \to H$, and $F \to G$, respectively, as in Corollary 1.10.5. Because the dimension of $T_f\,\pi_1$ is constant (as a function of $f \in F$), the mapping π_1 is, locally in F, a fibration with fiber dimension equal to the dimension of $\ker T_f\,\pi_1$. However, π_1 is invertible, so the fibers consist of single points and we conclude that π_1 is an immersion.

Because $\Phi \circ L(x) = L(\Phi(x)) \circ \Phi$ for any $x \in H$, it is sufficient to prove analyticity for Φ in a neighborhood of 1 in H, or in H°, so we may assume that H is connected. It follows that $H \times G$, and therefore F, is a countable union of compact subsets, because F is closed in $H \times G$.

We conclude that, if $\dim F < \dim H$, then $\pi_1(F)$ has measure 0 in H, in contradiction with $\pi_1(F) = H$. So $\dim F \geq \dim H$ and π_1 is actually a local diffeomorphism, and hence a diffeomorphism because π_1 is invertible. It finally follows that $\Phi = \pi_2 \circ \pi_1^{-1}$ is analytic: $H \to G$.

That π_1 is a local diffeomorphism, also can be proved as follows. $H = \pi_1(F)$ is a countable union of closed subsets. Since H is locally compact, Baire's theorem asserts that at least one of these sets contains an open subset of H; and the conclusion follows.

For a proof of Baire's theorem, suppose that H has a nonempty interior and is the union of a countable family F_j of closed subsets each having an empty interior. Then one can construct, by induction on i, a sequence of compact subsets $K_1 \supset K_2 \supset \ldots$, such that $K_i^{\text{int}} \neq \emptyset$ and:

$$(\bigcup_{j=1}^{i} \mathrm{F}_j) \cap K_i = \emptyset, \quad \text{for all} \quad i.$$

Because this implies that the intersection of the K_i is void, we get a contradiction. □

Note that, without any assumption on G, Corollary 1.10.8 is false: take $\dim H > 0$, $G = H$ with the discrete topology, and $\Phi =$ identity: $H \to G$. The graph of Φ is closed because Φ^{-1} is continuous: $G \to H$.

(1.10.9) Corollary. *Every continuous homomorphism between Lie groups is analytic.*

Proof. A continuous mapping maps the identity component into the identity component, so we may assume both groups to be connected. Also a continuous map has a closed graph. □

(1.10.10) Corollary. *Let G, H be Lie groups, H having only countably many connected components. Let Φ be an isomorphism of groups: $H \to G$ and continuous. Then Φ is an isomorphism of Lie groups: $H \to G$, that is, both Φ and Φ^{-1} are analytic.*

Proof. The assumptions imply that also G has only countably many components, so we can apply Corollary 1.10.8 to both Φ and Φ^{-1}. □

Example. let $H = (\mathbf{R}, +)$, and $G = \mathbf{GL}(V)$ with V a finite-dimensional vector space. Let h be a mapping: $(\mathbf{R}, +) \to \mathbf{GL}(V)$ such that:

(i) $h(t + s) = h(t)h(s)$, for all $t, s \in \mathbf{R}$;
(ii) if $t_n \to 0 \in \mathbf{R}$, and $h(t_n) \to A$ in $\mathbf{GL}(V)$, then $A = \mathrm{I}$.

Then the conclusion is that there is some $B \in \mathbf{L}(V, V)$ such that $h(t) = \mathrm{e}^{tB}$, for all $t \in \mathbf{R}$. The assumption (ii) can even be further weakened, because one can show that discontinuous homomorphisms: $(\mathbf{R}, +) \to \mathbf{GL}(V)$ must behave rather wildly.

1.11 Quotients

Let M be a set. An *action* of a group G on M is a homomorphism A from G to the group of bijective mappings: $M \to M$. Writing

$$(A(g))(x) = A(g, x) \qquad (g \in G, x \in M), \tag{1.11.1}$$

we can also describe the action of G on M as a mapping $A\colon G \times M \to M$ such that:

$$A(gh,x) = A\big(g, A(h,x)\big) \qquad (g,h \in G, x \in M). \tag{1.11.2}$$

If G is a Lie group and M a C^k manifold, with $k \geq 1$, then a C^k *action* of G on M is an action A that is C^k as a mapping: $G \times M \to M$. For a topological group G (see Section 11.5) and a topological space M a *continuous action* can be defined as one for which $A\colon G \times M \to M$ is continuous.

For each $x \in M$ the *orbit through* x for the action A is defined as the set:

$$A(G)(x) = G \cdot x = \{\, A(g)(x) \mid g \in G \,\}. \tag{1.11.3}$$

Here the second notation is related to the abbreviation:

$$A(g)(x) = g \cdot x \qquad (g \in G, x \in M), \tag{1.11.4}$$

which is quite customary (although it suppresses the action A).

(1.11.1) Lemma. *The relation $y \in G \cdot x$ is an equivalence relation in M, which partitions M into orbits.*

Proof. If 1 denotes the identity element of G, then $A(1)(x) = \text{identity}\,(x) = x$, so $x \in G \cdot x$. Secondly, if $y \in G \cdot x$, then $y = A(g)(x)$, for some $g \in G$. But then $x = A(g)^{-1}(y) = A(g^{-1})(y)$; hence $x \in G \cdot y$, showing that the relation is symmetric. Finally, if $y \in G \cdot x$ and $z \in G \cdot y$ then $y = g \cdot x$ and $z = h \cdot y$, for some $g, h \in G$; hence $z = h \cdot (g \cdot x) = (hg) \cdot x$, or $z \in G \cdot x$: the relation is transitive. □

If R is an equivalence relation in a topological space M, we denote by M/R the set of equivalence classes and by $\pi\colon M \to M/R$ the canonical projection which assigns to each $x \in M$ its equivalence class $\{\, y \in M \mid (x,y) \in R \,\}$. Defining a subset V of M/R to be open if and only if $\pi^{-1}(V)$ is open in M, we get a topology on M/R for which π is continuous. Actually this is the strongest topology on M/R with this property, and it is called the *quotient topology* on M/R.

(1.11.2) Lemma. *Let M be a topological space and R an equivalence relation in M. Then the quotient topology of M/R has the Hausdorff property if and only if R is a closed subset of $M \times M$.*

Proof. For any subset U of M, write $U^R = \pi^{-1}(\pi(U))$ for the set of $y \in M$ such that $(x,y) \in R$, for some $x \in U$. The lemma is an immediate consequence of the following statement: if U and V are subsets of M, then $(U \times V) \cap R = \emptyset$ if and only if $U^R \cap V^R = \emptyset$. (Apply this with U, and V, equal to an open neighborhood of x, and y, respectively, such that $(x,y) \notin R$.) In order to prove the latter statement, we observe that $U^R \cap V^R \neq \emptyset$ means that there exist $z \in M$, $x \in U$, $y \in V$ such that $(x,z) \in R$ and $(z,y) \in R$; and this is equivalent to $(x,y) \in R$.

The collection of orbits is called the *set-theoretic quotient* $G\backslash M$ of M under the action A of G on M. (The notation $G\backslash M$ is not very specific, because it does not mention the action A.) The surjective mapping:

$$\pi\colon x \mapsto G\cdot x\colon M \to G\backslash M \tag{1.11.5}$$

is called the *canonical projection*. Even for analytic actions of Lie groups on analytic manifolds it may happen that the quotient $G\backslash M$ cannot be provided with the structure of a Hausdorff topological space making the canonical projection continuous, see Remark 1.11.6 below. In certain algebraic situations one avoids this by taking as a quotient the space of closed orbits, see Section 14.7. However, in this section we will restrict ourselves to actions which are so nice that the set-theoretic quotient space can be provided with the structure of a smooth manifold, making the canonical projection into a smooth fibration, see Theorem 1.11.4 below.

The action is said to be *transitive* if M is an orbit, that is, if there is an $x \in M$ such that for each $y \in M$ we have $y = g\cdot x$, for some $g \in G$. This is equivalent to saying that there is only one orbit (which then must be equal to M), that is, for each $x, y \in M$ there is a $g \in G$ such that $y = g\cdot x$. In this case M is also called a *homogeneous space*.

A subset N of M is said to be *invariant under the action A* or *G-invariant* if $g\cdot y \in N$, whenever $y \in N$, $g \in G$. That is, $A(G\times N) \subset N$, and the restriction of A to $G\times N$ defines an action of G on N, called the *restriction to N of the action* of G on M. Note that N is invariant if and only if N is a union of orbits, thus identifying the collection of invariant subsets of M with the collection of all subsets of $G\backslash M$. The minimal nonvoid invariant subsets are the orbits; these are also precisely the nonvoid invariant subsets on which the (restriction of the) action is transitive.

For each $x \in M$,

$$G_x = \{\, g \in G \mid g\cdot x = x \,\} \tag{1.11.6}$$

is clearly a subgroup of G, called the *isotropy group* of the action at the point x, or the *stabilizer* of x under the action. (Again, this notation is not very specific.) The mapping:

$$A_x\colon g \mapsto A(g)(x)\colon G \to G\cdot x \tag{1.11.7}$$

is surjective, by the definition of $G\cdot x$. On the other hand, $A(h)(x) = A(g)(x)$ if and only if $g^{-1}h \in G_x$, that is, $h \in gG_x$. This shows that A_x is injective if and only if $G_x = \{1\}$; in this case the action is said to be *free at x*. The action is said to be *free* if it is free at x, for every $x \in M$, that is, if each orbit is in a bijective correspondence with G by means of the mappings A_x, with $x \in M$.

Let H be any subgroup of G. Then:

$$h \mapsto \mathrm{L}(h)\colon x \mapsto hx, \tag{1.11.8}$$

and,

$$h \mapsto \mathrm{R}(h)^{-1}\colon x \mapsto xh^{-1}, \tag{1.11.9}$$

define a free action of H on G, called the *action from the left*, and *the right*, respectively, of H on G. The orbits are the Hx (that is, the right cosets equal the right translates of H), and the xH, respectively, with $x \in G$ (that is, the left cosets equal the left translates of H in G). The corresponding quotient spaces are denoted by:

$$H\backslash G = \{\, Hx \mid x \in G \,\}, \tag{1.11.10}$$

and

$$G/H = \{\, xH \mid x \in G \,\}, \tag{1.11.11}$$

respectively.

The remark following (1.11.7) shows that the fibers of $A_x\colon G \to G \cdot x$ are just the cosets gG_x, for $g \in G$; and there is a unique bijective mapping:

$$B_x\colon G/G_x \xrightarrow{\sim} G \cdot x \tag{1.11.12}$$

such that $A_x = B_x \circ \pi$, if π denotes the canonical projection: $G \to G/G_x$.

As a further exercise with the definitions, we note that, for any subgroup H of G,

$$g \mapsto (xH \mapsto gxH) \qquad (g, x \in G), \tag{1.11.13}$$

defines a transitive action of G on G/H, with isotropy group at $1H$ equal to H. (The isotropy group at $xH \in G/H$ is equal to xHx^{-1}.) This shows that every subgroup H of G is equal to the isotropy group, at some point, of some action of G.

Finally, we recall that the product:

$$(xH) \cdot (yH) = (xy)H \qquad (x, y \in G), \tag{1.11.14}$$

in G/H is well-defined if and only if H is a *normal subgroup* of G, that is,

$$xHx^{-1} = H \qquad (x \in G). \tag{1.11.15}$$

This makes $G/H = H\backslash G$ into a group. The canonical projection $G \to G/H$ is a group homomorphism with kernel equal to H. Because the kernel of every group homomorphism is a normal subgroup, this shows that the normal subgroups are precisely the kernels of group homomorphisms from G to some other group G'. In Corollary 1.11.5 we shall see that if G is a Lie group, then the closed normal subgroups are precisely the kernels of homomorphisms of Lie groups $G \to G'$, with G' some other Lie group.

Until now we have only discussed some set-theoretical generalities about group actions. We now turn to their topological and smoothness properties. Recall that a continuous mapping Φ from a topological space U to a topological space V is said to be *proper* if $\Phi^{-1}(K)$ is compact in U for every compact subset K of V. (If U and V are Hausdorff, this implies that $\Phi(C)$ is closed in V for every closed subset C of U.) A continuous action of a topological group G on a topological space M is said to be a *proper action* if:

$$(1.11.16) \quad (g, x) \mapsto (g \cdot x, x) \quad \text{is a proper mapping} : G \times M \to M \times M.$$

(1.11.3) Lemma. *For a proper continuous action, the quotient topology on the orbit space has the Hausdorff property.*

Proof. According to Lemma 1.11.2 we have to prove that the orbit relation is closed. This is the case, because it is the image of $G \times M$ in $M \times M$ under the proper and continuous mapping $(g, x) \mapsto (g \cdot x, x)$.

(1.11.4) Theorem. *Let G be a Lie group, M a C^k manifold, for $k \geq 1$, and A a C^k action of G on M that is* **proper and free**. *Then the orbit space $G\backslash M$ has a unique structure of a C^k manifold, of dimension equal to* $\dim M - \dim G$, *with the following properties. If $\pi\colon M \to G\backslash M$ is the canonical projection $x \mapsto G \cdot x$, then for every $b \in G\backslash M$ there is an open neighborhood S in $G\backslash M$ and a C^k diffeomorphism:*

$$\tau\colon x \mapsto (\chi(x), s(x))\colon \pi^{-1}(S) \to G \times S,$$

such that, for $x \in \pi^{-1}(S)$, $g \in G$:

$$(1.11.17) \qquad \pi(x) = s(x) \quad \text{and} \quad \tau(g \cdot x) = (g\chi(x), s(x)).$$

The topology of $G\backslash M$ is equal to the quotient topology.

Proof. $\mathrm{T}_1(A_x)$ is injective: $\mathfrak{g} \to \mathrm{T}_x M$, for each $x \in M$. Indeed, let $X \in \mathfrak{g}$ satisfy $\mathrm{T}_1(A_x)(X) = 0$. Then:

$$\frac{d}{dt} \exp tX \cdot x = \left.\frac{d}{dh}\right|_{h=0} \exp tX \exp hX \cdot x = \mathrm{T}_x\big(A(\exp tX)\big) \circ \mathrm{T}_1(A_x)(X) = 0,$$

for all $t \in \mathbf{R}$. Hence $t \mapsto \exp tX \cdot x$ is constant; so it is equal to its value x at $t = 0$. The freeness of the action at x now implies that $\exp tX = 1$, for all $t \in \mathbf{R}$; differentiating this with respect to t at $t = 0$, yields $X = 0$.

Next, let S' be a C^k submanifold of M through x such that $\mathrm{T}_x S'$ is a linear complement of $\mathrm{T}_1(A_x)(\mathfrak{g})$ in $\mathrm{T}_x M$. By continuity it follows that there is an open neighborhood S of x in S', such that for all $y \in S$:

$$\mathrm{T}_1(A_y)(\mathfrak{g}) \oplus \mathrm{T}_y S = \mathrm{T}_y M.$$

Combined with the injectivity of $\mathrm{T}_1(A_y)$, it follows that:

$$B = A|_{G\times S}$$

has a bijective tangent map at $(1, y)$, for each $y \in S$. Using that $A(gh, y) = A(g)(A(h, y))$, for $g, h \in G$, $y \in S$, we get, differentiating this with respect to (h, y) at $h = 1$, $y \in S$, that:

$$\mathrm{T}_{(g,y)}\, B = \mathrm{T}_y\, A(g) \circ T_{(1,y)} B.$$

Now $A(g)$ is a diffeomorphism, so $\mathrm{T}_y\, A(g)$ is bijective; and the conclusion is that B has a bijective tangent map everywhere on $G \times S$. By the inverse function theorem it follows that B is a local C^k diffeomorphism from $G \times S$ onto an open subset U of M.

By sufficiently shrinking S, we can arrange that B is also injective; and then it is a C^k diffeomorphism: $G \times S \to U$. Indeed, if not, then there would be sequences y_j and $z_j \in S$, g_j and $h_j \in G$ such that:

$$\lim_{j\to\infty} y_j = x, \quad \lim_{j\to\infty} z_j = x, \quad g_j \cdot y_j = h_j \cdot z_j, \quad (g_j, y_j) \neq (h_j, z_j), \text{ for all } j.$$

This implies that $g_j \neq h_j$, for all j. Write $k_j = h_j^{-1} g_j$, so $k_j \neq 1$, $z_j = k_j \cdot y_j$. Because the $(k_j \cdot y_j, y_j) = (z_j, y_j)$ are contained in a compact subset of $M \times M$ (they converge to (x, x)), the properness of the action yields that the k_j are contained in a compact subset of G. Hence by passing to a suitable subsequence, we may assume that $\lim_{j\to\infty} k_j = k$, for some $k \in G$. By continuity of the action, $x = \lim_{j\to\infty} k_j \cdot y_j = k \cdot x$; so $k = 1$, the action being free at x. But now we get a contradiction with the injectivity of B on a suitable neighborhood of $(1, x)$ in $G \times S$.

Because:

$$A(g)\big(B(h, y)\big) = A(g) \circ A(h)(y) = A(gh)(y) = B(gh, y),$$

we see that $\tau = B^{-1}$ satisfies (1.11.17). We still have to prove that $G\backslash M$ has the structure of a C^k manifold, such that $y \mapsto G \cdot y$ is a C^k diffeomorphism from S onto an open subset of $G\backslash M$. Our task therefore is to show that if S, and x, are replaced by manifolds T, and $x' \in G \cdot x$, respectively, with the same properties, and if $V = A(G \times T)$, the corresponding G-invariant open subset of M, then:

$$U \cap V = A(G \times S_0) = A(G \times T_0),$$

for open subsets S_0, and T_0, of S, and T, respectively, and the mapping κ which assigns to $y \in S_0$ the unique element of $G \cdot y \cap T_0$ is a C^k diffeomorphism (see Fig.1.11.1).

But this follows directly: $B^{-1}(U \cap V)$ is open in $G \times S$ and invariant under the left action of G on the first factor, so it must be of the form $G \times S_0$ for some open subset S_0 of S. Similarly $U \cap V = A(G \times T_0)$ for an open subset T_0 of T. Now $B^{-1}(T_0)$ is a C^k submanifold of $G \times S_0$, intersecting each G-orbit $G \times \{y\}$, with $y \in S_0$, exactly once and transversally, because T_0 does so in

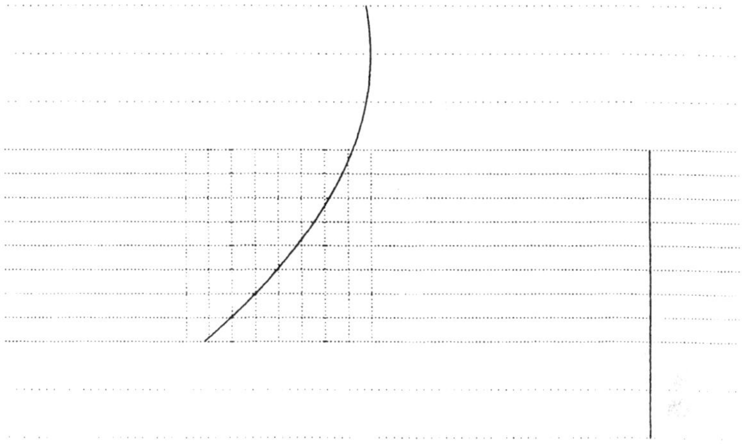

Fig. 1.11.1.

$U \cap V$. This shows that $B^{-1}(T_0)$ equals $\{ (\chi(y), y) \mid y \in S_0 \}$, the graph (in the wrong order) of a C^k mapping $\chi\colon S_0 \to G$, hence $\kappa\colon y \mapsto \chi(y) \cdot y\colon S_0 \to T_0$ is C^k as well. The same argument with the roles of S_0 and T_0 interchanged shows that κ^{-1} is also C^k. Clearly the topology of $G\backslash M$ is equal to the quotient topology and it follows from Lemma 1.11.3 that it is Hausdorff. □

The τ in Theorem 1.11.4 are the local trivializations making $\pi\colon M \to G\backslash M$ into a C^k fiber bundle; they also translate the action of G on M into the left action of G on the first component in $G \times S$. If $\sigma\colon \pi^{-1}(T) \to G \times T$ is another such trivialization then we have:

(1.11.18) $$\tau \circ \sigma^{-1}\colon (g, x) \mapsto (g\chi(x), x)\colon G \times (S \cap T) \to G \times (S \cap T),$$

for a C^k function $\chi\colon S \cap T \to G$, similarly as in the proof of Theorem 1.11.4. In the terminology of Definition A.4 in Appendix A a fiber bundle M with a Lie group G as fiber and retrivializations $\tau \circ \sigma^{-1}$ as in (1.11.18), is called a *principal fiber bundle with structure group G*. One verifies easily that the left G-actions on the first factor in the trivializations $G \times S$ induce an action of G on the principal fiber bundle that is proper and free, with the fibers as the orbits. In this way having a proper and free action of G on M is equivalent to saying that M is a principal fiber bundle with structure group G.

In Chapter 2 we shall drop the hypothesis that the action is free and investigate general proper actions, with applications to the structure of compact Lie groups. Later, in the study of noncompact Lie groups, we shall also have to deal with Lie group actions which are neither proper nor free.

Examples. For any Lie group G, the action $(x, X) \mapsto (\mathrm{Ad}\, x)(X) : G \times \mathfrak{g} \to \mathfrak{g}$ of G on $\mathfrak{g}$ is called the *adjoint action*. Lie group actions for which the transformations are linear transformations in a vector space, are precisely the *representations* of Definition 1.10.1. This is the reason why the adjoint action is also called the adjoint representation. Let us take a look at the simplest examples of Section 1.2. For $G = \mathbf{SO}(3, \mathbf{R})$ or $\mathbf{SU}(2)$, the orbits of the adjoint action are either points or two-dimensional spheres, the stabilizer groups G_x then are either the whole group G or a circle subgroup $T = S^1$ of G. It follows that $G \to G/T$ exhibits $\mathbf{SO}(3, \mathbf{R})$ and $\mathbf{SU}(2)$ as a principal circle bundle over the two-dimensional sphere S^2, both being nontrivial, that is, not equivalent to a Cartesian product. It is interesting to identify these bundles in the classification of all principal circle bundles over S^2, see Steenrod [1951]. For $G = \mathbf{SU}(2)$, we get the three-dimensional sphere S^3 as a principal circle bundle over S^2, which is the same as the action of the circle $\{ z \in \mathbf{C} \mid |z| = 1 \}$ as a multiplicative subgroup of $\mathbf{C} \setminus \{0\}$ on the unit sphere in $\mathbf{C}^2 \simeq \mathbf{R}^4$ by means of the complex multiplication by the scalars z. In differential topology this is known as the *Hopf fibration*; it occurs implicitly at many places in the classical literature. For $G = \mathbf{SL}(2, \mathbf{R})$, the adjoint orbits are diffeomorphic to a point, the punctured plane (a one-sheeted hyperboloid or a punctured half-cone) or the plane (a sheet of the two-sheeted hyperboloid). In this case the stabilizer groups are either G itself or $\mathbf{R} \times (\mathbf{Z}/2\mathbf{Z})$, or the circle. The circle fibration of $\mathbf{SL}(2, \mathbf{R})$ over the plane is trivial, but the fibration of $\mathbf{SL}(2, \mathbf{R})$ over the punctured plane is not: $\mathbf{SL}(2, \mathbf{R})$ is connected and the fibers are not connected.

For a Lie group G, the mapping $(g, x) \mapsto (gx, x)$ is bijective: $G \times G \to G \times G$. Since the inverse $(y, x) \mapsto (yx^{-1}, x)$ is continuous, it follows that the left action of G on G is proper and free. Similarly the right action of G on G is proper and free.

If H is a closed subgroup of G then it follows immediately that the left and the right action of H on G are proper and free as well. Because a closed subgroup of a Lie group is a Lie subgroup (Corollary 1.10.7) we therefore get:

(1.11.5) Corollary. *For any closed subgroup H of a Lie group G, there is a unique structure of an analytic manifold on $H\backslash G$, and G/H, making $G \to H\backslash G$, and $G \to G/H$, respectively, into an analytic principal fiber bundle with structure group H. The right, and left action, of G induces an analytic transitive action of G on $H\backslash G$, and G/H, respectively, with $G_{1H} = H$. Finally, if H is also a normal subgroup, then this analytic structure makes G/H into a Lie group and the canonical projection into a homomorphism of Lie groups.*

For a general C^k action of a Lie group G on a C^k manifold M, we can apply this to $H = G_x$, which is a closed subgroup, hence a Lie subgroup, of G because of the continuity of the mapping A_x (cf. (1.11.7)). Turning to the right action of G_x on G, we can identify G/G_x locally with the submanifolds of G occurring in the proof of Theorem 1.11.4 (replacing M, and G, by G, and G_x, respectively). We then see that the mapping B_x of (1.11.12) is a C^k immersion from the analytic manifold G/G_x into M, mapping G/G_x bijectively onto $G \cdot x$. This exhibits the orbit $G \cdot x$ as an immersed C^k submanifold of M, of dimension equal to $\dim G - \dim G_x$. Note also that, with the argument in the beginning of the proof of Theorem 1.11.4, the Lie algebra of G_x is equal to:

$$\mathrm{T}_1(G_x) = \mathfrak{g}_x = \ker \mathrm{T}_1(A_x), \tag{1.11.19}$$

whereas:

$$\mathrm{T}_x(G \cdot x) = \operatorname{im} \mathrm{T}_1(A_x) \xleftarrow{\sim} \mathfrak{g}/\mathfrak{g}_x. \tag{1.11.20}$$

Especially, a C^k homogeneous space is C^k diffeomorphic to an analytic manifold of the form G/H where H is a closed Lie subgroup of a Lie group G; this analytic structure is the unique one for which the action is analytic.

(1.11.6) *Remark.* If H is a nonclosed subgroup of G and G/H is provided with a topology for which the canonical projection: $G \to G/H$ is continuous, then a point xH, with $x \in H^{\mathrm{cl}} \setminus H$, is contained in any neighborhood of $1H$ in G/H. So G/H is not Hausdorff. An example is given by $H = \mathbf{Q}$ as a subgroup of the one-dimensional Lie group $(\mathbf{R}, +)$. Another example can be found at the end of Section 1.12.

We conclude this section with some more remarks about closed normal subgroups of Lie groups. A subalgebra $\mathfrak{h}$ of a Lie algebra $\mathfrak{g}$ is called an *ideal* in $\mathfrak{g}$ if:

$$[X, Y] \in \mathfrak{h} \qquad (X \in \mathfrak{g},\ Y \in \mathfrak{h}). \tag{1.11.21}$$

(1.11.7) Proposition. *Let H be a Lie subgroup of G with Lie algebra of H equal to $\mathfrak{h}$. Then the following assertions are equivalent.*

(i) H° *is normal in* G°.
(ii) $\operatorname{Ad} x(\mathfrak{h}) \subset \mathfrak{h}$, *for all* $x \in G^\circ$.
(iii) $\mathfrak{h}$ *is an ideal in* $\mathfrak{g}$.

If H is normal in G, then H° is normal in G. Finally, if $\mathfrak{h}$ is an ideal in $\mathfrak{g}$, then the group H generated by $\exp \mathfrak{h}$ is a connected Lie subgroup of G, normal in G°, and has Lie algebra equal to $\mathfrak{h}$.

Proof. If H° is normal in G°, then differentiating $h \mapsto (\mathbf{Ad}\, x)(h) \colon H \to H$ at $h = 1$ (for any $x \in G^\circ$) gives (ii). If (ii) holds, then differentiating $x \mapsto \operatorname{Ad} x(Y)$ at $x = 1$ (for any $Y \in \mathfrak{h}$) gives (iii). Finally if $X \in \mathfrak{g}$ and $\operatorname{ad} X(\mathfrak{h}) \subset \mathfrak{h}$, then $\mathrm{e}^{\operatorname{ad} X}(\mathfrak{h}) \subset \mathfrak{h}$, so $(\mathbf{Ad} \exp X)(\exp Y) = \exp(\mathrm{e}^{\operatorname{ad} X}(Y)) \in \exp \mathfrak{h}$, if $Y \in \mathfrak{h}$. Here we have used Theorem 1.5.2. Since H°, and G°, are generated by $\exp \mathfrak{h}$, and $\exp \mathfrak{g}$, respectively, this proves: (iii) $\Rightarrow$ (i). If $\Phi \colon H \to H$ is continuous, then it maps H° into H°; and applying this to the action of conjugation, we see that H° is normal in G if H is so. The last statement is obtained by combining Theorem 1.10.3 with (iii) $\Rightarrow$ (i). □

After this intermezzo on how to recognize from properties of its Lie algebra whether a Lie subgroup is normal, we now turn to the differential geometric properties of homomorphisms of Lie groups.

(1.11.8) Proposition. *Let $\Phi\colon H \to G$ be a homomorphism of Lie groups. Then $K = \ker \Phi$ is a closed normal Lie subgroup of H, with Lie algebra equal to $\ker \mathrm{T}_1 \Phi$. If π denotes the canonical projection: $H \to H/K$, then the unique homomorphism $\Psi\colon H/K \to G$ such that $\Phi = \Psi \circ \pi$, is an injective analytic immersion making $\Phi(H) = \Psi(H/K)$ into a Lie subgroup of G, with Lie algebra equal to the image of $\mathrm{T}_1 \Phi$. With this analytic structure on $\Phi(H)$, the mapping Φ is a principal fiber bundle: $H \to \Phi(H)$, with structure group equal to K. Finally, if H has only countably many components and Φ is surjective, then Φ is a principal fibration: $H \to G$, with structure group equal to K.*

Proof. Ψ is analytic because Φ is analytic and π is an analytic fibration (Corollary 1.11.5). Ψ is injective by construction. The argument in the proof of Theorem 1.11.4 that was used to show that $\mathrm{T}_1(A_x)$ is injective, now can be used to obtain that $\mathrm{T}_1 \Psi$ is injective; and $\mathrm{T}_h \Psi \circ \mathrm{T}_1 \mathrm{L}(h) = \mathrm{T}_1 \mathrm{L}(\Psi(h)) \circ \mathrm{T}_1 \Psi$ shows that $\mathrm{T}_h \Psi$ is injective, for all $h \in H$. Further $\mathrm{T}_1 K = \ker \mathrm{T}_1 \Phi$ follows from $\mathrm{T}_1 \Phi = \mathrm{T}_1 \Psi \circ \mathrm{T}_1 \pi$, $\ker \mathrm{T}_1 \pi = \mathrm{T}_1 K$ and the injectivity of $T_1\Psi$. For the last statement, we observe that the assumptions imply that Ψ is an isomorphism of Lie groups (Corollary 1.10.10). □

(1.11.9) Corollary. *Let G, H be Lie groups such that at least one of these has finitely many connected components. Let Φ, and Ψ, be an injective and continuous homomorphism: $G \to H$, and $H \to G$, respectively. Then $\Phi(G) = H$, $\Psi(H) = G$, and Φ and Ψ are isomorphisms of Lie groups.*

Proof. Proposition 1.11.8 implies that Φ and Ψ are analytic immersions, so $\dim G \leq \dim H \leq \dim G$; or $\dim G = \dim H$. But then $H_0 := \Phi(G)$, and $G_0 := \Psi(H)$, is open in H, and G, and therefore it is equal to the union of some of the connected components of H, and G, respectively. But now $\#(G/G^\circ) = \#(H_0/(H_0)^\circ) \leq \#(H/H^\circ) = \#(G_0/(G_0)^\circ) \leq \#(G/G^\circ)$ shows that all these numbers are equal, and finite. □

1.12 Connected Commutative Lie Groups

In this section we give a detailed description of the Lie groups of this simple type. We shall use the general theory developed up to this moment; nevertheless many of the proofs become much simpler, sometimes almost trivial in the commutative case.

(1.12.1) Theorem. *Let G be a connected Lie group with Lie algebra $\mathfrak{g}$. Then G is commutative if and only if $\mathfrak{g}$ is Abelian. If this is the case, then the exponential mapping is a surjective homomorphism of Lie groups: $(\mathfrak{g}, +) \to G$. Moreover $\Gamma = \ker(\exp)$ is a discrete (zero-dimensional, closed) subgroup of $(\mathfrak{g}, +)$, and the exponential mapping induces an isomorphism of Lie groups: $\mathfrak{g}/\Gamma \to G$.*

Proof. The first statement is Proposition 1.9.2. For the second statement we observe that, as $\mathrm{ad} = 0$, (1.6.2) becomes $\frac{dZ}{dt}(t) = X$; so $\mathfrak{g}_e^2 = \mathfrak{g} \times \mathfrak{g}$ and $\mu(X, Y) = X + Y$. Hence exp is a homomorphism of Lie groups: $(\mathfrak{g}, +) \to G$. Because $\mathrm{T}_0 \exp = \mathrm{I}$, its image is an open, hence also closed subgroup of G, which is equal to G because G is connected. For the remaining statements we refer to Proposition 1.11.8. □

From Example to Proposition 1.9.2. we learn that a nonconnected G may fail to be commutative and yet may have an Abelian $\mathfrak{g}$. We now investigate the possibilities for discrete subgroups Γ of $(\mathfrak{g}, +) \simeq (\mathbf{R}^n, +)$.

(1.12.2) Lemma. *Let H be a subgroup of $(\mathbf{R}, +)$. Then H is dense in $\mathbf{R}$, or $H = \{0\}$, or $H = \{\, na \mid n \in \mathbf{Z}\,\}$, for some $a > 0$.*

Proof. If $H \neq \{0\}$, then H contains elements $x > 0$ (note that $-x \in H$ if $x \in H$). For any $y \in \mathbf{R}$, let n be the largest integer k such that $k \leq \frac{y}{x}$. Then $nx \leq y < (n+1)x$, so $|y - nx| < x$. Noting that $nx \in H$ for all $n \in \mathbf{Z}$, we conclude that H is dense in $\mathbf{R}$ if $a := \inf\{\, x \in H \mid x > 0\,\}$ is equal to 0. If $a > 0$, then $a \in H$ and $H = \{\, na \mid n \in \mathbf{Z}\,\}$. Indeed, suppose $a \notin H$. Then, for any $\epsilon > 0$, there would exist $x \in H$ such that $a < x < a + \epsilon$, and by taking another $y \in H$ such that $a < y < x$, we would get $x - y \in H$ and $0 < x - y < \epsilon$. So, by starting out with $\epsilon < a$, we would arrive at a contradiction. Hence $a \in H$. For any $b \in H$, let n be the largest integer $\leq \frac{b}{a}$; accordingly $na \leq b < (n+1)a$. Then $0 \leq b - na < a$, and the definition of a now leads to $b - na \leq 0$, or $b - na = 0$. □

(1.12.3) Theorem. *Let V be a finite-dimensional vector space over $\mathbf{R}$ and Γ a discrete subgroup of $(V, +)$, with $\Gamma \neq \{0\}$. Then there exist an integer k, with $1 \leq k \leq \dim V$, and linearly independent vectors $e_1, \ldots, e_k \in V$, such that:*

$$\Gamma = \left\{ \sum_{j=1}^{k} n_j e_j \mid n_j \in \mathbf{Z} \right\}.$$

Proof. By induction on $\dim V$. Let $a \in \Gamma$, with $a \neq 0$, and let L be the linear subspace of V generated by a. Because $\Gamma \cap L$ is a discrete subgroup of $(L, +) \simeq (\mathbf{R}, +)$, we can apply Lemma 1.12.2 to obtain an element $b \in \Gamma \cap L$, with $b \neq 0$, such that:

$$\Gamma \cap L = \{ nb \mid n \in \mathbf{Z} \}.$$

Let Γ' be the image of Γ in V/L under the canonical projection: $V \to V/L$. If Γ' were not discrete, then there would exist a sequence x_j in Γ such that $x_j \notin L$, while $x_j + L$ converges to L in V/L; that is, $x_j - l_j \to 0$, for a sequence $l_j \in L$. Select $y_j \in \Gamma \cap L$ such that $l_j - y_j \in [0, b]$, the closed interval between 0 and b in L. Passing to a subsequence we may assume that $l_j - y_j$ converges to some $l \in L$. But then $x_j - y_j = (x_j - l_j) + (l_j - y_j)$ converges to l as well. Because $x_j, y_j \in \Gamma$, we get $x_j - y_j \in \Gamma$; and the discreteness of Γ now implies that $x_j - y_j = l$, for j sufficiently large. However $y_j \in L$ and $l \in L$ now leads to $x_j \in L$, for j sufficiently large, a contradiction. So Γ' is discrete in V/L, and $\dim(V/L) = \dim V - 1$. Therefore, by the induction hypothesis, either $\Gamma \subset L$ (and the theorem holds with $k = 1$) or there exist $e_1, \dots, e_{k-1} \in \Gamma$ satisfying: $e_1, \dots, e_{k-1}$ are linearly independent modulo L and, for each $x \in \Gamma$,

$$x + L = \sum_{j=1}^{k-1} n_j e_j + L,$$

for some $n_j \in \mathbf{Z}$. But this means that:

$$x - \sum_{j=1}^{k-1} n_j e_j \in \Gamma \cap L = \{ nb \mid n \in \mathbf{Z} \},$$

and the conclusion of the theorem holds with $b = e_k$. □

The subset Γ of V as described in Theorem 1.12.3 is called the *k-dimensional integral lattice in V generated by $e_1, \dots, e_k$*. Note that Γ is also generated by $f_1, \dots, f_k$ if and only if:

$$f_i = \sum_{j=1}^{k} m_{ji} e_j,$$

with $M = (m_{ij})$ an invertible matrix with **integral coefficients** such that M^{-1} has integral coefficients as well. The latter condition is equivalent to $\det M = \pm 1$; and this means that the k-dimensional volume of the *fundamental domain*:

$$F = \left\{ \sum_{j=1}^{k} x_j e_j \mid 0 \le x_j < 1 \right\} \tag{1.12.1}$$

does not depend on the choice of the set of generators of Γ. (If $k = n$, then $f \mapsto f + \Gamma$ sets up a bijection between F and V/Γ, and the volume of F is just the volume of the n-dimensional torus V/Γ, see Corollary 1.12.4 below.)

If G is a connected commutative Lie group, then the kernel of the exponential mapping is called the *integral lattice of G in the Lie algebra* $\mathfrak{g}$.

Let Γ be an integral lattice in V with generators $e_1, \dots, e_k$; choose $e_{k+1}, \dots, e_n$ such that $e_1, \dots, e_n$ is a basis of V, here $n = \dim V$. Then the mapping:

$$(1.12.2) \qquad (t_1 + \mathbf{Z}, \dots, t_k + \mathbf{Z}, t_{k+1}, \dots, t_n) \mapsto \sum_{j=1}^{n} t_j e_j + \Gamma,$$

is an isomorphism of Lie groups: $(\mathbf{R}/\mathbf{Z})^k \times \mathbf{R}^{n-k} \to V/\Gamma$.

(1.12.4) Corollary. *Every connected commutative Lie group G is isomorphic (as a Lie group) to the additive group:*

$$(\mathbf{R}/\mathbf{Z})^k \times \mathbf{R}^{n-k}, \quad \textit{where } n = \dim G \textit{ and } k = \dim(\textit{linear span of } \ker \exp).$$

In particular, every compact connected commutative Lie group is isomorphic (as a Lie group) to a standard torus $(\mathbf{R}/\mathbf{Z})^n$; this case occurs if and only if $k = n$. On the other hand, G is isomorphic to a vector space if and only if $k = 0$; and this condition is equivalent to $\ker \exp = \{0\}$.

A Lie subgroup H of $(\mathbf{R}/\mathbf{Z})^n$ is closed if and only if it is compact. This is the case if its Lie algebra $\mathfrak{h}$ is generated (as a vector space) by $\mathfrak{h} \cap \mathbf{Z}^n (\Leftrightarrow H^\circ$ is compact) and H/H° is finite. In turn, $\mathfrak{h}$ is generated by $\mathfrak{h} \cap \mathbf{Z}^n$ if and only if there are linearly independent $v_1, \dots, v_c \in \mathbf{Q}^n$ such that $< \mathfrak{h}, v_j > = 0$, for $j = 1, \dots, c$; here $c = n - \dim \mathfrak{h}$, the codimension of H in the torus. For example, the closure of the one-parameter subgroup $\{ \exp tX \mid t \in \mathbf{R} \}$, for given $X \in \mathfrak{g} \simeq \mathbf{R}^n$, is a compact connected subgroup of $(\mathbf{R}/\mathbf{Z})^n$ of dimension equal to $\dim_{\mathbf{Q}}(\mathbf{Q}X_1 + \cdots + \mathbf{Q}X_n)$; here the $X_j \in \mathbf{R}$ denote the coordinates of $X \in \mathbf{R}^n$ and $\mathbf{R}$ is viewed as (an infinite-dimensional) vector space over $\mathbf{Q}$. In particular, we have obtained the Approximation theorem of Kronecker [1884], stating that:

$$\{ tX + \mathbf{Z}^n \mid t \in \mathbf{R} \}$$

is dense in $\mathbf{R}^n/\mathbf{Z}^n$, if and only if $X_1, \dots, X_n$ are linearly independent over $\mathbf{Q}$. Here we have freely used the identification of $(\mathbf{R}/\mathbf{Z})^n$ with $\mathbf{R}^n/\mathbf{Z}^n$ and of exp with the canonical projection: $\mathbf{R}^n \to \mathbf{R}^n/\mathbf{Z}^n$. The subgroup $M = \{ nX + \mathbf{Z}^n \mid n \in \mathbf{Z} \}$ of $\mathbf{R}^n/\mathbf{Z}^n$ generated by $X + \mathbf{Z}^n$, is dense in $\mathbf{R}^n/\mathbf{Z}^n$, if and only if $1, X_1, \dots, X_n$ are linearly independent over $\mathbf{Q}$. Indeed,

$$\begin{aligned} & M \text{ is dense in } \mathbf{R}^n/\mathbf{Z}^n \\ \iff & \left\{ \begin{array}{c} (\{0\} \times \mathbf{R}^n)/\mathbf{Z}^{n+1} \text{ is contained in the closure of} \\ \{ tX' + \mathbf{Z}^{n+1} \mid t \in \mathbf{R} \} \text{ in } \mathbf{R}^{n+1}/\mathbf{Z}^{n+1} \end{array} \right\} \\ \iff & \{ tX' + \mathbf{Z}^{n+1} \mid t \in \mathbf{R} \} \text{ is dense in } \mathbf{R}^{n+1}/\mathbf{Z}^{n+1}. \end{aligned}$$

Here $X' = (1, X_1, \dots, X_n) \in \mathbf{R}^{n+1}$.

A Lie group G is called *monogenic* or *monothetic* if it contains an element x such that the subgroup generated by x is dense in G. A simple analysis shows that a Lie group G is monogenic if and only if it is either isomorphic to $\mathbf{Z}$, or if G° is monogenic and G/G° is isomorphic to $\mathbf{Z}/m\mathbf{Z}$, for some $m \in \mathbf{Z}_{>0}$. A connected Lie group (like G°) is monogenic if and only if it is commutative and compact, hence isomorphic to a torus. If G is a monogenic Lie group of dimension > 0, then the set of $x \in G$ such that $\{ x^n \mid n \in \mathbf{Z} \}$ is dense in G, has a complement that is a countable union of lower-dimensional Lie groups. In particular, this complement is meager and has $\dim G$-dimensional measure equal to 0. In this strong sense one can therefore say that "the generic element of G generates a dense subgroup of G" if G is monogenic and $\dim G > 0$.

1.13 Simply Connected Lie Groups

Let M be a connected smooth manifold and $x_0 \in M$. A *path in M starting at x_0* is a continuous curve $\gamma\colon [0,1] \to M$ such that $\gamma(0) = x_0$. The space $\mathrm{P} = \mathrm{P}(x_0, M)$ of all paths in M starting at x_0, will be provided with the topology of uniform convergence. The paths $\gamma, \gamma' \in \mathrm{P}(x_0, M)$ are called *equivalent*, notation $\gamma \sim \gamma'$, if there exists a *homotopy* from γ to γ' leaving the end points fixed, that is, a continuous curve:

$$s \mapsto \gamma_s\colon [0,1] \to \mathrm{P}(x_0, M) \quad \text{such that}$$
$$\gamma_0 = \gamma, \quad \gamma_1 = \gamma', \quad s \mapsto \gamma_s(1) \text{ is constant on } [0,1].$$

Of course this implies that $\gamma(1) = \gamma'(1)$. Note also that the continuity of $s \mapsto \gamma_s$ is equivalent to the continuity of the mapping:

$$\Gamma\colon (s,t) \mapsto \gamma_s(t)\colon [0,1] \times [0,1] \to M.$$

The relation $\gamma \sim \gamma'$ is an equivalence relation in $\mathrm{P}(x_0, M)$. For the transitivity one observes that if $s \mapsto \bar{\gamma}_s\colon [0,1] \to \mathrm{P}(x_0, M)$ is a homotopy from γ' to γ'' leaving the end points fixed, then $s \mapsto \bar{\bar{\gamma}}_s$, defined by:

$$\bar{\bar{\gamma}}_s = \gamma_{2s}, \qquad \text{for } 0 \leq s \leq \frac{1}{2},$$
$$\bar{\bar{\gamma}}_s = \bar{\gamma}_{2s-1}, \qquad \text{for } \frac{1}{2} \leq s \leq 1,$$

is a homotopy from γ to γ'' leaving the end points fixed. We shall write $[\gamma]$ for the equivalence class in P of $\gamma \in \mathrm{P}$, and we define $\widetilde{M}$ as the collection of equivalence classes in the path space P. Because $\gamma(1) = \gamma'(1)$ if $\gamma \sim \gamma'$, the mapping:

$$\tilde{\pi}\colon [\gamma] \to \gamma(1)\colon \widetilde{M} \to M$$

is well-defined; and it is surjective since M is pathwise connected.

If $\gamma \in \mathrm{P}(x_0, M)$ and $\gamma' \in \mathrm{P}(\gamma(1), M)$, then:

$$\gamma' \dashv\vdash \gamma = \text{“ first } \gamma, \text{ then } \gamma' \text{ ”} \in \mathrm{P}(x_0, M)$$

is defined by:

$$(1.13.1) \qquad \begin{aligned} (\gamma' \dashv\vdash \gamma)(t) &= \gamma(2t), && \text{if } 0 \leq t \leq \frac{1}{2}, \\ (\gamma' \dashv\vdash \gamma)(t) &= \gamma'(2t-1), && \text{if } \frac{1}{2} \leq t \leq 1. \end{aligned}$$

Note that we have already used this construction in $\mathrm{P}(\gamma, \mathrm{P}(x_0, M))$ in order to prove the transitivity of the equivalence relation in $\mathrm{P}(x_0, M)$. Clearly $\gamma' \dashv\vdash \gamma \sim \gamma_1' \dashv\vdash \gamma_1$ if $\gamma \sim \gamma_1$ and $\gamma' \sim \gamma_1'$, making:

$$[\gamma'] \dashv\vdash [\gamma] = [\gamma' \dashv\vdash \gamma]$$

well-defined.

In particular, this turns $\tilde{\pi}^{-1}(\{x_0\})$, the space of homotopy classes of loops starting and ending at x_0, into a group; it is called the *fundamental group* $\pi_1(M, x_0)$ *of* M *with base point* x_0. If x_1 is any other point in M, then we can choose $\epsilon \in \mathrm{P}(x_0, M)$ with $\epsilon(1) = x_1$; and $[\gamma] \mapsto [\epsilon] \dashv\vdash [\gamma]$ establishes a bijection: $\tilde{\pi}^{-1}(\{x_0\}) \to \tilde{\pi}^{-1}(\{x_1\})$. Also, $[\gamma] \mapsto [\epsilon] \dashv\vdash [\gamma] \dashv\vdash [\epsilon]^{-1}$ is an isomorphism of groups: $\pi_1(M, x_0) \to \pi_1(M, x_1)$; it is in this sense that the fundamental group is independent of the base point. However, the isomorphism is only independent of the choice of ϵ if the fundamental group is Abelian, as is easily verified; and in general this need not be the case at all. For example, the plane minus two points has fundamental group equal to the fundamental group of the figure 8, which in turn is isomorphic to the free group with two generators, according to the Van Kampen theorem, cf. Crowell and Fox [1963], p.67. Such free groups are extremely noncommutative!

The space M is said to be *simply connected* if its fundamental group is trivial, that is, if every loop can be shrunk to a point. (It is not hard to see that here it does not matter whether one fixes initial and end points of the loops or lets them free.) In other words, $\tilde{\pi}^{-1}(x_0) = [\underline{x}_0]$, if $\underline{x}_0$ denotes the constant curve at x_0; and in view of the remarks above this is equivalent to saying that $\tilde{\pi}\colon \widetilde{M} \to M$ is a bijection. The discussion of fundamental groups and universal coverings is usually presented in the framework of locally simply connected and pathwise connected topological spaces, see for instance Greenberg [1967], Sections 1–6 for more details. For our applications to Lie groups we shall need some more emphasis on what happens with these concepts in the smooth case.

(1.13.1) Theorem. *When provided with the quotient topology,* $\widetilde{M}$ *is simply connected. Moreover, there is a unique structure of a smooth manifold on* $\widetilde{M}$ *(with the same topology) making* $\tilde{\pi}$ *into a smooth fibration with discrete fibers. For each* $l \in \mathbf{Z}_{\geq 0}$*, the inclusion:*

$$\mathrm{P}(x_0, M) \cap \mathrm{C}^l([0,1], M) \to \mathrm{P}(x_0, M),$$

induces a homeomorphism of the quotients (providing the left hand side with the C^l *topology). Furthermore,* $\gamma, \gamma' \in \mathrm{P}(x_0, M) \cap \mathrm{C}^l([0,1], M)$ *are equivalent if and only if there is a smooth homotopy* $s \mapsto \gamma_s \colon [0,1] \to \mathrm{P}(x_0, M) \cap \mathrm{C}^l([0,1], M)$ *keeping the end points fixed, from* γ *to* γ'.

Proof. It will be convenient to use the smooth diffeomorphism λ, introduced in Lemma 11.1.1, from an open neighborhood Ω of $\Delta = \{ (x,y) \in M \times M \mid x = y \}$ in $M \times M$ onto an open neighborhood Θ of the zero section in $\mathrm{T}M$, such that:

$$\lambda(x,y) \in \mathrm{T}_x M, \quad \text{for all } (x,y) \in \Omega; \quad \lambda(x,x) = 0 \in \mathrm{T}_x M, \quad \text{for all } x \in M.$$

One can also arrange that $\Theta \cap \mathrm{T}_x M$ is convex in $\mathrm{T}_x M$, for each $x \in M$.

Suppose:

$$\gamma, \gamma' \in \mathrm{P}(x_0, M) \cap \mathrm{C}^l, \quad \gamma(1) = \gamma'(1), \quad (\gamma(t), \gamma'(t)) \in \Omega \qquad (t \in [0,1]).$$

Define γ_s by the relation $\lambda(\gamma(t), \gamma_s(t)) = s\lambda(\gamma(t), \gamma'(t))$, where we use the scalar multiplication by $s \in [0,1]$ in $\mathrm{T}_{\gamma(t)} M$. It follows that $s \mapsto \gamma_s$ is a smooth curve in the Banach manifold (see the beginning of Section 1.14) $\mathrm{P}(x_0, M) \cap \mathrm{C}^l$, with $\gamma_0 = \gamma$, $\gamma_1 = \gamma'$, and $s \mapsto \gamma_s(1)$ is constant. So we find an open neighborhood:

$$U_\gamma = \{ \gamma' \in \mathrm{P}(x_0, M) \mid (\gamma(t), \gamma'(t)) \in \Omega, \text{ for all } t \in [0,1] \}$$

of γ in $\mathrm{P}(x_0, M)$, while $\gamma' \sim \gamma$ by a smooth homotopy in $\mathrm{P}(x_0, M) \cap \mathrm{C}^l$ with fixed end points, whenever $\gamma, \gamma' \in \mathrm{P}(x_0, M) \cap \mathrm{C}^l$, $\gamma' \in U_\gamma$, $\gamma'(1) = \gamma(1)$. Because each $\gamma \in \mathrm{P}(x_0, M)$ can be uniformly approximated by smooth curves with the same endpoints, whereas each homotopy can be split into small steps, each of which can be replaced by smooth homotopies which are stationary at the ends so that they piece together to a smooth homotopy, the last two statements of the theorem follow.

Now, let $x_1 \in M$ and choose an open neighborhood V of x_1 in M such that $\{x_1\} \times V \subset \Omega$. Let $\gamma \in \mathrm{P}(x_0, M)$ be such that $\gamma(1) = x_1$. Take $0 < \delta \leq 1$ so small that $(\gamma(t), x) \in \Omega$, for all $1 - \delta \leq t \leq 1$, $x \in V$. For $x \in V$, define:

$$\begin{aligned} \gamma_x(t) &= \gamma(t), && \text{for } 0 \leq t \leq 1 - \delta, \\ \lambda(\gamma(t), \gamma_x(t)) &= \frac{t - 1 + \delta}{\delta} \lambda(\gamma(t), x), && \text{for } 1 - \delta \leq t \leq 1. \end{aligned}$$

This defines a smooth mapping:

$$\sigma \colon x \mapsto \gamma_x \colon V \to \mathrm{P}(x_0, M), \quad \text{such that } \gamma_x(1) = x, \quad \text{for all } x \in V.$$

Moreover $\gamma_{x_1} \sim \gamma$, because $\gamma_{x_1} \in U_\gamma$ and $\gamma_{x_1}(1) = x_1 = \gamma(1)$. So

$$s \colon x \mapsto [\gamma_x] \colon V \to \widetilde{M} \quad \text{satisfies } \tilde{\pi} \circ s = \text{ identity in } V, \quad s(x_1) = [\gamma].$$

In other words, for each $[\gamma] \in \tilde{\pi}^{-1}(\{x_1\})$ we have constructed a section s of $\tilde{\pi}$ on V, mapping x_1 to $[\gamma]$, as the composition of a smooth mapping $\sigma\colon V \to \mathrm{P}(x_0, M)$ with the canonical projection:

$$[\,]\colon \gamma \mapsto [\gamma]\colon \mathrm{P}(x_0, M) \to \widetilde{M}.$$

Now $s(V_0)$ is open in $\widetilde{M}$, for every open subset V_0 of V; and the continuity of $[\,]$ implies the continuity of s. So these s are homeomorphisms, and this shows that $\tilde{\pi}\colon \widetilde{M} \to M$ is a topological fibration with discrete fibers (see Definition A.1 in Appendix A). Now restrict the V to coordinate neighborhoods of smooth local coordinatizations κ of M, then the $(\kappa \circ \tilde{\pi})|_{s(V)}$ form an atlas which makes $\widetilde{M}$ into a smooth manifold (with the same coordinate changes in $\mathbf{R}^n$ as M) and $\tilde{\pi}\colon \widetilde{M} \to M$ into a smooth fibration with discrete fibers.

For each $\gamma \in \mathrm{P}(x_0, M)$, setting $\gamma_s(t) = \gamma(st)$, with $s, t \in [0,1]$, defines a homotopy from $\underline{x_0}$ to γ, so $s \mapsto [\gamma_s]$ is a continuous curve in $\widetilde{M}$ from $[\underline{x_0}]$ to $[\gamma]$. This shows that $\widetilde{M}$ is pathwise connected and therefore that $\tilde{\pi}\colon \widetilde{M} \to M$ is a covering (cf. Definition A.3). Note that now $\tilde{\pi}([\gamma_s]) = \gamma(s)$, for all $s \in [0,1]$.

In order to show that $\widetilde{M}$ is simply connected, we first recall that for **every** covering $\pi'\colon M' \to M$, one has the property of *unique lifting of curves*. That is, fix $x_0 \in M$ and $x_0' \in (\pi')^{-1}(\{x_0\}) \in M'$. Then, for any continuous curve $\gamma\colon [0,1] \to M$ with $x_0 = \gamma(0)$, there is a unique continuous curve $\gamma'\colon [0,1] \to M'$ such that $\gamma'(0) = x_0'$ and $\pi' \circ \gamma' = \gamma$. Moreover, the mapping $\gamma \mapsto \gamma'$ is continuous: $\mathrm{P}(x_0, M) \to \mathrm{P}(x_0', M')$. The proof is by "continuous induction on t", applied to $\gamma[0,t]$; the induction step is a local statement, where one uses that locally π' is a homeomorphism. In other words:

$$(\pi')_*\colon \gamma' \mapsto \pi' \circ \gamma'\colon \mathrm{P}(x_0', M') \to \mathrm{P}(x_0, M) \quad \text{is a homeomorphism,}$$

with the lifting as the inverse operation. Because $(\pi')_*$ maps equivalence classes onto equivalence classes, it induces a homeomorphism from $\widetilde{M'}$ onto $\widetilde{M}$, which will be denoted by $(\pi')_*$ as well. Note also that $(\pi')_*$ injectively maps the fiber $\pi_1(M', x_0')$ of $\widetilde{M'} \to M'$ over x_0' into the fiber $\pi_1(M, x_0)$ of $\widetilde{M} \to M$ over x_0.

Replacing in the results above π' by $\tilde{\pi}$, and x_0' by $[\underline{x_0}]$, we get that $(\tilde{\pi})_*\colon \widetilde{\widetilde{M}} \to \widetilde{M}$ is a homeomorphism. Reading through the definitions, we can write any element $[\Gamma]$ of $\widetilde{\widetilde{M}}$ as:

$$[s \mapsto [t \mapsto \Gamma(s,t)]], \quad \text{with } \Gamma\colon [0,1] \times [0,1] \to M \quad \text{continuous ;}$$
$$\Gamma(s,0) = x_0, \quad \text{for all } s; \quad \Gamma(0,t) = x_0, \quad \text{for all } t.$$

Now $(\tilde{\pi})_*([\Gamma]) = [s \mapsto \Gamma(s,1)]$, whereas the image of $[\Gamma]$ under the projection $\pi_{\widetilde{M}}\colon \widetilde{\widetilde{M}} \to \widetilde{M}$ is equal to $[t \mapsto \Gamma(1,t)]$. However, $s \mapsto \Gamma(s,1)$ is homotopic to $u \mapsto \Gamma(u,u)$ via $s \mapsto \Gamma(s, 1-\alpha+\alpha s)$, and $u \mapsto \Gamma(u,u)$ is homotopic to

$t \mapsto \Gamma(1,t)$ via $t \mapsto \Gamma(\beta + (1-\beta)t, t)$. So $(\tilde{\pi})_* = \pi_{\widetilde{M}}$, implying that $\pi_{\widetilde{M}}$ is bijective, or $\widetilde{M}$ is simply connected. □

A simply connected covering, like $\widetilde{M}$, is called a *universal covering*. Here the word "universal" refers to the property that for any covering $\pi' \colon M' \to M$ there is a covering $\psi \colon \widetilde{M} \to M'$ such that $\tilde{\pi} = \pi' \circ \psi \colon \widetilde{M} \to M$. Indeed, we can take for ψ the composition of the inverse of $(\pi')_* \colon \widetilde{M'} \to \widetilde{M}$, which is a homeomorphism because of the unique lifting of curves referred to in the proof of Theorem 1.13.1, with the covering: $\widetilde{M'} \to M'$.

We now turn to the special case of Lie groups.

(1.13.2) Proposition. *Let G be a connected Lie group. Then* $\mathrm{P}(1,G)$, *provided with the pointwise multiplication $(\gamma \cdot \gamma')(t) = \gamma(t)\gamma'(t)$, for $t \in [0,1]$, is a group, called the path group of G. Furthermore:*

$$\Lambda(G) = \{\gamma \in \mathrm{P}(1,G) \mid \gamma(1) = 1\}$$

is a closed normal subgroup of $\mathrm{P}(1,G)$, *called the loop group of G, while:*

$$\Lambda(G)^\circ = \{\gamma \in \mathrm{P}(1,G) \mid \gamma \sim \underline{1}\},$$

the path component of $[\underline{1}]$ in $\Lambda(G)$, is closed and normal in $\mathrm{P}(1,G)$ *as well. Now $\gamma' \sim \gamma$ in* $\mathrm{P}(1,G)$ *if and only if $\gamma' \in \gamma\Lambda(G)^\circ$, so that:*

$$\widetilde{G} = \mathrm{P}(1,G)/\Lambda(G)^\circ,$$

and $\widetilde{G}$ therefore becomes a group. Combined with its manifold structure, $\widetilde{G}$ is a Lie group and $\tilde{\pi} \colon \widetilde{G} \to G$ is a Lie group covering as defined before Corollary 1.10.5. On $\ker \tilde{\pi} = \pi_1(G,1)$, *the group structure of $\widetilde{G}$ coincides with that of the fundamental group. Every element of $\pi_1(G,1)$ commutes with every element of $\widetilde{G}$, in particular, $\pi_1(G,1)$ is commutative. Finally, all statements remain valid if* $\mathrm{P}(1,G)$ *is replaced by* $\mathrm{P}(1,G) \cap \mathrm{C}^l$, $\Lambda(G)$ *by* $\Lambda(G) \cap \mathrm{C}^l$ *and the equivalence is replaced by smooth homotopies with fixed end points in* $\mathrm{P}(1,G) \cap \mathrm{C}^l$, $l \in \mathbf{Z}_{\geq 0}$.

Proof. $\gamma \mapsto \gamma(1)$ is a continuous homomorphism: $\mathrm{P}(1,G) \to G$, with kernel $\Lambda(G)$, which therefore is closed and normal in $\mathrm{P}(1,G)$. A homotopy $s \mapsto \gamma_s$ from γ to γ' with fixed end points corresponds to a continuous curve $s \mapsto \gamma^{-1} \cdot \gamma_s$ in $\mathrm{P}(1,G)$ from 1 to $\gamma^{-1} \cdot \gamma'$, showing that $\widetilde{G} = \mathrm{P}(1,G)/\Lambda(G)^\circ$. The Lie group statements about $\widetilde{G}$ follow from the identification $\tilde{\pi} \colon [\gamma] \mapsto \gamma(1)$ of a neighborhood of 1 in $\widetilde{G}$ with a neighborhood of 1 in G, both with respect to the group structure and the manifold structure.

If $\gamma, \gamma' \in \mathrm{P}$ with $\gamma(1) = 1$, $\gamma'(1) = 1$, define (see Fig.1.13.1):

$$\gamma_s''(t) = \begin{cases} \gamma(\frac{2t}{2-s}) & \text{if } 0 \leq t \leq \frac{s}{2}, \\ \gamma(\frac{2t}{2-s})\gamma'(\frac{2t-s}{2-s}) & \text{if } \frac{s}{2} \leq t \leq 1 - \frac{s}{2}, \\ \gamma'(\frac{2t-s}{2-s}) & \text{if } 1 - \frac{s}{2} \leq t \leq 1. \end{cases}$$

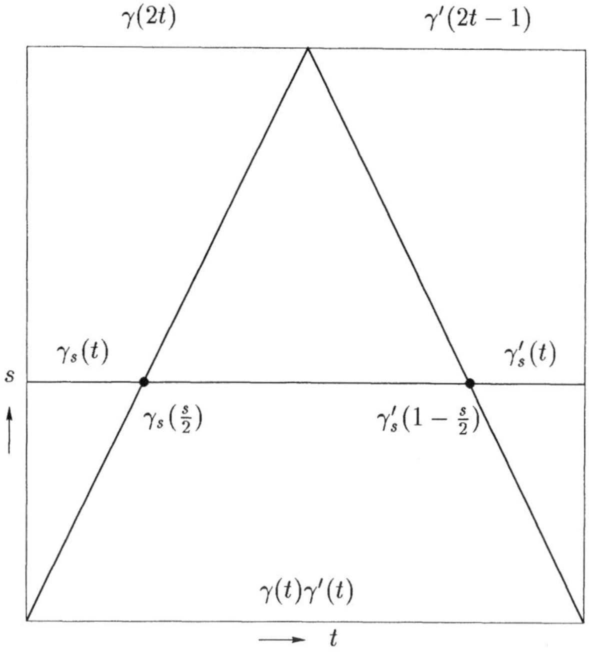

Fig. 1.13.1. $\gamma_s(t) = \gamma\left(\frac{t}{1-\frac{s}{2}}\right)$, $\quad \gamma'_s(t) = \gamma\left(\frac{t-\frac{s}{2}}{1-\frac{s}{2}}\right)$

Then γ''_s is a homotopy from $\gamma \cdot \gamma'$ to $\gamma' \dashv\!\vdash \gamma$, showing $[\gamma] \cdot [\gamma'] = [\gamma'] \dashv\!\vdash [\gamma]$ in $\pi_1(G, 1)$.

The last sentence follows from the corresponding remarks in Theorem 1.13.1, whereas the commuting properties follow from Lemma 1.13.3 below. □

(1.13.3) Lemma. *If G is a connected Lie group and C is a discrete normal subgroup of G, then every element of C commutes with every element of G.*

Proof. For every $c \in C$, the mapping $x \mapsto xcx^{-1}$ is continuous: $G \to C$. Because $\{c\}$ is open and closed in C, $G_c = \{\, x \in G \mid xcx^{-1} = c \,\}$ is open and closed in G. Because $1 \in G_c$, we have $G_c \neq \emptyset$; and therefore $G_c = G$, because G is connected. □

Consequently Lie group coverings $\pi' \colon G' \to G$ always arise by fixing a discrete subgroup C of:

$$\mathrm{Z}(G') = \{\, z \in G' \mid zx = xz, \text{ for all } x \in G' \,\},$$

the *center* of G', and then taking $G = G'/C$, and for π' the canonical projection: $G' \to G'/C$. In view of $\widetilde{G'} = \widetilde{G}$, cf. the remark preceding Proposition 1.13.2, we get $\widetilde{G}/\pi_1(G) \simeq G \simeq (\widetilde{G}/\pi_1(G'))/C$, hence $C \simeq \pi_1(G)/\pi_1(G')$.

Proposition 1.13.2 says that the universal covering of a Lie group is a Lie group covering, and this implies that every Lie group covering of G is an isomorphism if and only if G is simply connected. We have used this in the discussion of the examples of $\mathbf{SU}(2)$ and $\mathbf{SO}(3, \mathbf{R})$ following Corollary 1.10.5.

By differentiating with respect to the time parameter t, we now transport the study of $\mathrm{P}(1, G) \cap \mathrm{C}^1$ to the study of the space $\mathrm{P}(\mathfrak{g})$ of all continuous curves $\delta\colon [0,1] \to \mathfrak{g}$ in the Lie algebra $\mathfrak{g}$ of G. Provided with the supremum norm with respect to some norm in $\mathfrak{g}$, the space $\mathrm{P}(\mathfrak{g})$ is a Banach space, which will be referred to as the *path space* of $\mathfrak{g}$.

(1.13.4) Proposition. *Let G be a connected Lie group with Lie algebra $\mathfrak{g}$. Then the mapping:*

$$\mathrm{D} := \mathrm{D}^{\mathrm{R}}\colon \gamma \mapsto \left(t \mapsto (\mathrm{T}_1\, \mathrm{R}(\gamma(t)))^{-1} \frac{d\gamma}{dt}(t)\right) \tag{1.13.2}$$

is a homeomorphism from $\mathrm{P}(1, G) \cap \mathrm{C}^1$ *onto* $\mathrm{P}(\mathfrak{g})$. *For* $\delta \in \mathrm{P}(\mathfrak{g})$, *let:*

$$A_\delta \in \mathrm{C}^1([0,1], \mathbf{L}(\mathfrak{g}, \mathfrak{g}))$$

be the solution A of the differential equation, with initial condition:

$$\frac{dA}{dt}(t) = \mathrm{ad}\,\delta(t) \circ A(t), \quad A(0) = \mathrm{I}\colon \mathfrak{g} \to \mathfrak{g}. \tag{1.13.3}$$

Then, for all $\gamma, \gamma' \in \mathrm{P}(1, G) \cap \mathrm{C}^1$ *and* $t \in [0,1]$:

$$\mathrm{D}(\gamma \cdot \gamma')(t) = \mathrm{D}\,\gamma(t) + \mathrm{Ad}\,\gamma(t)(\mathrm{D}\,\gamma'(t)), \tag{1.13.4.a}$$

$$\mathrm{Ad}\,\gamma(t) = A_{\mathrm{D}\,\gamma}(t). \tag{1.13.4.b}$$

Finally, we have:

$$\mathrm{D}(\Lambda(G)^\circ \cap \mathrm{C}^1) = \mathrm{P}(\mathfrak{g})_0,$$

where $\mathrm{P}(\mathfrak{g})_0$ *is the set of the* $\delta \in \mathrm{P}(\mathfrak{g})$ *satisfying the following conditions:*

$$\exists \textit{ smooth mapping}: s \mapsto \delta_s\colon [0,1] \to \mathrm{P}(\mathfrak{g}) \quad \textit{with: } \delta_0 = 0, \quad \delta_1 = \delta,$$

$$\int_0^1 A_{\delta_s}(t)^{-1} \frac{\partial}{\partial s}\delta_s(t)\, dt = 0 \qquad (s \in [0,1]). \tag{1.13.5}$$

Proof. The statement that D is bijective: $\mathrm{P}(1, G) \cap \mathrm{C}^1 \to \mathrm{P}(\mathfrak{g})$ means that for every continuous curve $\delta\colon [0,1] \to \mathfrak{g}$, there is a unique C^1 solution curve $\gamma\colon [0,1] \to G$ of:

(1.13.6) $$\frac{d\gamma}{dt}(t) = \left(\delta(t)^{\mathrm{R}}\right)(\gamma(t)), \quad \gamma(0) = 1.$$

Here $\left(\delta(t)^{\mathrm{R}}\right)(x) = \left(\mathrm{T}_1\,\mathrm{R}(x)\right)(\delta(t))$, for all $x \in G$ (cf. (1.3.10)).

The mapping $t \mapsto \delta(t)^{\mathrm{R}}$ is continuous from $[0,1]$ to the space of analytic vector fields on G. From Appendix B we get for each $s \in [0,1]$, $x \in G$, a unique maximal solution:

$$t \mapsto \gamma(t,s,x) \colon I(s,x) \to G \quad \text{of}$$
$$\frac{d\gamma}{dt}(t,s,x) = \left(\delta(t)^{\mathrm{R}}\right)(\gamma(t,s,x)), \quad \gamma(s,s,x) = x.$$

Here $I(s,x)$ is an interval, open in $[0,1]$ and containing s. Defining the flow:

$$\Phi^{t,s} \colon x \mapsto \gamma(t,s,x) \quad \text{from time } s \text{ to time } t, \text{ on } G^{t,s} = \{\, x \in G \mid t \in I(s,x) \,\},$$

we have also the group property $\Phi^{t,s} \circ \Phi^{s,u} \subset \Phi^{t,u}$. Now, if $\frac{d\gamma}{dt}(t) = \left(\mathrm{T}_1\,\mathrm{R}(\gamma(t))\right)\delta(t)$ then, for any $x \in G$:

$$\begin{aligned}\frac{d}{dt}(\gamma(t)x) &= \left(\mathrm{T}_{\gamma(t)}\,\mathrm{R}(x)\right)\frac{d\gamma}{dt}(t) = \left(\mathrm{T}_{\gamma(t)}\,\mathrm{R}(x) \circ \mathrm{T}_1\,\mathrm{R}(\gamma(t))\right)(\delta(t))\\ &= \left(\mathrm{T}_1\left(\mathrm{R}(x) \circ \mathrm{R}(\gamma(t))\right)\right)(\delta(t)) = \left(\mathrm{T}_1\,\mathrm{R}(\gamma(t)x)\right)\delta(t).\end{aligned}$$

Hence $t \mapsto \gamma(t)x$ satisfies the same differential equation. This shows that $G^{t,s} = G$ and $\Phi^{t,s}(x) = \Phi^{t,s}(1)x$, for all $x \in G$. The group property implies that the $\Phi^{t,s}$, constructed first only for s,t close to each other, are globally defined on G, for all $s,t \in [0,1]$; in particular therefore the solution γ of (1.13.6) exists for all $t \in [0,1]$. The continuous dependence of the solution on the vector field implies that the inverse of D is continuous, so the first statement of the proposition is proved.

For (1.13.4.a), we compute:

$$\begin{aligned}\frac{d}{dt}(\gamma(t)\gamma'(t)) &= \left(\mathrm{T}_{\gamma(t)}\,\mathrm{R}(\gamma'(t))\right)\frac{d\gamma}{dt}(t) + \left(\mathrm{T}_{\gamma'(t)}\,\mathrm{L}(\gamma(t))\right)\frac{d\gamma'}{dt}(t)\\ &= \left(\mathrm{T}_{\gamma(t)}\,\mathrm{R}(\gamma'(t)) \circ \mathrm{T}_1\,\mathrm{R}(\gamma(t))\right)\,\mathrm{D}\,\gamma(t)\\ &\quad + \left(\mathrm{T}_{\gamma'(t)}\,\mathrm{L}(\gamma(t)) \circ \mathrm{T}_1\,\mathrm{R}(\gamma'(t))\right)\,\mathrm{D}\,\gamma'(t)\\ &= \left(\mathrm{T}_1\,\mathrm{R}(\gamma(t)\gamma'(t))\right)\left(\mathrm{D}\,\gamma(t) + \mathrm{Ad}\,\gamma(t)\,\mathrm{D}\,\gamma'(t)\right).\end{aligned}$$

In order to prove (1.13.4.b), observe that:

$$\frac{d}{dt}\,\mathrm{Ad}\,\gamma(t) = \left.\frac{d}{dh}\right|_{h=0}\mathrm{Ad}\left(\gamma(t+h) \circ \gamma(t)^{-1}\right) \circ \mathrm{Ad}\,\gamma(t) = \mathrm{ad}\,\mathrm{D}\,\gamma(t) \circ \mathrm{Ad}\,\gamma(t),$$
$$\mathrm{Ad}\,\gamma(0) = \mathrm{Ad}\,1 = \mathrm{I};$$

and this shows that $A \colon t \mapsto \mathrm{Ad}\,\gamma(t)$ satisfies (1.13.3) with $\delta = \mathrm{D}\,\gamma$.

If $u \mapsto \gamma_u$ is smooth: $[0,1] \to \mathrm{P}(1,G) \cap \mathrm{C}^1$ then, writing $\Phi_u^{t,s}$ for the flow from time s to time t, of the time-dependent vector field $\mathrm{D}\,\gamma_u(t)^{\mathrm{R}}$, we have the variational formula (cf. Appendix (B.10))

$$\frac{\partial}{\partial u}\gamma_u(1) = \int_0^1 \left(\mathrm{T}_{\gamma_u(s)}\,\Phi_u^{1,s}\right)\left(\frac{\partial}{\partial u}\,\mathrm{D}\,\gamma_u(s)^{\mathrm{R}}\right)(\gamma_u(s))\,ds.$$

Now

$$\begin{aligned}\mathrm{T}_{\gamma_u(s)}\,\Phi_u^{1,s} &= \mathrm{T}_{\gamma_u(s)}\left(\Phi_u^{1,0} \circ (\Phi_u^{s,0})^{-1}\right) = \mathrm{T}_1\,\Phi_u^{1,0} \circ (\mathrm{T}_1\,\Phi_u^{s,0})^{-1} \\ &= \mathrm{T}_1\,\mathrm{L}(\gamma_u(1)) \circ \mathrm{T}_1\,\mathrm{L}(\gamma_u(s))^{-1},\end{aligned}$$

according to Lemma 1.3.1. On the other hand,

$$\mathrm{D}\,\gamma_u(s)^{\mathrm{R}}(x) = \left(\mathrm{T}_1\,\mathrm{R}(x)\right)\,\mathrm{D}\,\gamma_u(s),$$

so the variational equation takes the form:

$$\frac{\partial}{\partial u}\gamma_u(1) = \mathrm{T}_1\,\mathrm{L}(\gamma_u(1)) \circ \int_0^1 \mathrm{Ad}\,\gamma_u(s)^{-1}\frac{\partial}{\partial u}\,\mathrm{D}\,\gamma_u(s)\,ds.$$

Thus $u \mapsto \gamma_u(1)$ is constant if and only if $\int_0^1 \mathrm{Ad}\,\gamma_u(s)^{-1}\frac{\partial}{\partial u}\,\mathrm{D}\,\gamma_u(s)\,ds = 0$. Again using $\mathrm{Ad}\,\gamma_u(s) = A_{\mathrm{D}\,\gamma_u}(s)$, we now obtain the last statement in the proposition. □

(1.13.5) *Remark.* We now give another **proof of Corollary 1.10.5**. Let H, and G, be a Lie group with Lie algebra $\mathfrak{h}$, and $\mathfrak{g}$, respectively, with H connected, and suppose that ϕ is a homomorphism of Lie algebras: $\mathfrak{h} \to \mathfrak{g}$. If Φ were a homomorphism of Lie groups: $H \to G$ such that $\mathrm{T}_1\,\Phi = \phi$, then, for each $\gamma \in \mathrm{P}(1,H) \cap \mathrm{C}^1$, the curve $\beta\colon t \mapsto \Phi(\gamma(t))$ would satisfy:

$$\begin{aligned}\frac{d\beta}{dt}(t) &= \left.\frac{d}{dh}\right|_{h=0} \Phi\left(\gamma(t+h)\gamma(t)^{-1}\gamma(t)\right) \\ &= \frac{d}{dh}\Phi\left(\gamma(t+h)\gamma(t)^{-1}\right)\Phi(\gamma(t)) = [\mathrm{T}_1\,\mathrm{R}(\beta(t))]\,\phi(\mathrm{D}\,\gamma(t)).\end{aligned}$$

Accordingly:

$$\mathrm{D}\,\beta(t) = \phi(\mathrm{D}\,\gamma(t)) \qquad (t \in [0,1]). \tag{1.13.7}$$

So, if $\gamma(1) = x$, then $\Phi(x)$ would be uniquely determined as the end point of $\mathrm{D}^{-1}(\phi \circ \mathrm{D}\,\gamma) \in \mathrm{P}(1,G)$.

Not yet knowing the existence of Φ, we nevertheless have the mapping:

$$\tilde{\phi}\colon \gamma \mapsto \mathrm{D}^{-1}(\phi \circ \mathrm{D}\,\gamma)\colon\ \mathrm{P}(1,H) \cap \mathrm{C}^1 \to \mathrm{P}(1,G) \cap \mathrm{C}^1\,. \tag{1.13.8}$$

The homomorphism property of ϕ implies that $\phi \circ \mathrm{ad}\,\delta(t) = \mathrm{ad}\,\phi(\delta(t)) \circ \phi$. Hence applying ϕ to (1.13.3), we get that $t \mapsto \phi \circ A_\delta(t)$ and $t \mapsto A_{\phi\circ\delta}(t) \circ \phi$ satisfy the same differential equation in $\mathbf{L}(\mathfrak{h},\mathfrak{g})$ with the same initial value ϕ.

Therefore $\phi \circ A_\delta(t) = A_{\phi\circ\delta}(t) \circ \phi$, for all $t \in [0,1]$. Applying ϕ to (1.13.4.a) we get:

$$\begin{aligned}\phi(\mathrm{D}(\gamma \cdot \gamma')(t)) &= \phi(\mathrm{D}\,\gamma(t)) + \phi \circ A_{\mathrm{D}\,\gamma}(t)(\mathrm{D}\,\gamma'(t))\\ &= (\phi \circ \mathrm{D}\,\gamma)(t) + A_{\phi\circ\mathrm{D}\,\gamma}(t)(\phi \circ \mathrm{D}\,\gamma'(t)).\end{aligned}$$

Hence $\tilde{\phi}$ is a homomorphism of path groups.

Because $A_{\phi\circ\delta}(t)^{-1} \circ \phi = \phi \circ A_\delta(t)^{-1}$, application of ϕ to (1.13.5) yields that $\delta \mapsto \phi \circ \delta$ maps $\mathrm{P}(\mathfrak{h})_0$ into $\mathrm{P}(\mathfrak{g})_0$, or $\tilde{\phi}(\Lambda(H)^\circ) \subset \Lambda(G)^\circ$. In view of Proposition 1.13.2 this means that the end point of $\mathrm{D}^{-1}(\phi \circ \mathrm{D}\,\gamma)$ remains the same if γ is replaced by an equivalent γ'. Consequently the map:

$$\widetilde{H} = \mathrm{P}(1,H) \cap \mathrm{C}^1 / \Lambda(H)^\circ \cap \mathrm{C}^1 \to \widetilde{G} = \mathrm{P}(1,G) \cap \mathrm{C}^1 / \Lambda(G)^\circ \cap \mathrm{C}^1$$

induced by $\tilde{\phi}$, followed by the covering map $\widetilde{G} \to G$, is a homomorphism of Lie groups $\widetilde{\Phi}\colon \widetilde{H} \to G$, such that $\mathrm{T}_1 \widetilde{\Phi} = \phi$. So if H is simply connected, $\widetilde{H} \xrightarrow{\sim} H$ and the desired homomorphism of Lie groups $\Phi\colon H \to G$ with $T_1\Phi = \phi$ exists.

(1.13.6) *Remark.* If H is not simply connected then the proof above shows that the desired Φ exists if and only if there is a set of generators $[\gamma_i] \in \pi_1(H,1)$ for which the end points of $\mathrm{D}^{-1}(\phi \circ \mathrm{D}\,\gamma_i)$ are equal to 1_G. According to Lemma 9.2.1, we may take the γ_i in a compact connected subgroup of H. It follows then from Corollary 3.9.6 that one can find finitely many commuting $X_i \in \mathfrak{h}$ such that $\gamma_i(t) = \exp tX_i$, with $t \in [0,1]$, define generators of $\pi_1(H,1)$. It follows that the homomorphism Φ exists if and only if $\exp \phi(X_i) = 1_G$, for these i.

For an example different from $H = \mathbf{SO}(3,\mathbf{R})$, discussed after Corollary 1.10.5, note that:

$$\left[t \mapsto \begin{pmatrix} \cos 2\pi t & -\sin 2\pi t \\ \sin 2\pi t & \cos 2\pi t \end{pmatrix} \middle| \; t \in [0,1] \right]$$

generates $\pi_1(\mathbf{SL}(2,\mathbf{R}),1)$. Thus $\phi\colon \mathfrak{sl}(2,\mathbf{R}) \to \mathfrak{g}$, a homomorphism of Lie algebras, extends to a homomorphism of Lie groups $\Phi\colon \mathbf{SL}(2,\mathbf{R}) \to G$, if and only if $\exp \phi \begin{pmatrix} 0 & -2\pi \\ 2\pi & 0 \end{pmatrix} = 1_G$. Fig.1.2.2 is a picture of the universal covering of $\mathbf{SL}(2,\mathbf{R})$, covering $\mathbf{SL}(2,\mathbf{R})$ in Fig.1.2.3, which in turn covers $\mathrm{Ad}\,\mathbf{SL}(2,\mathbf{R})$ in Fig.1.2.4.

(1.13.7) *Remark.* We give another **comment on Proposition 1.13.4.** The definition $\gamma_s(t) = \gamma(st)$ leads to a contraction $\gamma \mapsto \gamma_s$, with s running from 1 to 0, of $\mathrm{P}(x_0, M)$ to $\{\underline{x}_0\}$. Proposition 1.13.4 shows that for a connected Lie group G, the space $\mathrm{P}(1,G)$ is not only contractible, but even homeomorphic to the Banach space $\mathrm{P}(\mathfrak{g})$, while D conjugates the contraction above to the radial contraction in $\mathrm{P}(\mathfrak{g})$. The projection $\mathrm{P}(1,G) \to \widetilde{G}$ makes $\mathrm{P}(1,G)$ into a fiber bundle over $\widetilde{G}$, with connected fiber $\Lambda(G)^\circ = \Lambda(\widetilde{G})$, the loop space of $\widetilde{G}$.

For compact $\widetilde{G}$ of positive dimension, for instance $\widetilde{G} = \mathbf{SU}(2) = \widetilde{\mathbf{SO}(3,\mathbf{R})}$, such a fibration is hard to visualize. In fact, there exists no fibration of a finite-dimensional vector space with connected fibers over a compact manifold of positive dimension, cf. Borel [1950], Cor.2 on Th.3. Even stronger, Serre [1951], Ch.IV, has shown that the fibration of the contractible $\mathrm{P}(x_0, M)$ over $\widetilde{M}$ with $H_n(\widetilde{M}) \neq 0$, for $n > 0$, forces the fiber, the loop space $\Lambda(M)^\circ$, to have nonzero homology in infinitely many dimensions. For simply connected compact Lie groups, the infinitely many Betti numbers of the loop groups have been computed by Bott [1954], [1956], see also Bott and Samelson [1958]. The fact that $\mathrm{P}(\mathfrak{g})$ is contractible makes $\mathrm{P}(\mathfrak{g}) \to \widetilde{G}$ into the universal bundle for the group $\Lambda(\widetilde{G})$; the "big" group $\Lambda(\widetilde{G})$ therefore has the "small" $\widetilde{G}$ as its classifying space (see Steenrod [1951], Thm.19.4).

1.14 Lie's Third Fundamental Theorem in Global Form

The formulae in (1.13.4) show that $\mathrm{D}(\gamma \cdot \gamma') \in \mathrm{P}(\mathfrak{g})$ can be expressed entirely in terms of $\mathrm{D}\,\gamma$ and $\mathrm{D}\,\gamma' \in \mathrm{P}(\mathfrak{g})$ and the structure of the Lie algebra $\mathfrak{g}$. It is natural to ask whether for an arbitrarily given (finite-dimensional) Lie algebra $\mathfrak{g}$, this defines a Lie group structure on $\mathrm{P}(\mathfrak{g})$. Now, unless $\mathfrak{g} = 0$, the space $\mathrm{P}(\mathfrak{g})$ is an infinite-dimensional Banach space; so in order to formulate the answer, we have to introduce the notion of an infinite-dimensional *Banach Lie group*. This is defined as in Definition 1.1.1, but with the condition that it is a finite-dimensional manifold, replaced by the condition that it is an analytic Banach manifold, the group operations being analytic mappings. In turn, an analytic Banach manifold M is defined in the same fashion as an analytic finite-dimensional manifold, but with $\mathbf{R}^n$ appearing in the local coordinatizations replaced by a fixed Banach space E. In this case one says that the Banach manifold M is *modeled over* E. The generalities about Lie groups which are based on the implicit function theorem in Banach spaces (including the theory of ordinary differential equations in such spaces), remain valid for Banach Lie groups. This generalization does not hold for Theorem 1.10.6 and its corollaries, because of the use of local compactness in its proof.

Also the Banach manifold version of Theorem 1.11.4 and its corollaries need the additional assumption that for each $x \in M$ the linear subspace $\mathfrak{g} \cdot x = \{ \frac{d}{dt}\big|_{t=0} \exp tX \cdot x \mid X \in \mathfrak{g}\}$ of $\mathrm{T}_x\, M$ is closed and has a closed linear complement in $\mathrm{T}_x\, M$. This is true if G is finite-dimensional (let $v_1, \dots, v_n$ be a basis for $\mathfrak{g} \cdot x$, use the Hahn-Banach theorem to find continuous linear functionals $\lambda_1, \dots, \lambda_n$ on $\mathrm{T}_x\, M$ such that $\lambda_i(v_i) \neq 0$; then $\bigcap_{1 \leq i \leq n} \ker \lambda_i$ is a closed linear subspace complementary to $\mathfrak{g} \cdot x$), or if $\mathfrak{g} \cdot x$ is closed in $\mathrm{T}_x\, M$ with finite codimension, or if M is a Hilbert manifold and $\mathfrak{g} \cdot x$ is closed in $\mathrm{T}_x\, M$. For instance, Proposition 1.11.8 holds if G, H are Banach Lie groups, one of which is finite-dimensional.

After having established that $\mathrm{P}(\mathfrak{g})$ is a Banach Lie group, we will show that $\mathrm{P}(\mathfrak{g})_0$, defined as in the last statement of Proposition 1.13.4, is a closed normal subgroup and that $\mathrm{P}(\mathfrak{g})/\mathrm{P}(\mathfrak{g})_0$ is a simply connected Lie group with Lie algebra isomorphic to $\mathfrak{g}$. This can be viewed as the global version of Theorem 1.8.1, where we constructed only a local Lie group with a given Lie algebra $\mathfrak{g}$.

(1.14.1) Proposition. *Let $\mathfrak{g}$ be a finite-dimensional Lie algebra. For δ, $\delta' \in \mathrm{P}(\mathfrak{g})$, the path space of $\mathfrak{g}$, define the product $\delta \cdot \delta' \in \mathrm{P}(\mathfrak{g})$ by:*

$$(\delta \cdot \delta')(t) = \delta(t) + A_\delta(t)\, \delta'(t) \qquad (t \in [0,1]), \tag{1.14.1}$$

where A_δ is defined as in Proposition 1.13.4. This makes $\mathrm{P}(\mathfrak{g})$ into a Banach Lie group; its Lie algebra, $\mathrm{P}(\mathfrak{g})^{\mathrm{alg}}$, is the space $\mathrm{P}(\mathfrak{g})$, provided with the bracket:
(1.14.2)
$$[X,Y](t) = \frac{d}{dt}[\textstyle\int_0^t X(s)\, ds, \int_0^t Y(s)\, ds] \qquad (X,Y \in \mathrm{P}(\mathfrak{g}),\ t \in [0,1]).$$

Further, $\mathrm{P}(\mathfrak{g})_0$, defined as in the last sentence of Proposition 1.13.4, is a connected normal Banach Lie subgroup. Its Lie algebra is equal to:

$$\mathrm{P}(\mathfrak{g})_0^{\mathrm{alg}} = \{ X \in \mathrm{P}(\mathfrak{g})^{\mathrm{alg}} \mid \textstyle\int_0^1 X(t)\, dt = 0 \}, \tag{1.14.3}$$

the kernel of the surjective homomorphism of Lie algebras:

$$\mathrm{av}\colon X \mapsto \textstyle\int_0^1 X(t)\, dt\colon\ \mathrm{P}(\mathfrak{g})^{\mathrm{alg}} \to \mathfrak{g}. \tag{1.14.4}$$

Proof. Formula (1.13.3) means that $A_\delta(t)$ is the solution curve of the time-dependent vector field $(\mathrm{ad}\, \delta(t))^{\mathrm{R}}$ on $\mathbf{GL}(\mathfrak{g})$. This vector field is tangent to $\mathrm{Ad}\, \mathfrak{g}$, the adjoint group of the Lie algebra $\mathfrak{g}$ defined in Example 1.10.4. So its flow leaves $\mathrm{Ad}\, \mathfrak{g}$ invariant, and therefore:

$$A_\delta(t) \in \mathrm{Ad}\, \mathfrak{g} \subset \mathrm{Aut}\, \mathfrak{g} \qquad (t \in [0,1]). \tag{1.14.5}$$

Next we verify the identity:

$$A_{\delta \cdot \delta'}(t) = A_\delta(t) \circ A_{\delta'}(t) \qquad (t \in [0,1]). \tag{1.14.6}$$

Indeed,

$$\begin{aligned}\frac{d}{dt}(A_\delta(t) \circ A_{\delta'}(t)) &= \mathrm{ad}\, \delta(t) \circ A_\delta(t) \circ A_{\delta'}(t) + A_\delta(t) \circ \mathrm{ad}\, \delta'(t) \circ A_{\delta'}(t) \\ &= \mathrm{ad}\, \delta(t) \circ A_\delta(t) \circ A_{\delta'}(t) + \mathrm{ad}(A_\delta(t)\delta'(t)) \circ A_\delta(t) \circ A_{\delta'}(t) \\ &= \mathrm{ad}(\delta \cdot \delta')(t) \circ \big(A_\delta(t) \circ A_{\delta'}(t)\big),\end{aligned}$$

where we have used that $A_\delta(t) \circ \mathrm{ad}\, \delta'(t) = \mathrm{ad}(A_\delta(t)\delta'(t)) \circ A_\delta(t)$, because of $A_\delta(t) \in \mathrm{Aut}\, \mathfrak{g}$, and moreover (1.14.1). So $A(t) = A_\delta(t) \circ A_{\delta'}(t)$ satisfies (1.13.3) with δ replaced by $\delta \cdot \delta'$, hence (1.14.6) follows.

Now, for $\delta, \delta', \delta'' \in \mathrm{P}(\mathfrak{g})$:

$$\begin{aligned}((\delta \cdot \delta') \cdot \delta'')(t) &= (\delta \cdot \delta')(t) + A_{\delta \cdot \delta'}(t)\delta''(t) \\ &= \delta(t) + A_\delta(t)\delta'(t) + A_\delta(t) \circ A_{\delta'}(t)\delta''(t) \\ &= \delta(t) + A_\delta(t)(\delta' \cdot \delta'')(t) = (\delta \cdot (\delta' \cdot \delta''))(t),\end{aligned}$$

for all $t \in [0,1]$; and this shows the associativity of the product in $\mathrm{P}(\mathfrak{g})$. This makes $(\mathrm{P}(\mathfrak{g}), \cdot)$ into a group, with $\delta(t) = \underline{0}(t) \equiv 0$ as the identity element, and the inverse of $\delta \in \mathrm{P}(\mathfrak{g})$ given by:

$$(\delta^{-1})(t) = -A_\delta(t)^{-1}\delta(t) \qquad (t \in [0,1]). \tag{1.14.7}$$

The linear dependence of the right hand side of (1.13.3) on $\delta \in \mathrm{P}(\mathfrak{g})$ implies that $\delta \mapsto A_\delta$ is an analytic mapping: $\mathrm{P}(\mathfrak{g}) \to \mathrm{C}^1([0,1], \mathbf{GL}(\mathfrak{g}))$. In turn, this makes the product and the inversion in $\mathrm{P}(\mathfrak{g})$ into analytic operations, so $(\mathrm{P}(\mathfrak{g}), \cdot)$ is a Banach Lie group.

As usual for vector spaces, we identify $\mathrm{T}_{\underline{0}}\,\mathrm{P}(\mathfrak{g})$ with $\mathrm{P}(\mathfrak{g})$, but we distinguish the Lie algebra $\mathrm{P}(\mathfrak{g})^{\mathrm{alg}}$ from the Lie group $\mathrm{P}(\mathfrak{g})$ by denoting the elements of the former by $t \mapsto X(t)$. In order to compute the Lie bracket, we differentiate:

$$(\delta \cdot \delta' \cdot \delta^{-1})(t) = \delta(t) + A_\delta(t)\delta'(t) - A_\delta(t) \circ A_{\delta'}(t) \circ A_\delta(t)^{-1}\delta(t)$$

with respect to δ', at $\delta' = \underline{0}$, in the direction of $Y \in \mathrm{P}(\mathfrak{g})^{\mathrm{alg}}$. From the variational equations for (1.13.3) (see (B.10)) we read off that:

$$\left.\frac{d}{d\epsilon}\right|_{\epsilon=0} A_{\epsilon Y}(t) = \int_0^t (\mathrm{ad}\, Y)(s)\, ds, \tag{1.14.8}$$

which then leads to:

$$((\mathrm{Ad}\,\delta)Y)(t) = A_\delta(t)Y(t) - A_\delta(t) \circ \textstyle\int_0^t \mathrm{ad}\, Y(s)\, ds \circ A_\delta(t)^{-1} \circ \delta(t). \tag{1.14.9}$$

Differentiating this with respect to δ, at $\delta = \underline{0}$, in the direction of $X \in \mathrm{P}(\mathfrak{g})^{\mathrm{alg}}$, we find:

$$\begin{aligned}[X,Y](t) &= \textstyle\int_0^t \mathrm{ad}\, X(s)\, ds\, Y(t) - \int_0^t \mathrm{ad}\, Y(s)\, ds\, X(t) \\ &= [\textstyle\int_0^t X(s)\, ds, Y(t)] + [X(t), \int_0^t Y(s)\, ds] \\ &= \frac{d}{dt}[\textstyle\int_0^t X(s)\, ds, \int_0^t Y(s)\, ds],\end{aligned}$$

and this proves (1.14.2).

Noting that:

$$\begin{aligned}\textstyle\int_0^1 [X,Y](t)\, dt &= \int_0^1 \frac{d}{dt}[\textstyle\int_0^t X(s)\, ds, \int_0^t Y(s)\, ds]\, dt \\ &= [\textstyle\int_0^1 X(s)\, ds, \int_0^1 Y(s)\, ds],\end{aligned}$$

we see that the *averaging* in (1.14.4) is a continuous and surjective homomorphism of Lie algebras: $\mathrm{P}(\mathfrak{g})^{\mathrm{alg}} \to \mathfrak{g}$. It follows that the kernel $\mathrm{P}(\mathfrak{g})_0^{\mathrm{alg}}$ is a closed ideal in $\mathrm{P}(\mathfrak{g})^{\mathrm{alg}}$, and that averaging induces an isomorphism:

$$\mathrm{P}(\mathfrak{g})^{\mathrm{alg}} / \mathrm{P}(\mathfrak{g})_0^{\mathrm{alg}} \xrightarrow{\sim} \mathfrak{g}.$$

The connected Lie subgroup P_0 of $\mathrm{P}(\mathfrak{g})$ with Lie algebra equal to $\mathrm{P}(\mathfrak{g})_0^{\mathrm{alg}}$ can be characterized as the set of $\delta \in \mathrm{P}(\mathfrak{g})$ for which there exists a smooth curve $s \mapsto \delta_s \colon [0,1] \to \mathrm{P}(\mathfrak{g})$ such that $\delta_0 = \underline{0}$, $\delta_1 = \delta$ and:

$$\text{(1.14.10)} \qquad \mathrm{T}_{\underline{0}}\,\mathrm{L}(\delta_s)^{-1} \frac{d}{ds}\delta_s \in \mathrm{P}(\mathfrak{g})_0^{\mathrm{alg}} \qquad (s \in [0,1]).$$

In view of (1.14.1), (1.14.3), this reads as (1.13.5), showing that P_0 coincides with the set $\mathrm{P}(\mathfrak{g})_0$ defined in Proposition 1.13.4. This proves the last sentence in the proposition. □

In general it is a subtle problem to decide whether a connected Lie subgroup with Lie algebra equal to a given subalgebra, which exists according to Theorem 1.10.3, is a **closed** subgroup. In order to understand the situation for the subgroup $\mathrm{P}(\mathfrak{g})_0$ of $\mathrm{P}(\mathfrak{g})$, we introduce the analytic, $\mathfrak{g}$-valued 1-form ω on $\mathrm{P}(\mathfrak{g})$, defined by:

$$\text{(1.14.11)} \qquad \omega_\delta(X) = \mathrm{av}(\mathrm{T}_{\underline{0}}\,\mathrm{L}(\delta)^{-1} X) = \textstyle\int_0^1 A_\delta(t)^{-1} X(t)\, dt.$$

Thus $\omega_\delta(X^{\mathrm{L}}(\delta)) = \mathrm{av}(X)$, independently of δ, for every $X \in \mathrm{P}(\mathfrak{g})^{\mathrm{alg}}$. In view of (1.13.5), $\ker \omega_\delta = \mathrm{T}_\delta(\mathrm{P}(\mathfrak{g})_0)$, for each $\delta \in \mathrm{P}(\mathfrak{g})_0$.

In general, if ω is a 1-form and X, Y are vector fields on a manifold, then the exterior derivative $d\omega$ of ω is given by:

$$\text{(1.14.12)} \qquad d\omega(X,Y) = d(\omega(Y))(X) - d(\omega(X))(Y) + \omega([X,Y]),$$

cf. Spivak [1979], Theorem 7.13. Applying this to our ω, we get (cf. (1.10.3)):

$$\text{(1.14.13)} \qquad d\omega(X^{\mathrm{L}}, Y^{\mathrm{L}}) = -\,\mathrm{av}([X,Y]) = -[\mathrm{av}\,X, \mathrm{av}\,Y], \text{ constantly.}$$

According to (1.14.6), $\delta \mapsto A_\delta(1)$ is a homomorphism of Lie groups: $\mathrm{P}(\mathfrak{g}) \to \mathrm{Ad}\,\mathfrak{g}$; and because of (1.14.8) its tangent mapping at $\underline{0}$ is given by:

$$X \mapsto \left.\frac{d}{d\epsilon}\right|_{\epsilon=0} A_{\epsilon X}(1) = \int_0^1 (\mathrm{ad}\,X)(s)\, ds \colon \mathrm{P}(\mathfrak{g})^{\mathrm{alg}} \to \mathrm{ad}\,\mathfrak{g},$$

and this is a surjection. Hence, in view of the implicit function theorem in Banach spaces the kernel of $\delta \mapsto A_\delta(1)$,

$$\text{(1.14.14)} \qquad \mathrm{P}(\mathfrak{g})_1 = \{\delta \in \mathrm{P}(\mathfrak{g}) \mid A_\delta(1) = \mathrm{I}\},$$

is a smooth submanifold, hence a closed normal Banach Lie subgroup of $\mathrm{P}(\mathfrak{g})$. Moreover we obtain that the Lie algebra of $\mathrm{P}(\mathfrak{g})_1$ is equal to:

$$\begin{aligned} \mathrm{P}(\mathfrak{g})_1^{\mathrm{alg}} &= \{\, X \in \mathrm{P}(\mathfrak{g})^{\mathrm{alg}} \mid \textstyle\int_0^1 \operatorname{ad} X(t)\, dt = 0 \,\} \\ &= \{\, X \in \mathrm{P}(\mathfrak{g})^{\mathrm{alg}} \mid \operatorname{av} X \in \mathfrak{z} \,\}, \end{aligned} \tag{1.14.15}$$

where $\mathfrak{z} = \ker \operatorname{ad}$ is the *center* of $\mathfrak{g}$. From (1.14.3, 14) it follows that $\mathrm{P}(\mathfrak{g})_0 \subset (\mathrm{P}(\mathfrak{g})_1)^\circ$. Furthermore, from (1.14.11, 15) we see that $\omega|_{\mathrm{P}(\mathfrak{g})_1}$ is $\mathfrak{z}$-valued, and from (1.14.13) that $d\omega|_{\mathrm{P}(\mathfrak{g})_1} = 0$.

The latter equality suggests to construct a smooth mapping:

$$\phi\colon (\mathrm{P}(\mathfrak{g})_1)^\circ \to (\mathfrak{z}, +) \quad \text{with} \quad \phi(\underline{0}) = 0 \quad \text{such that} \quad d\phi = \omega.$$

Then $M = \phi^{-1}(\{0\})$ is a submanifold through $\underline{0}$ satisfying $\mathrm{T}_\delta\, M = \ker(d\phi)_\delta = \ker \omega_\delta = \mathrm{T}_\delta(\mathrm{P}(\mathfrak{g})_0)$, for each $\delta \in M$. That is, M is a closed integral manifold through $\underline{0}$ for the distribution $\ker \omega$, for which $\mathrm{P}(\mathfrak{g})_0$ is also an integral manifold; and in this fashion we shall obtain that $\mathrm{P}(\mathfrak{g})_0$ is a closed subgroup of $\mathrm{P}(\mathfrak{g})$.

Since $\operatorname{ad} \mathfrak{g}$ is the Lie algebra of the connected Lie group $\operatorname{Ad} \mathfrak{g}$, we can use Proposition 1.13.4 in order to identify $\mathrm{P}(\operatorname{ad} \mathfrak{g})$, and $\mathrm{P}(\operatorname{ad} \mathfrak{g})_0$, via the homeomorphism D^{-1} with $\mathrm{P}(1, \operatorname{Ad} \mathfrak{g}) \cap \mathrm{C}^1$, and $\Lambda(\operatorname{Ad} \mathfrak{g})^\circ \cap \mathrm{C}^1$, respectively, and then Proposition 1.13.2 to obtain:

$$\mathrm{P}(\operatorname{ad} \mathfrak{g}) / \mathrm{P}(\operatorname{ad} \mathfrak{g})_0 = \mathrm{P}(1, \operatorname{Ad} \mathfrak{g}) / \Lambda(\operatorname{Ad} \mathfrak{g})^\circ = \widetilde{\operatorname{Ad} \mathfrak{g}},$$

the universal covering group of the adjoint group of $\mathfrak{g}$. But in view of $\mathfrak{g}/\mathfrak{z} \simeq \operatorname{ad} \mathfrak{g}$ and (1.14.15) this implies:

$$\mathrm{P}(\mathfrak{g})/(\mathrm{P}(\mathfrak{g})_1)^\circ = (\mathrm{P}(\mathfrak{g})/\mathrm{P}(\mathfrak{z}))/((\mathrm{P}(\mathfrak{g})_1)^\circ/\mathrm{P}(\mathfrak{z})) = \widetilde{\operatorname{Ad} \mathfrak{g}}.$$

The Equations (1.14.13, 15) now show that $d\omega$ is equal to the pull-back $\pi^* \Omega$ of a 2-form Ω on $\mathrm{P}(\mathfrak{g})/(\mathrm{P}(\mathfrak{g})_1)^\circ = \widetilde{\operatorname{Ad} \mathfrak{g}}$ under the canonical projection $\pi\colon \mathrm{P}(\mathfrak{g}) \to \widetilde{\operatorname{Ad} \mathfrak{g}}$. Note that $(\mathrm{P}(\mathfrak{g})_1)^\circ$ is an open, hence closed subgroup of $\mathrm{P}(\mathfrak{g})_1$; it is a closed normal Banach Lie subgroup of $\mathrm{P}(\mathfrak{g})$. Because π is surjective and has surjective tangent map, Ω is uniquely determined and smooth. Because π is a homomorphism and $d\omega$ is left invariant, Ω is left invariant as well. So it is determined by its value at the identity element, where it is given by:

$$\Omega_1(\operatorname{ad} X, \operatorname{ad} Y) = -[\, X, Y \,] \qquad (X, Y \in \mathfrak{g}). \tag{1.14.16}$$

Also, $\pi^* d\Omega = d(\pi^* \Omega) = dd\omega = 0$, and using again the surjectivity of π and of $\mathrm{T}_0\, \pi$, we conclude that $d\Omega = 0$; so Ω defines a de Rham cohomology class $[\Omega] \in H^2(\widetilde{\operatorname{Ad} \mathfrak{g}}; \mathbf{R})$.

Now let $\delta\colon s \mapsto \delta_s$ be a continuous, piecewise C^1 curve: $[\,0,1\,] \to (\mathrm{P}(\mathfrak{g})_1)^\circ$ such that $\delta_0 = \underline{0} = \delta_1$. We now use that $\mathrm{P}(\mathfrak{g})$, as a Banach space, allows a radial contraction. Define $\epsilon\colon [\,0,1\,] \times [\,0,1\,] \to \mathrm{P}(\mathfrak{g})$ by $\epsilon(u, s) = u\delta_s$. We have $\epsilon(0, s) \equiv \underline{0}$, $\epsilon(u, 0) = \epsilon(u, 1) \equiv \underline{0}$, $\epsilon(1, s) = \delta_s$. However, $\delta_s \in (\mathrm{P}(\mathfrak{g})_1)^\circ$ implies that $\pi(\delta_s) \equiv 1$; so the mapping $A = \pi \circ \epsilon\colon [\,0,1\,] \times [\,0,1\,] \to \widetilde{\operatorname{Ad} \mathfrak{g}}$ maps the whole boundary of $[0,1] \times [0,1]$ to $\{1\}$, and therefore defines a homology class $[A] \in H_2(\widetilde{\mathrm{Ad}\,\mathfrak{g}}\,; \mathbf{Z})$. Now Stokes' theorem on the square yields:

$$
\begin{aligned}
\int_{[0,1]} \delta^*\omega &= \int_{[0,1]\times[0,1]} d(\epsilon^*\omega) = \int_{[0,1]\times[0,1]} \epsilon^* d\omega \\
&= \int_{[0,1]\times[0,1]} \epsilon^*\pi^*\Omega = \int_{[0,1]\times[0,1]} A^*\Omega \\
&= < [A], [\Omega] > .
\end{aligned}
\tag{1.14.17}
$$

But the following general fact about the topology of simply connected Lie groups implies that the right hand side is equal to zero.

(1.14.2) Theorem. *If G is a simply connected Lie group, then $H^2(G\,; \mathbf{R}) = 0$.*

Proof. Let $\pi_1 \colon (x,y) \mapsto x$ and $\pi_2 \colon (x,y) \mapsto y$ denote the two projections from $G \times G$ onto G. If ω_1 and ω_2 are C^∞ differential forms on G, then the exterior product $\pi_1^*\omega_1 \cdot \pi_2^*\omega_2$ is a C^∞ differential form on $G \times G$. This construction induces a bilinear mapping: $H^p(G\,;\mathbf{R}) \times H^q(G\,;\mathbf{R}) \to H^{p+q}(G \times G\,;\mathbf{R})$, which in turn induces a linear mapping

$$
K_{p,q} \colon H^p(G\,;\mathbf{R}) \otimes H^q(G\,;\mathbf{R}) \to H^{p+q}(G \times G\,;\mathbf{R}). \tag{1.14.18}
$$

The *Künneth formula* for the compactly supported de Rham cohomology $H_c^*(G\,;\mathbf{R})$, combined with the *Poincaré duality*

$$
H^p(G\,;\mathbf{R}) \cong (H_c^{n-p}(G\,;\mathbf{R}))^*, \quad n = \dim G,
$$

implies the following: $K_{p,q}$ is injective for each p, q; the images of the $K_{p,q}$, for the various p, q, are linearly independent; and $H^k(G \times G\,;\mathbf{R})$ is equal to the direct sum of the images of the $K_{p,k-p}$, for $0 \le p \le k$, if for each p either $H^p(G\,;\mathbf{R})$ or $H^{k-p}(G\,;\mathbf{R})$ is finite-dimensional, see Greub, Halperin, Vanstone [1972], Chap.V, §6.

Now $H^0(G\,;\mathbf{R}) = \mathbf{R}$ and $H^1(G\,;\mathbf{R}) = 0$, as the space of homomorphisms: $\pi_1(G,1) \to (\mathbf{R},+)$, so we get:

$$
H^2(G \times G\,;\mathbf{R}) = \pi_1^* H^2(G\,;\mathbf{R}) \oplus \pi_2^* H^2(G\,;\mathbf{R}). \tag{1.14.19}
$$

The multiplication $\mu \colon (x,y) \mapsto xy \colon G \times G \to G$ induces the ring homomorphism $\mu^* \colon H^\bullet(G\,;\mathbf{R}) \to H^\bullet(G \times G\,;\mathbf{R})$. If $\sigma \in H^2(G\,;\mathbf{R})$ then, by (1.14.19), we obtain:

$$
\mu^*(\sigma) = \pi_1^*\sigma_1 + \pi_2^*\sigma_2, \tag{1.14.20}
$$

for $\sigma_1, \sigma_2 \in H^2(G\,;\mathbf{R})$. Writing $\iota_1 \colon x \mapsto (x,1)$, $\iota_2 \colon y \mapsto (1,y)$, from G to $G \times G$, we have:

$$
\begin{aligned}
\mu \circ \iota_1 &= \pi_1 \circ \iota_1 = \mu \circ \iota_2 = \pi_2 \circ \iota_2 = \text{identity in } G, \\
\pi_2 \circ \iota_1 &= \pi_1 \circ \iota_2 = \text{projection } : G \to \{1\}.
\end{aligned}
$$

So applying ι_1^*, and ι_2^*, to $\mu^*(\sigma)$, we get $\sigma = \sigma_1$, and $\sigma = \sigma_2$, respectively; so (1.14.20) becomes:

(1.14.21) $$\mu^*(\sigma) = \pi_1^*(\sigma) + \pi_2^*(\sigma) \qquad (\sigma \in H^2(G\,;\mathbf{R})).$$

Taking the m-th power, and using that H^{even} is commutative, we find:

$$\begin{aligned}\mu^*(\sigma^m) = (\mu^*(\sigma))^m &= \sum_{k=0}^{m} \binom{m}{k} (\pi_1^*(\sigma))^k \cdot (\pi_2^*(\sigma))^{m-k} \\ &= \sum_{k=0}^{m} \binom{m}{k} K_{k,m-k}(\sigma^k \otimes \sigma^{m-k}).\end{aligned}$$

Obviously $\sigma^m = 0$, if $2m > \dim G$.

If $\sigma^m = 0$, then the Künneth formula implies that $\sigma^k \otimes \sigma^{m-k} = 0$, for all $0 \le k \le m$. In particular, $\sigma \otimes \sigma^{m-1} = 0$; hence $\sigma = 0$ or $\sigma^{m-1} = 0$, so $\sigma^{m-1} = 0$ in any case. By downward induction on m we arrive at $\sigma = 0$, for every $\sigma \in H^2(G\,;\mathbf{R})$. □

Applying Theorem 1.14.2 with G replaced by $\widetilde{\operatorname{Ad}\mathfrak{g}}$, we conclude that $\int_{[0,1]} \delta^*\omega = 0$, for every closed, continuous, piecewise C^1 loop δ in $(\mathrm{P}(\mathfrak{g})_1)^\circ$. So, if $\alpha \in (\mathrm{P}(\mathfrak{g})_1)^\circ$ is given, then the value $\phi(\alpha) := \int_{[0,1]} \delta^*\omega$, where $s \mapsto \delta_s$ is a C^1 curve: $[0,1] \to (\mathrm{P}(\mathfrak{g})_1)^\circ$ such that $\delta_0 = \underline{0}$, $\delta_1 = \alpha$, actually does not depend on the choice of δ; and it defines a $\mathfrak{z}$-valued smooth function ϕ on $(\mathrm{P}(\mathfrak{g})_1)^\circ$, such that $\phi(\underline{0}) = 0$ and $d\phi = \omega$.

Consider now, for fixed $\delta' \in (\mathrm{P}(\mathfrak{g})_1)^\circ$, the mapping:

$$\delta \mapsto \phi(\delta \cdot \delta') - \phi(\delta) \colon (\mathrm{P}(\mathfrak{g})_1)^\circ \to (\mathfrak{z}, +).$$

Its derivative vanishes since $\omega = d\phi$ is left invariant; and because $\phi(\underline{0}) = 0$, it is equal to the constant map $\delta \mapsto \phi(\delta')$. Thus ϕ is a homomorphism of Lie groups $(\mathrm{P}(\mathfrak{g})_1)^\circ \to (\mathfrak{z}, +)$. Hence $\ker\phi$ is a closed normal Banach Lie subgroup of $(\mathrm{P}(\mathfrak{g})_1)^\circ$, with Lie algebra equal to $\mathrm{P}(\mathfrak{g})_0^{\text{alg}}$. Therefore $\mathrm{P}(\mathfrak{g})_0$ is equal to $(\ker\phi)^\circ$, which is closed in $\mathrm{P}(\mathfrak{g})$, because $(\ker\phi)^\circ$ is an open, hence closed subgroup in $\ker\phi \subset (\mathrm{P}(\mathfrak{g})_1)^\circ \subset \mathrm{P}(\mathfrak{g})$, where each inclusion is closed.

Now $\mathrm{P}(\mathfrak{g})/\mathrm{P}(\mathfrak{g})_0$ is a Banach Lie group with Lie algebra equal to $\mathrm{P}(\mathfrak{g})^{\text{alg}}/\mathrm{P}(\mathfrak{g})_0^{\text{alg}}$, which is isomorphic to $\mathfrak{g}$ via averaging. So there is a Lie group G with Lie algebra $\mathfrak{g}$, and Propositions 1.13.2, 4 then say that $\mathrm{P}(\mathfrak{g})/\mathrm{P}(\mathfrak{g})_0$ is canonically isomorphic to the universal covering $\widetilde{G}$ of G.

We finally note that, because of (1.13.4.b) and (1.14.5), the center $\mathrm{Z}(\widetilde{G})$ of $\widetilde{G}$ is equal to $\ker(\operatorname{Ad}\colon \widetilde{G} \to \operatorname{Ad}\mathfrak{g})$; and using (1.14.14) we obtain $\mathrm{Z}(\widetilde{G}) = \mathrm{P}(\mathfrak{g})_1/\mathrm{P}(\mathfrak{g})_0$. Furthermore $(\mathrm{P}(\mathfrak{g})_1)^\circ/\mathrm{P}(\mathfrak{g})_0$ projects onto $(\mathrm{P}(\mathfrak{g})_1)^\circ/\ker\phi \cong (\mathfrak{z}, +)$, with discrete fiber $(\ker\phi)/(\ker\phi)^\circ$; but because $\mathfrak{z}$ is simply connected this is an isomorphism. Therefore the identity component of $\mathrm{Z}(\widetilde{G})$ is a connected commutative Lie group that is isomorphic to a vector group. But according to Corollary 1.12.4 the exponential map is an isomorphism for such groups. That is, we have proved the following result.

(1.14.3) Theorem. *Let $\mathfrak{g}$ be a finite-dimensional Lie algebra. Then there exists a simply connected Lie group $\widetilde{G}$ with Lie algebra equal to $\mathfrak{g}$. The restriction of the exponential mapping: $\mathfrak{g} \to \widetilde{G}$ to the center $\mathfrak{z}$ of $\mathfrak{g}$ induces an isomorphism from $(\mathfrak{z}, +)$ onto the identity component of the center of $\widetilde{G}$.*

(1.14.4) *Remark.* By construction $< [A], [\Omega] > \in \mathfrak{z}$. On the other hand, (1.14.16) shows that Ω takes its values in $\{ [X, Y] \mid X, Y \in \mathfrak{g} \}$; so $< [A], [\Omega] > \in [\mathfrak{g}, \mathfrak{g}]$, the vector space generated by the $[X, Y]$, with $X, Y \in \mathfrak{g}$. The space $[\mathfrak{g}, \mathfrak{g}]$ is an ideal in $\mathfrak{g}$, called the *derived Lie algebra* of $\mathfrak{g}$. If $\mathfrak{z} \cap [\mathfrak{g}, \mathfrak{g}] = 0$, then we would get $< [A], [\Omega] > = 0$, for a simpler reason than Theorem 1.14.2.

On the other hand, if $\mathfrak{z} \cap [\mathfrak{g}, \mathfrak{g}] = 0$, then taking $\mathfrak{g}_1$ equal to the preimage, under the projection $\mathfrak{g} \mapsto \mathfrak{g}/[\mathfrak{g}, \mathfrak{g}]$, of a complementary subspace to $\mathfrak{z} + [\mathfrak{g}, \mathfrak{g}]$ in the Abelian Lie algebra $\mathfrak{g}/[\mathfrak{g}, \mathfrak{g}]$, we get $\mathfrak{g} = \mathfrak{g}_1 \oplus \mathfrak{z}$, and $\mathfrak{g}_1$ is a Lie subalgebra of $\mathfrak{g}$. Then $\operatorname{ad}|_{\mathfrak{g}_1}$ is an isomorphism of Lie algebras: $\mathfrak{g}_1 \to \operatorname{ad} \mathfrak{g}$, and $\widetilde{G} = \widetilde{\operatorname{Ad} \mathfrak{g}} \times (\mathfrak{z}, +)$ immediately gives a simply connected Lie group with Lie algebra isomorphic to $\mathfrak{g}$.

(1.14.5) *Remark.* For an infinite-dimensional Banach Lie algebra $\mathfrak{g}$, the conclusion of Theorem 1.14.3 need not be true, as is demonstrated by the following counterexample due to Van Est and Korthagen [1964]. Its point of departure is the observation, with $G_0 = \Lambda(\mathbf{SU}(2))$ denoting the loop group of $\mathbf{SU}(2)$, that the second homotopy group $\pi_2(G_0)$ is isomorphic to $\pi_3(\mathbf{SU}(2)) \cong \mathbf{Z}$, in contrast with Theorem 1.14.2.

If $\mathfrak{g}$ is the Lie algebra of a Banach Lie group G, then the restriction ϕ to $(\mathrm{P}(\mathfrak{g})_1)^\circ$ of the homomorphism: $\mathrm{P}(\mathfrak{g}) \to \widetilde{G} \cong \mathrm{P}(\mathfrak{g}) / \mathrm{P}(\mathfrak{g})_0$ takes its values in the identity component of the center of $\widetilde{G}$, which we identify with $(\mathfrak{z}, +)/\Gamma$ where $\mathfrak{z}$ is the center of $\mathfrak{g}$ and $\Gamma = \ker \exp \colon \mathfrak{z} \to \widetilde{G}$. On the other hand, $d\omega = \phi$, so (1.14.17) takes its values in the discrete subgroup Γ of $(\mathfrak{z}, +)$.

Because $\mathbf{SU}(2)$ is simply connected, the Lie algebra $\mathfrak{g}_0$ can be identified with $\mathrm{P}(\mathfrak{su}(2))_0^{\mathrm{alg}}$, the Lie algebra consisting of the $X \in \mathrm{C}^0([0,1], \mathfrak{su}(2))$ such that $\int_0^1 X(t)\, dt = 0$. On $\mathfrak{g}_0$, consider the antisymmetric bilinear form τ defined by:

$$\tau(X, Y) = \int_0^1 \operatorname{tr}\left(\left(\int_0^t X(s)\, ds \circ Y(t) \right) \right) dt \qquad (X, Y \in \mathfrak{su}(2)). \tag{1.14.22}$$

Now provide $\mathfrak{g} = \mathbf{R} \times \mathfrak{g}_0$ with the bracket, for $a_j \in \mathbf{R}$, $X_j \in \mathfrak{g}_0$:

$$[(a_1, X_1), (a_2, X_2)] = (\tau(X_1, X_2), [X_1, X_2]). \tag{1.14.23}$$

One verifies that $\mathfrak{g}$ is a Lie algebra, with one-dimensional center $\mathfrak{z} = \mathbf{R} \times \{0\}$.

In order to show that $[\Omega] \neq 0$, we only need to look at the $\mathfrak{z}$-part of Ω which, according to (1.14.16), is the left invariant 2-form on $\widetilde{\operatorname{Ad} \mathfrak{g}} \cong G_0$ that at 1 is equal to $-\tau$.

Because $(\mathrm{T}_{\underline{0}}\,\mathrm{L}(\delta)^{-1}X)(t) = A_\delta(t)^{-1}X(t)$ we get, upon writing $\delta = \mathrm{D}^2\,\gamma = \frac{\partial\gamma}{\partial t}\circ\gamma^{-1}$ and letting γ depend on t, u, v:

$$\begin{aligned}\mathrm{T}_{\underline{0}}\,\mathrm{L}(\delta)^{-1}\circ\frac{\partial\delta}{\partial u} &= \gamma^{-1}\circ\Big(\frac{\partial^2\gamma}{\partial u\partial t}\circ\gamma^{-1} - \frac{\partial\gamma}{\partial t}\circ\gamma^{-1}\circ\frac{\partial\gamma}{\partial u}\circ\gamma^{-1}\Big)\circ\gamma \\ &= \gamma^{-1}\circ\frac{\partial^2\gamma}{\partial u\partial t} - \gamma^{-1}\circ\frac{\partial\gamma}{\partial t}\circ\gamma^{-1}\circ\frac{\partial\gamma}{\partial u} \\ &= \frac{\partial}{\partial t}\Big(\gamma^{-1}\circ\frac{\partial\gamma}{\partial u}\Big).\end{aligned}$$

So we have to verify whether we can make:

$$(1.14.24)\qquad -\iiint \mathrm{tr}\Big(\gamma^{-1}\circ\frac{\partial\gamma}{\partial u}\circ\frac{\partial}{\partial t}\Big(\gamma^{-1}\circ\frac{\partial\gamma}{\partial v}\Big)\Big)\,dt\,du\,dv$$

nonzero, for some 3-cycle $\gamma\colon (t,u,v)\mapsto\gamma(t,u,v)\in\mathbf{SU}(2)$.

Now it is convenient to use the identification of $\mathbf{SU}(2)$, and $\mathfrak{su}(2)$, in (1.2.19), and (1.2.23), with the unit sphere, and a 3-dimensional linear subspace, respectively, in the space $\mathbf{H}$ of quaternions. For $X\in S$, the unit sphere in $\mathfrak{su}(2)$, write:

$$\gamma(t,X) = \exp(2\pi tX) = \cos 2\pi t\,\mathrm{I} + \sin 2\pi t\,X.$$

The derivative of this with respect to X in the direction of $Y\in\mathrm{T}_X\,S$ is equal to $\sin 2\pi t\,Y$, so the integrand of (1.14.24) gets the form:

$$-\,\mathrm{tr}\Big[(\cos 2\pi t\,\mathrm{I} - \sin 2\pi t\,X)\sin 2\pi t\,Y\frac{d}{dt}\big((\cos 2\pi t\,\mathrm{I} - \sin 2\pi t\,X)\sin 2\pi t\,Z\big)\Big],$$

for $Y, Z\in\mathrm{T}_X\,S$. Integrating this over t, for $t\in[0,1]$, we get $\pi\,\mathrm{tr}(YXZ)$. Because, for:

$$X = \begin{pmatrix} i & 0\\ 0 & -i\end{pmatrix},\quad Y = \begin{pmatrix} 0 & \beta\\ -\bar\beta & 0\end{pmatrix},\quad Z = \begin{pmatrix} 0 & \gamma\\ -\bar\gamma & 0\end{pmatrix},$$

one has $\mathrm{tr}(YXZ) = 2\det(\beta,\gamma)$, for $\beta,\gamma\in\mathbf{C}\cong\mathbf{R}^2$, we see that (1.14.24) is equal to $2\pi 4\pi > 0$.

In order to obtain a nondiscrete subgroup $\Gamma = \{< c, [\Omega] > \mid c\in H_2(\widetilde{\mathrm{Ad}\,\mathfrak{g}};\mathbf{Z})\}$ of $(\mathfrak{z}, +)$, let:

$$0\to\mathfrak{z}_i\to\mathfrak{g}_i\to\mathfrak{g}_{0,i}\to 0$$

be two central extensions $(\mathfrak{z}(\mathfrak{g}_{0,i}) = 0)$ like in (1.14.23) with $\mathbf{R}\times\{0\}$ replaced by $\mathfrak{z}_i$ and with $\mathfrak{g}_{0,i}$ the Lie algebra of the Banach Lie groups $G_{0,i}$. Then $\mathfrak{z} = \mathfrak{z}_1\times\mathfrak{z}_2$, $\mathfrak{g} = \mathfrak{g}_1\times\mathfrak{g}_2$, $\mathfrak{g}_0 = \mathfrak{g}_{0,1}\times\mathfrak{g}_{0,2}$ is a central extension $0\to\mathfrak{z}\to\mathfrak{g}\to\mathfrak{g}_0\to 0$ like in (1.14.23), with $\Gamma = \Gamma_1\times\Gamma_2\subset\mathfrak{z}_1\times\mathfrak{z}_2$. Now any linear mapping $\lambda\colon\mathfrak{z}\to\mathfrak{z}'$ induces a central extension $0\to\mathfrak{z}'\to\mathfrak{g}'\to\mathfrak{g}_0\to 0$ as in (1.14.23), with $\Gamma' = \lambda(\Gamma) = \lambda(\Gamma_1\times\Gamma_2)$. So it suffices for instance to take two copies of the previous example with $G_0 = \Lambda(\mathbf{SU}(2))$ and then take $\lambda\colon (\tau_1, \tau_2) \mapsto \tau_1 + c\tau_2$ with $c \in \mathbf{R} \setminus \mathbf{Q}$, to get a Γ' which lies dense in $\mathbf{R}$.

Banach Lie groups which are non-Hausdorff, as the $\mathrm{P}(\mathfrak{g})/\mathrm{P}(\mathfrak{g})_0$ above, have also been considered by, for instance, Plaisant [1980].

1.15 Exercises

(1.1) Exercise. The complex classical Lie groups are:

$$
\begin{aligned}
&\text{(i)} \quad \mathbf{SL}(n+1,\mathbf{C}) &&= \{ A \in \mathbf{L}_{\mathbf{C}}(\mathbf{C}^{n+1}, \mathbf{C}^{n+1}) \mid \det A = 1 \};\\
&\text{(ii)} \quad \mathbf{SO}(2n+1,\mathbf{C}) &&= \{ A \in \mathbf{SL}(2n+1,\mathbf{C}) \mid {}^tAA = \mathrm{I} \};\\
&\text{(iii)} \quad \mathbf{Sp}(n,\mathbf{C}) &&= \{ A \in \mathbf{GL}(2n,\mathbf{C}) \mid {}^tAJ_nA = J_n \};\\
&\text{(iv)} \quad \mathbf{SO}(2n,\mathbf{C}) &&= \{ A \in \mathbf{SL}(2n,\mathbf{C}) \mid {}^tAA = \mathrm{I} \}.
\end{aligned}
$$

In (iii) $J_n = \begin{pmatrix} 0 & \mathrm{I}_n \\ -\mathrm{I}_n & 0 \end{pmatrix}$, and $\mathbf{Sp}(n,\mathbf{C})$ is called the complex symplectic group.

Verify that these are complex-analytic Lie groups, and prove that the corresponding complex classical Lie algebras are given by:

$$
\begin{aligned}
&\text{(i)} \quad A_n = \mathfrak{sl}(n+1,\mathbf{C}) &&= \{ X \in \mathbf{L}_{\mathbf{C}}(\mathbf{C}^{n+1}, \mathbf{C}^{n+1}) \mid \operatorname{tr} X = 0 \};\\
&\text{(ii)} \quad B_n = \mathfrak{so}(2n+1,\mathbf{C}) &&= \{ X \in \mathbf{L}_{\mathbf{C}}(\mathbf{C}^{2n+1}, \mathbf{C}^{2n+1}) \mid {}^tX + X = 0 \};\\
&\text{(iii)} \quad C_n = \mathfrak{sp}(n,\mathbf{C}) &&= \left\{ \begin{pmatrix} X_1 & X_2 \\ X_3 & -{}^tX_1 \end{pmatrix} \,\middle|\, X_i \in \mathbf{L}_{\mathbf{C}}(\mathbf{C}^n, \mathbf{C}^n), \right.\\
& && \qquad\qquad\qquad\qquad \left. {}^tX_2 = X_2,\ {}^tX_3 = X_3 \right\};\\
&\text{(iv)} \quad D_n = \mathfrak{so}(2n,\mathbf{C}) &&= \{ X \in \mathbf{L}_{\mathbf{C}}(\mathbf{C}^{2n}, \mathbf{C}^{2n}) \mid {}^tX + X = 0 \}.
\end{aligned}
$$

Check that these Lie algebras, in fact, are closed under the Lie bracket of matrices, and that they are invariant under conjugation by elements in the corresponding Lie group. (A Lie algebra $\mathfrak{g}$ in a particular family is specified by an integer subscript n, which is the dimension of a subalgebra of diagonal matrices in $\mathfrak{g}$, relative to a suitably chosen basis for the vector space.)
The corresponding compact classical Lie groups are

$$
\begin{aligned}
&\text{(i)} \quad \mathbf{SU}(n+1);\\
&\text{(ii)} \quad \mathbf{SO}(2n+1);\\
&\text{(iii)} \quad \mathbf{Sp}(n) = \{ A \in \mathbf{U}(2n) \mid {}^tAJ_nA = J_n \};\\
&\text{(iv)} \quad \mathbf{SO}(2n).
\end{aligned}
$$

Verify that they are real-analytic Lie groups, and determine the corresponding Lie algebras.

(1.2) Exercise. Let $k = \mathbf{R}$ or $\mathbf{C}$. Show that a Lie algebra of dimension 2 over k is either Abelian or possesses a basis $\{X, Y\}$ such that $[X, Y] = X$. Let U be the noncommutative Lie group of dimension 2 given in Section 1.2 with Lie algebra $\mathfrak{u}$. Show that $X = \begin{pmatrix} 0 & 1 \\ 0 & 0 \end{pmatrix}$ and $Y = \begin{pmatrix} -\frac{1}{2} & 0 \\ 0 & \frac{1}{2} \end{pmatrix}$ form such a basis for $\mathfrak{u}$.

(1.3) Exercise. Let $U \subset \mathbf{GL}(n, \mathbf{R})$ be the subgroup of upper triangular matrices with positive eigenvalues. Show that the map:

$$(a, u) \mapsto au \colon \mathbf{O}(n, \mathbf{R}) \times U \to \mathbf{GL}(n, \mathbf{R})$$

is a diffeomorphism of analytic manifolds (this is the content of the Gram-Schmidt procedure). Deduce that $\mathbf{GL}(n, \mathbf{R})$ is analytically diffeomorphic with $\mathbf{O}(n, \mathbf{R}) \times \mathbf{R}^{\frac{1}{2}n(n+1)}$. Let $P \subset \mathbf{GL}(n, \mathbf{R})$ be the **set** of positive definite symmetric matrices. Show that multiplication induces an analytic diffeomorphism $\mathbf{O}(n, \mathbf{R}) \times P \to \mathbf{GL}(n, \mathbf{R})$.
Hint: if $A \in \mathbf{GL}(n, \mathbf{R})$ then ${}^tAA \in P$, hence ${}^tAA = R^2$ for some $R \in P$, while $AB^{-1} \in \mathbf{O}(n, \mathbf{R})$.

(1.4) Exercise. The mapping $\exp\colon \mathfrak{gl}(n, \mathbf{C}) \to \mathbf{GL}(n, \mathbf{C})$ is surjective.
Hint: Apply the theory of the Jordan canonical form to write a conjugate of $A \in \mathbf{GL}(n, \mathbf{C})$ as $D(\mathrm{I} + U)$ with $D \in \mathbf{GL}(n, \mathbf{C})$ diagonal and U nilpotent, and use $\log(\mathrm{I} + U) = \sum_{k=0}^{\infty} \frac{(-1)^k}{k+1} U^{k+1}$, which is a terminating series.

(1.5) Exercise. Consider $x \in \mathbf{R}^3$ as a column vector and define:

$$X = \begin{pmatrix} 0 & {}^tx \\ x & 0 \end{pmatrix} \in \mathbf{L}_{\mathbf{R}}(\mathbf{R}^4, \mathbf{R}^4).$$

Find recurrence relations for X^{2n} and X^{2n+1} where $n \geq 0$, and prove:

$$e^{sX} = \begin{pmatrix} \cosh s|x| & \frac{\sinh s|x|}{|x|}\, {}^tx \\ \frac{\sinh s|x|}{|x|}\, x & \mathrm{I} + \frac{\cosh s|x| - 1}{|x|^2}\, x\, {}^tx \end{pmatrix}, \quad \text{for } s \in \mathbf{R}.$$

(1.6) Exercise. We define the Heisenberg group H^n for $n \geq 1$ as follows. As an analytic manifold $H^n = \mathbf{R}^{2n+1}$. We denote elements in H^n by (t_i, q_i, p_i) with $t_i \in \mathbf{R}$ and q_i, $p_i \in \mathbf{R}^n$, and we denote the standard inner product on $\mathbf{R}^n$ by $\cdot$. Then we define multiplication in H^n by:

$$(t_1, q_1, p_1)(t_2, q_2, p_2) = (t_1 + t_2 + \frac{1}{2}(p_2 \cdot q_1 - p_1 \cdot q_2), q_1 + q_2, p_1 + p_2,).$$

Note that $(q_1, p_1) J_n {}^t(q_1, p_1) = p_2 \cdot q_1 - p_1 \cdot q_2$ with J_n as in Exercise 1.1. Verify that H^n is a Lie group. Determine the center of H^n. Let $\widetilde{H}^n$ be $\mathbf{R}^{2n+1}$ with the multiplication:

$$(t_1, q_1, p_1)(t_2, q_2, p_2) = (t_1 + t_2 + p_1 \cdot q_2, q_1 + q_2, p_1 + p_2).$$

Prove that $\widetilde{H}^n$ is a Lie group, and that $(t, q, p) \mapsto (t + \frac{1}{2} p \cdot q, q, p) \colon H^n \to \widetilde{H}^n$ is an isomorphism of Lie groups.
Show that for every right invariant vector field v on H^n there exist $T \in \mathbf{R}$, Q and $P \in \mathbf{R}^n$ such that v considered as a mapping from $\mathbf{R}^{2n+1}$ into itself is given by:

$$v(t, q, p) = v_{(T,Q,P)}(t, q, p) = (T + \frac{1}{2}(p \cdot Q - P \cdot q), Q, P).$$

Prove that $\mathrm{D}\, v_{(T,Q,P)}(t, q, p)(\delta t, \delta q, \delta p) = \frac{1}{2}(\delta p \cdot Q - P \cdot \delta q, 0, 0)$ and deduce

$$[\, v_{(0,Q,0)}, v_{(0,0,P)} \,](t, q, p) = (P \cdot Q, 0, 0),$$

all other commutators being zero, where the Lie bracket $[\, v, w \,]$ of two vector fields v and w is defined in (1.8.3). In particular, the Lie algebra $\mathfrak{h}^n$ of H^n is $\mathbf{R}^{2n+1}$; and denoting by T, Q_i and P_i the standard basis vectors in $\mathbf{R}^{2n+1}$, its Lie algebra structure is given by $[\, Q_i, P_i \,] = T$ for $1 \le i \le n$, all other Lie brackets being zero.

(1.7) Exercise. Let $\mathfrak{g}$ be a Lie algebra with center $\{0\}$. Then the center Z' of the group $G' = \mathrm{Ad}\, \mathfrak{g}$ of inner automorphisms of $\mathfrak{g}$ consists of $\{\mathrm{I}\}$.
Hint: Write Ad' for the adjoint representation of G'. Now G' has Lie algebra $\mathfrak{g}' = \mathrm{ad}\, \mathfrak{g}$. The mapping $\Phi \colon g'Z' \to \mathrm{Ad}'\, g' \colon G'/Z' \to \mathrm{Ad}'\, \mathfrak{g}'$ is an isomorphism of Lie groups. On the other hand, $\mathrm{ad} \colon \mathfrak{g} \to \mathfrak{g}'$ is an isomorphism of Lie algebras, and hence $\Psi \colon g \mapsto \mathrm{ad} \circ \mathrm{Ad}\, g \circ \mathrm{ad}^{-1} \colon G' \to \mathrm{Ad}'\, \mathfrak{g}'$ is an isomorphism of Lie groups too. Verify $\Psi(e^{\mathrm{ad}\, X}) = \mathrm{ad} \circ e^{\mathrm{ad}\, X} \circ \mathrm{ad}^{-1} = \mathrm{Ad}'(e^{\mathrm{ad}\, X}) \in \mathrm{Ad}'\, \mathfrak{g}'$, for $X \in \mathfrak{g}$, and consider $\Psi^{-1} \circ \Phi \colon G'/Z' \to G'$.

(1.8) Exercise. Prove that the center Z of $\mathbf{SO}(n, \mathbf{R})$ is $\{\mathrm{I}\}$ or $\{\mathrm{I}, -\mathrm{I}\}$ for n odd or even, respectively.
Hint: Let $\mathcal{L}$ be the collection of linear subspaces of $\mathbf{R}^n$ of dimension 1 or 2, respectively, and consider the induced action of $\mathbf{SO}(n, \mathbf{R})$ on $\mathcal{L}$. For every $L \in \mathcal{L}$ there exists $A \in \mathbf{SO}(n, \mathbf{R})$ such that L is the unique fixed point of A. For any $C \in Z$ we have $AC(L) = C(L)$, and thus $C(L) = L$.

(1.9) Exercise. Let G be a Lie group and H a closed subgroup. Prove that G is connected if H and the analytic manifold G/H are connected. Show that the Lie groups $\mathbf{SO}(n, \mathbf{R})$ for $n \ge 2$, and $\mathbf{SU}(n)$ and $\mathbf{Sp}(n)$ for $n \ge 1$, from Exercise 1.1 act transitively on the respective unit spheres S^{n-1}, S^{2n-1} and S^{4n-1} of $\mathbf{R}^n$, $\mathbf{C}^n$ and $\mathbf{C}^{2n}$. Deduce the existence of analytic diffeomorphisms between these spheres and the analytic manifolds $\mathbf{SO}(n, \mathbf{R})/\,\mathbf{SO}(n-1, \mathbf{R})$, $\mathbf{SU}(n)/\,\mathbf{SU}(n-1)$ and $\mathbf{Sp}(n)/\,\mathbf{Sp}(n-1)$ respectively. Prove that $\mathbf{SO}(n, \mathbf{R})$, $\mathbf{SU}(n)$ and $\mathbf{Sp}(n)$ are connected for $n \ge 1$.

(1.10) Exercise. Let G be a Lie group, M a C^k manifold, for $k \geq 1$, and A a C^k action of G on M, and let A_x be as in (1.11.7). Prove that for every $x \in M$ the map A_x has a constant rank, and if this rank equals m, then: (i) G_x is a Lie subgroup of dimension $\dim G - m$; (ii) for some neighborhood U of the identity in G the set $U \cdot x$ is a C^k submanifold of dimension m in M; (iii) if the orbit is a C^k submanifold of M, then it has dimension m. Now assume that G is compact. Prove that every orbit is a C^k submanifold of M.
Hint: It is sufficient to prove this in a neighborhood of a point $x \in M$. Let U be as in (ii). The orbit $G \cdot x$ is a union of the disjoint sets $U \cdot x$ and $C \cdot x$, where $C = G \setminus UG_x$. Since C is compact, the set $C \cdot x$ is closed in M. Thus the intersection of $G \cdot x$ with the open neighborhood $M \setminus C \cdot x$ of x in M is a submanifold.

(1.11) Exercise. Let A be a transitive differentiable action of a Lie group G on a connected differentiable manifold M. Prove (i) the group G^0 also acts transitively on M; (ii) $G/G^0 \simeq G_x/(G_x \cap G^0)$ for every $x \in M$; (iii) if G_x is connected for a point $x \in M$, then G is connected.

(1.12) Exercise. Let G be a Lie group and V a finite-dimensional vector space, and suppose that $\Phi\colon G \to \mathbf{GL}(V)$ is a homomorphism of Lie groups. Define the semidirect product $G \ltimes V$ of G and V to be the Lie group whose underlying manifold is the Cartesian product $G \times V$ and whose multiplication is given by

$$(g_1, v_1)(g_2, v_2) = (g_1 g_2,\ v_1 + \Phi(g_1) v_2), \quad \text{for } g_1, g_2 \in G, v_1, v_2 \in V.$$

Show $(g, v)^{-1} = (g^{-1},\ -\Phi(g^{-1})v)$. Verify that the elements in $G \ltimes V$ of the form $(g, 0)$, and $(1, v)$, form a subgroup of $G \ltimes V$ which is isomorphic as a Lie group with G, and V, respectively. The subgroup isomorphic to V is a normal subgroup, and the homomorphism Φ can be reconstructed from the multiplication through $(g, 0)(1, v)(g^{-1}, 0) = (1,\ \Phi(g)v)$. Show that the Lie algebra of $G \ltimes V$ has the Lie bracket, for X_1, $X_2 \in \mathfrak{g}$, Y_1, $Y_2 \in V$

$$[\,(X_1, Y_1), (X_2, Y_2)\,] = ([\,X_1, X_2\,],\ [\,Y_1, Y_2\,] + \phi(X_1)Y_2 - \phi(X_2)Y_1).$$

Here $\phi = \mathrm{T}_1 \Phi\colon \mathfrak{g} \to \mathbf{L}(V, V)$. Prove that the Euclidean group, consisting of all isometries of $\mathbf{R}^n$, is the semidirect product $\mathbf{SO}(n, \mathbf{R}) \ltimes \mathbf{R}^n$, with $\mathbf{SO}(n, \mathbf{R})$ acting on $\mathbf{R}^n$ as group of rotations.

(1.13) Exercise. Let H be a one-parameter subgroup of a Lie group G. Show that the closure of H in G is compact and thus a torus, if H is not closed in G.
Hint: Use Corollary 1.10.7 to reduce to the case that G is connected and commutative.

(1.14) Exercise. Let G be a compact connected complex-analytic Lie group of dimension n. Prove that G is commutative and is isomorphic to $\mathbf{C}^n/\Gamma$, where Γ is an integral lattice in $\mathbf{C}^n$ with $2n$ generators (over $\mathbf{R}$).
Hint: The adjoint representation $G \to \operatorname{Ad}\mathfrak{g}$ is trivial since complex-analytic functions on a compact connected manifold are constant.

(1.15) Exercise. Let H be the Lie group whose underlying manifold is $\mathbf{R}^4$ and whose multiplication is given by:

$$(x_1, y_1, z_1, t_1)(x_2, y_2, z_2, t_2) = (x_1 + x_2 + z_1t_2, y_1 + y_2 + \alpha z_1t_2, z_1 + z_2, t_1 + t_2),$$

where $\alpha \in \mathbf{R}$ is irrational. Let $D \subset H$ be the discrete central subgroup consisting of the elements $(p, q, 0, 0)$, with p and $q \in \mathbf{Z}$. Let $G = H/D$. Show that the commutator subgroup $[G, G]$ is not closed in G. Here $[G, G]$ is the subgroup of G generated by all elements of the form $[g, h] = ghg^{-1}h^{-1}$. ($G/[G, G]$ is Abelian and $[G, G]$ is contained in every normal subgroup K such that G/K is Abelian.)
Now suppose that G is a simply connected Lie group. Prove that any normal connected subgroup H of G is necessarily closed in G.
Hint: Let $\mathfrak{g}$ be the Lie algebra of G and $\mathfrak{h} \subset \mathfrak{g}$ the ideal corresponding to H. If B is a connected Lie group whose Lie algebra is isomorphic to $\mathfrak{g}/\mathfrak{h}$, prove that there is a continuous homomorphism $G \to B$ having H for the connected component of the identity of its kernel.

(1.16) Exercise. Let G be a connected Lie group. An automorphism of G is an automorphism of the underlying group structure of G which is also a diffeomorphism of the underlying manifold. These automorphisms form a group, denoted by $\operatorname{Aut} G$. Prove (i) the mapping $\mathrm{T}\colon \Phi \to \mathrm{T}_1\Phi$ is an injective homomorphism of groups $\operatorname{Aut} G \to \operatorname{Aut}\mathfrak{g}$; (ii) if G is simply connected, the map T is an isomorphism; and by transport of structure via T^{-1} we obtain a Lie group structure on $\operatorname{Aut} G$; (iii) if $G = \widetilde{G}/C$, where $\widetilde{G}$ is the simply connected universal covering group of G and C is a discrete subgroup of the center of $\widetilde{G}$, then $\operatorname{Aut} G$ may be identified with the subgroup of $\operatorname{Aut}\widetilde{G}$ consisting of automorphisms Φ such that $\Phi(C) = C$.

(1.17) Exercise. Verify that $U = \{\, x \in \mathbf{R} \mid |x| < \frac{1}{2} \,\}$ is a local Lie group with (commutative) multiplication and inversion, respectively, given by:

$$\mu(x, y) = \frac{2xy - x - y}{xy - 1}, \qquad \iota(x) = \frac{x}{2x - 1}.$$

Prove $\kappa\colon \{\, x \in \mathbf{R} \mid x < 1 \,\} \to \mathbf{R}$ with $\kappa(x) = \frac{x}{x-1}$ satisfies $\kappa(x + y) = \mu(\kappa(x), \kappa(y))$ and $\kappa(-x) = \iota(\kappa(x))$, where defined. Deduce that μ and ι are the usual addition and inversion on $\mathbf{R}$ in a local chart for $\mathbf{R}$.
Now prove the following general result using Theorem 1.14.3. Let $U \subset \mathfrak{g}$ be a local Lie group with multiplication μ and inversion ι. Then there exists a

simply connected Lie group G and a coordinate chart $\kappa\colon V \to U$ where V is a neighborhood of the identity in G, such that:

$$\kappa(1) = 0, \qquad \kappa(xy) = \mu(\kappa(x), \kappa(y)), \qquad \kappa(x^{-1}) = \iota(\kappa(x)),$$

whenever $x,\ y \in V$.

1.16 Notes

The theory of Lie groups was created by S. Lie in the 1870's and, with the collaboration of Engel, he wrote a comprehensive account of his work in the three-volume book Lie [1888-93]. This also contains many detailed references to the history of the subject at that time; further information is contained in the obituaries written by Engel [1900], [1935]. Lie's main sources of inspiration were geometry and partial differential equations. In geometry, Klein and Lie in close cooperation had observed the importance of the group of transformations that leave invariant a given geometric structure. It led Klein [1872] to the formulation of his "Erlanger Programm" of studying geometries by means of the properties of the corresponding automorphism groups. For Lie the interesting point was that in many cases the transformations in the groups depend, in an analytic or even algebraic fashion, on a finite number of continuous parameters, in the same way as the transformations that he had met in his study of certain partial differential equations. The transformations that leave invariant the form of a differential equation, map solutions to solutions; and it was a natural idea for Lie to investigate to what extent the groups of such transformations can play a similar fundamental role for the differential equations as the Galois groups do for algebraic equations.

For Lie, an n-parameter group is always a group of transformations in some manifold; in the terminology of this chapter: an analytic action of a Lie group G on an analytic manifold M, as defined in Section 1.11. In fact, any abstract group G acts on itself by left and right multiplications respectively, and because these actions are simply transitive, G can be identified with these transformation groups. We also used these actions systematically in this chapter and therefore followed Lie closely in his point of view. As a consequence, the infinitesimal transformations (the elements of $\mathfrak{g} = \mathrm{T}_1 G$) are always thought of by Lie as vector fields on the manifold on which G acts, the 1-parameter subgroups of G are the flows of these vector fields, etc.

Although his discussion of examples like the projective linear group indicate that he considers groups as global objects, Lie in his general theory is never very specific about domains of definition and the reader has to sort out for himself which proofs lead to local results only. For instance, the statement that every finite-dimensional Lie algebra of vector fields leads to a Lie group of transformations, Lie's "second fundamental theorem", (considered by himself to be the most important one of his "three fundamental theorems") in general is only valid in a local form like Theorem 1.8.3. Similarly the extension of a homomorphism of Lie algebras to a homomorphism of Lie groups is done by Lie only in a local form and not as in Corollary 1.10.5 or Remark 1.13.5. It was much later that Weyl [1925], pp.290-291, observed that in general a representation of the Lie algebra is tangent to a representation of the group, only if the group is simply connected, so that one has to pass to the universal covering group, whose existence seems to have been evident to him. Universal coverings and fundamental groups were introduced by Poincaré [1895] in the context of complex analysis, but the proof that the universal covering of a Lie group is a (Lie) group is due to Schreier [1926].

The name "Lie group" was introduced by É. Cartan [1930-II]. The emphasis on a Lie group being a manifold, with its global topological aspects, is due to Weyl [1925] who also mentioned the name "Lie algebra" (coined by Jacobson) in 1934. The realization in Section 1.2 of $\mathbf{SU}(2)$ as the three-dimensional sphere in $\mathbf{R}^4$ and the two-fold covering: $\mathbf{SU}(2) \to \mathbf{SO}(3, \mathbf{R})$ is due to Hamilton [1847]. The picture of $\mathbf{SL}(2, \mathbf{R})$ we learned from Atiyah, p.29. That Lie did not only think local is illustrated by his introduction of the identity component G° as the group generated by $\exp \mathfrak{g}$; he also proves that G° is a normal subgroup of G and that all cosets xG° have the same dimension. (He understood $\pi_0(G)$, but not yet $\pi_1(G)$.)

Some more translations from Volume I of Lie's book:

continuirlich	=	connected,
discontinuirlich	=	discrete (p.3),
r-gliedrig	=	depending analytically on r complex or real parameters (p.15),
wesentlich	=	effective action (p.12),
ähnlich	=	the actions are conjugate by a diffeomorphism, (p.24)
Gruppe	=	semigroup (p.3),

Lie makes a point that many of his statements are valid without assuming the presence of the identity or inverse transformations.

kanonische Form	=	logarithmic coordinates (p.171),
Zusammensetzung	=	the constants c_{ijk} in $[X_i, X_j] = \sum_k c_{ijk} X_k$, with X_i a basis in $\mathfrak{g}$, an abstract Lie algebra (p.94, p.289)
gleichzusammengesetzt	=	homomorphic Lie algebras (p.291),
holoëdrisch isomorph	=	isomorphic (p.293),
meroëdrisch isomorph	=	homomorphic image of (p.293).

The Jacobi identity occurs in Jacobi [1836/37], p.348, for the Poisson brackets, cf. Lemma 13.1.2. It was formulated for arbitrary vector fields by Lie [1888], p.94, as an easy consequence of viewing the Lie bracket as the commutator of the corresponding first order partial differential operators. The formula $\exp X = \sum_{k=0}^{\infty} \frac{1}{k!} X^k$ occurs in Lie, ibidem, p.51, with the interpretation of the vector field X as a linear (differential) operator acting on a vector space of analytic functions. If X is a linear vector field, then restriction to linear coordinate functions yields (1.4.1), which occurs already in Laguerre [1867]. An amusing description of $\exp X$ as obtained by "iterating X infinitely often" can be traced to Jordan [1868/69], p.243. In the concrete form: $\exp X = \lim_{n \to \infty} h_n^{[t_n]}$, this idea appears in the proof of Theorem 1.10.6, which is due to von Neumann [1927], for H_0 a closed subgroup of $\mathbf{GL}(V)$. The observation that $\mathbf{GL}(V)$ can be replaced by any Lie group G without any changes in the proof, is due to É. Cartan [1930-II], and the form presented here to Chevalley [1946]. Corollary 1.10.9 can be traced to von Neumann [1929].

Formula (1.5.1) for the tangent map of exp occurs in F. Schur [1891], and even more explicitly in Poincaré [1899], who wrestles with the singularities and complex-analytic extensions of the inverse of exp. Our presentation of Section 1.6, together with the conclusion that any C^2 Lie group is analytic, very closely follows F. Schur [1891], [1893]. Campbell [1897/98], Baker [1905] and Hausdorff [1906] made further remarks on the recursive identities which express the Taylor expansion at the origin of the product in logarithmic coordinates, in terms of the Lie algebra structure; and it has become customary to refer to this as the *formula of Campbell-Baker-Hausdorff*. The explicit Formula (1.7.3) was found by Dynkin [1950]. Instead of adding the names of Schur (in front) and Dynkin (at the back), we propose here to call it just "(the Taylor expansion of) the product in logarithmic coordinates".

Lie's "third fundamental theorem" is the statement that every abstract Lie algebra occurs as a Lie algebra of analytic vector fields. In [1888-93], Vol.III, p.598 we can read that he first tried to prove this via the adjoint representation, but then realized that there are problems with the center. His subsequent proof in Vol.II, Kap.17 (see Proposition 13.5.3 for this proof) is in terms of what nowadays is called the Lie-Poisson structure of $\mathfrak{g}^*$, cf. Weinstein [1983]. The explicit realization by means of the X^{R} in (1.6.2, 5), cf. Lemma 1.8.2, is due to F. Schur [1891]. In Vol.III, p.598 Lie stresses that it should be true that every finite-dimensional Lie algebra $\mathfrak{g}$ is isomorphic to a Lie subalgebra of $\mathfrak{gl}(V)$, for some finite-dimensional vector space V; and then Theorem 1.10.3 can be used to get a Lie subgroup of $\mathbf{GL}(V)$ with Lie algebra isomorphic to $\mathfrak{g}$. This was proved in 1935 by Ado (see Ado [1947]) but the proof requires quite a bit of structure theory of Lie algebras. É. Cartan [1930-I], [1936] constructed a simply connected Lie group with Lie algebra $\mathfrak{g}$ as a central extension of $\widetilde{\mathrm{Ad}\,\mathfrak{g}}$, and in the process found a topological obstruction that is equivalent to the $[\Omega] \in H^2(\widetilde{\mathrm{Ad}\,\mathfrak{g}}\,; \mathbf{R})$ introduced before Theorem

1.14.2. We like to view our construction in Section 1.14 as a further clarification of Cartan's.

The idea of the proof of Theorem 1.14.2 is due to H. Hopf [1941], who used it to unravel the whole ring structure of the cohomology of Lie groups. It is interesting to note here that when de Rham [1931] developed his cohomology theory, É. Cartan [1929-II] immediately used it in order to reduce the computation of the cohomology of compact symmetric spaces to a Lie algebra computation. See also Section 9.3.

The idea of avoiding the global problems with the exponential map by looking at the relation between arbitrary curves in $\mathfrak{g}$ and G is known, see, for instance, Loos [1971]. Banach Lie groups were introduced in Birkhoff [1938]. A lot of information on loop groups is contained in Pressley and Segal [1986].

The mapping $\mathrm{D}^{\mathrm{R}}\colon \gamma \mapsto \left(t \mapsto \left(\mathrm{T}_1\,\mathrm{R}(\gamma(t))\right)^{-1}\frac{d\gamma}{dt}(t)\right)$ in Formula (1.13.2) is the inverse of a *product-integral* introduced by Volterra [1887] (in the case of finite-dimensional transformations). His Teorema 5.I corresponds to our Formula (1.13.4.a).

The concept of principal fiber bundles (and the definition of bundles in general) has been developed by Ehresmann in the early 1940's, for the basic theory see Steenrod [1951], Section 8. Theorem 1.11.4 however is due to Gleason [1950], for compact groups and with the emphasis on topological questions, but the transfer to our situation is clear. The analytic structure on G/H, cf. Corollary 1.11.5 and Proposition 1.11.8, is older, see for instance Mayer and Thomas [1935]. Often these properties are treated as rather obvious, which they are basically.

We thank A. Borel for historical comments and J. Leslie for warning us to be more careful in our treatment of Banach Lie groups.

References for Chapter One

Ado, I.D. (1947) The representation of Lie algebras by matrices. Uspekhi Mat. Nauk. **2** (1947) 159-173 = Amer. Math. Soc. Transl. (1) **9** (1962) 308-327

Atiyah, M.F. Characters of Semi-simple Lie Groups. Mimeographed notes. University of Oxford = Collected Works, Vol. 4, pp. 489-557. Clarendon Press, Oxford 1988

Baker, H.F. (1905) Alternants and continuous groups. Proc. London Math. Soc. (2) **3** (1905) 24-47

Birkhoff, G. (1938) Analytical groups. Trans. Amer. Math. Soc. **43** (1938) 61-101

Borel, A. (1950) Impossibilité de fibrer une sphère par un produit de sphères. C.R. Acad. Sci. Paris Sér. I. Math. **231** (1950) 943-945 = Œuvres Collected Papers, Vol. I, pp. 67-69. Springer-Verlag, Berlin Heidelberg New York Tokyo 1983

Bott, R. (1954) On torsion in Lie groups. Proc. Nat. Acad. Sci. U. S. A. **40** (1954) 586-588 = Collected Papers, Vol. 1, pp. 123-125. Birkhäuser, Boston 1994

Bott, R. (1956) An application of the Morse theory to the topology of Lie-groups. Bull. Soc. Math. France **84** (1956) 251-281 = Collected Papers, Vol. 1, pp. 159-189. Birkhäuser, Boston 1994

Bott, R., Samelson, H. (1958) Applications of the theory of Morse to symmetric spaces. Amer. J. Math. **80** (1958) 964-1029 = Collected Papers, Vol. 1, pp. 327-393. Birkhäuser, Boston 1994

Campbell, J.E. (1897/98) On a law of combination of operators bearing on the theory of continuous transformation groups. Proc. London Math. Soc. (1) **28** (1897) 381-390; **29** (1898) 14-32

Cartan, É. (1929-II) Sur les invariants intégraux de certains espaces homogènes clos et les propriétés topologiques de ces espaces. Ann. Soc. Pol. Math. **8** (1929) 181-225 = Œuvres Complètes, Partie I, pp. 1081-1125. Gauthier-Villars, Paris 1952

Cartan, É. (1930-I) Le troisième théorème fondamental de Lie. C.R. Acad. Sci. Paris Sér. I. Math. **190** (1930) 914-916, 1005-1007 = Œuvres Complètes, Partie I, pp. 1143-1148. Gauthier-Villars, Paris 1952

Cartan, É. (1930-II) La théorie des groupes finis et continus et l'Analysis situs. Mémorial Sc. Math. **42** (1930) = Œuvres Complètes, Partie I, pp. 1165-1225. Gauthier-Villars, Paris 1952

Cartan, É. (1936) La Topologie des Groupes de Lie. Paris, Hermann = Œuvres Complètes, Partie I, pp. 1307-1330. Paris, Gauthier-Villars 1952

Chevalley, C. (1946) Theory of Lie Groups: I. Princeton Math. Series No.8. Princeton University Press, Princeton

Crowell, R.H., Fox, R.H. (1963) Introduction to Knot Theory. Ginn, Boston

Dynkin, E. (1950) Normed Lie algebras and analytic groups. Uspekhi Mat. Nauk. **5** (1950) 135-186 = Amer. Math. Soc. Transl. (1) **9** (1962) 470-534

Eilenberg, S. (1944) Singular homology theory. Ann. of Math. **45** (1944) 407-447

Engel, F. (1900) Sophus Lie. Jahresber. Deutsch. Math.-Verein. **8** (1900) 30-46

Engel, F. (1935) Friedrich Schur. Jahresber. Deutsch. Math.-Verein. **45** (1935) 1-31

van Est, W.T., Korthagen, Th.J. (1964) Non-enlargible Lie algebras. Nederl. Akad. Wetensch. Proc. Ser. A **67** (1964) 15-31 = Indag. Math. **26** (1964) 15-31

Euler, L. (1770/71) Problema algebraicum ob affectiones prorsus singulares memorabile. Novi Comment. Acad. Sci. Petropolitanæ**15** (1770/1771) 74-106 = Opera Omnia 6, 1 Ser., pp. 287-315. Teubner, Leipzig 1921

Freudenthal, H. (1941) Die Topologie der Lieschen Gruppen als algebraisches Phänomen, I. Ann. of Math. **42** (1941) 1051-1074

Gleason, A.M. (1950) Spaces with a compact Lie group of transformations. Proc. Amer. Math. Soc. **1** (1950) 35-43; 826

Greenberg, M. (1967) Lectures on Algebraic Topology. W.A. Benjamin, New York Amsterdam

Greub, W., Halperin, S., Vanstone, R. (1972) Connections, Curvature and Cohomology, Volume I. Academic Press, New York San Francisco London

Guillemin, V., Pollack, A. (1974) Differential Topology. Prentice-Hall, Englewood Cliffs

Hamilton, W.R. (1847) On quaternions, or on a new system of imaginaries in algebra: with some geometrical illustrations. Proc. Roy. Irish Acad. **3** (1847) 1-16 = Mathematical Papers, Vol. 3, pp. 355-362. Cambridge University Press, Cambridge 1967

Hausdorff, F. (1906) Die symbolische Exponentialformel in der Gruppentheorie. Berichte der Köngl. Sächsischen Gesellschaft der Wiss. zu Leipzig (Math.-Phys. Klasse) **58** (1906) 19-48

Hopf, H. (1941) Über die Topologie der Gruppen-Mannigfaltigkeiten und ihre Veralgemeinerungen. Ann. of Math. **42** (1941) 22-52 = Selecta, pp. 119-151. Springer-Verlag, Berlin Göttingen Heidelberg New York 1964

Jacobi, C.G.J. (1836/37) Über diejenigen Probleme der Mechanik in welchen eine Kräftefunction existirt und über die Theorie der Störungen. Gesammelte Werke, Second Ed., Band 5, pp. 217-395. Chelsea Publishing Cy., New York 1969

Jordan, C. (1868/69) Mémoire sur les groupes de mouvements. Annali di Math. **2** (1868-1869) 167-215, 322-345 = Œuvres, Tome 4, pp. 231-302. Gauthier-Villars, Paris 1964

Klein, F. (1872) Vergleichende Betrachtungen über neuere geometrische Forschungen (Das Erlanger Programm). Math. Ann. **43** (1893) 63-100 = Gesammelte Math. Abh., Band 1, pp. 460-497. Julius Springer, Berlin 1921

Kronecker, L. (1884) Näherungsweise ganzzahlige Auflösung linearer Gleichungen. Monatsber. Königlich. Preuss. Akad. Wiss. zu Berlin **1884**, pp. 1179-1193, 1271-1299 = Werke, Band 3:1, pp. 47-109. Teubner, Leipzig 1899

Laguerre, E.N. (1867) Sur le calcul des systèmes linéaires. J. Ecole Polytechnique **62** (1867) = Œuvres, Second Ed., Tome 1, pp. 221-267. Chelsea Publishing Cy., New York 1972

Lie, S. (1888-1893) Theorie der Transformationsgruppen. Unter Mitwirkung von Prof. dr. F. Engel. Teubner, Leipzig 1888, 1890, 1893

Loos, O. (1971) Lie transformation groups of Banach manifolds. J. Differential Geom. **5** (1971) 175-185

Mayer, W., Thomas, T.Y. (1935) Foundations of the theory of Lie groups. Ann. of Math. **36** (1935) 770-822

von Neumann, J. (1927) Zur Theorie der Darstellungen kontinuierlicher Gruppen. Sitzungsber. Preuss. Akad. Wiss., Phys.-Math. Kl. **1927** pp. 76-90 = Collected Works, Vol. I, pp. 134-148. Pergamon Press, Oxford 1961

von Neumann, J. (1929) Über die analytischen Eigenschaften von Gruppen linearer Transformationen und ihrer Darstellungen. Math. Z. **30** (1929) 3-42 = Collected Works, Vol. I, pp. 509-548. Pergamon Press, Oxford 1961

Plaisant, M. (1980) Q-variétés banachiques. Application à l' intégrabilité des algèbres de Lie. C.R. Acad. Sci. Paris Sér. A-B. **280** (1980) 185-188

Poincaré, H. (1895) Analysis situs. J. Ecole Polytechnique **1** (1895) 1-121 = Œuvres, Tome 6, pp. 193-288. Gauthier-Villars, Paris 1953

Poincaré, H. (1899) Sur les groupes continus. Cambridge Philos. Trans. **18** (1899) 220-255 = Œuvres, Tome 3, pp. 173-212. Gauthier-Villars, Paris 1934

Pressley, A., Segal, G. (1986) Loop Groups. Oxford University Press, Oxford

de Rham, G. (1931) Sur l'Analysis situs des variétés à n dimensions. J. Math. Pures Appl. (9) **10** (1931) 115-200

Schreier, O. (1926) Abstrakte kontinuierliche Gruppen. Abh. Math. Sem. Univ. Hamburg **4** (1926) 15-32

Schur, F. (1891) Zur Theorie der endlichen Transformationsgruppen. Math. Ann. **38** (1891) 263-286

Schur, F. (1893) Ueber den analytischen Character der eine endliche continuirliche Transformationsgruppe darstellenden Functionen. Math. Ann. **41** (1893) 509-538

Serre, J-P. (1951) Homologie singulière des espaces fibrés. Applications. Ann. of Math. **54** (1951) 425-505 = Œuvres Collected Papers, Vol. I, pp. 24-104. Springer-Verlag, Berlin Heidelberg New York Tokyo 1986

Spivak, M. (1979) A Comprehensive Introduction to Differential Geometry, Volume One, 2nd Ed. Publish or Perish, Wilmington

Steenrod, N. (1951) The Topology of Fibre Bundles. Princeton Math. Series, No.14. Princeton University Press, Princeton

Volterra, V. (1887) Sui fondamenti della teoria delle equazioni differenziali lineari. Mem. Soc. It. Sc. Ser. 3ª **6** (1887) 1-107 = Opere Matematiche, Vol. I, pp. 209-290. Accademia Nazionale dei Lincei, Roma 1954

Walter, W. (1985) Analysis, I. Grundwissen Mathematik, Bd.3. Springer-Verlag, Berlin Heidelberg New York Tokyo

Weinstein, A. (1983) Sophus Lie and symplectic geometry. Exposition. Math. **1** (1983) 95-96

Weyl, H. (1925/26) Theorie der Darstellung kontinuierlicher halb-einfacher Gruppen durch lineare Transformationen, I, II, III, Nachtrag. Math. Z. **23** (1925) 271-309; **24** (1926) 328-376, 377-395, 789-791 = Gesammelte Abh., Band II, pp. 543-647. Springer-Verlag, Berlin Heidelberg New York 1968

Yamabe, H. (1950) On an arcwise connected subgroup of a Lie group. Osaka Math. J. **2** (1950) 13-14

Chapter 2

Proper Actions

2.1 Review

In this section we recall the main results of Section 1.11 on general group actions, adding some further comments.

An *action* of a group G on a space M is a homomorphism A from G to the group of transformations in M. If G is a Lie group and M a (real-analytic) manifold, then the action is *of class* C^k if the mapping $(g, x) \mapsto g \cdot x := A(g)(x) \colon G \times M \to M$ is of class C^k, for $(1 \leq k \leq \omega)$. Here C^ω denotes the real-analytic mappings. The *orbit* $G \cdot x$ *through* $x \in M$ is defined to be the image of the mapping $A_x \colon g \mapsto A(g)(x) \colon G \to M$.

The space M is partitioned into orbits; the set of orbits is called the *quotient* $G\backslash M$ of M under the action of G on M, and $\pi \colon x \mapsto G \cdot x \colon M \to G\backslash M$ denotes the *canonical projection*. The *quotient topology* on $G\backslash M$ is the one for which V is open in $G\backslash M$ if and only if the G-invariant subset $\pi^{-1}(V)$ is open in M. In general the quotient topology need not be Hausdorff: $\{ G \cdot x \}$ is closed in $G\backslash M$ if and only if $G \cdot x$ is closed in M; and it easy to find examples with nonclosed orbits. The action $(t, (x_1, x_2)) \mapsto (x_1 + tx_2, x_2)$ of $(\mathbf{R}, +)$ on $\mathbf{R}^2$ is an example with all orbits being closed. For each x_1, $x_1' \in \mathbf{R}$, any two invariant neighborhoods of $(x_1, 0)$ and $(x_1', 0)$ intersect each other, making the quotient topology a non-Hausdorff topology. We recall from Lemma 1.11.2 that the quotient topology is Hausdorff if and only if the *orbit relation* $\{ (x, y) \in M \times M \mid y \in G \cdot x \}$ is a closed subset of $M \times M$.

Recall that an action A is is said to be *proper* if the mapping $(g, x) \mapsto (g \cdot x, x)$ is proper: $G \times M \to M \times M$. According to Lemma 1.11.3 a proper and continuous action leads to a Hausdorff quotient space $G\backslash M$. It is the main purpose of this section to show that for proper C^k actions of Lie groups on manifolds, a quite detailed description of the orbit structure can be given; in particular, the orbit space is a locally finite union of C^k manifolds, pieced together in a nice way.

The main result in Section 1.11 is Theorem 1.11.4, stating that for a proper C^k action of G on M that is *free*, that is, $g \cdot x \neq x$ whenever $x \in M$, $g \in G$, $g \neq 1$, the quotient $G\backslash M$ has a unique structure of a C^k manifold such that $\pi\colon M \to G\backslash M$ is a C^k fiber bundle of a special kind, called a *principal fiber bundle* (see Definition A.4 in Appendix A).

If H is a closed Lie subgroup of G, then the right action $h \mapsto (g \mapsto gh^{-1})$ of H on G is proper and free; therefore the quotient, now denoted by G/H, is a real-analytic manifold, and $G \to G/H$ is a real-analytic principal fibration. The left action $\mathrm{L}_{G/H}\colon g \mapsto (xH \mapsto gxH)$ is a real-analytic action of G on G/H that is *transitive*, that is, there is only one orbit, G/H.

We now return to the general C^k action of G on M. Then:

$$G_x := \{\, g \in G \mid A(g)(x) = x \,\},$$

the *isotropy* or *stabilizer group* at x, is a closed, and therefore a Lie subgroup of G. The mapping $A_x\colon G \to M$ induces a bijective mapping $B_x\colon G/G_x \to G{\cdot}x$; it is a C^k immersion, exhibiting the orbit as an immersed C^k submanifold of M. Moreover, B_x intertwines (cf. Definition 2.2.2 below) the left action of G on G/G_x with the action of G on $G{\cdot}x$, that is, $A(g)|_{G\cdot x} = B_x \circ \mathrm{L}_{G/G_x}(g) \circ B_x^{-1}$, for every $g \in G$.

In particular, each transitive action can be identified with the left action of G on G/H for some closed Lie subgroup H of G; this fact reduces the theory of transitive actions to the structure theory of Lie groups.

Write $\alpha_x = \mathrm{T}_1\, A_x\colon \mathfrak{g} \to \mathrm{T}_x\, M$ for the *infinitesimal action* at x, then the Lie algebra of G_x is equal to $\mathfrak{g}_x = \ker \alpha_x$, see (1.11.19). We conclude this review with the following general description, which however is local both in M and in G.

(2.1.1) Lemma. *Let A be a C^k action $(k \geq 1)$ of the Lie group G on the manifold M. For $x_0 \in M$, let S be a C^k submanifold of M through x_0 such that:*

$$\mathrm{T}_{x_0}\, M = \alpha_{x_0}(\mathfrak{g}) \oplus \mathrm{T}_{x_0}\, S, \tag{2.1.1}$$

and let C be a C^k submanifold of G through 1 such that:

$$\mathfrak{g} = \mathfrak{g}_{x_0} \oplus \mathrm{T}_1\, C. \tag{2.1.2}$$

Then there is an open neighborhood C_0, and S_0, of 1, and x_0, in C, and S, respectively, such that $A_0 = A|_{C_0 \times S_0}$ is a C^k diffeomorphism from $C_0 \times S_0$ onto an open neighborhood M_0 of x_0 in M.

Proof. The tangent mapping of $A|_{C\times S}$ at $(1, x_0)$ is equal to

$$(X, v) \mapsto \alpha_{x_0}(X) + v\colon\; \mathrm{T}_1\, C \times \mathrm{T}_{x_0}\, S \to \mathrm{T}_{x_0}\, M.$$

Suppose the image vector $\alpha_{x_0}(X) + v$ is equal to 0. The fact $\alpha_{x_0}(\mathfrak{g}) \cap \mathrm{T}_{x_0}\, S = 0$ then implies that $\alpha_{x_0}(X) = 0$ and $v = 0$. Next $\ker \alpha_{x_0} \cap \mathrm{T}_1\, C = 0$ implies that $X = 0$. So $\mathrm{T}_{(1, x_0)}\, A|_{C\times S}$ is injective. On the other hand:

$$\begin{aligned}\dim(C \times S) &= \dim C + \dim S \\ &= (\dim \mathfrak{g} - \dim \ker \alpha_{x_0}) + (\dim \mathrm{T}_{x_0} M - \dim \alpha_{x_0}(\mathfrak{g})) \\ &= \dim \mathrm{T}_{x_0} M = \dim M.\end{aligned}$$

Hence $\mathrm{T}_{(1,x_0)} A|_{C\times S}$ is bijective and we can apply the inverse mapping theorem. □

Remarks. The set $\alpha_{x_0}(\mathfrak{g})$ is equal to the tangent space at x_0 of the orbit $G\cdot x_0$ through x_0, the latter viewed as an immersed submanifold of M. Condition (2.1.1) says that S intersects $G \cdot x_0$ transversally and has complementary dimension.

The mapping $\pi_2 \circ A_0^{-1}\colon M_0 \to S_0$, where π_2 is the projection $C_0 \times S_0 \to S_0$ onto the second factor, is a (trivial) C^k fibration, whose fibers are submanifolds of orbits. To be precise, $c \mapsto c\cdot s$ is a C^k diffeomorphism from C_0 onto the fiber over $s \in S_0$. In particular, all (local) orbits of neighborhoods of x_0, as in Lemma 2.1.1 intersect S_0 near x_0 transversally. In the next section it will be shown that, if the action is proper at x_0, then S_0 can be chosen such that the orbits near x_0 intersect S_0 in orbits for the action of G_{x_0}, the isotropy group at x_0. However, in general such a nice description of the intersections of the nearby orbits with S_0 is not possible.

The nearby local orbits intersect S_0 in isolated points if and only $\dim \mathfrak{g}_x = \dim \mathfrak{g}_{x_0}$ for all x near x_0. Note that $\mathfrak{g}_{x_0} = \{ X \in \mathfrak{g} \mid \alpha_{x_0}(X) = 0 \}$ implies that $\dim \mathfrak{g}_x \leq \dim \mathfrak{g}_{x_0}$ for all x near x_0. In the special case that $\mathfrak{g}_{x_0} = 0$, that is, if the action is infinitesimally (locally) free at x_0, then C_0 is an open neighborhood of 1 in G. In the local identification of M with $C_0 \times S_0$, the local action of $g \in G$ then consists of left multiplication by G only on the first factor. Theorem 1.11.4 is a global version of this, but it needs the much stronger assumption of a globally free and proper action.

It is also clear that Lemma 2.1.1 remains true for locally defined actions of a local Lie group; this in turn is given by its infinitesimal action, which is a finite-dimensional Lie algebra $\mathfrak{g}$ of vector fields on M. If $G = (\mathbf{R}, +)$, and $\dim \mathfrak{g} = 1$, respectively, the action is the flow of a vector field X on M. The condition that $\mathfrak{g}_{x_0} = 0$ means that $X(x_0) \neq 0$, and Lemma 2.1.1 is the "flow box theorem" stating that in suitable local coordinates the flow after time t is equal to the translation $(x_1, x_2, \dots, x_n) \mapsto (x_1 + t, x_2, \dots, x_n)$ over t in the first variable. In this situation the manifold S is also called a *local Poincaré section* for the vector field X; condition (2.1.1) then just expresses that:

(2.1.3) $$\dim S = \dim M - 1 \quad \text{and} \quad X(x_0) \notin \mathrm{T}_{x_0} S.$$

The condition that the globally defined action of $G = (\mathbf{R}, +)$ is globally free would mean that $X(x) \neq 0$ for all $x \in M$, and that X has no periodic solutions. Requiring in addition that the action is proper would yield that M is a principal $(\mathbf{R}, +)$-fiber bundle over an $(n-1)$-dimensional manifold N. Because the fibers are affine spaces, one can use partitions of unity over

N in order to average local sections to a global section $s\colon N \to M$. Since a principal fiber bundle with a global section is trivial, we get that M is diffeomorphic to $\mathbf{R} \times N$ (or $\mathbf{R} \times S$, with $S = s(N)$), in such a way that the action of $\mathbf{R}$ is just translation in the first factor. Such simple behavior is rarely seen for flows of vector fields.

2.2 Bochner's Linearization Theorem

Let M be a real-analytic finite-dimensional manifold and K a compact topological group, acting continuously on M by means of C^k transformations $(1 \leq k \leq \omega)$. In Theorem 11.5.1 it is proved that this implies that the action A is a continuous homomorphism from K to the topological group $\mathrm{Diff}^k(M)$ of C^k diffeomorphisms of M. The following theorem says that, near a fixed point, the action of K can be identified with the linear action of a closed Lie subgroup of the orthogonal group, acting on a ball around the origin in a Euclidean space.

(2.2.1) Theorem. *Let A be a continuous homomorphism from a compact group K to $\mathrm{Diff}^k(M)$, with $k \geq 1$ and let $x_0 \in M$, $A(k)(x_0) = x_0$, for all $k \in K$. Then there exists a K-invariant open neighborhood U of x_0 in M and a C^k diffeomorphism χ from U onto an open neighborhood V of 0 in $\mathrm{T}_{x_0} M$, such that:*

$$\chi(x_0) = 0, \quad \mathrm{T}_{x_0}\chi = \mathrm{I}\colon \mathrm{T}_{x_0} M \to \mathrm{T}_{x_0} M \tag{2.2.1}$$

and:

$$\chi(A(k)(x)) = \mathrm{T}_{x_0} A(k)\chi(x) \quad (k \in K,\ x \in U). \tag{2.2.2}$$

Proof. For every open neighborhood U' of x_0 in M there exists a K-invariant open neighborhood U of x_0 in M that is contained in U'. Indeed, because $A\colon (k, x) \mapsto A(k)(x)\colon K \times M \to M$ is continuous at (k, x_0) and $A(k, x_0) = x_0$, there exist for each $k \in K$ open neighborhoods $W(k)$ of k in K, and $U(k)$ of x_0 in M, respectively, such that $A(W(k) \times U(k)) \subset U'$. Because K is compact, there is a finite subset F of K such that $\bigcup_{k \in F} W(k) = K$ and it follows that $A(K \times U'') \subset U'$, where $U'' = \bigcap_{k \in F} U(k)$, an open neighborhood of x_0 in M. Now $U = A(K \times U'') = \bigcup_{k \in K} A(k)(U'')$ is the desired K-invariant neighborhood of x_0 in M.

Let $\tilde{\chi}$ be a C^k mapping from an open neighborhood $\tilde{U}$ of x_0 in M, to $\mathrm{T}_{x_0} M$, such that $\tilde{\chi}(x_0) = 0$ and $T_{x_0}\tilde{\chi} =$ identity in $\mathrm{T}_{x_0} M$; the existence of such $\tilde{\chi}$ is obvious. Restricting $\tilde{\chi}$ suitably, we can arrange that $\tilde{U}$ is K-invariant. Now:

$$(k, \chi) \mapsto \mathrm{T}_{x_0} A(k) \circ \chi \circ A(k)^{-1}$$

is a continuous representation of K in the space of C^k mappings $\chi: \tilde{U} \to \mathrm{T}_{x_0} M$ such that $\chi(x_0) = 0$, which is a complete locally convex topological vector space. Applying the averaging principle of Section 4.2, we can form:

$$\bar{\chi} = \int_K \mathrm{T}_{x_0} A(k) \circ \tilde{\chi} \circ A(k)^{-1} \, dk,$$

or, more explicitly,

$$\bar{\chi}(x) = \int_K \mathrm{T}_{x_0} A(k) \, \tilde{\chi}(A(k^{-1})(x)) \, dk \quad (x \in \tilde{U}).$$

Because the average is a fixed vector, we have:

$$\mathrm{T}_{x_0} A(k) \circ \bar{\chi} \circ A(k)^{-1} = \bar{\chi} \quad (k \in K),$$

that is, $\bar{\chi}$ satisfies (2.2.2) on $\tilde{U}$. Moreover:

$$\mathrm{T}_{x_0} (\mathrm{T}_{x_0} A(k) \circ \tilde{\chi} \circ A(k)^{-1}) = \mathrm{T}_{x_0} A(k) \circ \mathrm{T}_{x_0} \tilde{\chi} \circ \mathrm{T}_{x_0} A(k)^{-1} = \mathrm{I},$$

for all $k \in K$, so $\mathrm{T}_{x_0} \bar{\chi} =$ identity on $\mathrm{T}_{x_0} M$. By the inverse mapping theorem there is an open neighborhood U' of x_0 in M such that $\bar{\chi}|_{U'}$ is a C^k diffeomorphism from U' onto an open neighborhood of 0 in $\mathrm{T}_{x_0} M$. Restricting further $\bar{\chi}$ to a K-invariant open neighborhood U of x_0 in U' we get the desired χ. □

If g is an arbitrary inner product on $\mathrm{T}_{x_0} M$, then:

$$\bar{g} = \int_K \mathrm{T}_{x_0} A(k)^* g \, dk$$

is an inner product on $\mathrm{T}_{x_0} M$ that is invariant under the tangent action of K on $\mathrm{T}_{x_0} M$. In other words,

$$K' = \{ \mathrm{T}_{x_0} A(k) \mid k \in K \}$$

is a compact, and hence closed Lie subgroup of the orthogonal group of the Euclidean space $E = (\mathrm{T}_{x_0} M, \bar{g})$. Let B be an open ball around 0 in E that is contained in V, with V as in Theorem 2.2.1. Then B is K'-invariant, so, in view of (2.2.2), $\chi^{-1}(B)$ is a K-invariant open neighborhood of x_0 in U on which K acts as described in the sentence preceding Theorem 2.2.1.

(2.2.2) Definition. *If A, and B, are actions of a group G on a space X, and Y, respectively, then one says that a mapping $\Phi\colon X \to Y$ intertwines A with B, or is G-equivariant: $X \to Y$, if:*

$$\Phi \circ A(g) = B(g) \circ \Phi \quad (g \in G). \tag{2.2.3}$$

This means that $A(g)$, for each $g \in G$, maps each fiber of Φ onto a fiber of Φ; one also sometimes says that the action A covers the action B with respect to the mapping Φ.

If G is a Lie group and A, and B, is a C^k action of G on the manifold X, and Y, respectively, then Φ is an *equivalence of* C^k *actions* if Φ is a C^k diffeomorphism: $X \to Y$, intertwining A with B; the actions A and B are said to be C^k *equivalent* if there exists an equivalence of C^k actions between A and B.

In this terminology, Theorem 2.2.1 says that the action of K, restricted to a suitable K-invariant open neighborhood U in M of the fixed point x_0, is equivalent to the linear tangent action of K on $\mathrm{T}_{x_0} M$, restricted to an open neighborhood of 0 in $\mathrm{T}_{x_0} M$. Indeed, (2.2.2) is just (2.2.3) with G, g, Φ, $B(g)$ replaced by K, k, χ, $\mathrm{T}_{x_0} A(k)$, respectively.

2.3 Slices

(2.3.1) Definition. *Let $A\colon G \times M \to M$ be a C^k action ($k \geq 1$) of a Lie group G on a manifold M. A C^k slice at $x_0 \in M$ for the action A is a C^k submanifold S of M through x_0 such that, in the notation of Lemma 2.1.1:*

(i) $\mathrm{T}_{x_0} M = \alpha_{x_0}(\mathfrak{g}) \oplus \mathrm{T}_{x_0} S$; *and* $\mathrm{T}_x M = \alpha_x(\mathfrak{g}) + \mathrm{T}_x S$, $(x \in S)$;
(ii) S *is* G_{x_0}*-invariant;*
(iii) if $x \in S$, $g \in G$, *and* $A(g)(x) \in S$, *then* $g \in G_{x_0}$.

It follows that the identity mapping: $S \to M$ induces a bijective mapping, even a homeomorphism: $G_{x_0} \cdot x \mapsto G \cdot x$, from the space $G_{x_0} \backslash S$ of G_{x_0}-orbits in S onto an open neighborhood of $G \cdot x_0$ in the space $G \backslash M$ of G-orbits in M. Note that the action of G_{x_0} on S has x_0 as a fixed point, by definition.

(2.3.2) Definition. *The action A is said to be proper at x_0 if for every sequence x_j in M, and g_j, in G such that $\lim_{j \to \infty} x_j = x_0$, and $\lim_{j \to \infty} g_j \cdot x_j = x_0$, respectively, there is a subsequence $j = j(k)$ such that $g_{j(k)}$ converges in G as $k \to \infty$.*

If G is not compact, one can find a sequence of compact subsets K_j of G such that g_j has no convergent subsequence whenever $g_j \notin K_j$, for all j. Using this one obtains that the action is proper at x_0 if and only if there exists a neighborhood U of x_0 in M such that $\{ g \in G \mid A(g)(U) \cap U \neq \emptyset \}$ has a compact closure in G. Note that properness of the action at x_0 implies that G_{x_0} is a compact subgroup of G.

In the case of dynamical systems, that is, if $G = (\mathbf{R}, +)$ or $G = (\mathbf{Z}, +)$, then a point $x_0 \in M$ at which the action is not proper is called a *nonwandering point*, in the terminology of Birkhoff (1927).

(2.3.3) Theorem. *Let A be a C^k action ($k \geq 1$) of the Lie group G on the manifold M, and suppose that the action is proper at $x_0 \in M$. Then there exists a C^k slice S at x_0 for the action A.*

Proof. As already remarked above, $K = G_{x_0}$ is a compact subgroup of G, and it has x_0 as a fixed point by definition. Applying Section 2.2, we find a G_{x_0}-invariant open neighborhood U of $x_0 \in M$ with the following properties. U can be identified, by means of a C^k diffeomorphism χ such that $\chi(x_0) = 0$, $\mathrm{T}_{x_0} \chi = \mathrm{I}$, with an open ball B_ϵ of radius $\epsilon > 0$ around the origin in $\mathrm{T}_{x_0} M$; and the action of G_{x_0} on U is identified with the tangent action of G_{x_0} in B_ϵ. Moreover the action is by orthogonal linear transformations with respect to a suitable inner product on $\mathrm{T}_{x_0} M$.

Because $A(k)^{-1}(x_0) = x_0$, if $k \in G_{x_0}$, $g \in G$, we obtain:

$$A(k)\big(A(g)(x_0)\big) = A(k) \circ A(g) \circ A(k)^{-1}(x_0) = A(kgk^{-1})(x_0).$$

Differentiating this relation with respect to $g \in G$ at $g = 1$ in the direction of $X \in \mathfrak{g}$, we get:

$$\mathrm{T}_{x_0} A(k)\big(\alpha_{x_0}(X)\big) = \alpha_{x_0}\big(\mathrm{Ad}\, k(X)\big) \quad (k \in G_{x_0},\ X \in \mathfrak{g}). \tag{2.3.1}$$

This implies that the tangent action of G_{x_0} on $\mathrm{T}_{x_0} M$ leaves the tangent space $\alpha_{x_0}(\mathfrak{g})$ of the G-orbit through x_0 invariant. The orthogonal complement $\alpha_{x_0}(\mathfrak{g})^\perp$ of $\alpha_{x_0}(\mathfrak{g})$ in $\mathrm{T}_{x_0} M$ with respect to the G_{x_0}-invariant inner product in $\mathrm{T}_{x_0} M$ is G_{x_0}-invariant, and so is the C^k submanifold:

$$S_\epsilon := \chi^{-1}\big(\alpha_{x_0}(\mathfrak{g})^\perp \cap B_\epsilon\big)$$

of M through x_0. This proves (ii). Clearly $\mathrm{T}_{x_0} S_\epsilon = \alpha_{x_0}(\mathfrak{g})^\perp$, hence $\mathrm{T}_{x_0} M = \alpha_{x_0}(\mathfrak{g}) \oplus \mathrm{T}_{x_0} S_\epsilon$. Because for any C^1 submanifold S of M the set $\{ x \in S \mid \mathrm{T}_x M = \alpha_x(\mathfrak{g}) + \mathrm{T}_x S \}$ is open in S, we get (i) by taking $S = S_\epsilon$, for $\epsilon > 0$ sufficiently small.

Suppose (iii) does not hold for any $S = S_\epsilon$, with $\epsilon > 0$. Then, for a given $\epsilon > 0$, there are sequences x_j in S_ϵ, and g_j in G, such that $\lim_{j \to \infty} x_j = x_0$, $\lim_{j \to \infty} A(g_j)(x_j) = x_0$, and moreover $A(g_j)(x_j) \in S_\epsilon$, but $g_j \notin G_{x_0}$, for all j. Because of the properties of the action at x_0 we may assume, by passing to a subsequence, that $\lim_{j \to \infty} g_j = g$ in G. Using the continuity of the action at (g, x_0), we get that $A(g)(x_0) = x_0$, or $g \in G_{x_0}$. Replace g_j by $g^{-1}g_j$, then we may assume that $\lim_{j\to\infty} g_j = 1$, still having $g_j \notin G_{x_0}$, for all j.

In the notation of Lemma 2.1.1, an application of the inverse mapping theorem yields the existence of open neighborhoods C_0, V, and W of 1 in C, G_{x_0}, and G, respectively, such that $(c, k) \mapsto ck$ is a diffeomorphism from $C_0 \times V$ onto W. Thus we write $g_j = c_j\, k_j$ with $c_j \in C_0$, $k_j \in V$, then the conditions $g_j \notin G_{x_0}$, $k_j \in G_{x_0}$ imply that $c_j \neq 1$, for all j. On the other hand, the G_{x_0}-invariance of S_ϵ and $x_j \in S_\epsilon$ imply that $A(k_j)(x_j) \in S_\epsilon$, for all j. Now apply Lemma 2.1.1 with S_0 replaced by S_ϵ, for sufficiently small ϵ; the conditions that $A(g)(x_j) = A(c_j)\big(A(k_j)(x_j)\big) \in S_\epsilon$ and $A(k_j)(x_j) \in S_\epsilon$ then imply $c_j = 1$, a contradiction. □

2.4 Associated Fiber Bundles

In the Tube theorem 2.4.1 it will be shown that, in a suitable G-invariant neighborhood of any orbit $G{\cdot}x$, the action is equivalent to a standard one that is constructed in terms of the Lie group G, the stabilizer group G_x (a closed Lie subgroup of G), and the tangent representation of G_x on $\mathrm{T}_x\, M/\mathrm{T}_x(G \cdot x)$. This construction can be described in the framework of *associated fiber bundles*; we shall start by defining this useful general concept.

Let X and Y be C^k manifolds and let H be a Lie group acting in a C^k fashion both on X and Y. The action of $h \in H$ on X will be denoted by $x \mapsto x \cdot h^{-1}$, and the one on Y by $y \mapsto h \cdot y$. Furthermore assume that the action of H on X is proper and free, so that the orbit space X/H is a C^k manifold of dimension equal to $\dim X - \dim H$, and $X \to X/H$ is a principal fiber bundle with structure group H, cf. Theorem 1.11.4 and the remarks thereafter.

Under these conditions, the action of H on $X \times Y$, defined by:

$$(2.4.1) \qquad (h, (x, y)) \mapsto (x \cdot h^{-1}, h \cdot y) \quad \big(h \in H, (x, y) \in X \times Y\big),$$

is proper and free as well; for this it suffices to look at what happens with the first component. The quotient manifold is a C^k manifold, which will be denoted by:

$$(2.4.2) \qquad X \times_H Y = \big\{\, \{\, (x \cdot h^{-1}, h \cdot y) \mid h \in H \,\} \bigm| (x, y) \in X \times Y \,\big\},$$

and $X \times Y \to X \times_H Y$ is another principal fiber bundle with structure group H. The projection $X \times Y \to X$ onto the first factor induces a mapping $X \times_H Y \to X/H$, the unique one which makes the diagram

$$\begin{array}{ccc} X \times Y & \longrightarrow & X \times_H Y \\ \big\downarrow & & \big\downarrow \\ X & \longrightarrow & X/H \end{array}$$

Fig. 2.4.1.

commutative. The claim is that $X \times_H Y \to X/H$ is a C^k fiber bundle over X/H with fiber equal to Y; this will be called the *fiber bundle over X/H with fiber Y, associated to the principal fiber bundle $X \to X/H$ with structure group H and using the action of H on Y*. Note that the notation $X \times_H Y$ is not completely informative, because it does not specify the actions of H on X and Y, respectively. If necessary, one may attach labels to X and Y, respectively, which specify these H-actions.

In order to prove the claim we recall that in the proof of Theorem 1.11.4 the C^k structure on X/H was obtained by taking C^k submanifolds S of X such that $(h, s) \mapsto s \cdot h^{-1}$ is a C^k diffeomorphism from $H \times S$ onto an open subset U of X, and by taking the mappings $s \mapsto s \cdot H$ from S to U/H as the inverses of the local coordinate charts. Therefore $(h, (s, y)) \mapsto (s \cdot h^{-1}, h \cdot y)$ is a C^k diffeomorphism from $H \times (S \times Y)$ onto $U \times Y$; and the mapping $S \times Y \to U \times_H Y$ which assigns to (s, y) its H-orbit in $U \times Y$ is the inverse of a local coordinate chart in $X \times_H Y$.

Now let G be another Lie group, with an action $(g, x) \mapsto g \cdot x$ on X that **commutes** with the action $(h, x) \mapsto x \cdot h^{-1}$ of H on X, that is, $x \mapsto g \cdot x$ commutes with $x \mapsto x \cdot h^{-1}$ for every $g \in G$ and $h \in H$, or:

$$g \cdot (x \cdot h^{-1}) = (g \cdot x) \cdot h^{-1} \quad (g \in G,\, x \in X,\, h \in H). \tag{2.4.3}$$

The equivalent formulation of this condition is that $((g, h), x) \mapsto (g \cdot (x \cdot h^{-1}))$ is an action of $G \times H$ on X. In the case of two commuting actions, the custom to write one action as a left multiplication and the other as a right multiplication, makes the commutativity of the action look like an associative law in (2.4.3). Compare this with the actions of left and right multiplication of a group on itself, cf. (1.3.3 – 5).

From (2.4.3) it follows that the action A_X of G on X maps H-orbits onto H-orbits, so it defines a unique action $A_{X/H}$ of G on X/H covered by A_X with respect to the projection $X \to X/H$. Using the local coordinate charts for X/H as in the proof of Theorem 1.11.4, one verifies immediately that $A_{X/H}$ is a C^k action on X/H.

The action $(g, (x, y)) \mapsto (g \cdot x, y)$ of G on $X \times Y$ also commutes with the action (2.4.1) of H on $X \times Y$, so it covers a unique action of G on $X \times_H Y$ with respect to the projection $X \times Y \to X \times_H Y$. This action is C^k and in fact all the arrows in Diagram 2.4.1 are C^k fibrations and intertwine the respective actions of G on X, $X \times_H Y$ and X/H.

An interesting special case occurs if H is a closed (hence Lie) subgroup of G, $X = G$, and if we let G and H act on G by means of left and right multiplications, respectively. In this case G acts transitively on the base space G/H of the fibration $G \times_H Y \to G/H$ (with fiber Y). Conversely, let $\psi : B \to Z$ be a C^k fibration, intertwining a C^k action of G on B with a **transitive** C^k action of G on Z; this is called a *homogeneous G-bundle.* For some $z \in Z$, write $Y = \psi^{-1}(\{z\})$, the fiber in B over z, which is a closed C^k submanifold of B. Also, $H = G_z$, the stabilizer of z in G, is a closed subgroup of G, which acts on Y. A straightforward verification shows that the mapping $(g, y) \mapsto g \cdot y : G \times Y \to B$ induces a C^k diffeomorphism: $G \times_H Y \to B$. In the diagram:

$$\begin{array}{ccc} G \times_H Y & \xrightarrow{\;\sim\;} & B \\ \downarrow & & \downarrow \\ G/H & \xrightarrow{\;\sim\;} & Z \end{array}$$

Fig. 2.4.2.

all arrows are C^k fibrations intertwining the respective G-actions, and the horizontal ones are C^k diffeomorphisms. This shows that each homogeneous G-bundle is equivalent to one of the form $G \times_H Y \to G/H$, for a suitable closed Lie subgroup H of G and C^k action of H on a manifold Y, the fiber of the original bundle.

The action of G on $Z \xrightarrow{\sim} G/H$ is proper if and only if $G_z = H$ is a compact subgroup of G; for the "if" part use that the canonical projection $G \to G/H$ is a proper mapping if H is compact. In this case the action of G on the homogeneous G-bundle $G \times_H Y$ is also proper.

Another interesting special case occurs when $Y = E$ is a finite-dimensional vector space on which H acts by linear transformations (a "linear representation of H", in the terminology of Chapter 4). Then each fiber of $X \times_H E \to X/H$ has a unique structure of a vector space for which $e \mapsto \{(x \cdot h^{-1}, h \cdot e) \mid h \in H\}$ is a linear mapping from E to the fiber over xH, for each $x \in X$. This makes $X \times_H E$ into a C^k vector bundle over X/H, called the *associated vector bundle over X/H with fiber E, defined by the given representation of H in E.*

We now come to the standard model for proper actions in G-invariant neighborhoods, as announced in the beginning of this section.

(2.4.1) Tube Theorem. *Let A be a C^k action of a Lie group G on a manifold M, proper at $x_0 \in M$. Then there exists a G-invariant open neighborhood U of x_0 in M such that the G-action in U is C^k equivalent to the action of G on $G \times_{G_{x_0}} B$. Here B is an open G_{x_0}-invariant neighborhood* *of* 0 *in* $\mathrm{T}_{x_0} M/\alpha_{x_0}(\mathfrak{g})$, *on which* G_{x_0} *acts linearly, via the tangent action* $k \mapsto \mathrm{T}_{x_0} A(k)$ *modulo* $\alpha_{x_0}(\mathfrak{g})$.

Proof. Let S be a slice at x_0 for the action A, as in Theorem 2.3.3. Write $\widetilde{A}$ for the restriction of the mapping $A\colon G \times M \to M$ to $G \times S$. Then $\mathrm{T}_{(g,x)} \widetilde{A}$ is surjective for each $(g, x) \in G \times S$. This is true for $g = 1$, because $\mathrm{T}_x M = \alpha_x(\mathfrak{g}) + \mathrm{T}_x S$, for every $x \in S$. It then follows for arbitrary $g \in G$ by differentiating $A(gh, x) = A(g)(A(h, x))$ with respect to $(h, x) \in G \times S$ at $h = 1$. As a consequence, $U = \widetilde{A}(G \times S) = A(G \times S)$ is an open G-invariant neighborhood of x_0 in M.

Now let (g, x), (h, y) be in $G \times S$, with $A(g, x) = A(h, y)$. Then we have that $A(h^{-1}g)(x) = y$, and by the slice property it follows that $k = h^{-1}g \in G_{x_0}$. That is, $A(g, x) = A(h, y)$ if and only if $(h, y) = (gk^{-1}, A(k)(x))$, for some $k \in G_{x_0}$; for the "if" part we also use that S is G_{x_0}-invariant. So there is a unique mapping $\Phi\colon G \times_{G_{x_0}} S \to U$ such that $\widetilde{A} = \Phi \circ \pi$, where π denotes the fibration $G \times S \to G \times_{G_{x_0}} S$, and Φ is bijective.

Using the usual local coordinate charts in $G \times_{G_{x_0}} S$ we see that Φ is a C^k mapping. It has surjective tangent mappings, because $\mathrm{T}_{(g,x)} \widetilde{A} = \mathrm{T}_{\pi(g,x)} \Phi \circ \mathrm{T}_{(g,x)} \pi$ is surjective for every $(g, x) \in G \times S$. Because:

$$\dim G \times_{G_{x_0}} S = \dim G/G_{x_0} + \dim S = \dim \alpha_{x_0}(\mathfrak{g}) + \dim S = \dim M,$$

Φ has bijective tangent mappings, so Φ^{-1} is C^k in view of the inverse mapping theorem. That is, Φ is a C^k equivalence between the G-actions on $G \times_{G_{x_0}} S$ and on U, respectively.

Finally note that the action of G_{x_0} on S is C^k equivalent to the tangent action of G_{x_0} on an open neighborhood of 0 in $\alpha_{x_0}(\mathfrak{g})^\perp \cong \mathrm{T}_{x_0} M/\alpha_{x_0}(\mathfrak{g})$, see the construction of the slice in the proof of Theorem 2.3.3. □

Observe that $\Phi^{-1}\colon U \to G \times_{G_{x_0}} B$, followed by the projection $G \times_{G_{x_0}} B \to G/G_{x_0}$, defines a G-equivariant C^k fibration $U \to G/G_{x_0}$, for which the orbit $G \cdot x_0$ is a global section. Also notice that the properness of the G-action on $G \times_{G_{x_0}} B$ implies that the action of G on the G-invariant open neighborhood U of x_0 in M is proper. Apparently "proper at x_0" is equivalent to "proper on a G-invariant neighborhood".

2.5 Smooth Functions on the Orbit Space

Let G be a Lie group acting properly on the manifold M. This latter condition is equivalent to: the action is proper at x for each $x \in M$, and the topology of the orbit space $G\backslash M$ is Hausdorff.

Indeed, the properness of the action implies that the orbit relation $\{ (x, g \cdot x) \in M \times M \mid x \in M, g \in G \}$ is closed in $M \times M$, which in turn is equivalent to the Hausdorff property for $G \backslash M$, see Lemma 1.11.3. Now conversely suppose that $x_j \to x$ and $g_j \cdot x_j \to y$ in M as $j \to \infty$, for sequences x_j, and g_j, in M, and G, respectively. The closedness of the orbit relation implies that $y = g \cdot x$, for some $g \in G$; and then $(g^{-1} g_j) \cdot x_j = g^{-1} \cdot (g_j \cdot x_j) \to g^{-1} \cdot (g \cdot x) = x$, as $j \to \infty$, because of the continuity of the action. Now the properness of the action at x implies that a subsequence of the $g^{-1} g_j$ converges and then the corresponding subsequence of the g_j converges as well.

An example of an action which is proper at all points without being proper, is obtained by taking the flow of the vector field $(x, y) \mapsto (\frac{x^2+y^2}{x^2+1}, 0)$ on $M = \mathbf{R}^2 \setminus \{(0,0)\}$. Note that every invariant neighborhood of $(-1, 0)$ intersects every invariant neighborhood of $(1, 0)$, whereas $(-1, 0)$ and $(1, 0)$ do not belong to the same orbit; and therefore the quotient space has no Hausdorff topology.

Because the canonical projection $\pi \colon M \to G \backslash M$ is continuous and maps open subsets of M onto open subsets of $G \backslash M$, the orbit space $G \backslash M$ is locally compact. Also it is locally pathwise connected, even locally contractible. In order to describe the connected components of $G \backslash M$, we note that, for each $g \in G$, the transformation $A(g)$ maps any connected component C of M diffeomorphically onto a connected component C' of M. Furthermore, $A(g)(C) = C$ if $g \in G^\circ$; so we get a natural action of the discrete group G/G° on the discrete space $\pi_0(M)$ of connected components of M. For each connected component $\underline{C}$ of $G \backslash M$, the set $\pi^{-1}(\underline{C})$ is equal to the union of the sets C in a G/G°-orbit in $\pi_0(M)$, and $\underline{C} = \pi(C)$, for any such C. The group $G_{(C)} = \{ g \in G \mid A(g)(C) = C \} = \{ g \in G \mid A(g)(C) \cap C \neq \emptyset \}$ is open and closed in G and acts on C; and $\underline{C}$ can be identified with $G_{(C)} \backslash C$.

Assuming from now on that M is paracompact, we have that each connected component C is equal to the union of a countable collection of compact subsets C_i. Hence $\underline{C} = \pi(C) = \bigcup_i \pi(C_i)$, and $\pi(C_i)$ is compact because of the continuity of π; and this shows that $G \backslash M$ **is paracompact**. Here we have used the theorem that a Hausdorff, locally compact space is paracompact if and only if it is the disconnected union of spaces, each of which is a union of countably many compact subsets, cf. Bourbaki [1951], §9, No.10, Th.5. The "if-part" of this criterion has been used before in Theorem 1.9.1 to prove that every group is paracompact.

Although in general $G \backslash M$ is not a smooth manifold (in the sequel we shall see in more detail how close we can get), it is natural to call a function f on an open subset V of $G \backslash M$ to be *of class* C^k if and only if $\pi^* f = f \circ \pi$ is a function of class C^k on M. These functions $\pi^* f$ are precisely the C^k functions ϕ on M that are constant on the G-orbits; or $\phi(A(g)(x)) = \phi(x)$, for all $x \in M$, $g \in G$, or $A(g)^* \phi = \phi$, for each $g \in G$. These are the *G-invariant C^k functions on M*. In turn this means that $A^* \phi = \pi_2^* \phi$, where A

is the action map: $G \times M \to M$ and $\pi_2 \colon G \times M \to M$ the projection onto the second factor. The space of G-invariant C^k functions on M is denoted by $\mathrm{C}^k(M)^G$, and the gist of the remarks above is that π^* is an isomorphism from $\mathrm{C}^k(G\backslash M)$ onto $\mathrm{C}^k(M)^G$, more or less by definition.

Replacing M by $U = A(G \times S)$, where S is a slice for the G-action at x_0 as in the proof of Theorem 2.4.1, we get that the C^k function ϕ on U is G-invariant if and only if $A^*\phi|_{(G\times S)} = \pi_2^*\psi$, for a C^k function ψ on S which is G_{x_0}-invariant. In this way not only $G\backslash U$ is homeomorphic to $G_{x_0}\backslash S$, but also the space of C^k functions on $V = G\backslash U$ gets canonically identified with the space of C^k functions on $G_{x_0}\backslash S$.

Moreover the G_{x_0}-action on S is C^k equivalent to the restriction to an open neighborhood B of 0 in $E = \mathrm{T}_{x_0} M/\alpha_{x_0}(\mathfrak{g})$ of the tangent action of G_{x_0} on $\mathrm{T}_{x_0} M$ modulo $\alpha_{x_0}(\mathfrak{g})$; and the latter is by orthogonal linear transformations with respect to some invariant inner product in E. In particular, any function of the distance to the origin is G_{x_0}-invariant, and we can find such a function χ of class C^∞ such that $\chi \geq 0$, $\chi(0) > 0$ and the support of χ is contained in any given neighborhood of 0 in B. Transporting this to $V \subset G\backslash M$ and extending the resulting function by 0 to $G\backslash M$, we see that for every $\underline{x}_0 \in G\backslash M$ and every neighborhood V of $\underline{x}_0$ in $G\backslash M$ there exists a function f of class $\mathrm{C}^{\min(k,\infty)}$ on $G\backslash M$, such that $f \geq 0$, $f(\underline{x}_0) > 0$ and the support of f is contained in V. (Notice that the previous argument doesn't apply in the real-analytic category.) In combination with the paracompactness of $G\backslash M$, we have proved the following

(2.5.1) Lemma. *For every open covering $\mathcal{V}$ of $G\backslash M$ there is a partition of unity $\{f_j\}$ on $G\backslash M$, of class $\mathrm{C}^{\min(k,\infty)}$ and subordinate to $\mathcal{V}$. That is, each f_j is a $\mathrm{C}^{\min(k,\infty)}$ function on $G\backslash M$, $f_j \geq 0$, and the support* $\mathrm{supp}(f_j)$ *of f_j is contained in some $V_j \in \mathcal{V}$. Moreover, the* $\mathrm{supp}(f_j)$ *form a locally finite family of compact subsets of $G\backslash M$, and $\sum f_j = 1$ on $G\backslash M$.*

Such partitions of unity can be used for piecing together G-invariant structures that are defined in G-invariant neighborhoods in M, to global ones in M. The structures should belong to a category where one can take arbitrary convex linear combinations. As an application we give the

(2.5.2) Proposition. *Let G be a Lie group acting properly and in a C^k fashion on the paracompact manifold M, with $1 \leq k \leq \infty$. Then M has a G-invariant Riemannian structure g of class C^{k-1}.*

Proof. Let H be a compact subgroup of G acting by linear transformations on a finite-dimensional vector space E. Taking an arbitrary left invariant Riemannian structure α on G (in view of the left trivialization $\mathcal{L}^{-1} \colon \mathrm{T}\, G \to G \times \mathfrak{g}$ with $\mathcal{L}$ as in (1.3.11), this amounts to the choice of an arbitrary inner product on $\mathfrak{g}$), and some inner product β on E, we get a left G-invariant Riemannian structure $\gamma = \alpha \times \beta$ on $G \times E$. Averaging over the action

$(h, (g, e)) \mapsto (gh^{-1}, h \cdot e)$ of H on $G \times E$, we get a real-analytic, H-invariant Riemannian structure $\overline{\gamma}$ on $G \times E$, which still is left G-invariant, because the left G-action commutes with the action of H on $G \times E$. On $G \times_H E$ there is a unique Riemannian structure δ such that each tangent mapping of the canonical projection $G \times E \to G \times_H E$ is an orthogonal linear transformation from the orthogonal complement of the tangent space of the fiber onto the tangent space of $G \times_H E$. The structure δ is well-defined. Indeed, two points in the same fiber are mapped to each other by the action of an $h \in H$; and its tangent mapping maps the tangent spaces of the orbits, and hence their orthogonal complements, onto such ones, and by orthogonal linear transformations. It is also not hard to verify that δ is real-analytic and G-invariant.

In view of Theorem 2.4.1 the G-invariant real-analytic Riemannian structures on the $G \times_{G_{x_0}} B$ are mapped by the C^k equivalence to G-invariant Riemannian structures on U of class C^{k-1}. These then can be pieced together to a G-invariant Riemannian structure g of class C^{k-1} on M, by means of a G-invariant partition of unity subordinate to the U's, of class C^k (C^{k-1} is sufficient). □

Conversely, Theorem 11.7.7 says that if g is a Riemannian structure of class C^{k-1} on a paracompact manifold M with finitely many connected components, then the group I of isometries of the corresponding metric space is equal to the group of automorphisms of (M, g), and is a finite-dimensional Lie group with countably many components. Its action on M is proper and of class C^k, and its Lie group topology coincides with the C^k topology on $I \subset \mathrm{Diff}^k(M)$ and also with the topology of pointwise convergence. Here $k > 2$; if $k = 1$, and $k = 2$, we have to add the condition that g is Hölder continuous, and that the first-order derivatives of g are Hölder continuous, respectively. Any closed subgroup G of I is then also a Lie group acting properly and in a C^k fashion on M.

On the other hand, if g is the G-invariant structure of the Proposition, and Φ_j is a sequence in G such that $A(\Phi_j)$ converges pointwise in M to a mapping $\Phi \colon M \to M$, then the properness of the action implies that a subsequence of the Φ_j converges in G, to some $\Psi \in G$. Because this implies that $A(\Phi_j)$ converges pointwise to $A(\Psi)$, we get that $\Phi = A(\Psi)$. In other words, $A(G)$ is a closed subgroup of the isometry group I for the Riemannian structure g. As usual we assume that the action is effective ($\ker A = \{1\}$), by passing to the Lie group $G/\ker A$, if necessary. In this sense we have that, for $3 \leq k \leq \infty$ and for a paracompact manifold M with finitely many components, that **proper effective C^k actions of Lie groups** and **closed subgroups of isometries for C^{k-1} Riemannian structures** can be regarded as the same topics.

2.6 Orbit Types and Local Action Types

We keep our standing assumption that A is a proper C^k action of a Lie group G on a manifold M.

(2.6.1) Definition. *We say that $x, y \in M$, and $G \cdot x, G \cdot y \in G\backslash M$,* are of *the same type, with notation $x \sim y$, and $G \cdot x \sim G \cdot y$, respectively, if there exists a G-equivariant bijection from $G \cdot x$ to $G \cdot y$. We say that x* dominates *y, and that $G \cdot x$* dominates *$G \cdot y$, with notation: $y \lesssim x$, and $G \cdot y \lesssim G \cdot x$, respectively, if there is a G-equivariant mapping from $G \cdot x$ to $G \cdot y$.*

Clearly, $\sim$ is an equivalence relation in M, and $G\backslash M$; we denote the equivalence classes, sometimes simply called the orbit types *in M, and $G\backslash M$, respectively, by:*

$$M_x^{\sim} = \{\, y \in M \mid y \sim x \,\}, \quad G\backslash M_{G\cdot x}^{\sim} = \{\, G \cdot y \in G\backslash M \mid G \cdot y \sim G \cdot x \,\}.$$

On the other hand, $\lesssim$ is a pre-order in M, and $G\backslash M$, respectively; and we write:

$$M_x^{\lesssim} = \{\, y \in M \mid y \lesssim x \,\}, \quad G\backslash M_{G\cdot x}^{\lesssim} = \{\, G \cdot y \in G\backslash M \mid G \cdot y \lesssim G \cdot x \,\}.$$

We start with some direct observations about these notions.

(2.6.2) Lemma.

(i) We have $x \sim y$ if and only if G_x is conjugate to G_y within G, that is, $G_y = g^{-1} G_x g := \{\, g^{-1} h g \in G \mid h \in G_x \,\}$, for some $g \in G$.

(ii) Also $y \lesssim x$ if and only G_x is conjugate, within G, to a subgroup of G_y. That is, $g^{-1} G_x g \subset G_y$, for some $g \in G$.

(iii) If Φ is an G-equivariant mapping: $G \cdot x \to G \cdot y$, then $G_x \subset G_{\Phi(x)}$ and $\Phi = B_{\Phi(x)} \circ \pi \circ B_x^{-1}$. Here π is the real-analytic, G-equivariant fibration $\pi \colon g G_x \mapsto g G_{\Phi(x)} \colon G/G_x \to G/G_{\Phi(x)}$, with fiber $G_{\Phi(x)}/G_x$ induced by $A_x \colon g \mapsto g \cdot x \colon G \to G \cdot x$. In particular, Φ automatically is a C^k fibration with fiber diffeomorphic to $G_{\Phi(x)}/G_x$.

(iv) Finally, $x \sim y$ if and only if $x \lesssim y$ and $y \lesssim x$.

Proof. Φ is a G-equivariant mapping: $G \cdot x \to G \cdot y$ if and only if $\Phi(x) \in G \cdot y$ and $\Phi(g \cdot x) = g \cdot \Phi(x)$, for all $g \in G$. If $z = \Phi(x)$, there exists a mapping $\Phi \colon G \cdot x \to M$ such that $\Phi(g \cdot x) = g \cdot z$, for all $g \in G$, if and only if $g \cdot x = g' \cdot x$ implies $g \cdot z = g' \cdot z$. That is, $g^{-1} g' \in G_x$ implies $g^{-1} g' \in G_z$, for all $g, g' \in G$; and in turn this is equivalent to $G_x \subset G_z = G_{\Phi(x)}$. Next observe that, if $z = g \cdot y$, then $h \in G_z \Leftrightarrow hg \cdot y = h \cdot z = z = g \cdot y \Leftrightarrow g^{-1} h g \in G_y \Leftrightarrow h \in g G_y g^{-1}$, or:

$$G_{g \cdot y} = g G_y g^{-1} \quad (g \in G,\ y \in M). \tag{2.6.1}$$

These observations prove (iii), (ii), and also (i). Indeed, a G-equivariant mapping $\Phi\colon G\cdot x \to G\cdot y$, according to (iii), is injective, and hence bijective, if and only if $G_{\Phi(x)} = G_x$.

Only in the proof of (iv) we use the properness of the action, via the compactness of the stabilizer groups, which implies that these stabilizer groups have finitely many connected components. Combine (i), (ii), and Corollary 1.11.9. □

(2.6.3) Definition. *For any subgroup H of G, we define the set of fixed points for H in M as:*

$$M^H = \{\, y \in M \mid h\cdot y = y,\ \text{for all } h \in H \,\}. \tag{2.6.2}$$

For any continuous action, M^H is a closed subset of M. Note that if H is compact, acting in a C^k fashion on M, for $k \geq 1$, then Bochner's linearization theorem 2.2.1 implies that locally M^H is a C^k submanifold of M. That is, each connected component of M^H is a closed submanifold of M, but different connected components of M^H can have different dimensions.

(2.6.4) Lemma.

(i) For any $x \in M$, we have $M_x^{\lesssim} = G\cdot(M^{G_x})$. Moreover, $M_x^{\lesssim}$, and $G\backslash M_x^{\lesssim}$, is a closed subset of M, and $G\backslash M$, respectively.

(ii) If S is a C^k slice at $x \in M$ for the G-action on M, then:

$$M_x^{\sim} \cap G\cdot S = G\cdot (S^{G_x}) = M_x^{\lesssim} \cap G\cdot S.$$

(Note that $G\cdot S$ is a G-invariant open neighborhood of x in M.) Restricting S to a suitable G-invariant open neighborhood of x in M, we get that $M_x^{\sim} \cap G\cdot S$ is a closed C^k submanifold of $G\cdot S$, which is G-equivariantly C^k diffeomorphic to $G/G_x \times (S^{G_x})$. Here S^{G_x} is a closed C^k submanifold of S of dimension equal to $\dim(\mathrm{T}_x\, M/\alpha_x(\mathfrak{g}))^{G_x}$.

Proof. (i) $y \in M_x^{\lesssim}$ is equivalent to the existence of a $g \in G$ such that $G_x \subset g G_y g^{-1} = G_{g\cdot y}$ cf. (2.6.1). On the other hand, $G_x \subset G_z$ just means that $z \in M^{G_x}$. This proves that $M_x^{\lesssim} = G\cdot(M^{G_x})$.

Now suppose that $y_j \in M_x^{\lesssim}$ and $\lim_{j\to\infty} y_j = y$. Let T be a slice at $y \in M$. Then we can replace y_j by $y_j' \in G\cdot y_j \cap T$ such that still $\lim_{j\to\infty} y_j' = y$. Now $y_j' \in M_x^{\lesssim}$ as well, and $y_j' \in T$ implies that $G_{y_j'} \subset G_y$. We find a $g_j \in G$ such that $g_j G_x g_j^{-1} \subset G_{y_j'} \subset G_y$, hence $y \in M_x^{\lesssim}$. This proves that $M_x^{\lesssim}$ is closed. Note that the properness of the action immediately implies that $G\cdot F$ is closed if F is compact, or if F is closed and G is compact. However, it is easy to find proper actions of noncompact Lie groups G and closed subsets F of M such that $G\cdot F$ is not closed.

(ii) If $s \in S$, then $s \in M_x^{\lesssim}$ if and only if $G_x \subset gG_sg^{-1}$, for some $g \in G$. However, $s \in S$ implies that $G_s \subset G_x$, so $G_x \subset gG_sg^{-1} \subset gG_xg^{-1}$. Because G_x is a compact Lie subgroup of gG_xg^{-1} of the same dimension and with the same number of connected components as gG_xg^{-1}, it follows that $G_x = gG_sg^{-1} = gG_xg^{-1}$, hence $G_s = G_x$ and so $s \in S^{G_x}$. Also $M_x^{\lesssim} \cap G \cdot S = M_x^{\sim} \cap G \cdot S$.

Applying Bochner's linearization theorem 2.2.1 to the G_x-action on S and shrinking S suitably, we get that S^{G_x} is C^k diffeomorphic to the intersection with an open ball around the origin, of the fixed point set in $\mathrm{T}_x M/\alpha_x(\mathfrak{g})$ for the tangent action of G_x. Obviously this fixed point set is a linear subspace of $\mathrm{T}_x M/\alpha_x(\mathfrak{g})$. Therefore $G \times S^{G_x}$ is a closed C^k submanifold of $G \times S$, invariant under the (right) action of G_x; and hence $G \times_{G_x} S^{G_x} \cong G/G_x \times S^{G_x}$ is a closed C^k submanifold of $G \times_{G_x} S$. Passing to M via the G-equivariant C^k diffeomorphism: $G \times_{G_x} S \to G \cdot S$ of the Tube theorem 2.4.1, we get that the image $G \cdot (S^{G_x})$ is a closed C^k submanifold of $G \cdot S$. □

(2.6.5) Definition. *The elements $x, y \in M$, and $G \cdot x, G \cdot y \in G\backslash M$, respectively, are said to be of the same local type, with the notation: $x \approx y$ and $G \cdot x \approx G \cdot y$, if there is a G-equivariant C^k diffeomorphism Φ from an open G-invariant neighborhood U of x in M onto an open G-invariant neighborhood V of y in M.*

Clearly this defines an equivalence relation $\approx$ in M, and $G\backslash M$, respectively; it is finer than $\sim$, that is, each $\sim$-equivalence class is partitioned into $\approx$-equivalence classes. These equivalence classes sometimes will be called *local action types* in M, and $G\backslash M$, respectively, and denoted by:

$$M_x^{\approx} = \{y \in M \mid y \approx x\}, \quad G\backslash M_{G \cdot x}^{\approx} = \{G \cdot y \in G\backslash M \mid G \cdot y \approx G \cdot x\}.$$

Note that the Tube theorem 2.4.1 shows that $x \approx y$ if and only $x \sim y$ and the actions of G_x, and G_y, on $\mathrm{T}_x M/\alpha_x(\mathfrak{g})$, and $\mathrm{T}_y M/\alpha_y(\mathfrak{g})$, respectively, are equivalent via a linear intertwining isomorphism, that is, as representations, cf. Section 2.4.

(2.6.6) Definition. *For any subgroup H of G, the* normalizer *of H in G is defined as $\mathrm{N}(H) = \mathrm{N}_G(H) = \{g \in G \mid gHg^{-1} = H\}$. It is the largest subgroup of G containing H as a normal subgroup. It is closed in G if H is closed in G.*

(2.6.7) Theorem.

(i) Each local action type is an open and closed subset of the corresponding orbit type.

(ii) The set $M_x^{\approx} \cap M^{G_x}$ is open in M^{G_x}, and a locally closed C^k submanifold of M (that is, all its connected components have the same dimension), and it is also $\mathrm{N}(G_x)$-invariant.

(iii) *The canonical projection:* $M \to G\backslash M$ *maps* $M_x^{\approx} \cap M^{G_x}$ *onto* $G\backslash M_{G.x}^{\approx}$, *its fibers in* $M_x^{\approx} \cap M^{G_x}$ *are the orbits for the* $\mathrm{N}(G_x)$*-action on* $M_x^{\approx} \cap M^{G_x}$. *The action of* $\mathrm{N}(G_x)/G_x$ *on* $M_x^{\approx} \cap M^{G_x}$ *is proper and free; hence there is a unique structure of a* C^k *manifold on* $G\backslash M_{G.x}^{\approx}$ *for which* $\pi \colon M_x^{\approx} \cap M^{G_x} \to G\backslash M_{G.x}^{\approx}$ *is the corresponding principal fibration with structure group* $\mathrm{N}(G_x)/G_x$.

(iv) $M_x^{\approx}$ *is a locally closed* G*-invariant* C^k *submanifold of* M, *and the* G*-action induces a* G*-equivariant* C^k *diffeomorphism from the associated fiber bundle* $G/G_x \times_{\mathrm{N}(G_x)/G_x} (M_x^{\approx} \cap M^{G_x})$ *onto* $M_x^{\approx}$. *Further* $\pi \colon M_x^{\approx} \to M_{G.x}^{\approx}$ *is a* C^k *fiber bundle with fiber* G/G_x.

(v) *We have:*

$$\begin{aligned}
\dim M_x^{\approx} &= \dim G - \dim G_x + \dim(\mathrm{T}_x M/\alpha_x(\mathfrak{g}))^{G_x} \\
&= \dim G - \dim \mathrm{N}(G_x) + \dim(M_x^{\approx} \cap M^{G_x}); \\
\dim G\backslash M_{G.x}^{\approx} &= \dim(\mathrm{T}_x M/\alpha_x(\mathfrak{g}))^{G_x} \\
&= \dim(M_x^{\approx} \cap M^{G_x}) - \dim G_x + \dim \mathrm{N}(G_x).
\end{aligned}$$

Proof. (i) According to the Tube theorem 2.4.1 the set S^{G_x} consists of elements of the same local action type. Therefore Lemma 2.6.4.(ii) shows that each local action type is open in its orbit type, and hence closed as the complement of the union of the other local action types in the same orbit type.

(ii) That $M_x^{\approx} \cap M^{G_x}$ is open in M^{G_x} also follows from Lemma 2.6.4.(ii). Further locally M^{G_x} is a C^k submanifold of M as follows from Bochner's linearization theorem. Indeed, for $K = G_x$ and any $x_0 \in M^{G_x}$, this theorem implies that near x_0 it is a closed C^k submanifold of dimension equal to $\dim(\mathrm{T}_{x_0} M)^{G_x}$, where $(\mathrm{T}_{x_0} M)^{G_x}$ is the vector space of the $v \in \mathrm{T}_{x_0} M$ such that $\mathrm{T}_{x_0} A(g)\, v = v$, for all $g \in G_x$. Note that if $x_0 \in M_x^{\sim} \cap M^{G_x}$, then $G_x = G_{x_0}$; and if $x_0 \in M_x^{\approx} \cap M^{G_x}$, then $M^{G_{x_0}} = M^{G_x}$ is, near x_0, a submanifold of M of the same dimension as M^{G_x} is near x.

If $g \in G$, then $g \cdot x \in M^{G_x}$ if and only if $h \cdot g \cdot x = g \cdot x$, or $g^{-1} hg \cdot x = x$ or $g^{-1} hg \in G_x$, for all $h \in G_x$; that is $g^{-1} G_x g = G_x$, or $g \in \mathrm{N}(G_x)$.

If $y \in M_x^{\sim} \cap M^{G_x}$ and $g \in G$, then $G_y = G_x$ shows that $g \cdot y \in M^{G_x}$ if and only if $g \in \mathrm{N}(G_x)$. Note also that $g \cdot y \in M_x^{\approx}$ if $y \in M_x^{\approx}$, so we have proved both that $M_x^{\approx} \cap M^{G_x}$ is $\mathrm{N}(G_x)$-invariant and that the fibers of $\pi : M \to G\backslash M$ restricted to $M_x^{\approx} \cap M^{G_x}$ are the $\mathrm{N}(G_x)$-orbits in $M_x^{\approx} \cap M^{G_x}$.

(iii) Because $M_x^{\lesssim} = G \cdot (M^{G_x})$ (Lemma 2.6.4.(i)), we get $M_x^{\sim} = G \cdot (M_x^{\sim} \cap M^{G_x})$ and $M_x^{\approx} = G \cdot (M_x^{\approx} \cap M^{G_x})$; and this implies that π is surjective: $M_x^{\approx} \cap M^{G_x} \to G\backslash M_{G\cdot x}^{\approx}$. The properness of the $\mathrm{N}(G_x)$-action on $M_x^{\approx} \cap M^{G_x}$ follows from the properness of the G-action on M and the fact that $\mathrm{N}(G_x)$ is a closed subgroup of G. Because $G_y = G_x$, for any $y \in M_x^{\sim} \cap M^{G_x}$, the action of the Lie group $\mathrm{N}(G_x)/G_x$ on $M_x^{\approx} \cap M^{G_x}$ is proper, free, and of class C^k.

(vi) That $M_x^{\approx}$ is a locally closed C^k submanifold of M follows from Lemma 2.6.4.(ii), by observing that $M_x^{\approx} \cap G \cdot S = M_x^{\sim} \cap G \cdot S$. The equivalence with $G/G_x \times_{\mathrm{N}(G_x)/G_x} (M_x^{\approx} \cap M^{G_x})$ follows as in the proof of the Tube theorem 2.4.1. Note that here $\mathrm{N}(G_x)/G_x$ acts on G/G_x by $(hG_x, gG_x) \mapsto gh^{-1}G_x$, since $G_x h = hG_x$, for all $h \in \mathrm{N}(G_x)$.

(v) This follows from combining Lemma 2.6.4.(ii) with (iv) above. □

Remarks. $G/G_x \times_{\mathrm{N}(G_x)/G_x} (M_x^{\approx} \cap M^{G_x}) \cong M_x^{\approx}$ is a C^k fiber bundle over $G/\,\mathrm{N}(G_x)$ with fiber $M_x^{\approx} \cap M^{G_x}$, the G-action on it covering the G-action on $G/\,\mathrm{N}(G_x)$ by left multiplications. However, because the action of $\mathrm{N}(G_x)/G_x$ on $M_x^{\approx} \cap M^{G_x}$ is also proper and free, $M_x^{\approx}$ is also a C^k fiber bundle over $(\mathrm{N}(G_x)/G_x)\backslash(M_x^{\approx} \cap M^{G_x}) \cong G\backslash M_{G\cdot x}^{\approx}$, with fiber G/G_x; of course, this is just the fibration of $M_x^{\approx}$ into its G-orbits. Here the G-action covers the trivial action of G on $G\backslash M_{G\cdot x}^{\approx}$. The common fibers of the intersections of the two fibrations are the $\mathrm{N}(G_y)/G_y$-orbits in $M_x^{\approx} \cap M^{G_y} = M_y^{\approx} \cap M^{G_y}$, for $y \in M_x^{\approx}$.

Different local orbit types in a given orbit type can have different dimensions. This is one of the reasons why we preferred to formulate Theorem 2.6.7.(ii)–(v) for local action types rather than for orbit types.

2.7 The Stratification by Orbit Types

Theorem 2.6.7.(iv) shows that the local action types partition M into locally closed C^k submanifolds, each of which is C^k fibered by G-orbits. In this section we shall study the local properties of this partioning.

Consider the action of a compact Lie group H acting on a Euclidean space E by means of orthogonal linear transformations. To start with, $E_0^{\sim} = E^H = \{\, v \in E \mid h{\cdot}v = v, \text{ for all } h \in H \,\}$ is a linear subspace of E, which is H-invariant. Hence F, the orthogonal complement of E^H, is H-invariant as well. The linear isomorphism $+\colon (e, f) \mapsto e{+}f\colon E^H \times F \to E$ is H-equivariant if we let act H on $E^H \times F$ by $(h, (e, f)) \mapsto (e, h{\cdot}f)$, for $h \in H$, $(e, f) \in E^H \times F$. Note that $F_0^{\sim} = F^H = \{0\}$. In order to study the H-action on $F \setminus \{0\}$, we observe that the unit sphere $\Sigma = \{\, f \in F \bigm| \|f\| = 1 \,\}$ in F is H-invariant and that in turn the real-analytic diffeomorphism $p\colon (r, f) \mapsto r\,f\colon \mathbf{R}_{>0} \times \Sigma \to F \setminus \{0\}$ is H-equivariant if we let act H on $\mathbf{R}_{>0} \times \Sigma$ by $(h, (r, f)) \mapsto (r, h \cdot f)$, for $h \in H$, $(r, f) \in \mathbf{R}_{>0} \times \Sigma$. It follows that the orbit types for the action on the H-invariant open subset $E \setminus E_0^{\sim}$ are the sets of the form $E^H + p(\mathbf{R}_{>0} \times T)$, where T runs over the orbit types for the H-action on Σ.

Next let H be a compact subgroup of a Lie group G and let π denote the projection: $G \times E \to G \times_H E$. Each G-orbit in $G \times_H E$ meets $\pi(\{1\} \times E)$; and $\pi(1, e) = g \cdot \pi(1, e) = \pi(g, e)$ if and only if $(1, e) = (gh^{-1}, h \cdot e)$, that is $g = h$ and $h{\cdot}e = e$, for some $h \in H$. In other words, $G_{\pi(1,e)} = H_e$; or the orbit types in $G \times_H E$ are the sets of the form $\pi(G \times \mathcal{T})$, where $\mathcal{T}$ runs over the orbit types for the action of H in E. In view of the previous paragraph, the orbit types in $G\times_H E$ therefore are the sets of the form $\pi\big(G\times(E^H+p(\mathbf{R}_{>0}\times T))\big)$, where T runs over the orbit types for the action of H on Σ.

Using the Tube theorem 2.4.1 and Theorem 2.6.7.(i) we obtain the following proposition by induction on the dimension of the manifold on which the Lie group acts properly.

(2.7.1) Proposition. *For a proper C^1 action of a Lie group G on a manifold M, there are locally only finitely many distinct orbit types with locally only finitely many connected components. In particular, one also has locally only finitely many distinct local action types. If M is compact, then there are only finitely many distinct connected components of orbit types, and the same is true if M is a finite-dimensional vector space on which G acts by linear transformations.*

Any connected component of an orbit type for the action of G on $G\times_H E$ that is not the orbit type of $\pi(G\times\{0\})$, has dimension equal to $\dim G - \dim H + \dim E^H + 1 + \dim T'$, where T' is a connected component of an orbit type T for the action of H on Σ. Because the dimension of the orbit type of $\pi(G\times\{0\})$ is equal to $\dim G - \dim H + \dim E^H$, we get, using again the Tube theorem 2.4.1 and Theorem 2.6.7.(i)

(2.7.2) Proposition. *For each $x\in M$ there is a G-invariant open neighborhood U of x in M such that, for each $y\in U\setminus M_x^{\approx}$:*

$$\dim M_y^{\approx} = \dim M_x^{\approx} + 1 + \dim T, \tag{2.7.1}$$

where T is some local action type for the tangent action of G_x on the unit sphere in the orthogonal complement of $\alpha_x(\mathfrak{g}) + (\mathrm{T}_x M)^{G_x}$ in $\mathrm{T}_x M$, with respect to a given G_x-invariant inner product on $\mathrm{T}_x M$. In the same vein:

$$\dim G\backslash M_y^{\approx} = \dim G\backslash M_x^{\approx} + 1 + \dim G_x\backslash T. \tag{2.7.2}$$

In particular, $\dim M_y^{\approx} > \dim M_x^{\approx}$ and $\dim G\backslash M_y^{\approx} > \dim G\backslash M_x^{\approx}$, for all $y\in U\setminus M_x^{\approx}$. Note also that $x \lesssim y$, hence $\dim G.y \geq \dim G.x$, for all y sufficiently close to x. In particular, the estimates above show that "smaller orbits cannot compensate for being small by massive appearance, not even in the orbit space".

(2.7.3) Definition. *A C^k stratification of the manifold M is a locally finite partition of M into locally closed connected C^k submanifolds M_i $(i\in I)$ of M, called the strata of the stratification, such that the following is satisfied. For each $i\in I$ the closure M_i^{cl} of M_i in M is equal to $M_i\cup\bigcup_{j\in I_i} M_j$, where $I_i\subset I\setminus\{i\}$, and $\dim M_j < \dim M_i$, for each $j\in I_i$.*

The stratification is said to be a Whitney stratification if the following conditions (a) and (b) are met.

(a) *For each $i \in I$, $j \in I_i$ and each sequence x_n in M_i such that* $\lim_{n\to\infty} x_n = x \in M_j$ *and* $\lim_{n\to\infty} \mathrm{T}_{x_n} M_i = L$ *in the Grassmann bundle of* $\mathrm{T}\,M$, *we have* $\mathrm{T}_x M_j \subset L$.

(b) *If x_n is a sequence as in (a) and y_n is a sequence in the limit stratum M_j, such that $\lim y_n = x$ and $y_n \neq x_n$, for all n, then each limit of the one-dimensional subspaces $\mathbf{R} \cdot \lambda(x_n, y_n)$ of $\mathrm{T}_{x_n} M$, for $n \to \infty$, is contained in L. Here λ is a diffeomorphism from an open neighborhood of the diagonal in $M \times M$ to an open neighborhood of the zero section of $\mathrm{T}\,M$, as in Lemma 11.1.1. Clearly the set of limit lines does not depend on the choice of λ.*

(2.7.4) Theorem. *The connected components of the orbit types in M form a Whitney stratification of M.*

Proof. Propositions 2.7.1 and 2.7.2 together imply that the connected components of the orbit types form a stratification.

For the verification of the Whitney conditions (a) and (b) we may assume that $M = G \times_H E$, $x = \pi(1, 0)$, as in the discussion preceding Proposition 2.7.1. The orbit types are obtained by applying the G-action to the H-orbit types in E. Let $\mathfrak{c}$ be a linear subspace of $\mathfrak{g}$ complementary to $\mathfrak{h} = \mathrm{T}_1 H$, and $\mathfrak{c}_0$ an open neighborhood of 0 in $\mathfrak{c}$ such that we can use the inverse of $(X, e) \mapsto \pi(\exp X, e)\colon \mathfrak{c}_0 \times E \to M$ as a local chart around x. In this chart, the orbit types are of the form $\mathfrak{c}_0 \times U$, where U is an H-orbit type in E. We may assume that x_n is not of the same orbit type as x; so $x_n = \pi(\exp X_n, e_n + r_n{\cdot}f_n)$ with $X_n \in \mathfrak{c}_0$, $\lim_{n\to\infty} X_n = 0$, $e_n \in E^H$, $\lim_{n\to\infty} e_n = 0$, $r_n > 0$, $\lim_{n\to\infty} r_n = 0$ and $f_n \in T'$, a connected component of an orbit type for the H-action on Σ. In the chart, the orbit type M_j of x can be identified with $\mathfrak{c}_0 \times E^H$ and the orbit type M_i of the x_n with $\mathfrak{c}_0 \times (E^H + \mathbf{R}_{>0} \cdot T')$. So $\mathrm{T}_{x_n} M_i = \mathfrak{c} \times (E^H + \mathbf{R}{\cdot} f_n + \mathrm{T}_{f_n} T')$. Hence any limit position of the $\mathrm{T}_{x_n} M_j$, for $n \to \infty$, certainly contains $\mathfrak{c} \times E^H = \mathrm{T}_x M_j$. This proves (a).

For (b), write $y_n = \pi(\exp Y_n, \widetilde{e}_n)$, with $Y_n \in \mathfrak{c}_0$, $\lim_{n\to\infty} Y_n = 0$, $\widetilde{e}_n \in E^H$, $\lim_{n\to\infty} \widetilde{e}_n = 0$. By passing to a subsequence we may assume that f_n converges, in the compact set Σ, to $f \in \Sigma$. It follows that $L = \lim_{n\to\infty} \mathrm{T}_{x_n} M_i = \mathfrak{c} \times (E^H + \mathbf{R} \cdot f + \Lambda)$, where $\Lambda = \lim_{n\to\infty} \mathrm{T}_{f_n} T'$. In the chart, we have $(X_n, e_n + r_n f_n) - (Y_n, \widetilde{e}_n) = (X_n - Y_n, (e_n - \widetilde{e}_n) + r_n f_n)$, and the lines through these vectors have their limit positions l always contained in $\mathfrak{c} \times (E^H + \mathbf{R}{\cdot} f) \subset L$. The proof is complete (and did not even need an induction on the dimension of M). □

(2.7.5) *Remarks.* We feel that the stratification by orbit types has even more special properties than general Whitney stratifications.

The orbit space $G\backslash M$ is stratified by the connected components of the orbit types (cf. Theorem 2.6.7. We define the dimension of a connected component of $G\backslash M$ as the maximum of the dimensions of its strata). But in order to say that this is a Whitney stratification, we have to embed $G\backslash M$ at least locally in a smooth manifold. For convenience, we assume in the following discussion that $k = \infty$.

Locally the space of smooth (C^∞) functions on $G\backslash M$, identified in Section 2.5 with $\mathrm{C}^\infty(M)^G$, can be identified with $\mathrm{C}^\infty(B)^K$, where $K = G_x$ and B is a K-invariant open neighborhood of 0 in $E = \mathrm{T}_x M/\alpha_x(\mathfrak{g})$. The action of K on E is the one induced by the tangent action $k \mapsto \mathrm{T}_x A(k)$ of $K = G_x$ on $\mathrm{T}_x M$.

The theorem of Schwarz [1975], applied to this representation of the compact group K on E, states that $\mathrm{C}^\infty(E)^K$ contains finitely many polynomials $p_1, \dots, p_k$, which may be chosen homogeneous of degree $m_1, \dots, m_k > 0$, such that every $f \in \mathrm{C}^\infty(E)^K$ can be written as $f = \phi \circ p$, for some $\phi \in \mathrm{C}^\infty(\mathbf{R}^k)$. Here p denotes the mapping $x \mapsto (p_1(x), \dots, p_k(x)) \colon E \to \mathbf{R}^k$.

The conic structure of multiplication by $r > 0$ in E, which maps K-orbits to K-orbits, is intertwined by p with the action of $r > 0$ in $\mathbf{R}^k$ given by: $(y_1, \dots, y_k) \mapsto (r^{m_1} y_1, \dots, r^{m_k} y_k)$. (The invariance of $p(E)$ under such an action is said to be a quasihomogeneous structure on $p(E)$.) In view of the compactness of the unit sphere in E, this leads to the properness of the mapping $p\colon E \to \mathbf{R}^k$; and it follows that $p(E)$ is a closed subset of $\mathbf{R}^k$.

The mapping p induces a homeomorphism: $K\backslash E \to p(E)$. Let X be a local action type of the K-action on E. Using the local description of the action in the Tube theorem 2.4.1 and of the orbit types preceding Proposition 2.7.1, one sees that, for each $x \in X$, there exist K-invariant C^∞ functions $f_1, \dots, f_r$ near x such that $df_1, \dots, df_r$ are linearly independent and $r = \dim K\backslash X$. Combined with the theorem of Schwarz, this implies that $\operatorname{rank} \mathrm{T}_x(p|_X) = \dim K\backslash X$, for all $x \in X$. In turn this implies that $p(X)$ is a smooth submanifold of $\mathbf{R}^k$, diffeomorphic to $K\backslash X$ under the homeomorphism: $K\backslash E \to p(E)$. Because of the properness of p, the $p(X)$ form a stratification of $p(E) \subset \mathbf{R}^k$.

Using that the action of K on E is polynomial (cf. Corollary 14.6.2), one can prove that the strata X are semialgebraic. According to the Tarski-Seidenberg theorem (cf. Hörmander [1983], Appendix A.2), which says that the image of semi-algebraic sets under polynomial mappings is semi-algebraic, one gets that both $p(E)$ and the $p(X)$'s are semi-algebraic. Łojasiewicz [1965], p.150–153 has proved that every locally-finite partition of a semi-analytic set into semi-analytic sets can be refined into a Whitney stratification. So, in particular, the stratification of $p(E)$ by the $p(X)$'s can be refined to a Whitney stratification of $p(E)$. We have not investigated the question whether the $p(X)$'s themselves automatically form a Whitney stratification.

2.8 Principal and Regular Orbits

A part of the information about the orbit type stratification can be encoded in its *directed graph* $\mathcal{T}$. The vertices of $\mathcal{T}$ are the connected components of the orbit types in $G\backslash M$ and we draw an arrow from $A \in \mathcal{T}$ to $B \in \mathcal{T}$, notation: $A \to B$, if $B \subset A^{\mathrm{cl}}$; in words, if elements of A converge to elements of B. Clearly the connected components in the graph correspond to the connected components of the orbit space $G\backslash M$.

If $A \to B$, then $A \gtrsim B$ (Lemma 2.6.4.(ii)). If also $A \neq B$, then according to Proposition 2.7.2 we have $\dim A > \dim B$ and $\dim \pi^{-1}(A) > \dim \pi^{-1}(B)$, where π denotes the projection: $M \to G\backslash M$. The latter shows that $A \to B$ defines a partial ordering in $\mathcal{T}$ (that is, if $A \to B$ and $B \to C$ then $A \to C$, and $A = B$ if and only if $A \to B$ and $B \to A$), in which any chain at most has $1 + \dim G\backslash M$ many elements. In pictures, one usually only draws the arrows between immediate successors. Note that the **minimal** elements in the graph $\mathcal{T}$ are precisely the connected components of orbit types in $G\backslash M$ and their preimages in M that are **closed** in $G\backslash M$, and M, respectively.

Example. The graph

B A C

Fig. 2.8.1.

is the directed graph of the action of $\mathbf{SO}(3,\mathbf{R})$ on the two-dimensional sphere in $\mathbf{R}^3$: A is the orbit type of the circular orbits, and B and C are the fixed points, the North and the South Pole, respectively. (For the action of $\mathbf{SO}(3,\mathbf{R})$ on $\mathbf{R}^3$, the vertices B and C would get identified since the vertical axis is a single orbit type of fixed points.) The action of $\mathbf{SO}(3,\mathbf{R})$ on itself by conjugation (cf. Section 1.2.A) has the same directed graph: A is the collection of conjugacy classes of rotations through an angle between 0 and π, whereas B is the conjugacy class of the rotation through π and C is the identity. Although of quite a different nature, the action of $\mathbf{SU}(2)$ on itself by conjugation (cf. Section 1.2.B) also has this directed graph, with A the collection of two-dimensional conjugacy classes and B, C corresponding to the elements $\pm\mathrm{I}$ in $\mathbf{SU}(2)$.

(2.8.1) Definition. *For a proper* C^k *action* $(k \geq 1)$ *of the Lie group* G *on the manifold* M*, the orbit* $G \cdot x$*, with* $x \in M$*, is said to be a principal orbit if its local action type* $M_x^{\approx}$ *is open in* M*, that is, if it belongs to a*

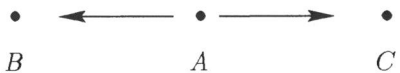

maximal element of the directed graph $\mathcal{T}$. We write $M^{\text{princ}} = \{x \in M \mid G \cdot x$ is a principal orbit $\}$, and $G\backslash M^{\text{princ}} \subset G\backslash M$ for the set of principal orbits.

Clearly M^{princ} is an open G-invariant subset of M, projecting onto the open subset $G\backslash M^{\text{princ}}$ of the orbit space $G\backslash M$. Moreover, M^{princ} is also dense in M as the complement of the union of the, locally finitely many, orbit types of positive codimension in M. The last statement in Theorem 2.6.7.(iv) shows that the open and closed subsets $M_x^{\approx}$ of M^{princ} are open G-invariant subsets of M fibered by G-orbits. These are maximal in the sense that no $x \in M \setminus M^{\text{princ}}$ has a G-invariant open neighborhood that is fibered by G-orbits. Indeed, each nearby orbit intersects a slice at x in some point y, and then $G_y \subset G_x$. If $\dim G \cdot y = \dim G \cdot x$, then $\dim G_y = \dim G_x$; or G_x/G_y is finite and $G \cdot y \cong G/G_y \to G/G_x \cong G \cdot x$ is a fibration with finite fiber G_x/G_y. This contradicts local triviality of the orbit structure near x, unless $G_y = G_x$, for all $y \in S$ near x. In terms of the description preceding Proposition 2.7.1 we have $x \in M^{\text{princ}}$ if and only if the space F occurring there is equal to $\{0\}$.

It is the main purpose of this section to show that, in each connected component of $G\backslash M$, the set $G\backslash M^{\text{princ}}$ is connected, cf. Theorem 2.8.5 below. In other words, **each connected component of the graph $\mathcal{T}$ has a unique maximal element**. If M is connected, this means that there is **only one principal orbit type**.

In the sequel, it will be convenient to use a somewhat coarser partition of $G\backslash M$ than the one into orbit types.

(2.8.2) Definition. *Two elements $x, y \in M$ are said to be of the same infinitesimal type, notation: $x \underset{\text{inf}}{\sim} y$, if there exists $g \in G$ such that $\operatorname{Ad} g^{-1}(\mathfrak{g}_y) = \mathfrak{g}_x$, (or $g^{-1}(G_y)^{\circ} g = G_x^{\circ}$). One says that y dominates x infinitesimally, notation: $x \underset{\text{inf}}{\lesssim} y$, if there exists $g \in G$ such that $\operatorname{Ad} g^{-1}(\mathfrak{g}_y) \subset \mathfrak{g}_x$, (or $g^{-1}(G_y)^{\circ} g \subset G_x^{\circ}$).*

Because $\mathfrak{g}_x$ is the Lie algebra of G_x and $\operatorname{Ad} g^{-1}(\mathfrak{g}_y)$ is the Lie algebra of $g^{-1}(G_y)g$, we have $x \sim y \Rightarrow x \underset{\text{inf}}{\sim} y$, and $x \lesssim y \Rightarrow x \underset{\text{inf}}{\lesssim} y$; so the infinitesimal version is a coarsening of the partition, and ordering, respectively by orbit types.

Furthermore, if S is a slice at x and $y \in S$, then $G_y \subset G_x$ and hence $\mathfrak{g}_y \subset \mathfrak{g}_x$. So all nearby orbits are dominating infinitesimally. Also $y \underset{\text{inf}}{\sim} x$ if and only if $\mathfrak{g}_y = \mathfrak{g}_x$. The intersection with S of the infinitesimal type of x therefore is equal to the common set of zeroes in S of the vector fields $\alpha(X)$, for $X \in \mathfrak{g}_x$. In a Bochner linearization these vector fields are all linear and this set of zeroes is a linear subspace. As in Theorem 2.6.7.(iv), one obtains that the set $M_x^{\underset{\text{inf}}{\sim}}$ of elements of the same infinitesimal type as x locally is a closed C^k submanifold of M. As in Proposition 2.7.2, one gets that the dimension of nearby different infinitesimal types is strictly larger, both in M and in the orbit space $G\backslash M$. As in Theorem 2.7.4, the infinitesimal types form a Whitney stratification of M.

(2.8.3) Definition. *The set $G \cdot x$ is said to be a regular orbit if the dimension of the orbits $G \cdot y$ is constant (not strictly larger), for all y near x. That is, if x is in the interior of its infinitesimal type; or, if $G \cdot x$ belongs to a maximal element of the directed graph $\mathcal{T}^{\mathrm{inf}}$ defined by the infinitesimal type. Elements on regular orbits will be called regular points for the action of G on M; and we shall denote the set of these points by M^{reg}, and the set of regular orbits by $G\backslash M^{\mathrm{reg}} \subset G\backslash M$.*

Because $\dim G \cdot y = \dim \mathfrak{g} - \dim \mathfrak{g}_y$ and $\{\, y \in M \mid \dim \mathfrak{g}_y > r \,\}$ is a closed subset of M for every $r \in \mathbf{Z}_{>0}$, we get that $G \cdot x$ is a regular orbit if and only if $\dim \mathfrak{g}_x \leq \dim \mathfrak{g}_y$, or $\dim \mathfrak{g}_x = \dim \mathfrak{g}_y$, for all y near x. Almost by definition M^{reg} is a dense open subset of M. It is G-invariant and projects onto the dense open subset $G\backslash M^{\mathrm{reg}}$ of $G\backslash M$. Clearly $M^{\mathrm{princ}} \subset M^{\mathrm{reg}}$, and both M^{princ} and M^{reg} are unions of local action types.

We now consider the orbit type strata of codimension 1 in M.

(2.8.4) Lemma. *Suppose that the Lie group G acts properly and in a C^k fashion on the manifold M, for $k \geq 1$. Let $x \in M$ and $\dim M_x^{\approx} = \dim M - 1$. Then $M_x^{\approx} \subset M^{\mathrm{reg}}$ and $\dim_{\mathbf{R}} F = 1$, where F is the subspace appearing in the description preceding Proposition 2.7.1, while G_x acts on F as the group $\mathbf{O}(1, \mathbf{R}) = \{1, -1\}$. The orbits near $G \cdot x$ are fibered over $G \cdot x$, with fibers consisting of two elements; if G is connected then these are two-fold coverings.*

Proof. Because $\dim M_x^{\approx} = \dim \alpha_x(\mathfrak{g}) + \dim(\mathrm{T}_x M/\alpha_x(\mathfrak{g}))^{G_x}$ (cf. Theorem 2.6.7.(v)), we have $\dim M_x^{\approx} = \dim M - 1$ if and only if F, which is the orthogonal complement of $(\mathrm{T}_x M/\alpha_x(\mathfrak{g}))^{G_x}$ in $\mathrm{T}_x M/\alpha_x(\mathfrak{g})$ is one-dimensional. Because G_x acts in a nontrivial way on F by orthogonal linear transformations, it acts as $\{1, -1\}$ on F. This implies the last statement, and in turn it shows that $\dim G \cdot y = \dim G \cdot x$, for y near x, or $x \in M^{\mathrm{reg}}$. □

Note that near x, the hypersurface $M_x^{\approx}$ disconnects M into two half spaces, but that G_x interchanges the two half spaces. It also follows that near $G \cdot x$ the orbit space $G\backslash M$ can be viewed as a C^k manifold with boundary (a "half space"), where $G \cdot x$ is a point on the codimension-one boundary. This is even so in terms of the Remarks 2.7.5: near $G \cdot x$ the orbit space can be identified with $(\mathrm{T}_x M/\alpha_x(\mathfrak{g}))^{G_x} \times (\{1, -1\}\backslash F)$, so we can take $p_1, \dots, p_{k-1}$ to be a basis of the linear forms on the first factor, and $p_k = f^2$, if f denotes the f-coordinate. So p maps $E = (\mathrm{T}_x M/\alpha_x(\mathfrak{g}))^{G_x} \times F$ onto the half space $\{\, (y_1, \dots, y_k) \in \mathbf{R}^k \mid y_k \geq 0 \,\}$. Thereby $S \cap M_x^{\approx}$ gets mapped to a neighborhood of 0 in the boundary $\{ (y_1, \dots , y_k) \in \mathbf{R}^k \mid y_k = 0 \}$, and $S \cap M^{\mathrm{princ}}$ to a neighborhood of 0 in the interior $\{ (y_1, \dots , y_k) \in \mathbf{R}^k \mid y_k > 0 \}$.

(2.8.5) Principal Orbit Theorem. *Suppose that the Lie group G acts properly and in a C^1 fashion on the manifold M. Then $M \setminus M^{\mathrm{reg}}$ is equal to the union of local orbit types (strata for the stratification described in Section 2.7) of codimension ≥ 2 in M. For every connected component M° of M, the subset $M^{\mathrm{reg}} \cap M^\circ$ is connected, open and dense in M°. Each connected component $(G\backslash M)^\circ$ of $G\backslash M$ contains only one principal orbit type, which is a connected, open and dense subset of it.*

Proof. Let x, $y \in M^{\mathrm{reg}} \cap M^\circ$ and let $\gamma \colon [0,1] \to M^\circ$ be a continuous curve in M° such that $\gamma(0) = x$, $\gamma(1) = y$. For each z in the compact subset $K = \gamma([0,1])$ of M, there is an open neighborhood U_z of z in M that meets only finitely many strata. There is a finite subset K_0 of K such that $K \subset \bigcup_{z \in K_0} U_z := U$, and the open neighborhood U of K in M meets only finitely many strata as well. A transversality principle (cf. Guillemin and Pollack [1974], p.70) yields the existence of a C^1 curve $\widetilde{\gamma}$ in U, running from x to y that is transversal to all the finitely many strata in U. But transversality means missing all strata of codimension ≥ 2. Hence Lemma 2.8.4 implies $\widetilde{\gamma}([0,1]) \subset M^{\mathrm{reg}} \cap M^\circ$; and therefore $M^{\mathrm{reg}} \cap M^\circ$ is connected.

Now let x, $y \in M^{\mathrm{princ}}$ such that $G \cdot x$, $G \cdot y$ are in the same connected component Γ of $G\backslash M$. By replacing y by a suitable element of $G \cdot y$, we may assume that x and y are in the same connected component M° of M. By the previous argument we can connect x and y with a C^1 curve γ' from x to y that misses all strata of codimension ≥ 2 and intersects the codimension-one strata T only transversally. It follows also that the $t_i \in [0,1]$ such that $z_i = \widetilde{\gamma}(t_i) \in T$ form a discrete and hence finite subset of $[0,1]$. Using the half space description of $G\backslash M$ near $G \cdot \widetilde{\gamma}(t_i)$ following Lemma 2.8.4, we see that $t \mapsto G \cdot \widetilde{\gamma}(t)$ can be replaced by a ("shorter") C^1 curve δ from $G \cdot x$ to $G \cdot y$ that runs entirely in $(G\backslash M)^{\mathrm{princ}}$, see Fig.2.8.2.
This shows that $(G\backslash M)^\circ \cap G\backslash M^{\mathrm{princ}}$ is connected and, since it is partitioned into principal open orbits, it consists of a single principal orbit type. □

The smallest nonvoid, open and closed G-invariant subsets are the preimages of the connected components of $G\backslash M$ or, equivalently, the sets $G \cdot M^\circ$ where M° is a connected component of M. By restricting the discussion to those subsets, we may assume that $G\backslash M$ is connected; and according to Theorem 2.8.5 this implies that there is only one principal orbit type. (It might be another procedure to restrict the attention to a connected component M° of M, if thereby one replaces the group G by $G_{(M^\circ)} = \{ g \in G \mid A(g)(M^\circ) = M^\circ \}$, which is an open subgroup of G. The group G acts transitively on the set of connected components in $G \cdot M^\circ$; hence this set can be identified with $G/G_{(M^\circ)}$. The actions of the $G_{((M^\circ)')}$

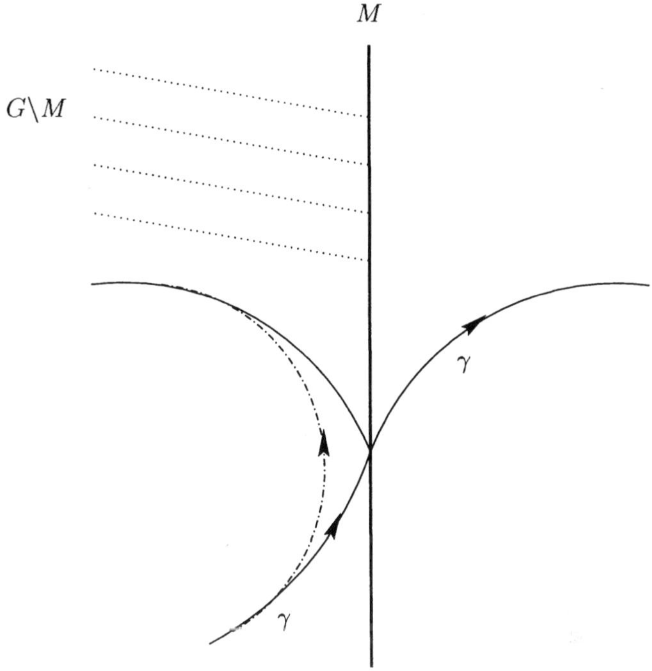

Fig. 2.8.2.

on the various connected components $(M^\circ)'$ of $G \cdot M^\circ$ are intertwined by the elements of G.)

(2.8.6) Corollary. *Suppose that G acts properly and in a C^k $(k \geq 1)$ fashion on M, and assume that $G\backslash M$ is connected. Let $x \in M^{\mathrm{princ}}$ and write C for the connected component of x in $M^{\mathrm{princ}} \cap M^{G_x}$. Then:*

(i) $M^{\mathrm{princ}} = M_x^{\approx} = M_x^{\sim}$.
(ii) The set H, the union of the connected components of M^{G_x} that meet M^{princ}, is a closed C^k submanifold of M, which contains $\mathrm{N}(G_x) \cdot C = M^{\mathrm{princ}} \cap M^{G_x}$ as an open subset.
(iii) $G_{(C)} = \{\, g \in G \mid A(g)(C) = C \,\}$ induces a G-equivariant C^k diffeomorphism $G/G_x \times_{G_{(C)}/G_x} C \overset{\sim}{\to} M^{\mathrm{princ}}$.
(iv) If C^{cl} denotes the closure of C, then $C^{\mathrm{cl}} \subset H$ and $G \cdot C^{\mathrm{cl}} = M$.

Proof. (i) We have $M^{\mathrm{princ}} = M_x^{\approx}$, because there is only one principal orbit type. If $y \in M$, $G_y = G_x$ and $y \notin M^{\mathrm{princ}}$, then in a slice at y there exists $z \in M^{\mathrm{princ}}$ such that $G_z \Subset G_y = G_x$, since M^{princ} is open and dense in M.

But then $z \notin M_x^{\approx}$, in contradiction with $M^{\mathrm{princ}} = M_x^{\approx}$. This implies that $M_x^{\sim} = M_x^{\approx}$.

(ii) As observed after Eq. 2.6.2, each connected component of M^{G_x} is a closed C^k submanifold of M. If π denotes the canonical projection: $M \to G\backslash M$, then Theorem 2.6.7.(iii) yields that $\pi(M^{\mathrm{princ}} \cap M^{G_x}) = G\backslash M^{\mathrm{princ}}$. According to Theorem 2.8.5 this set is connected; and the fibers of $\pi\colon M^{\mathrm{princ}} \cap M^{G_x} \to G\backslash M^{\mathrm{princ}}$ are the $\mathrm{N}(G_x)/G_x$-orbits in $M^{\mathrm{princ}} \cap M^{G_x}$. This shows that $\mathrm{N}(G_x) \cdot C = M^{\mathrm{princ}} \cap M^{G_x}$; it also follows that all connected components of $M^{\mathrm{princ}} \cap M^{G_x}$, and hence also of H, have the same dimension, equal to $\dim C$.

(iii) According to Theorem 2.6.7.(iii) we have $g \in G_{(C)}$ if and only if $A(g)(y) \in C$, for some $y \in C$. Also $G_{(C)} \subset \mathrm{N}(G_x)$. Because $\mathrm{N}(G_x) \cdot y \subset H$ and C is an open subset of H, we get $G_{(C)}$ is open in $\mathrm{N}(G_x)$. Theorem 2.6.7.(iv) implies that $M^{\mathrm{princ}} = G \cdot (M^{\mathrm{princ}} \cap M^{G_x}) = G \cdot G \cdot C = G \cdot C$. Now the G-equivariant C^k diffeomorphism $G/G_x \times_{G_{(C)}/G_x} C \overset{\sim}{\to} M^{\mathrm{princ}}$ is obtained using the Tube theorem 2.4.1.

(iv) Let $y \in M$. Using the description preceding Proposition 2.7.1, and the fact that M^{princ} is dense in M, we can find a continuous curve $\gamma\colon [\,0, \epsilon\,] \to M$, with $\epsilon > 0$, such that $\gamma(0) = y$ and $\gamma(r) \in M^{\mathrm{princ}}$, $G_{\gamma(r)} = G_{\gamma(\epsilon)}$, for all $r \in \,]\,0, \epsilon\,]$. Because $G \cdot C = M^{\mathrm{princ}}$, there exists a $g \in G$ such that $g \cdot \gamma(\epsilon) \in C$. In particular, $G_{g\cdot\gamma(r)} = gG_{\gamma(r)}g^{-1} = gG_{\gamma(\epsilon)}g^{-1} = G_{g\cdot\gamma(\epsilon)} = G_x$, for all $r \in \,]\,0, \epsilon\,]$. That is $g \cdot \gamma(r) \in M^{\mathrm{princ}} \cap M^{G_x}$, for all $r \in \,]\,0, \epsilon\,]$. Because $g \cdot \gamma(\epsilon) \in C$ and C is a connected component of $M^{\mathrm{princ}} \cap M^{G_x}$, it follows that $g \cdot \gamma(r) \in C$, for all $r \in \,]\,0, \epsilon\,]$ and $g \cdot y = \lim_{r \searrow 0} g \cdot \gamma(r) \in C^{\mathrm{cl}}$. This proves that $y \in G \cdot C^{\mathrm{cl}}$. □

The purpose of Theorem 2.6.7 was to obtain reductions to proper and free actions of $\mathrm{N}(G_x)/G_x$ on $M_x^{\approx} \cap M^{G_x}$; if $x \in M^{\mathrm{princ}}$, then $G \cdot (M_x^{\approx} \cap M^{G_x})$ is open in M. The aim of Corollary 2.8.6 is to combine this with connectivity statements. If the action is free at some point $x \in M$, we get that $G_x = \{1\}$, $\mathrm{N}(G_x) = G$ and $M^{G_x} = M$; and the reductions are not very substantial in that case. For the action by conjugation of a compact Lie group on itself, one has another extreme case, where $\mathrm{N}(G_x)/G_x$, the *Weyl group*, is finite. In this situation, $G_{(C)}/G_x$ is even much smaller and often trivial. See Chapter 3.

For general proper G-actions, it might be interesting to investigate the relationship between the $\mathrm{N}(G_x)/G_x$-orbit types in H and the intersections with H of the G-orbit types in M. And also, whether the restriction to C^{cl} of the G-orbit types leads to Whitney stratification of C^{cl}, and what can be said about the fibers of the projection $\pi\colon C^{\mathrm{cl}} \to G\backslash M$.

For the action by conjugation of a compact Lie group on itself, we shall return to these questions in Section 3.1. There we will also use the following variation on Theorem 2.6.7 and Corollary 2.8.6. For any Lie subalgebra $\mathfrak{h}$ of $\mathfrak{g}$, we write

$$M^{\mathfrak{h}} = \{\, y \in M \mid \alpha_y(X) = 0, \text{ for all } X \in \mathfrak{h} \,\}, \tag{2.8.1}$$

the set of zeroes of the infinitesimal actions of the $X \in \mathfrak{h}$, that is, the fixed point set of the connected Lie subgroup H of G with Lie algebra equal to $\mathfrak{h}$. Also:

$$\text{(2.8.2)} \qquad \mathrm{N}_G(\mathfrak{h}) = \{ g \in G \mid \mathrm{Ad}\, g(\mathfrak{h}) = \mathfrak{h} \},$$

the normalizer of $\mathfrak{h}$ in G, i.e., the normalizer of H in G. Clearly $\mathrm{N}_G(\mathfrak{h})$ is a closed Lie subgroup of G. Furthermore, if H is contained in a compact subgroup K of G, then the Bochner linearization theorem 2.2.1 shows that $M^{\mathfrak{h}}$ is a locally closed subset of M, locally equal to a C^k submanifold, for which the dimensions of different connected components may be different.

(2.8.7) Proposition. *Suppose that G acts properly and in a C^k ($k \geq 1$) fashion on M, and assume that $G\backslash M$ is connected. Let $x \in M^{\mathrm{reg}}$ and write C' for the connected component of x in $M^{\mathrm{reg}} \cap M^{\mathfrak{g}_x}$, and H' for the union of the connected components of $M^{\mathfrak{g}_x}$ that meet M^{reg}. Then:*

(i) We have $M^{\mathrm{reg}} = \widetilde{M_x^{\mathrm{inf}}}$.
(ii) The set H' is a closed C^k submanifold of M and is $\mathrm{N}_G(\mathfrak{g}_x)$-invariant. Further $M^{\mathrm{reg}} \cap M^{\mathfrak{g}_x}$ is open and dense in H', and equal to the set of regular points for the $\mathrm{N}_G(\mathfrak{g}_x)/(G_x)^{\circ}$-action on H'. The G-action induces a G-equivariant C^k diffeomorphism from the associated fiber bundle:

$$G/(G_x)^{\circ} \times_{\mathrm{N}_G(\mathfrak{g}_x)/(G_x)^{\circ}} (M^{\mathrm{reg}} \cap M^{\mathfrak{g}_x})$$

onto M^{reg}. Also $(\mathrm{N}_G(\mathfrak{g}_x)/(G_x)^{\circ})\backslash M^{\mathrm{reg}} \cap M^{\mathfrak{g}_x} \cong G\backslash M^{\mathrm{reg}}$.
(iii) $\mathrm{N}_G(\mathfrak{g}_x) \cdot C' = M^{\mathrm{reg}} \cap M^{\mathfrak{g}_x}$; and $G_{(C')} := \{ g \in G \mid A(g)(C') = C' \}$ is an open subgroup of $\mathrm{N}_G(\mathfrak{g}_x)$, and the action $A \colon G \times C' \to M$ induces a G-equivariant C^k diffeomorphism:

$$G/(G_x)^{\circ} \times_{G_{(C')}/(G_x)^{\circ}} C' \xrightarrow{\sim} M^{\mathrm{reg}}.$$

(iv) If $(C')^{\mathrm{cl}}$ denotes the closure of C', then $(C')^{\mathrm{cl}} \subset H'$ and $G \cdot (C')^{\mathrm{cl}} = M$.

Proof. (i) If $y \in M$, $\mathfrak{g}_y = \mathfrak{g}_x$ and $y \notin M^{\mathrm{reg}}$, then in a slice at y there exists $z \in M^{\mathrm{reg}}$ such that G_z is a subgroup of G_y with $\dim G_z < \dim G_y$. But then $\mathfrak{g}_z \Subset \mathfrak{g}_y = \mathfrak{g}_x$, in contradiction with the fact that there is only one regular orbit type.

(ii)–(iv) Now $y \in M^{\mathfrak{g}_x}$ means that $\mathfrak{g}_x \subset \mathfrak{g}_y$. If $g \in G$, then $g \cdot y \in M^{\mathfrak{g}_x}$ if and only if $\mathfrak{g}_x \subset \mathfrak{g}_{g \cdot y} = \mathrm{Ad}\, g(\mathfrak{g}_y)$; and this automatically holds if g^{-1}, and hence g, belongs to $\mathrm{N}_G(\mathfrak{g}_y)$. Conversely, if $\mathfrak{g}_x \subset \mathrm{Ad}\, g(\mathfrak{g}_y)$ and $y \in M^{\mathrm{reg}}$, hence $\dim \mathfrak{g}_x = \dim \mathfrak{g}_y$, then we have $\mathfrak{g}_x = \mathrm{Ad}\, g(\mathfrak{g}_y)$, or $\mathrm{Ad}\, g^{-1}(\mathfrak{g}_x) = \mathfrak{g}_y = \mathfrak{g}_x$, or $g \in \mathrm{N}_G(\mathfrak{g}_x)$. The statements in (ii) now follow, compare also with the proof of Corollary 2.8.6.(ii)–(iv). □

(2.8.8) Corollary. *Assume that $G\backslash M$ is connected and also, for $x \in M^{\mathrm{princ}}$, that G_x is connected. Then the submanifold H in Corollary 2.8.6 is equal to the submanifold H' in Proposition 2.8.7, and further $\mathrm{N}_G(\mathfrak{g}_x) = \mathrm{N}_G(G_x)$,*

$M^{\mathfrak{g}_x} = M^{G_x}$, *$M^{\text{princ}} \cap M^{\mathfrak{g}_x}$ is dense in H, $C \subset C' \subset H$, $G_{(C)} = G_{(C')} \subset$ $\mathrm{N}(G_x)$.*

Proof. Let $y \in M^{\text{reg}} \cap M^{\mathfrak{g}_x}$, let S be a slice at y. Then $G_z \subset G_y$, for $z \in S$. Taking Lie algebras we get that $\mathfrak{g}_z \subset \mathfrak{g}_y$; and using that $y \in M^{\text{reg}}$, we conclude that $\mathfrak{g}_z = \mathfrak{g}_y = \mathfrak{g}_x$. Hence $z \in M^{\mathfrak{g}_x}$, and we have proved that $S \subset M^{\mathfrak{g}_x}$. Because S contains elements of M^{princ}, arbitrarily close to y, we have proved that $M^{\text{princ}} \cap M^{\mathfrak{g}_x}$ is dense in $M^{\text{reg}} \cap M^{\mathfrak{g}_x}$, and in turn this set was dense in H'. This also proves that H, which contains $M^{\text{princ}} \cap M^{\mathfrak{g}_x} = M^{\text{princ}} \cap M^{G_x}$, is dense in H' and hence is equal to H'. If $A(g)(C) = C$, then $A(g)(C') \supset A(g)(C) = C$ meets C'; and therefore $A(g)(C) = C'$. This proves that $G_{(C)} \subset G_{(C')}$. □

2.9 Blowing Up

If E is a finite-dimensional vector space over $\mathbf{R}$, then the *projective space* $\mathbf{P}(E)$ is defined as the space of one-dimensional linear subspaces l of E. As is well-known, this is a compact real-analytic manifold; it also can be identified with the unit sphere in E (with respect to any inner product in E) with antipodal points identified. If E is a C^k vector bundle over a manifold X, then $\mathbf{P}(E)$ will be the bundle over X such that $(\mathbf{P}(E))_x$, the fiber over x, is defined as $\mathbf{P}(E_x)$. Then $\mathbf{P}(E)$ is a C^k bundle over X in a natural way. If X is a C^k submanifold of a manifold M, then the *normal bundle* $\mathrm{N}\,X$ of X in M is the bundle over X such that $(\mathrm{N}\,X)_x = \mathrm{T}_x M / \mathrm{T}_x X$, for each $x \in X$; $\mathrm{N}\,X$ is a C^{k-1} vector bundle over X in a natural way.

Now assume also that X is closed in M. Then the *blowup of M along X* is a C^{k-1} manifold W, together with a C^{k-1} mapping $\beta\colon W \to M$ that is defined in the following way. In local coordinates we may assume that M is an open neighborhood of 0 in $\mathbf{R}^n = \mathbf{R}^p \times \mathbf{R}^q$, and that $X = \{\,(y,z) \in M \mid z = 0\,\}$. In this case we set $W = \{\,(y,z,l) \in M \times \mathbf{P}(\mathbf{R}^q) \mid z \in l\,\}$ and $\beta(y,z,l) = (y,z)$, for $(y,z,l) \in W$. Then W is a closed real-analytic n-dimensional submanifold of $M \times \mathbf{P}(\mathbf{R}^q)$. Near $(y,z,l) \in W$ with $z \neq 0$ this is obvious; and near $(y^0, 0, \mathbf{R}\cdot v^0)$, with $(y^0, 0) \in X$ and, say $v_i^0 = 1$, it is given by $\{\,(y,z,\mathbf{R}\cdot v) \mid v_i = 1, z_j = z_i v_j,$ for all $j \neq i\,\}$. Further $\beta\colon W \to M$ has to be a real-analytic mapping such that its restriction to $\beta^{-1}(M \backslash X)$ is a real-analytic diffeomorphism onto $M \backslash X$. On the other hand, $\beta^{-1}(X) = X \times \mathbf{P}(\mathbf{R}^q) \subset W$, and the restriction of β to $\beta^{-1}(X)$ is equal to the projection of $X \times \mathbf{P}(\mathbf{R}^q)$ onto the first factor X.

A change of such coordinates is a local C^k diffeomorphism $\Phi\colon (y,z) \mapsto (\eta(y,z), \zeta(y,z))$ such that $\zeta(y,0) = 0$. On $\beta^{-1}(M \backslash X)$, the mapping $\beta^{-1} \circ \Phi \circ \beta$ is equal to $\Psi\colon (y,z,\mathbf{R}\cdot z) \mapsto (\eta(y,z), \epsilon(y,z), \mathbf{R}\cdot \epsilon(y,z))$. It is easily verified that the choice $\Psi(y,0,l) = (\eta(y,0), 0, \frac{\partial \epsilon}{\partial z}(y,0)(l))$ defines an extension of Ψ to a C^{k-1} diffeomorphism of the W's. These coordinate changes together define a global manifold W, conjointly with a proper C^{k-1} mapping $\beta: W \to M$ such that:

(2.9.1) $\beta|_{\beta^{-1}(M\setminus X)}: \beta^{-1}(M \setminus X) \to M \setminus X$ is an analytic diffeomorphism,

and

(2.9.2) $\quad \beta^{-1}(X) = \mathbf{P}(\mathrm{N}\, X), \quad \beta|_{\beta^{-1}(X)}$ is the projection $\mathbf{P}(\mathrm{N}\, X) \to X$.

Actually, in order to get (2.9.1), we identify $W \setminus \beta^{-1}(X)$ as a manifold with the open subset $M \setminus X$ of M, gluing to it the open neighborhood of $\beta^{-1}(X)$ in W constructed above via the projections $(y, z, l) \mapsto (y, z)$ in the local coordinates.

The mapping β "blows down" the projective spaces $\mathbf{P}((\mathrm{N}\, X)_x)$ to the points x.

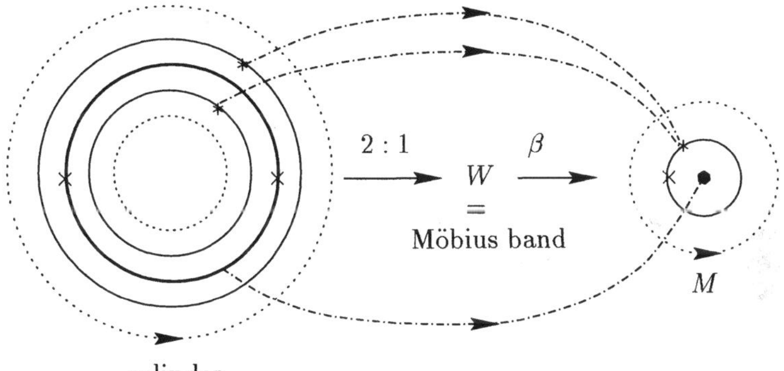

Fig. 2.9.1.

Now assume that in addition X is invariant under a proper C^k action of a Lie group G on M. Then this action induces a C^{k-1} action of G on the tangent bundle $\mathrm{T}_X\, M$ of M along X, on the tangent bundle $\mathrm{T}\, X$ of X, and therefore also on the normal bundle $\mathrm{N}\, X$ of X in M. In turn this induces a C^{k-1} action of G on $\mathbf{P}(\mathrm{N}\, X) = \beta^{-1}(X)$.

The mapping β intertwines the G-action on $M \setminus X$ with a unique action of G on $W \setminus \beta^{-1}(X) = \beta^{-1}(M \setminus X)$. Together with the previously defined G-action on $\beta^{-1}(X)$ we get a proper C^{k-1} action of G on W, intertwined by β with the G-action on M.

In $W \setminus \beta^{-1}(X)$ we obviously get the same orbit types as in $M \setminus X$, but for $l \in \mathbf{P}((\mathrm{N}\, X)_x)$ we have $G_l = \{ g \in G_x \mid \mathrm{T}_x\, A(g)v = v \}$, for any $v \in (\mathrm{T}_x\, X)^{\perp}$ such that $v \neq 0$ and $v + \mathrm{T}_x\, X \in l$. Here we take the orthogonal complement with respect to any G_x-invariant inner product in $\mathrm{T}_x\, M$. Also, $\mathfrak{g}_l = \{ X \in \mathfrak{g}_x \mid \alpha_x(X)v = 0 \}$, for any v as above. In view of the description preceding Proposition 2.7.1 the latter equation leads to:

(2.9.1) Proposition. *Let τ be a* **minimal** *nonmaximal element of the directed graph T' of the connected components of the infinitesimal types in $G\backslash M$ of the proper C^k action of G on M. Let $X = \pi^{-1}(\tau)$ be the corresponding closed G-invariant C^k submanifold of M and $\beta\colon W \to M$ the blowup of M along X. Then the directed graph of the connected components of the infinitesimal types in $G\backslash W$ is equal to $T' \setminus \{\tau\}$. The one of $G\backslash\beta^{-1}(M)$ is equal to the set of the $\sigma \in T'$ such that $\sigma \neq \tau$, $\sigma \to \tau$, or $\tau \subset \sigma^{\mathrm{cl}}$.*

We also can blow up M simultaneously along all $\pi^{-1}(\tau)$, with τ minimal and nonmaximal in T'. If l denotes the maximal length of a chain in T', then, if $k \geq l$, after $l-1$ such blowups we end up with a manifold V, a proper C^{k-l+1} map $\delta\colon V \to M$, and a proper C^{k-l+1}-action of G on V such that:

$$\delta|_{\delta^{-1}(M^{\mathrm{reg}})}\colon \delta^{-1}(M^{\mathrm{reg}}) \to M^{\mathrm{reg}} \text{ is an analytic diffeomorphism;} \tag{2.9.3}$$

$$\delta^{-1}(M \setminus M^{\mathrm{reg}}) \text{ is a union of disjoint closed } \mathrm{C}^{k-l+1} \text{ submanifolds of positive codimension in } V; \tag{2.9.4}$$

(2.9.5) *V consists entirely of regular points for the action of G on V.*

In analogy with the corresponding definition in Algebraic Geometry a mapping $\delta\colon V \to M$ as in Eq. (2.9.3–4) can be called a (real) *desingularization*. However, it should be emphasized that here the desingularization has been obtained by successive application of blowups along closed smooth submanifolds; and this gives a much better control than in the case of an arbitrary desingularization.

Applying now Proposition 2.8.7.(ii) to the action of G on V, we get, for any $x \in V$, that we have a G-equivariant diffeomorphism:

$$G/(G_x)^\circ \times_{\mathrm{N}_G(\mathfrak{g}_x)/(G_x)^\circ} V^{\mathfrak{g}_x} \xrightarrow{\sim} V;$$

and the $\mathrm{N}_G(\mathfrak{g}_x)/(G_x)^\circ$-orbit structure on $V^{\mathfrak{g}_x}$ can be identified with the G-orbit structure on V. In particular, because $V^{\mathfrak{g}_x}$ consists of regular points for the $\mathrm{N}_G(\mathfrak{g}_x)/(G_x)^\circ$-action, the stabilizer groups in $\mathrm{N}_G(\mathfrak{g}_x)/(G_x)^\circ$ of all $y \in V^{\mathfrak{g}_x}$ are finite. Hence we obtain a quotient:

$$(\mathrm{N}_G(\mathfrak{g}_x)/(G_x)^\circ)\backslash V^{\mathfrak{g}_x} \cong G\backslash V$$

that has the structure of a V-*manifold* in the sense of Satake [1956] (at least, if there are no codimension-one orbit types). Further $G\backslash V$ is a desingularization of $G\backslash M$ in a straightforward way.

We could also blow up M along $X = \pi^{-1}(\tau)$, where τ is a minimal, nonmaximal element of the directed graph $\mathcal{T}$ of the connected components of the orbit types in $G\backslash M$. Then the directed graph of $G\backslash W$ is equal to $\mathcal{T} \setminus \{\tau\}$, unless there exist a $v \in (\mathrm{T}_x X)^\perp$, $v \neq 0$, and a $g \in G_x$ such that $\mathrm{T}_x A(g)\, v = -v$. In the latter case the directed graph of $G\backslash W$ is equal to

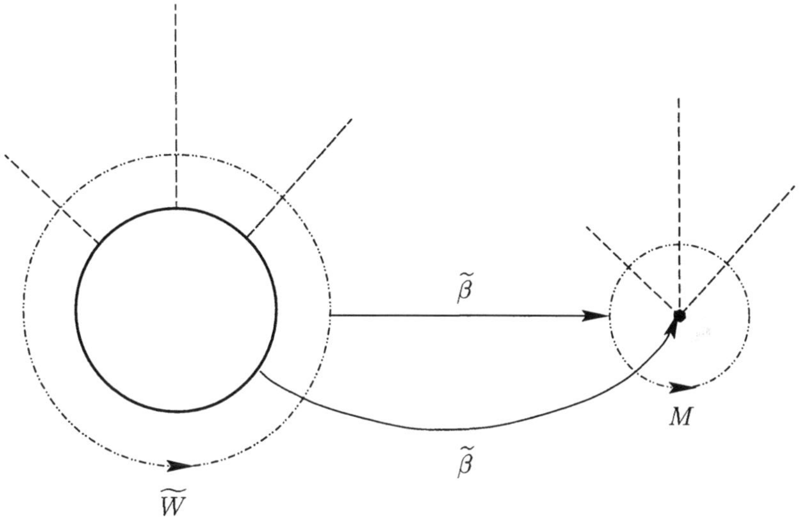

Fig. 2.9.2.

$\mathcal{T}$ with τ replaced by some other element, and no improvement has been obtained. If such an obstruction never occurs, then blowing up repeatedly along the minimal types, in many cases leads, after finitely many steps, to a desingularization $\delta \colon V \to M$ with a proper action of G on V such that $V^{\text{princ}} = V$. According to Theorem 2.6.7 we then have, for any $x \in V$, that the action of $\mathrm{N}(G_x)/G_x$ on V^{G_x} is proper and free; its orbit space then is a smooth manifold which can be identified with $G \backslash V$. Here and in the sequel we assume for simplicity that $G \backslash M$ is connected. The mapping $\delta \colon V \to M$ induces a desingularization $G \backslash V \to G \backslash M$, which now really deserves its name because $G \backslash V$ is smooth whereas $G \backslash M$ usually will have singularities.

If the "obstruction of antipodal action of G_x" occurs, then it looks natural to replace the blowing up with projective spaces by "blowing up along X with spheres". One way to describe this is to replace, in the beginning of this section, $\mathbf{P}(E)$ by the unit sphere:

$$\mathbf{S}(E) = \{ v \in E \mid < v, v > = 1 \} \tag{2.9.6}$$

with respect to any inner product in E, and the set W, and the mapping β, respectively, by:

$$\widetilde{W} = X \times \mathbf{S}(\mathbf{R}^q) \times [0, \epsilon[, \quad \widetilde{\beta}(y, s, r) = (y, rs), \tag{2.9.7}$$

(in local coordinates).

The resulting blow up is a manifold $\widetilde{W}$ with boundary and a C^{k-1} map $\widetilde{\beta}\colon \widetilde{W} \to M$ which is an analytic diffeomorphism: $\widetilde{W} \setminus \widetilde{\beta}^{-1}(M) \to M \setminus X$, whereas $\widetilde{\beta}^{-1}(M) = \partial\widetilde{W} = \mathbf{S}(\mathrm{N}\,X)$ and $\widetilde{\beta}\colon \widetilde{\beta}^{-1}(M) \to X$ is identified with the projection from the normal sphere bundle $\mathbf{S}(\mathrm{N}\,X)$ onto X. Note also that there is a natural map $\widetilde{W} \to W$ which is a two-fold covering: $\widetilde{\beta}^{-1}(M) \to \beta^{-1}(M)$.

At the price of having to allow manifolds with boundary, the directed graph of the G-action on $\widetilde{W}$ now is equal to $T \setminus \{\tau\}$. At a later stage the obstruction of antipodal actions of isotropy groups could occur at the boundary; in this case the blowup with spheres leads to a "manifold with corners" (cf. Borel and Serre [1973] for the formal definition and some applications of this concept.) At the end of the process one may get corners at which a number of smooth boundary manifolds meet in general position. So the general result (for $G\backslash M$ connected) is that after finitely many successive blowups along minimal types, blowing up with spheres only in the case of antipodal actions of isotropy groups, one often ends up with a manifold V with corners whereupon G acts properly and freely. This then leads to a principal fiber bundle over an orbit space $G\backslash V$ which also is a manifold with corners. The desingularization $\delta\colon V \to M$ induces therefore a desingularization $G\backslash V \to G\backslash M$ of the orbit space $G\backslash M$ by a manifold with corners.

It goes without saying that the blowups of this section, apart from their intrinsic conceptual interest, especially are useful in proving properties that are "invariant under blowing up".

2.10 Exercises

(2.1) Exercise. Let $k = \mathbf{R}$ or $\mathbf{C}$. Verify that the usual action of the matrix group $G = \mathbf{SL}(2,k)$ on k^2 induces an action on the projective line over k which for $x \in k$ is given by:

$$\begin{pmatrix} a_{11} & a_{12} \\ a_{21} & a_{22} \end{pmatrix} \cdot x = \frac{a_{11}x + a_{12}}{a_{21}x + a_{22}}.$$

That is, the elements of G act as linear fractional transformations on k. The infinitesimal action of $X \in \mathfrak{g}$ on k is given by $X(x) = \frac{d}{dt}\big|_{t=0} \exp tX \cdot x$. Set:

$$X = \begin{pmatrix} 0 & 1 \\ 0 & 0 \end{pmatrix}, \quad H = \begin{pmatrix} 1 & 0 \\ 0 & -1 \end{pmatrix}, \quad Y = \begin{pmatrix} 0 & 0 \\ 1 & 0 \end{pmatrix} \in \mathfrak{g}.$$

Prove $[H,X] = 2X$, $[H,Y] = -2Y$ and $[X,Y] = H$, and furthermore:

$$X(x) = 1, \qquad H(x) = 2x, \qquad Y(x) = -x^2, \quad \text{for } x \in k.$$

Prove that the mapping from $\mathfrak{g}$ to the Lie algebra of vector fields on k which sends $X \in \mathfrak{g}$ to the vector field $x \mapsto X(x)$ on k, in fact, is a homomorphism of Lie algebras. (For $k = \mathbf{R}$, it was shown by Lie that the Lie algebra of vector fields on $\mathbf{R}$ given above, and its subalgebras generated by the vector fields X, and X and H, respectively, are precisely all the finite-dimensional Lie algebras of vector fields on $\mathbf{R}$ up to local similarity.)

(2.2) Exercise. Let P_n be the linear space of homogeneous polynomials of degree n in two variables z_1 and z_2. Verify that we have an action of the Lie group $G = \mathbf{SL}(2, \mathbf{C})$ on P_n if we set:

$$(A \cdot f)(z) = f(zA), \text{ for } A = \begin{pmatrix} a_{11} & a_{12} \\ a_{21} & a_{22} \end{pmatrix} \in G, \quad f \in P_n, \quad z = (z_1, z_2).$$

Here $zA = (a_{11}z_1 + a_{21}z_2, a_{12}z_1 + a_{22}z_2)$ is defined through the usual matrix multiplication. The infinitesimal action of $X \in \mathfrak{g}$ on P_n is the linear operator $f \mapsto X \cdot f \colon P_n \to P_n$ given by:

$$(X \cdot f)(z) = \left.\frac{d}{dt}\right|_{t=0} (\exp tX \cdot f)(z) = \left.\frac{d}{dt}\right|_{t=0} f(z \exp tX).$$

Define X, Y and $H \in \mathfrak{g}$ as in Exercise 2.1. Prove:

$$X \cdot f = z_1 \frac{\partial f}{\partial z_2}, \quad Y \cdot f = z_2 \frac{\partial f}{\partial z_1}(z), \quad H \cdot f = z_1 \frac{\partial f}{\partial z_1} - z_2 \frac{\partial f}{\partial z_2}.$$

Determine the matrices of the nilpotent operators X and Y and of the semisimple H in $\mathbf{L}(P_n, P_n)$ with respect to the basis $\{\, z \mapsto \binom{n}{j} z_1^{n-j} z_2^j \mid 0 \le j \le n \,\}$ for P_n.

(2.3) Exercise. Let A be a C^k action ($k \ge 1$) of a connected Lie group G on a connected manifold M. Show that the action A is transitive if and only if the infinitesimal actions $\alpha_x = \mathrm{T}_1 A_x \colon \mathfrak{g} \to \mathrm{T}_x M$ are surjective for every $x \in M$. **Hint**: Note that every G-orbit is open in M if the condition is satisfied.

(2.4) Exercise. (i) Suppose we have a proper and free C^k action of a Lie group H on a C^k manifold X, and let V be a finite-dimensional vector space on which H acts by linear transformations. Consider the associated vector bundle $X \times_H V = E \to M = X/H$ with fiber V. Assume we have a C^k action of a Lie group G on X which commutes with the action of H on X, and consider the induced action of G on E, and on M. Denote by $\Gamma(E)$ the (generally speaking, infinite-dimensional) linear space of all C^k sections $s \colon M \to E$, and by $\Gamma(E)^G$ the linear subspace of sections which are invariant under the action of G on $\Gamma(E)$ given by $(g \cdot s)(x) = g \cdot s(g^{-1}x)$, for $g \in G$ and $x \in M$. Now let $x \in M$ be arbitrary but fixed, let $E_x \simeq V$ be the fiber over X, and define $\mathrm{ev} \colon \Gamma(E) \to V$ by $\mathrm{ev}(s) = s(x)$. Verify that ev is H-equivariant and that $\mathrm{ev} \colon \Gamma(E)^G \to V^H$ is an isomorphism of vector spaces.

(ii) If $E' \to M$ is another associated vector bundle with fiber V', then the vector space $\mathrm{Hom}_G(E', E)$ consisting of morphisms of associated vector bundles which are linear on the fibers, is linearly isomorphic with $\mathrm{Hom}_H(V', V)$.
(iii) Finally, suppose H is a closed subgroup of G and $X = G$, and let G, and H, act on G by left, and right multiplications, respectively. Then to the action of H on V there corresponds a linear action of G on the space of sections $\Gamma(E)$; this is said to be the representation of G induced by the representation of the subgroup H. Let W be an arbitrary vector space on which G acts by linear transformations. Now apply the result in (ii) with E' equal to the trivial bundle $M \times W$ to deduce the following version of Frobenius reciprocity (see Theorem 4.7.1):

$$\mathrm{Hom}_G(W, \Gamma(E)) = \mathrm{Hom}_H(W, V).$$

(2.5) Exercise. Show that there is an infinite number of orbit types for the action in $\mathbf{R}^3$ of the Lie group consisting of the matrices

$$\begin{pmatrix} 1 & a & b \\ 0 & 1 & 0 \\ 0 & 0 & 1 \end{pmatrix}, \quad \text{for } a, b \in \mathbf{R}.$$

(2.6) Exercise. Suppose we have a proper C^k action of a Lie group G on a C^k manifold M, and that all orbits of G in M are isomorphic, say to G/H. Then the canonical projection $M \to G\backslash M$ is the projection of a differentiable fiber bundle with fiber G/H and structure group $\mathrm{N}_G(H)/H$.

(2.7) Exercise. Let G be a Lie group and $H \subset N \subset G$ two closed subgroups where H is normal in N. Prove that the natural mapping $G/H \to G/N$ is the projection of an analytic principal bundle, with structure group N/H, whose right action on G/H is given by $gH \cdot nH = gnH$, for $g \in G$ and $n \in N$. Now consider the analytic G-homogeneous space $M = G/H$, and set $N = \mathrm{N}_G(H)$, the normalizer of H in G. Prove that $\mathrm{Aut}_G\, M$, the group of all G-equivariant analytic diffeomorphisms of M, is isomorphic with N/H. Show that the orbit of $x \in M$ under the action of $\mathrm{Aut}_G\, M$ equals $\{\, y \in M \mid G_y = G_x \,\}$.

(2.8) Exercise. Suppose we have a proper C^k action of a Lie group G on a C^k manifold M, and let $x \in M$. Show that $M_x^{\sim}$ and $M_x^{\sim} \cap M^{G_x}$ are locally closed C^k submanifolds of M. Show that there is a unique structure of a C^k manifold on $G\backslash M_{G.x}^{\sim}$ for which $\pi\colon M_x^{\sim} \cap M^{G_x} \to G\backslash M_{G.x}^{\sim}$ is a principal fibration with structure group $\mathrm{N}(G_x)/G_x$. Prove that the mapping $(g, x) \mapsto g \cdot x\colon G \times (M_x^{\sim} \cap M^{G_x}) \to M$ induces a G-equivariant C^k diffeomorphism $G/G_x \times_{\mathrm{N}(G_x)/G_x} (M_x^{\sim} \cap M^{G_x}) \to M_x^{\sim}$.

2.11 Notes

The concept of associated fiber bundles seems to be due to Serre [1958]. But Bredon [1972] calls $X \times_H Y$ a "twisted product" if one drops the condition that $X \to X/H$ is a principal fibration. The last sentence in Proposition 2.7.2 is borrowed from Jänich [1968].

The theory in Section 2.9 is a real version of the desingularizations introduced in a complex-algebraic setting by Kirwan [1985], in order to be able to apply her results in Kirwan [1984].

References for Chapter Two

Birkhoff, G. (1927) Dynamical Systems. Amer. Math. Soc. Colloq. Publ., Vol. IX. Amer. Math. Soc., New York

Borel, A., Serre, J-P. (1973) Corners and arithmetic groups. Comment. Math. Helv. **48** (1973) 436-491 = Œuvres Collected Papers, Vol. III, pp. 244-299. Springer-Verlag, Berlin Heidelberg New York Tokyo 1983

Bourbaki, N. (1951) Éléments de Mathématique. Topologie Générale, Chap. 1 et 2. Hermann, Paris

Bredon, G.E. (1972) Introduction to Compact Transformation Groups. Academic Press, New York London

Guillemin, V., Pollack, A. (1974) Differential Topology. Prentice-Hall, Englewood Cliffs

Hörmander, L. (1983) The Analysis of Linear Partial Differential Operators, Vol. II. Springer-Verlag, Berlin Heidelberg New York Tokyo

Jänich, K. (1968) Differenzierbare G-Mannigfaltigkeiten. Lecture Notes in Math. **59**. Springer-Verlag, Berlin Heidelberg New York

Kirwan, F.C. (1984) Cohomology of Quotients in Symplectic and Algebraic Geometry. Math. Notes No. 31. Princeton University Press, Princeton

Kirwan, F.C. (1985) Partial desingularisations of quotients of nonsingular varieties and their Betti numbers. Ann. of Math. **122** (1985) 41-85

Łojasiewicz, S. (1965) Ensembles Semi-Analytiques. Mimeographed notes. Inst. Hautes Études Sci., Bures-sur-Yvette

Satake, I. (1956) On a generalization of the notion of manifold. Proc. Nat. Acad. Sci. U. S. A. **42** (1956) 359-363

Schwarz, G.W. (1975) Smooth functions invariant under the action of a compact Lie group. Topology **14** (1975) 63-68

Serre, J-P. (1958) Espaces fibrés algébriques. In: Séminaire C. Chevalley: Anneaux de Chow et Applications. Secrétariat Mathématique, Paris

Chapter 3

Compact Lie Groups

3.0 Introduction

Throughout this chapter, G will be a compact Lie group. It acts on itself by means of conjugation; we shall write:

$$\mathbf{Ad}\, g(x) = gxg^{-1} \quad \big((g, x) \in G \times G\big). \tag{3.0.1}$$

Because G is compact, this action is proper. Applying the general principles of Chapter 2, we shall obtain a detailed description of the structure of G. This includes the basic theorems about connected, compact Lie groups, viz.:

(i) The **maximal torus theorem**, stating that every element of G is conjugate to an element of a maximal torus. Also, all maximal tori are conjugate to each other and equal to the centralizer of an element of principal orbit type. See Theorem 3.7.1.

(ii) The **Weyl covering theorem** and the **Weyl integration theorem**, concerned with the mapping Γ from $G/T \times T$ onto G, defined by the action of conjugation. See Theorems 3.7.2 and 3.14.1.

(iii) The computation of the **fundamental group** of G. See, for instance, Proposition 3.11.1.

In several respects, the action of conjugation of a compact Lie group on itself enjoys special properties when compared to general actions of compact Lie groups. Sometimes therefore one could give direct, easy proofs, without having to refer to the general theory. Still, even in these cases, we found it instructive to treat the structure theory of compact Lie groups in the framework of general proper group actions.

Another disadvantage of proceeding from the more general theory to this special situation, is that the same principle may appear several times under slightly different guises, which may make orientation more difficult. For this reason it might be helpful to start by browsing through this chapter, in order to see how the final results, say for connected, compact Lie groups, look like.

3.1 Centralizers

For the action by conjugation of a compact Lie group G on itself, the general theory of proper actions of Chapter 2 immediately leads to a number of interesting conclusions.

(a) (See Sections 2.1 and 2.5.) The orbit space $(\mathbf{Ad}\, G)\backslash G$, the space of conjugacy classes in G, is a compact, Hausdorff topological space. Functions on it are identified with the functions on G that are constant on the conjugacy classes, that is:

$$f(gxg^{-1}) = f(x) \quad (x, g \in G). \tag{3.1.1}$$

These are also called the *class functions* on G. The space of C^k functions $(1 \leq k \leq \omega)$ on the orbit space is defined as the space of C^k class functions on G.

(b) In the Sections 2.2–4, the basic tool is the concept of a *slice* at any point x, a submanifold S through x whose tangent space is complementary to the tangent space of the orbit. It is also invariant under the action of the stabilizer group G_x of x, and the action of G near x can be completely described in terms of G_x and the linear action of G_x on $\mathrm{T}_x\, S$.

For the action by conjugation of G on itself, the stabilizer or isotropy group of $x \in G$ is equal to the *centralizer*:

$$G_x = \mathrm{Z}_G(x) := \{\, g \in G \mid gx = xg \,\} \tag{3.1.2}$$

of x in G. Then G_x is a closed, hence Lie subgroup of G, with Lie algebra equal to:

$$\mathrm{T}_1(G_x) = \mathfrak{g}_x := \mathfrak{z}_{\mathfrak{g}}(x) := \ker(\mathrm{Ad}\, x - \mathrm{I}), \tag{3.1.3}$$

the *centralizer of x in* $\mathfrak{g}$. Indeed, $Y \in \mathfrak{g}$ belongs to the Lie algebra of G_x if and only if:

$$\exp tY = x(\exp tY)x^{-1} = \exp \mathrm{Ad}x(tY) = \exp t\, \mathrm{Ad}\, x(Y), \quad \text{for all } t \in \mathbf{R};$$

and this is equivalent to $\mathrm{Ad}\, x(Y) = Y$.

However, because G acts on itself, G_x can also be seen as a subset of the manifold on which G acts. As such, it is a closed real-analytic submanifold, which passes through the point x, that is:

$$x \in G_x. \tag{3.1.4}$$

The consequences of this are quite remarkable.

(3.1.1) Proposition.

(i) *Near x, the set G_x is a slice at x for the action by conjugation, and locally it is the only one. The tangent space to the orbit is given by:*

$$\alpha_x(\mathfrak{g}) := \mathrm{T}_x(\mathbf{Ad}\, G(x)) = \mathrm{T}_1\, \mathbf{R}(x)\big(\mathrm{im}(\mathrm{Ad}\, x - \mathrm{I})\big), \tag{3.1.5}$$

and the tangent space to the slice by:

$$\mathrm{T}_x(G_x) = \mathrm{T}_1\, \mathbf{R}(x)(\mathfrak{g}_x) = \mathrm{T}_1\, \mathbf{R}(x)\big(\ker(\mathrm{Ad}\, x - \mathrm{I})\big). \tag{3.1.6}$$

(ii) *The logarithmic chart at x intertwines the action of G_x in G, near x, with the adjoint representation of G_x in $\mathfrak{g}$, near the origin.*

Proof. The mapping $g \mapsto \mathbf{R}(x)^{-1}(\mathbf{Ad}\, g(x)) = gxg^{-1}x^{-1}$ is real-analytic, maps 1 to 1, and has tangent mapping at 1 equal to $\mathrm{I} - \mathrm{Ad}\, x$; and this proves (3.1.5). On the other hand, (3.1.6) follows from (3.1.3).

If $Y \neq 0$, $Y \in \mathrm{im}(\mathrm{Ad}\, x - \mathrm{I}) \cap \ker(\mathrm{Ad}\, x - \mathrm{I})$, then $Y = (\mathrm{Ad}\, x - \mathrm{I})(Z)$, and $(\mathrm{Ad}\, x - \mathrm{I})^2(Z) = 0$. It follows that:

$$(\mathrm{Ad}\, x)^k(Z) = (\mathrm{I} + (\mathrm{Ad}\, x - \mathrm{I}))^k(Z) = (\mathrm{I} + k(\mathrm{Ad}\, x - \mathrm{I}))(Z) = Z + kY.$$

This is unbounded as $k \to \infty$, in contradiction with the fact that the $(\mathrm{Ad}\, x)^k$ belong to the compact subgroup $\mathrm{Ad}\, G_x$ of $\mathbf{GL}(\mathfrak{g})$. So the image and the kernel of $\mathrm{Ad}\, x - \mathrm{I}$ have zero intersection, and because these spaces have complementary dimensions, $\mathfrak{g}$ is equal to their direct sum. Applying the linear isomorphism $\mathrm{T}_1\, \mathbf{R}(x) \colon \mathfrak{g} \to \mathrm{T}_x\, G$, we get that $\mathrm{T}_x\, G$ is equal to the direct sum of $\mathrm{T}_x(\mathbf{Ad}\, G(x))$ and $\mathrm{T}_x(G_x))$.

Because G_x, being a group, is **Ad**-invariant, Theorem 2.3.3 shows that there is an open neighborhood U of x in G such that $G_x \cap U$ is a real-analytic slice at x for the action by conjugation.

Now let $\log_x$ be the logarithmic chart at x that is the inverse of the analytic diffeomorphism $X \mapsto (\exp X)x$, defined on an open, $\mathrm{Ad}\, G_x$-invariant neighborhood V of 0 in $\mathfrak{g}$, and mapping onto an open neighborhood U of x in G. For $g \in G_x$,

$$g(\exp X\, x)g^{-1} = (g \exp X g^{-1})\,(gxg^{-1}) = \exp \mathrm{Ad}\, g(X)\, x;$$

so $\log_x$ intertwines the action of G_x in U with the adjoint action of G_x in V. In other words, here the Bochner linearization of Theorem 2.2.1 is realized by the logarithmic chart.

We now prove the uniqueness in (i). In the logarithmic chart, the slice G_x is equal to $\mathfrak{g}_x \cap V$. With the complementary subspace:

$$\mathfrak{q} = \mathrm{im}(\mathrm{Ad}\, x - \mathrm{I}) = \mathrm{T}_1\, \mathbf{R}(x)^{-1}\big(\mathrm{T}_x(\mathbf{Ad}\, G(x))\big),$$

any C^1 slice S at x is locally of the form $\{\, Y + q(Y) \mid Y \in \mathfrak{g}_x \cap V_0 \,\}$, where V_0 is an open neighborhood of 0 in V, and q is a C^1 mapping from $\mathfrak{g}_x \cap V_0$ to $\mathfrak{q}$ such that $q(0) = 0$.

The G_x-invariance of S gives that, for every $Y \in \mathfrak{g}_x \cap V_0$, $g \in G_x$,

$$\operatorname{Ad} g(Y) + \operatorname{Ad} g(q(Y)) = \operatorname{Ad} g(Y + q(Y)) = Z + q(Z),$$

for some $Z \in \mathfrak{g}_x \cap V_0$. As $\mathfrak{g}_x$ and $\mathfrak{q}$ are $\operatorname{Ad} G_x$-invariant, and $\mathfrak{g}_x \cap \mathfrak{q} = 0$, this means $\operatorname{Ad} g(Y) = Z$ and $\operatorname{Ad} g(q(Y)) = q(Z)$; or $\operatorname{Ad} g(q(Y)) = q(\operatorname{Ad} g(Y))$, for all $Y \in \mathfrak{g}_x \cap V_0$, $g \in G_x$. Applying this to $g = x \in G_x$ (!), we see that $\operatorname{Ad} x(q(Y)) = q(Y)$, or $q(Y) \in \mathfrak{g}_x \cap \mathfrak{q} = 0$. This shows that $q = 0$, or $S = \mathfrak{g}_x \cap V_0$. □

Combining now the theory of Sections 2.2-4 with Proposition 3.1.1, we get the following conclusions. In an open, conjugacy invariant neighborhood of x in G, the action of conjugation is analytically equivalent to the G-action on $G \times_{G_x} U$, where U is an $\operatorname{Ad} G_x$-invariant open neighborhood of 0 in $\mathfrak{g}_x$, and the action of G_x on U is the adjoint one. The C^k class functions near x are identified (in the logarithmic chart and via restriction to $\mathfrak{g}_x$) with the $\operatorname{Ad} G_x$-invariant functions of class C^k on $\mathfrak{g}_x$ near 0, or with the G_x-class functions of class C^k on G_x, near the identity element of G_x.

In the sequel, we shall not only use centralizers in G and $\mathfrak{g}$, respectively, of single elements of G, but also of subgroups. Furthermore, we shall need the centralizers in G and $\mathfrak{g}$, respectively, of elements in $\mathfrak{g}$. Below we have collected the definitions of all the combinations that occur later on.

For any subset H of any Lie group G,

$$G_H := \mathrm{Z}_G(H) := \bigcap_{h \in H} G_h = \{\, g \in G \mid gh = hg, \text{ for all } h \in H \,\} \tag{3.1.7}$$

is called the *centralizer of* H *in* G. Here H is regarded as a subset of the manifold on which G acts. However, considering H as a subset of the group which acts, $\mathrm{Z}_G(H)$ can also be recognized as the set of fixed points of H, or

$$G_H = \mathrm{Z}_G(H) = G^H, \tag{3.1.8}$$

in the notation of (2.6.2).

$\mathrm{Z}_G(H)$ is a closed, hence Lie subgroup of G, with Lie algebra equal to:

$$\begin{aligned} \mathfrak{z}_{\mathfrak{g}}(H) :&= \bigcap_{h \in H} \mathfrak{g}_h = \{\, X \in \mathfrak{g} \mid \operatorname{Ad} h(X) = X, \text{ for all } h \in H \,\} \\ &= \mathfrak{g}^{\operatorname{Ad} H}, \end{aligned} \tag{3.1.9}$$

called the *centralizer of* H *in* $\mathfrak{g}$.

If $H = G$, then:

$$\mathrm{Z}(G) := \mathrm{Z}_G(G) = \{\, g \in G \mid gx = xg, \text{ for all } x \in G \,\} \tag{3.1.10}$$

is called the *center* of G. Its Lie algebra is equal to:

$$\mathfrak{z}(G) := \mathfrak{z}_{\mathfrak{g}}(G), \tag{3.1.11}$$

which is called the *center of G in $\mathfrak{g}$*.

Note that if H is a **connected** Lie subgroup of G with Lie algebra equal to $\mathfrak{h}$, then:

$$\mathrm{Z}_G(H) = \mathrm{Z}_G(\mathfrak{h}) := \{\, g \in G \mid \mathrm{Ad}\, g\,(Y) = Y, \text{ for all } Y \in \mathfrak{h}\,\}, \tag{3.1.12}$$

the *centralizer of $\mathfrak{h}$ in G*. Furthermore,

$$\mathfrak{z}_{\mathfrak{g}}(H) = \mathfrak{z}_{\mathfrak{g}}(\mathfrak{h}) := \{\, X \in \mathfrak{g} \mid [\, X, Y\,] = 0, \text{ for all } Y \in \mathfrak{h}\,\}, \tag{3.1.13}$$

is said to be the *centralizer of $\mathfrak{h}$ in $\mathfrak{g}$*.

If G itself is connected, then:

$$\mathrm{Z}(G) = \ker \mathrm{Ad}, \quad \text{with } \mathrm{Ad}\colon G \to \mathbf{GL}(\mathfrak{g}), \tag{3.1.14}$$

and the Lie algebra of the center $\mathrm{Z}(G)$ of G is equal to:

$$\begin{aligned} \mathfrak{z} := \mathfrak{z}(\mathfrak{g}) := \{\, X \in \mathfrak{g} \mid [\, X, Y\,] = 0, \text{ for all } Y \in \mathfrak{g}\,\} = \ker \mathrm{ad}, \\ \text{with } \mathrm{ad}\colon \mathfrak{g} \to \mathbf{L}(\mathfrak{g}, \mathfrak{g}), \end{aligned} \tag{3.1.15}$$

the *center* of $\mathfrak{g}$, which we have met before in (1.14.15).

(c) G has a Riemannian structure which is left and right invariant, and therefore also conjugacy invariant. These Riemannian structures on G are real-analytic, and the mapping $\beta \mapsto \beta_1$ is a bijection from the space of these onto the space of $\mathrm{Ad}\, G$-invariant inner products on $\mathfrak{g}$. The latter are obtained by averaging any inner product on $\mathfrak{g}$ over $\mathrm{Ad}\, G$. The proof is immediate, although one could also argue that it coincides with the proof of Proposition 2.5.2, when applied to the left-right action of $G \times G$ on G.

(d) In Section 2.6 the central theme is the study of the *orbit types*:

$$G_x^{\sim} = \{\, y \in G \mid G_y \text{ is conjugate to } G_x \text{ in } G\,\}.$$

From the description of the action of conjugation near x after Proposition 3.1.1, it follows that this action is completely determined, up to equivalence, by the group G_x (which determines the adjoint action of G_x on its own Lie algebra $\mathfrak{g}_x$). It follows that **the orbit types $G_x^{\sim}$ are equal to the local action types $G_x^{\approx}$** of Section 2.6. In the further description of the local action types, the fixed point set for the action of G_x, and the normalizer $\mathrm{N}(G_x)$ of G_x play an important part. About these we have some more consequences of (3.1.4):

(3.1.2) Lemma.

(i) *For every $x \in G$, the fixed point set $\mathrm{Z}_G(G_x)$ of G_x in G is contained in G_x, so it is equal to the center $\mathrm{Z}(G_x)$ of the group G_x.*

(ii) *The normalizer $\mathrm{N}(G_x)$ of G_x in G has the same dimension as G_x, so $\mathrm{N}(G_x)/G_x$ is a finite group.*

Proof. (i) $x \in G_x$ implies that $\mathrm{Z}_G(G_x) \subset \mathrm{Z}_G(\{x\}) = G_x$.

(ii) An element $g \in \mathrm{N}(G_x)$ close to 1 maps $x \in G_x$ to an element in G_x close to x, because of the continuity of the action. Because G_x, near x, is a slice at x by Proposition 3.1.1.(i), we conclude that $g \in G_x$. This proves that $\dim \mathrm{N}(G_x) = \dim G_x$. Because $\mathrm{N}(G_x)$, as a closed subgroup of G, is compact, $\mathrm{N}(G_x)/G_x$ is compact and discrete, hence finite. □

Theorem 2.6.7 now leads to the following conclusions. To begin with, the orbit type $G_x^\sim$ is a locally closed, real-analytic submanifold of G, of codimension equal to the dimension of $G_x/\mathrm{Z}(G_x)$.

Furthermore, **the fixed point set $\mathrm{Z}(G_x)$ of G_x in G is an Abelian group** (a compact Lie subgroup of G_x, hence of G). Now $G_x^\sim \cap \mathrm{Z}(G_x)$ is an open subset of $\mathrm{Z}(G_x)$, containing x. The finite group $\mathrm{N}(G_x)/G_x$ acts freely on $G_x^\sim \cap \mathrm{Z}(G_x)$, and the quotient is naturally identified with $(\mathbf{Ad}\, G)\backslash G_x^\sim$, the orbit type in the orbit space. In this way the latter gets the structure of a real-analytic manifold, of dimension equal to $\dim \mathrm{Z}(G_x)$.

The action of G induces a G-equivariant analytic diffeomorphism:

$$G/G_x \times_{\mathrm{N}(G_x)/G_x} (G_x^\sim \cap \mathrm{Z}(G_x)) \xrightarrow{\sim} G_x^\sim .$$

The partitioning of $G_x^\sim$ into conjugacy classes defines a real-analytic fibration: $G_x^\sim \to (\mathbf{Ad}\, G)\backslash G_x^\sim$, with fiber equal to G/G_x.

(e) From Section 2.7, combined with the compactness of the manifold G on which G acts, we get that there are only finitely many orbit types, each of these having only finitely many connected components. These constitute a Whitney stratification in G.

If $y \in (G_x^\sim)^{\mathrm{cl}} \setminus G_x^\sim$, a "boundary point" of a stratum in $G_x^\sim$, then G_x is conjugate (in G) to a subgroup of G_y. Moreover, $\dim \mathrm{Z}(G_y) < \dim \mathrm{Z}(G_x)$, and therefore a fortiori:

$$\operatorname{codim} G_y^\sim = \dim G_y/\mathrm{Z}(G_y) > \dim G_x/\mathrm{Z}(G_x) = \operatorname{codim} G_x^\sim .$$

(For (f)–(h), see Section 2.8)

(f) By definition and by (d), $x \in G$ is of *principal orbit type* if $G_x^\sim$ is open in G. The set of these is denoted by G^{princ}. We have $x \in G^{\mathrm{princ}}$ if and only if $\dim \mathrm{Z}(G_x) = \dim G_x$, or $G_x/\mathrm{Z}(G_x)$ is a finite group, or **the adjoint representation of G_x on its Lie algebra $\mathfrak{g}_x$ is trivial**. This implies that $\mathfrak{g}_x$ is Abelian, and that $(G_x)^\circ = (\mathrm{Z}(G_x))^\circ$.

$G \setminus G^{\text{princ}}$ is a closed subset of G. It is equal to the disjoint union of finitely many locally closed, connected, real-analytic submanifolds of G (the connected components of the other orbit types), of codimension ≥ 1 in G.

In the union $'G$ of all connected components of G that meet a given conjugacy class, there is precisely one principal orbit type. In other words, the directed graph of orbit types in $'G$ is connected, and has a unique maximal element. In particular, there is only one principal orbit type in G°, the identity component of G. This implies that all G_x, for $x \in G^{\text{princ}} \cap G^{\circ}$, are conjugate to each other.

(g) An element x is said to be a *regular* element of G if all nearby orbits have the same dimension. That is, $\dim \mathfrak{g}_y = \dim \mathfrak{g}_x$, for all y near x in G; this in turn is equivalent to the condition that $\mathfrak{g}_x$ **is Abelian**. Indeed, let S be a slice at x, which is a neighborhood of x in G_x. If $y \in S$, then $g \in G_y$ implies that $\mathbf{Ad}\, g(y) = y \in S$, hence $g \in G_x$ (this is part of the definition of a slice). Or $G_y \subset G_x$; and in turn this implies that $\mathfrak{g}_y \subset \mathfrak{g}_x$. Because all elements in G near x are conjugate to an element of S, the conclusion is that x is regular if and only if $\mathfrak{g}_x \subset \mathfrak{g}_y$, for all $y \in G_x$ sufficiently close to x. If $y = zx$, this means that $\operatorname{Ad} z = \mathrm{I}$ on $\mathfrak{g}_x$ for all $z \in G_x$ near 1, which clearly is equivalent to the condition that $\mathfrak{g}_x$ is Abelian.

The set G^{reg} of regular elements in G is open. The complement $G^{\text{sing}} = G \setminus G^{\text{reg}}$, the set of *singular elements* in G, is equal to the disjoint union of finitely many connected, locally closed, real-analytic submanifolds of G of codimension ≥ 2. These are the *infinitesimal orbit types*, the partitioning of G into them is a coarsening of the partitioning into orbit types. They also define a Whitney stratification in G. In particular, $G^{\text{reg}} \cap C$ is connected for each connected component C of G. In Lemma 3.9.1 we shall see that, if G is connected, then the codimensions above are actually ≥ 3, which is the basic ingredient in the computation of the fundamental group of G.

The open set G^{reg} contains the open subset G^{princ}. The orbits in $G^{\text{reg}} \setminus G^{\text{princ}}$ are called *exceptional*; they have the same dimension as the nearby principal orbits. That is, if $x \in G^{\text{reg}} \setminus G^{\text{princ}}$, and $y \in G^{\text{princ}}$ belongs to the slice at x, then G_y is a subgroup of G_x of the same dimension; so these groups only differ by number of their connected components, but they have the same identity component. The injection $G_y \to G_x$ induces a finite covering $G/G_y \to G/G_x$, where G/G_y, and G/G_x, can be identified with the orbit through y, and x, respectively.

(h) Let $x \in G^{\text{princ}}$, write C for the connected component of x in $G^{\text{princ}} \cap \mathrm{Z}(G_x)$. Also, write $'G^{\text{princ}} = G^{\text{princ}} \cap {'G}$, where $'G$ is as in (f). Combining (d) and Corollary 2.8.6, it follows that every element of $'G^{\text{princ}} \cap \mathrm{Z}(G_x)$ is conjugate, by means of an element of $\mathrm{N}(G_x)$, to an element of C.

Furthermore, the group $\mathrm{N}(C) := \{g \in G \mid gCg^{-1} = C\}$ is an open subgroup of $\mathrm{N}(G_x)$ containing G_x. The finite group $\mathrm{N}(C)/G_x$ acts freely on C, the orbit space is in bijective correspondence with the set of conjugacy classes in $'G^{\mathrm{princ}}$. The action of G induces a G-equivariant analytic diffeomorphism: $G/G_x \times_{\mathrm{N}(C)/G_x} C \xrightarrow{\sim} {'G^{\mathrm{princ}}}$.

Finally, **each element of $'G$ is conjugate to an element of C^{cl}, the closure of C in G, which is contained in $\mathrm{Z}(G_x)$**

(3.1.3) Proposition. *Let G be connected, $x \in G^{\mathrm{reg}}$. Then:*

(i) $T := (G_x)^\circ$ is a **torus** *in G, that is, a connected, compact, Abelian subgroup of G.*

(ii) Each element of G is conjugate to an element of T, and actually to an element in the closure of any connected component C of $T \cap G^{\mathrm{princ}}$.

(iii) $x \in T$.

Proof. Assertion (i) follows from the description of regular elements in (g).

(ii) Let $y \in G^{\mathrm{princ}}$ be in the slice at x. From the description in (g) and (f) we get $\mathfrak{g}_x = \mathfrak{g}_y$; hence $T = (G_x)^\circ = (G_y)^\circ = \mathrm{Z}(G_y)^\circ$, because $G_y/\mathrm{Z}(G_y)$ is finite if y is of principal orbit type. Now let C be any connected component of $\mathrm{Z}(G_y) \cap G^{\mathrm{princ}}$. Then $1 \in G$ is conjugate to an element of C^{cl}, and observing $z = 1$ if z is conjugate to 1, we get that $1 \in C^{\mathrm{cl}}$. Because C^{cl} is a connected subset of $\mathrm{Z}(G_y)$, and contains 1, it is contained in the identity component T of $\mathrm{Z}(G_y)$.

(iii) Applying (ii) to x itself, we get an element $g \in G$ such that $x \in gTg^{-1}$. Now $g^{-1}xg \in \mathrm{Z}(G_x)$ already implies that $g \in \mathrm{N}(G_x)$, that is $\mathbf{Ad}\, g$ leaves G_x invariant. One may verify this directly, it is an instance of the general principle that for any group action we have that g^{-1} maps x to a fixed point of G_x if and only if $g \in \mathrm{N}(G_x)$. Now $g \in \mathrm{N}(G_x)$ in turn implies that $\mathbf{Ad}\, g$ leaves T invariant. Now we have $x \in \mathbf{Ad}\, g\,(T) = T$. □

Remark. In Corollary 3.3.2, we shall actually see that y is of principal orbit type in a connected, compact Lie group G, if and only if G_y is a torus.

(3.1.4) Corollary. *For any connected, compact Lie group G with Lie algebra $\mathfrak{g}$, the exponential mapping is surjective: $\mathfrak{g} \to G$.*

Proof. Take T as in Proposition 3.1.3. Then $T = \exp \mathfrak{t}$, if $\mathfrak{t}$ denotes the Lie algebra of T, cf. Theorem 1.12.1. It follows that every element y of G is of the form $y = g \exp X g^{-1} = \exp \mathrm{Ad}\, g\,(X)$, for some $g \in G$, $X \in \mathfrak{t}$. □

Remark. Let $\mathfrak{h}$ be a Lie subalgebra of $\mathfrak{g}$ which happens to be the Lie algebra of a **closed** subgroup H of G. Then $\exp \mathfrak{h}$ is a compact subset of G, and also a subgroup, because it is equal to H°. In particular, $\exp \mathfrak{g}_x = (G_x)^\circ$ for all $x \in G$. Note that for a general Lie subalgebra $\mathfrak{h}$ of the Lie algebra of a Lie group G, the subset $\exp \mathfrak{h}$ is neither closed in G, nor a subgroup of G.

3.2 The Adjoint Action

Analogous conclusions as in Section 3.1 can be drawn for the adjoint action Ad of G on its Lie algebra $\mathfrak{g}$. Actually the situation is somewhat simpler here, due to the fact that $\mathfrak{g}$ is a vector space, on which moreover G acts by means of linear transformations.

(a) The orbit space $(\operatorname{Ad} G)\backslash\mathfrak{g}$ is a Hausdorff topological space. If we delete the origin, it has a natural *conic structure*, that is, a proper and free action of the multiplicative group $\mathbf{R}_{>0}$. Functions on it are the $\operatorname{Ad} G$-invariant functions on $\mathfrak{g}$, they are called the *class functions on* $\mathfrak{g}$ *with respect to the group* G. Note that if G is connected, then $\operatorname{Ad} G$ is equal to the Lie subgroup of $\mathbf{GL}(\mathfrak{g})$ generated by $e^{\operatorname{ad}\mathfrak{g}}$, the *adjoint group* of $\mathfrak{g}$, which is denoted by $\operatorname{Ad}\mathfrak{g}$, cf. Example 1.10.4. In this case we simply talk about the *class functions of* $\mathfrak{g}$. On $\mathfrak{g}$ we can also talk about the polynomial class functions, in addition to their C^k counterparts. This can be used to turn $(\operatorname{Ad} G)\backslash\mathfrak{g}$ into a real affine algebraic variety, see Chapter 14.

The action of G by conjugation in a suitable conjugacy-invariant open neighborhood of 1 in G is equivalent to the adjoint action of G on a corresponding $\operatorname{Ad} G$-invariant open neighborhood of 0 in $\mathfrak{g}$, via the logarithmic chart. By means of multiplications by nonzero scalars, which commute with the adjoint action, the adjoint action can completely be identified with the conjugation action near 1 in G. However, in the other direction, the action of conjugation in G not always can be recovered completely from the adjoint representation, not even if G is connected. See the discussion in Section 3.4 of the examples of $\mathbf{SO}(3,\mathbf{R})$ and $\mathbf{SU}(2)$. The local equivalence implies that centralizers in G of elements in G near 1 are equal to those of elements in $\mathfrak{g}$ near 0. This is formulated somewhat more precisely in the following lemma; actually the compactness of G does not play any role in it.

(3.2.1) Lemma. *Let U be an $\operatorname{Ad} G$-invariant subset of $\mathfrak{g}$ on which the exponential mapping is injective. Then $G_X = G_{\exp X}$, for all $X \in U$. In particular, $G_{\exp X} = G_{tX}$, if $t \in \mathbf{R}\setminus\{0\}$ is sufficiently small.*

Proof. Let $X \in U$, $g \in G$. Then $g\exp X g^{-1} = \exp(\operatorname{Ad} g\,(X))$ is equal to $\exp X$, if and only if $\operatorname{Ad} g\,(X) = X$. The second statement follows from $G_X = G_{tX}$, if $t \in \mathbf{R}\setminus\{0\}$. □

(b) We now turn to the description of the slices at $X \in \mathfrak{g}$. The analogue of (3.1.4) for the adjoint representation is:

$$X \in \mathfrak{g}_X, \quad \text{or } [X, X] = 0. \tag{3.2.1}$$

See (1.1.20), one may also view this as a consequence of the existence of an Abelian Lie subgroup of G with Lie algebra equal to $\mathbf{R}\cdot X$, cf. Theorem 1.3.2.

In view of the remarks above about the relation between the adjoint action and the action of conjugation in G, Proposition 3.1.1 immediately implies:

(3.2.2) Proposition. *For every $X \in \mathfrak{g}$, we have $\mathrm{T}_X(\mathrm{Ad}\, G\,(X)) = \mathrm{im}(\mathrm{ad}\, X)$. Near X, the Lie subalgebra $\mathfrak{g}_X = \mathrm{T}_1(G_X) = \ker(\mathrm{ad}\, X)$ is a slice at X for the adjoint representation of G, and it is locally the only one.*

The theory of Sections 2.2–4 now leads to the following conclusions. In an open, $\mathrm{Ad}\, G$-invariant neighborhood of X in $\mathfrak{g}$, the adjoint action is analytically equivalent to the G-action on $G \times_{G_X} U$, where U is an $\mathrm{Ad}\, G_X$-invariant neighborhood of 0 in $\mathfrak{g}_X$, and the action of G_X on $\mathfrak{g}_X$ is the adjoint one. The C^k class functions near X are identified with the $\mathrm{Ad}\, G_X$-invariant C^k functions on $\mathfrak{g}_X$, near 0.

(c) The Lie algebra $\mathfrak{g}$ has an $\mathrm{Ad}\, G$-invariant inner product.

(d) From Lemma 3.1.2, we get:

(3.2.3) Lemma.
(i) For every $X \in \mathfrak{g}$, the fixed point set $\mathfrak{z}_{\mathfrak{g}}(G_X)$ of $\mathrm{Ad}\, G_X$ in $\mathfrak{g}$ is contained in $\mathfrak{g}_X$, so is equal to $\mathfrak{z}(G_X)$, the Lie algebra of $\mathrm{Z}(G_X)$.
(ii) The normalizer $\mathrm{N}(G_X)$ of G_X in G has the same dimension as G_X, and $\mathrm{N}(G_X)/G_X$ is a finite group.

This time, Theorem 2.6.7 takes the following form. For each $X \in \mathfrak{g}$, the orbit type:

$$\mathfrak{g}_X^{\sim} = \{\, Y \in \mathfrak{g} \mid \mathfrak{g}_Y = \mathrm{Ad}\, g(\mathfrak{g}_X), \ \text{for some } g \in G \,\}$$

is equal to the local action type $\mathfrak{g}_X^{\approx}$, according to (b). It is a locally closed real-analytic submanifold of $\mathfrak{g}$, of codimension equal to $\dim \mathfrak{g}_X/\mathfrak{z}(G_X)$.

The set $\mathfrak{g}_X^{\sim} \cap \mathfrak{z}(G_X)$ is an open subset of the **Abelian** Lie subalgebra $\mathfrak{z}(G_X)$ of $\mathfrak{g}_X$, containing the point X. The finite group $\mathrm{N}(G_X)/G_X$ acts freely on it, and the quotient is naturally identified with $(\mathrm{Ad}\, G)\backslash\mathfrak{g}_X^{\sim}$; hence the orbit type stratum in the orbit space is a real-analytic manifold of dimension equal to $\dim \mathfrak{z}(G_X)$.

The adjoint action of G induces a G-equivariant analytic diffeomorphism:

$$G/G_X \times_{\mathrm{N}(G_X)/G_X} (\mathfrak{g}_X^{\sim} \cap \mathfrak{z}(G_X)) \xrightarrow{\sim} \mathfrak{g}_X^{\sim}.$$

The partitioning of $\mathfrak{g}_X^{\sim}$ into $\mathrm{Ad}\, G$-orbits defines a real-analytic fibration:

$$\mathfrak{g}_X^{\sim} \to (\mathrm{Ad}\, G)\backslash\mathfrak{g}_X^{\sim},$$

with fiber G/G_X.

(e) Using the identification of the orbit types with those near the origin, by means of the conic structure, we obtain from Section 2.7 the following results. There are only finitely many orbit types in $\mathfrak{g}$, and each of these has only finitely many connected components; they define a Whitney stratification in $\mathfrak{g}$.

If $X, Y \in \mathfrak{g}$, $Y \in (\mathfrak{g}_X^{\sim})^{\mathrm{cl}} \setminus \mathfrak{g}_X^{\sim}$, then $\mathrm{Ad}\, u(\mathfrak{g}_X) \subset \mathfrak{g}_Y$, for some $u \in G$, and $\dim \mathfrak{z}(G_Y) < \dim \mathfrak{z}(G_X)$, and therefore a fortiori $\mathrm{codim}\, \mathfrak{g}_Y^{\sim} = \dim G_Y/\mathfrak{z}(G_Y) > \dim G_X/\mathfrak{z}(G_X) = \mathrm{codim}\, \mathfrak{g}_X^{\sim}$.

(For (f)–(h), see Section 2.8.)

(f) The element $X \in \mathfrak{g}$ is of *principal orbit type for the adjoint action* if and only if $\mathfrak{g}_X^{\sim}$ is open, or $\dim \mathfrak{z}(G_X) = \dim \mathfrak{g}_X$, or *the adjoint representation of G_X on its Lie algebra $\mathfrak{g}_X$ is trivial.* This implies that $\mathfrak{g}_X$ is Abelian.

Because $\mathfrak{g}$ is connected, there is only one principal orbit type in $\mathfrak{g}$, denoted by $\mathfrak{g}^{\mathrm{princ}}$. This implies that **all** G_X, for $X \in \mathfrak{g}^{\mathrm{princ}}$ **are conjugate to each other**. The complement of $\mathfrak{g}^{\mathrm{princ}}$ in $\mathfrak{g}$ is closed and composed of finitely many locally closed, connected, real-analytic submanifolds of $\mathfrak{g}$ (the connected components of the other orbit types), of codimension ≥ 1 in $\mathfrak{g}$.

(g) X is a *regular element of* $\mathfrak{g}$, if all nearby adjoint orbits have the same dimension, that is, $\dim \mathfrak{g}_Y = \dim \mathfrak{g}_X$, for all Y near X in $\mathfrak{g}$. Using the slice $\mathfrak{g}_X$ at X, we obtain as in 3.1 (g) that X is regular in $\mathfrak{g}$ if and only if $\mathfrak{g}_X$ **is Abelian**. Note that this condition is formulated in terms of the Lie algebra structure of $\mathfrak{g}$ only, the reason being that regularity is defined in terms of the infinitesimal action, ad, of Ad. The set $\mathfrak{g}^{\mathrm{reg}}$ of regular elements in $\mathfrak{g}$ is connected, because the vector space $\mathfrak{g}$ is connected. In Corollary 3.3.2, we shall actually see that $\mathfrak{g}^{\mathrm{reg}} = \mathfrak{g}^{\mathrm{princ}}$, whereas Theorem 3.8.3.(ii) yields that the singular strata in $\mathfrak{g}$ have codimension ≥ 3.

(h) Let $X \in \mathfrak{g}^{\mathrm{princ}}$, and let $\mathfrak{c}$ be the connected component of X in $\mathfrak{g}^{\mathrm{princ}} \cap \mathfrak{z}(G_X)$. Then $\mathrm{Ad}\, \mathrm{N}(G_X)\,(\mathfrak{c}) = \mathfrak{g}^{\mathrm{princ}} \cap \mathfrak{z}(G_X)$.

Furthermore, $\mathrm{N}(\mathfrak{c}) := \{g \in G \mid \mathrm{Ad}\, g(\mathfrak{c}) = \mathfrak{c}\}$ is an open subgroup of $\mathrm{N}(G_X)$ containing G_X, so that $\mathrm{N}(\mathfrak{c})/G_X$ is a finite group which acts freely on $\mathfrak{c}$; the quotient space can be identified with $\mathrm{Ad}\, G\backslash\mathfrak{g}^{\mathrm{princ}}$. The adjoint action of G induces a G-equivariant analytic diffeomorphism: $G/G_X \times_{\mathrm{N}(\mathfrak{c})/G_X} \mathfrak{c} \xrightarrow{\sim} \mathfrak{g}^{\mathrm{princ}}$.

Finally, $\mathfrak{g} = \mathrm{Ad}\, G(\mathfrak{c}^{\mathrm{cl}})$.

3.3 Connectedness of Centralizers

In general it is a nontrivial problem to decide whether the set of solutions of certain equations is connected, and this applies also to the centralizer groups appearing in Sections 3.1 and 3.2. We shall prove here that, in a connected, compact Lie group, the centralizer of any element in the Lie algebra is connected.

(3.3.1) Theorem. *Let G be a connected, compact Lie group. Then:*

(i) $x \in (G_x)^\circ$*, for every* $x \in G$.
(ii) G_X *is connected, for every* $X \in \mathfrak{g}$.
(iii) G_S *is connected, for every torus* $S \subset G$.

Proof. (i) Because, near x, the set G_x is a slice at x, there exist elements $y \in G_x \cap G^{\text{princ}}$ near x. Note that $G_y \subset G_x$, because any $g \in G$ that maps an element in the slice (in this case y) back into the slice, belongs to G_x. But then, according to Proposition 3.1.3.(iii), we have $y \in (G_y)^\circ \subset G_y \subset G_x$, so $y \in (G_x)^\circ$. Taking the limit we get $x \in (G_x)^\circ$.

(ii) Let $z \in G_X$, that is, $X \in \mathfrak{g}_z$, the Lie algebra of $(G_z)^\circ$; and we also have that $z \in (G_z)^\circ$ by (i). Hence, if we can show that $z \in G_X \cap (G_z)^\circ$ implies $z \in (G_X)^\circ$, then we are done. That is, it suffices to show that if G is a connected, compact Lie group and $X \in \mathfrak{g}$, then $z \in (G_X)^\circ$ for every $z \in G_X \cap \mathrm{Z}(G)$.

According to Lemma 3.2.1, $G_X = G_{tX} = G_{\exp tX}$, if $t \in \mathbf{R} \setminus \{0\}$ and t is sufficiently small. Near $x = \exp tX$, the set G_x is a slice at x; therefore we can find $y \in G_X \cap G^{\text{princ}}$ (near x). Note that $G_y \subset G_X$, hence $(G_y)^\circ \subset (G_X)^\circ$. In view of Proposition 3.1.3.(ii), there exists $g \in G$ such that $gzg^{-1} \in (G_y)^\circ$, but $z = gzg^{-1}$, because $z \in \mathrm{Z}(G)$.

(iii) If S is a torus in G with Lie algebra $\mathfrak{s}$, then there exist $X \in \mathfrak{s}$ such that $\{ \exp tX \mid t \in \mathbf{R} \}$ is dense in S (cf. the comments after Corollary 1.12.4), hence $G_S = G_X$, which is connected according to (ii). □

Remark. For a quite different proof of Theorem 3.3.1, see Remark (b) after Corollary 3.8.4.

(3.3.2) Corollary. *Let G be a connected, compact Lie group with Lie algebra $\mathfrak{g}$. Then $\mathfrak{g}^{\text{princ}} = \mathfrak{g}^{\text{reg}}$; that is, there are no exceptional orbits in $\mathfrak{g}$. Also, if $x \in G$, then G_x is a torus if and only if $x \in G^{\text{princ}}$. All these tori G_x, $x \in G^{\text{princ}}$, are conjugate to each other.*

Proof. If X is regular, then $\mathfrak{g}_X$ is Abelian, so G_X, which is connected, is a torus. But then its adjoint representation is trivial, or X is of principal orbit type.

Now let U be an $\mathrm{Ad}\, G$-invariant neighborhood of 0 in $\mathfrak{g}$ on which the exponential mapping is injective. Then $V = \exp U$ is a neighborhood of 1 in G, and because G^{princ} is dense in G, there exists $x \in V \cap G^{\text{princ}}$. If we write $x = \exp X$, Lemma 3.2.1 yields that $G_x = G_X$ is connected, hence a torus because we know already that $\mathfrak{g}_x$ is Abelian if $x \in G^{\text{princ}}$. However, because there is only one principal orbit type, cf. 3.1.(f), we have G_y is conjugate to

G_x, and therefore a torus as well, for every $y \in G^{\text{princ}}$.

If conversely $x \in G$ and G_x is a torus, then clearly the adjoint representation of G_x on its Lie algebra is trivial, which implies that $x \in G^{\text{princ}}$. □

Theorem 3.3.1 leads to several simplifications of the results in Sections 3.1 and 3.2, for connected G. For instance, if $X \in \mathfrak{g}$, then $\mathfrak{z}(G_X) = \mathfrak{z}(\mathfrak{g}_X)$. Also:

$$\mathrm{N}(G_X) = \mathrm{N}(\mathfrak{g}_X) := \{\, y \in G \mid \mathrm{Ad}\, y(\mathfrak{g}_X) = \mathfrak{g}_X \,\},$$

the *normalizer of* $\mathfrak{g}_X$ *in* G. The groups $\mathrm{N}(G_X)/G_X$ (and $\mathrm{N}(G_x)/G_x$, for $x \in G^{\text{princ}}$) coincide with the component groups of $\mathrm{N}(G_X)$ (and $\mathrm{N}(G_x)$, respectively, for $x \in G^{\text{princ}}$).

For $X \in \mathfrak{g}^{\text{reg}}$, the quotient $\mathrm{N}(T)/T$ is called the *Weyl group* of the Abelian subalgebra $\mathfrak{t} = \mathfrak{g}_X$, here $T = \exp \mathfrak{t} = G_X$. We will return to this in the Sections 3.5, 3.7, and 3.10, where the use of a little more linear algebra leads to a considerable further clarification.

3.4 The Group of Rotations and its Covering Group

In $\mathbf{SO}(3, \mathbf{R})$, cf. Section 1.2.(a), the conjugacy classes of all elements $x \neq 1$ are two-dimensional, so all these elements are regular. Let R_a denote the rotation about the vertical axis through the angle $a \in \,]\,0, \pi\,]$. If the element $x \in \mathbf{SO}(3, \mathbf{R})$ commutes with R_a, then it has to leave invariant the vertical axis, that is the eigenspace of R_a for the eigenvalue 1. On it, it acts as the identity or as the reflection -1. In the horizontal plane, x then acts as a rotation, or as an orientation-reversing orthogonal linear transformation. As is easily verified, the second case is excluded if $a < \pi$; the centralizer of R_a in $\mathbf{SO}(3, \mathbf{R})$ then is equal to the circle group T of all rotations about the vertical axis. It follows that all R_a with $a \in \,]\,0, \pi\,[$ are of principal orbit type.

The normalizer $\mathrm{N}(T)$ of T in $\mathbf{SO}(3, \mathbf{R})$ consists of all $x \in \mathbf{SO}(3, \mathbf{R})$ that leave the vertical axis invariant; then it is equal to the disjoint union of T and sT, where we can take for s the element that sends the standard basis vectors e_1, e_2, e_3 to e_2, e_1, $-e_3$, respectively. $\mathrm{N}(T)$ has therefore two connected components, the Weyl group $\mathrm{N}(T)/T$ is the commutative group of two elements. Because the centralizer of R_π is equal to $\mathrm{N}(T)$, the element R_π is not of principal orbit type, its conjugacy class therefore is the only exceptional orbit. In Section 1.2.(a) we described $\mathbf{SO}(3, \mathbf{R})^{\text{princ}}$ as an open ball with the center deleted, and fibered by the conjugacy classes, which are the concentric spheres. Also the conjugacy class of R_π was described as a real projective plane; the fact that it is even topologically different from a sphere causes that it cannot take part in the fibration by spheres, and this confirms again the exceptional nature of the conjugacy class of R_π.

N(T) itself is a quite interesting noncommutative group, with commutative identity component T. On T, the conjugation action of N(T) is by a reflection, the orbits consist of two points, except for the fixed points R_0 and R_π. On the other hand, using that $R_a \cdot s = s \cdot R_a$ and $s^2 = 1$, we get $R_a \cdot (s \cdot R_b) \cdot R_a^{-1} = s \cdot R_{b-2a}$, so $R_a \in T$ acts on $s \cdot T$ as a translation over $-2a$. That is, the connected component $s \cdot T$ is a single orbit for the action by conjugation. This may also serve as a warning that in nonconnected compact Lie groups, the regular orbits can easily have different dimensions in different connected components of the group.

If we take $G = \mathbf{SU}(2)$, then $G^{\text{princ}} = G^{\text{reg}} = \mathbf{SU}(2) \setminus \{-1, 1\}$, which is fibered by the conjugacy classes (2-dimensional spheres). Both for $\mathbf{SO}(3, \mathbf{R})$ and for $\mathbf{SU}(2)$, the orbit space is homeomorphic to a segment on the real axis, for $\mathbf{SO}(3, \mathbf{R})$ however, one of the end points represents an exceptional orbit, whereas for $\mathbf{SU}(2)$ both endpoints represent singular orbits (actually elements of the center).

The Lie algebras of $\mathbf{SO}(3, \mathbf{R})$ and $\mathbf{SU}(2)$ are isomorphic, the adjoint group acts on them as the rotation group. Clearly the regular = principal set is equal to the complement of the origin, and fibered by concentric spheres. The orbit space is homeomorphic to $[0, \infty[$, with the conic structure on $]0, \infty[$. It is also easy here to determine for which elements in the Lie algebra the conclusion in Lemma 3.2.1 does not hold.

3.5 Roots and Root Spaces

Let $\mathfrak{g}$ be the Lie algebra of a compact Lie group G. It will be convenient to study $\operatorname{ad} X \in \mathbf{L}(\mathfrak{g}, \mathfrak{g})$, for $X \in \mathfrak{g}$, by passing to the complexification $\mathfrak{g}_\mathbf{C} = \mathfrak{g} \otimes_\mathbf{R} \mathbf{C}$ of $\mathfrak{g}$. Note that $\mathfrak{g}_\mathbf{C} = \mathfrak{g} + i\mathfrak{g}$ (as vector spaces over $\mathbf{R}$), and that $\mathfrak{g}_\mathbf{C}$ is a complex Lie algebra, with the unique complex bilinear mapping $[\,.\,,.\,] \colon \mathfrak{g}_\mathbf{C} \times \mathfrak{g}_\mathbf{C} \to \mathfrak{g}_\mathbf{C}$ that extends the Lie bracket of $\mathfrak{g}$.

Every $A \in \mathbf{L}(\mathfrak{g}, \mathfrak{g})$ has a unique complex linear extension: $\mathfrak{g}_\mathbf{C} \to \mathfrak{g}_\mathbf{C}$, which will also be denoted by A. Conversely, a map $A \in \mathbf{L}_\mathbf{C}(\mathfrak{g}_\mathbf{C}, \mathfrak{g}_\mathbf{C})$ arises in this way, if and only if $A(\mathfrak{g}) \subset \mathfrak{g}$.

Now let $X \in \mathfrak{g}$. For every $t \in \mathbf{R}$, we have: $\mathrm{e}^{t \operatorname{ad} X} = \mathrm{e}^{\operatorname{ad} tX} \in \operatorname{Ad} G$, which is a compact subset of $\mathbf{L}(\mathfrak{g}, \mathfrak{g})$, as the image of the compact group G under the continuous mapping Ad. It follows that the $\mathrm{e}^{t \operatorname{ad} X}$, for $t \in \mathbf{R}$, form a bounded subset of $\mathbf{L}(\mathfrak{g}, \mathfrak{g})$; and the same holds for the subset of $\mathbf{L}_\mathbf{C}(\mathfrak{g}_\mathbf{C}, \mathfrak{g}_\mathbf{C})$ formed by the complex linear extensions.

The Jordan normal form theorem for $\operatorname{ad} X \in \mathbf{L}_\mathbf{C}(\mathfrak{g}_\mathbf{C}, \mathfrak{g}_\mathbf{C})$ says that $\mathfrak{g}_\mathbf{C}$ can be written as the direct sum of complex linear subspaces $\mathfrak{g}_j$ with the following properties. Each $\mathfrak{g}_j$ is invariant under $\operatorname{ad} X$, and on it, $\operatorname{ad} X$ is of the form:

$$\operatorname{ad} X|_{\mathfrak{g}_j} = c_j \, \mathrm{I} + N_j, \tag{3.5.1}$$

where $c_j \in \mathbf{C}$ and $N_j \in \mathbf{L}_{\mathbf{C}}(\mathfrak{g}_j, \mathfrak{g}_j)$ is nilpotent, that is, $(N_j)^m = 0$ for some $m \in \mathbf{Z}_{>0}$. For any linear mapping of the form $L = c\mathrm{I} + N$, with $c \in \mathbf{C}$ and $N^m = 0$, we have the formula:

$$\mathrm{e}^{tL} = \mathrm{e}^{tc} \sum_{k=0}^{m-1} \frac{t^k}{k!} N^k. \tag{3.5.2}$$

Assuming that the vector space on which L acts is nonzero, we have $\ker N \neq 0$. On $\ker N$, the operator e^{tL} acts as multiplication by e^{tc}; so if $\{ \mathrm{e}^{tL} \mid t \in \mathbf{R} \}$ is bounded, then c must be purely imaginary. Multiplying by the numbers e^{-tc}, for $t \in \mathbf{R}$, which remain bounded, we see by downward induction on k that also necessarily $N^k = 0$, for $k = m, \ldots, 1$, or: $N = 0$. We have proved:

(3.5.1) Lemma. *For each X in the Lie algebra $\mathfrak{g}$ of a compact Lie group, $\mathrm{ad}\, X \in \mathbf{L}_{\mathbf{C}}(\mathfrak{g}_{\mathbf{C}}, \mathfrak{g}_{\mathbf{C}})$ is diagonalizable, with only purely imaginary eigenvalues.*

If $X, Y \in \mathfrak{g}$, and $[X, Y] = 0$, then $\mathrm{ad}\, X$ and $\mathrm{ad}\, Y$ commute, because ad is a homomorphism: $\mathfrak{g} \to \mathbf{L}(\mathfrak{g}, \mathfrak{g}) \subset \mathbf{L}_{\mathbf{C}}(\mathfrak{g}_{\mathbf{C}}, \mathfrak{g}_{\mathbf{C}})$, cf. (1.1.22). This implies that $\mathrm{ad}\, Y$ leaves invariant the eigenspaces of $\mathrm{ad}\, X$. Any of these, say E, then decomposes further as the direct sum of the eigenspaces of $\mathrm{ad}\, Y|_E$. In this way $\mathfrak{g}_{\mathbf{C}}$ is the direct sum of the complex linear subspaces which are the common eigenspaces of $\mathrm{ad}\, X$ and $\mathrm{ad}\, Y$.

Assume now that $\mathfrak{t}$ is an *Abelian subspace* of $\mathfrak{g}$, that is, $\mathfrak{t}$ is a real linear subspace of $\mathfrak{g}$ such that $[X, Y] = 0$ whenever $X, Y \in \mathfrak{t}$. Applying the result above successively to a basis of $\mathfrak{t}$, we then get a direct sum decomposition of $\mathfrak{g}_{\mathbf{C}}$ into finitely many complex linear subspaces $\mathfrak{g}_j \neq 0$, and real linear functions $\alpha_j : \mathfrak{t} \to \mathbf{C}$ (actually taking purely imaginary values only), such that:

$$[X, Y] = \alpha_j(X) Y, \quad \text{for every } X \in \mathfrak{t},\ Y \in \mathfrak{g}_j. \tag{3.5.3}$$

The construction is such that $\alpha_j \neq \alpha_k$, whenever $j \neq k$.

A more intrinsic notation is obtained by defining, for every $\mathbf{R}$-linear function $\alpha \colon \mathfrak{t} \to \mathbf{C}$, the "common eigenspace for $\mathfrak{t}$":

$$\mathfrak{g}_\alpha := \mathfrak{g}_{\alpha^{\mathfrak{t}}} := \{ Y \in \mathfrak{g}_{\mathbf{C}} \mid [X, Y] = \alpha(X) Y, \quad \text{for all } X \in \mathfrak{t} \}. \tag{3.5.4}$$

These complex linear subspaces of $\mathfrak{g}_{\mathbf{C}}$ are nonzero for only finitely many $\alpha \in (\mathfrak{t}^*)_{\mathbf{C}}$ (and then actually $\alpha(\mathfrak{t}) \subset i\mathbf{R}$), and $\mathfrak{g}_{\mathbf{C}}$ is equal to the direct sum of the nonzero $\mathfrak{g}_\alpha$. Also,

$$\overline{\mathfrak{g}_\alpha} = \mathfrak{g}_{-\alpha}, \tag{3.5.5}$$

where $^-$ denotes the complex conjugation $X + iY \mapsto X - iY$, for $X, Y \in \mathfrak{g}$. In particular, $\mathfrak{g}_{-\alpha}$ has the same dimension as $\mathfrak{g}_\alpha$; and the space $\mathfrak{g}_0$, which contains $\mathfrak{t}$, is invariant under complex conjugation, that is, $\mathfrak{g}_0 = (\mathfrak{g}_0 \cap \mathfrak{g}) + i(\mathfrak{g}_0 \cap \mathfrak{g})$.

$\mathfrak{t}$ is said to be a *maximal Abelian subspace* of $\mathfrak{g}$ if it is an Abelian subspace of $\mathfrak{g}$, and $\mathfrak{s} = \mathfrak{t}$ for every Abelian subspace $\mathfrak{s}$ of $\mathfrak{g}$ such that $\mathfrak{t} \subset \mathfrak{s}$. Because $\mathfrak{t} + \mathbf{R} \cdot Y$ is an Abelian subspace of $\mathfrak{g}$ if $Y \in \mathfrak{g}_0 \cap \mathfrak{g}$, this amounts to requiring that $\mathfrak{t}$ is an Abelian subspace of $\mathfrak{g}$ such that:

$$\mathfrak{g}_0 \cap \mathfrak{g} = \mathfrak{t}, \quad \mathfrak{g}_0 = \mathfrak{t} \oplus i\mathfrak{t} := \mathfrak{h}. \tag{3.5.6}$$

The space $\mathfrak{h}$ then is a maximal Abelian subspace of $\mathfrak{g}_{\mathbf{C}}$, such that $\operatorname{ad} X \in \mathbf{L}_{\mathbf{C}}(\mathfrak{g}_{\mathbf{C}}, \mathfrak{g}_{\mathbf{C}})$ is diagonalizable for every $X \in \mathfrak{h}$. Such a subspace is said to be a *Cartan subalgebra* of the complex Lie algebra $\mathfrak{g}_{\mathbf{C}}$, see Section 5.4. The $\alpha \in (\mathfrak{t}^*)_{\mathbf{C}}$ extend to complex linear forms on $\mathfrak{h}$, also denoted by α; and $[X, Y] = \alpha(X)Y$, for every $X \in \mathfrak{h}$, $Y \in \mathfrak{g}_\alpha$. In other words, we may identify $(\mathfrak{t}^*)_{\mathbf{C}}$ with $\mathfrak{h}^*$, the complex dual of $\mathfrak{h}$.

The $\alpha \in \mathfrak{h}^*$ such that $\mathfrak{g}_\alpha \neq 0$ **and** $\alpha \neq 0$ are called the *roots*, or *root forms* of the Cartan subalgebra $\mathfrak{h}$, and the $\mathfrak{g}_\alpha$ are the corresponding *root spaces.* The set of roots will be denoted by R. (The terminology comes from the computation of $\alpha(X)$ as the root of the eigenvalue equation $\det(\operatorname{ad} X - c\mathrm{I}) = 0$, $c \in \mathbf{C}$.)

Combined with (3.5.5), the direct sum decomposition:

$$\mathfrak{g}_{\mathbf{C}} = \mathfrak{h} \oplus \sum_{\alpha \in R} \mathfrak{g}_\alpha, \tag{3.5.7}$$

has as a real counterpart:

$$\mathfrak{g} = \mathfrak{t} \oplus \sum_{\alpha \in P} (\mathfrak{g}_\alpha \oplus \mathfrak{g}_{-\alpha}) \cap \mathfrak{g}. \tag{3.5.8}$$

Here $(\mathfrak{g}_\alpha \oplus \mathfrak{g}_{-\alpha}) \cap \mathfrak{g}$, the real part of $\mathfrak{g}_\alpha \oplus \mathfrak{g}_{-\alpha}$, has real dimension equal to $2 \dim_{\mathbf{C}} \mathfrak{g}_\alpha$; and we need only to sum over a subset P of R which contains one of each pair α, $-\alpha$ in R.

The $i^{-1}\alpha$, $\alpha \in R$, are nonzero real linear forms on $\mathfrak{t}$, so the $\ker \alpha \cap \mathfrak{t}$, with $\alpha \in R$, are real hyperplanes in $\mathfrak{t}$, called the *root hyperplanes* in $\mathfrak{t}$. An element $X \in \mathfrak{t}$ is said to be *regular* in $\mathfrak{t}$, if $\alpha(X) \neq 0$, for all roots α, that is, if it belongs to:

$$\mathfrak{t}^{\mathrm{reg}} := \mathfrak{t} \setminus \bigcup_{\alpha \in R} \ker \alpha, \tag{3.5.9}$$

the complement in $\mathfrak{t}$ of the finitely many root hyperplanes. The elements of $\mathfrak{t}$ which lie on a root hyperplane are said to be *singular*. It is easy to verify that $\mathfrak{t}^{\mathrm{reg}} = \mathfrak{t} \cap \mathfrak{g}^{\mathrm{reg}}$, cf. Theorem 3.7.1.(ii), which explains this terminology. From the description in Corollary 3.10.3 of the action of the Weyl group $W = \operatorname{Ad} \mathrm{N}(\mathfrak{t})|_{\mathfrak{t}} = \mathrm{N}(T)/T$ on $\mathfrak{t}$, it follows also that $\mathfrak{t}^{\mathrm{reg}}$ is also equal to the set where W acts freely; that is, $\mathfrak{t}^{\mathrm{reg}}$ is the principal orbit type for the action of W on $\mathfrak{t}$. (Warning: for the action of a finite group, every point is regular; so here it is essential to use the more refined concept of orbit types.)

Now let $\mathfrak{c}$ be a connected component of $\mathfrak{t}^{\mathrm{reg}}$. For each $\alpha \in R$, the real-valued continuous function $i^{-1}\alpha|_{\mathfrak{t}}$ has no zeros in $\mathfrak{c}$, so the connectedness of $\mathfrak{c}$ implies that it is either everywhere > 0 or everywhere < 0 on $\mathfrak{c}$. That is,

$$R = P \cup (-P), \quad P \cap (-P) = \emptyset, \tag{3.5.10}$$

if we take:

$$P = P(\mathfrak{c}) := \{\, \alpha \in R \mid i^{-1}\alpha > 0 \text{ on } \mathfrak{c} \,\}, \tag{3.5.11}$$

and if we write $-P = \{-\alpha \mid \alpha \in P\}$. Conversely, if P is a subset of R satisfying (3.5.10), then the set:

$$\mathfrak{c} = \mathfrak{c}(P) := \{\, X \in \mathfrak{t} \mid i^{-1}\alpha(X) > 0, \text{ for all } \alpha \in P \,\} \tag{3.5.12}$$

is an open, convex polyhedral cone in $\mathfrak{t}$, contained in $\mathfrak{t}^{\mathrm{reg}}$. It is nonvoid, and therefore equal to a connected component of $\mathfrak{t}^{\mathrm{reg}}$, if and only if the convex cone in $i\mathfrak{t}^*$ generated by P is *proper*, that is, if P satisfies:

$$\sum_{\alpha \in P} c_\alpha \alpha = 0, \quad c_\alpha \geq 0, \text{ for all } \alpha \in P \Rightarrow c_\alpha = 0, \text{ for all } \alpha \in P. \tag{3.5.13}$$

A subset P of R, satisfying (3.5.10) and (3.5.13), is said to be a *choice of positive roots*. The connected components of $\mathfrak{t}^{\mathrm{reg}}$ are called the *Weyl chambers* in $\mathfrak{t}$. We have proved:

(3.5.2) Lemma. *Each of the relations (3.5.11), and (3.5.12), respectively, defines a bijective correspondence between the set of choices P of positive roots in R and the set of Weyl chambers $\mathfrak{c}$ in $\mathfrak{t}$.*

If one has fixed the choice of a Weyl chamber $\mathfrak{c}$, then it is customary to write:

$$P = R^+, \quad \mathfrak{c} = \mathfrak{t}^+, \tag{3.5.14}$$

motivated by (3.5.11), (3.5.12), respectively.

3.6 Compact Lie Algebras

As an intermezzo, we give an algebraic characterization of those Lie algebras which arise as the Lie algebra of a compact Lie group. Because the results of this section will not be used in the remainder of Chapter 3, the reader may also pass directly to the next section.

For our purposes, a convenient tool is the bilinear form κ, defined on any Lie algebra $\mathfrak{g}$ by:

$$\kappa(X,Y) = \operatorname{tr}(\operatorname{ad} X \circ \operatorname{ad} Y), \quad X, Y \in \mathfrak{g}. \tag{3.6.1}$$

Then κ is called the *Killing form* of $\mathfrak{g}$; and it will be denoted by $\kappa_{\mathfrak{g}}$ or $\kappa_{\mathfrak{g}}^{\mathbf{R}}$, if there is danger of confusion. This bilinear form is **symmetric**, because in general $\operatorname{tr}(A \circ B) = \operatorname{tr}(B \circ A)$, for linear endomorphisms A and B. Because the real trace of a linear mapping is equal to the complex trace of its complex linear extension, the Killing form extends to a complex bilinear form on $\mathfrak{g}_{\mathbf{C}}$ by means of the formula:

$$\kappa(X,Y) = \operatorname{tr}_{\mathbf{C}}(\operatorname{ad} X \circ \operatorname{ad} Y), \quad X, Y \in \mathfrak{g}_{\mathbf{C}}, \tag{3.6.2}$$

where now $\operatorname{ad} X \circ \operatorname{ad} Y$ is considered as an element of $\mathbf{L}_{\mathbf{C}}(\mathfrak{g}_{\mathbf{C}}, \mathfrak{g}_{\mathbf{C}})$. Note however that the Killing form of $\mathfrak{g}_{\mathbf{C}}$, if we consider $\mathfrak{g}_{\mathbf{C}}$ as a Lie algebra over $\mathbf{R}$, is equal to $2 \operatorname{Re} \kappa_{\mathfrak{g}}^{\mathbf{R}}$; this because the trace of multiplication by $c \in \mathbf{C}$, considered as a real linear mapping: $\mathbf{C} \to \mathbf{C}$, is equal to $2 \operatorname{Re} c$.

Now let $\mathfrak{g}$ satisfy the conclusion of Lemma 3.5.1, and let $\mathfrak{t}$ be a maximal Abelian subspace of $\mathfrak{g}$. Then, for $X \in \mathfrak{t}$,

$$\kappa(X,X) = \operatorname{tr}((\operatorname{ad} X)^2) = \sum_{\alpha \in R} (\alpha(X))^2 \dim_{\mathbf{C}} \mathfrak{g}_\alpha, \tag{3.6.3}$$

cf. (3.5.7) and (3.5.4). Because $\alpha(X) \in i\mathbf{R}$, for all $\alpha \in R$, it follows that $\kappa(X,X) \le 0$; and $\kappa(X,X) = 0$ if and only if $\alpha(X) = 0$ for all $\alpha \in R$. This in turn is the case if and only if X belongs to $\mathfrak{z}$, the center of $\mathfrak{g}$. For any $X \in \mathfrak{g}$, we have $[X, X] = 0$, so $\mathbf{R} \cdot X$ is an Abelian subspace of $\mathfrak{g}$; and this in turn is contained in a maximal Abelian subspace $\mathfrak{t}$ of $\mathfrak{g}$. The conclusion is therefore that κ is negative semi-definite on $\mathfrak{g}$, and $\ker \kappa = \mathfrak{z}$.

It is also obvious that:

$$\kappa_{\mathfrak{g}} = \pi^* \kappa_{\mathfrak{g}/\mathfrak{z}}, \tag{3.6.4}$$

if π denotes the canonical projection: $\mathfrak{g} \to \mathfrak{g}/\mathfrak{z}$. In particular, the Killing form of $\mathfrak{g}/\mathfrak{z}$ is negative definite in our situation.

For the next lemma, we recall from Example 1.10.4 that the automorphism group $\operatorname{Aut} \mathfrak{g}$ of any Lie algebra $\mathfrak{g}$ is a closed Lie subgroup of $\mathbf{GL}(\mathfrak{g})$; with Lie algebra equal to the space $\operatorname{Der} \mathfrak{g}$ of *derivations* of $\mathfrak{g}$, the linear mappings $D \colon \mathfrak{g} \to \mathfrak{g}$ such that:

$$D([X,Y]) = [D(X), Y] + [X, D(Y)], \quad \text{for all } X, Y \in \mathfrak{g}. \tag{3.6.5}$$

This can also be written as:

$$\operatorname{ad} D(X) = [D, \operatorname{ad} X] \text{ in } \mathbf{L}(\mathfrak{g}, \mathfrak{g}), \quad \text{for all } X \in \mathfrak{g}. \tag{3.6.6}$$

We also recall that the adjoint group $\operatorname{Ad} \mathfrak{g}$ of $\mathfrak{g}$ is the connected Lie subgroup of $\operatorname{Aut} \mathfrak{g}$ with Lie algebra $\operatorname{ad} \mathfrak{g} \subset \operatorname{Der} \mathfrak{g}$.

(3.6.1) Lemma. *Let $\mathfrak{g}$ be a Lie algebra with a nondegenerate Killing form κ; this implies that $\mathfrak{z} = 0$. Then* $\operatorname{ad}\mathfrak{g} = \operatorname{Der}\mathfrak{g}$, *and* $\operatorname{Ad}\mathfrak{g} = (\operatorname{Aut}\mathfrak{g})^{\circ}$, *a closed Lie subgroup of* $\mathbf{GL}(\mathfrak{g})$. *If moreover κ is definite, then* $\operatorname{Aut}\mathfrak{g}$, *and therefore also* $\operatorname{Ad}\mathfrak{g}$, *is compact; and κ is negative definite.*

Proof. Let $D \in \operatorname{Der}\mathfrak{g}$. Because κ is nondegenerate, the linear form $Y \mapsto \operatorname{tr}(D \circ \operatorname{ad} Y)$ on $\mathfrak{g}$ can be written as $Y \mapsto \kappa(X, Y)$, for a unique $X \in \mathfrak{g}$. Write $D_0 = D - \operatorname{ad} X$, so that now $D_0 \in \operatorname{Der}\mathfrak{g}$ and $\operatorname{tr}(D_0 \circ \operatorname{ad} Y) = 0$, for all $Y \in \mathfrak{g}$. Then, if $Y, Z \in \mathfrak{g}$, we have:

$$\begin{aligned}\kappa(D_0(Y), Z) &= \operatorname{tr}(\operatorname{ad} D_0(Y) \circ \operatorname{ad} Z) = \operatorname{tr}((D_0 \circ \operatorname{ad} Y - \operatorname{ad} Y \circ D_0) \circ \operatorname{ad} Z) \\ &= \operatorname{tr}(D_0 \circ [\operatorname{ad} Y, \operatorname{ad} Z]) = \operatorname{tr}(D_0 \circ \operatorname{ad}[Y, Z]) = 0,\end{aligned}$$

where we used (3.6.6) and $\operatorname{tr}(\operatorname{ad} Y \circ D_0 \circ \operatorname{ad} Z) = \operatorname{tr}(D_0 \circ \operatorname{ad} Z \circ \operatorname{ad} Y)$. Because this holds for every $Z \in \mathfrak{g}$, and κ is nondegenerate, we get $D_0(Y) = 0$ for all $Y \in \mathfrak{g}$, or $D_0 = 0$, or $D = \operatorname{ad} X$.

This proves that $\operatorname{ad}\mathfrak{g} = \operatorname{Der}\mathfrak{g}$, and therefore that $\operatorname{Ad}\mathfrak{g} = (\operatorname{Aut}\mathfrak{g})^{\circ}$.

If $\Phi \in \operatorname{Aut}\mathfrak{g}$, $X \in \mathfrak{g}$, then $\operatorname{ad}\Phi(X) = \Phi \circ \operatorname{ad} X \circ \Phi^{-1}$, and because the trace is a conjugacy invariant function on $\mathbf{L}(\mathfrak{g}, \mathfrak{g})$, it follows that $\operatorname{Aut}\mathfrak{g}$ leaves the Killing form κ invariant. That is, $\operatorname{Aut}\mathfrak{g}$ is a closed subgroup of the orthogonal group of $\mathfrak{g}$ with respect to κ. Because the latter group is compact if κ is definite, the remaining statements of the lemma are clear now. □

In the following theorem we also encounter the *derived Lie algebra* $[\mathfrak{g}, \mathfrak{g}]$ of $\mathfrak{g}$; it is defined as the linear span of the $[X, Y]$, for $X, Y \in \mathfrak{g}$, and we have met it before in 1.14.4.

(3.6.2) Theorem. *Let $\mathfrak{g}$ be a finite-dimensional Lie algebra over* $\mathbf{R}$. *Then the following conditions are equivalent:*

(i) *$\mathfrak{g}$ is equal to the Lie algebra of a compact Lie group G.*
(ii) $\operatorname{Ad}\mathfrak{g}$ *is compact.*
(iii) *For each $X \in \mathfrak{g}$, the $e^{t \operatorname{ad} X}$, with $t \in \mathbf{R}$, form a bounded subset of* $\mathbf{L}(\mathfrak{g}, \mathfrak{g})$.
(iv) *For each $X \in \mathfrak{g}$, we have* $\operatorname{ad} X \in \mathbf{L}_{\mathbf{C}}(\mathfrak{g}_{\mathbf{C}}, \mathfrak{g}_{\mathbf{C}})$ *is diagonalizable and has only purely imaginary eigenvalues.*
(v) *The Killing form of $\mathfrak{g}$ is negative semi-definite, and its kernel is equal to $\mathfrak{z}$, the center of $\mathfrak{g}$.*
(vi) *There exists an inner product β on $\mathfrak{g}$ such that, for all $X, Y, Z \in \mathfrak{g}$, we have $\beta([X, Y], Z) + \beta(Y, [X, Z]) = 0$.*
(vii) *$\mathfrak{g} = \mathfrak{g}' \oplus \mathfrak{z}$, for a Lie subalgebra $\mathfrak{g}'$ of $\mathfrak{g}$, on which the Killing form is negative definite.*

Finally, if $\mathfrak{g}'$ is as in (vii), then $[\mathfrak{g}, \mathfrak{g}] = \mathfrak{g}'$.

Proof. The assertions (i) ⇒ (ii) ⇒ (iii) ⇔ (iv) ⇒ (v) have already been dealt with.

Now assume that (v) holds. Let $Z \in \mathfrak{g}$ be orthogonal to $[\mathfrak{g}, \mathfrak{g}]$, with respect to the Killing form. That is, $0 = \kappa([X, Y], Z) = \kappa(X, [Y, Z])$, for all $X, Y \in \mathfrak{g}$; or $[Y, Z] \in \mathfrak{z}$ for all $Y \in \mathfrak{g}$. But then

$$[Z, [X, Y]] = [[Z, X], Y] + [X, [Z, Y]] = 0,$$

for all $X, Y \in \mathfrak{g}$; and this implies that $\operatorname{ad} Z \circ \operatorname{ad} X = 0$. Hence, taking the trace, we see $\kappa(Z, X) = 0$, for all $X \in \mathfrak{g}$, or $Z \in \mathfrak{z}$. Using (3.6.4) and the fact that $\kappa_{\mathfrak{g}/\mathfrak{z}}$ is nondegenerate, we get that $\pi([\mathfrak{g}, \mathfrak{g}]) = \mathfrak{g}/\mathfrak{z}$, or $\mathfrak{g} = [\mathfrak{g}, \mathfrak{g}] + \mathfrak{z}$.

The mapping

$$\pi : \phi \mapsto (X + \mathfrak{z} \mapsto \phi(X) + \mathfrak{z}) : \operatorname{ad} \mathfrak{g} \to \operatorname{ad}(\mathfrak{g}/\mathfrak{z})$$

is a surjective homomorphism of Lie algebras. We have a similar mapping $\Pi : \operatorname{Ad} \mathfrak{g} \to \operatorname{Ad}(\mathfrak{g}/\mathfrak{z})$, which is a homomorphism of Lie groups, and $T_1 \Pi = \pi$. So $\Pi(\operatorname{Ad} \mathfrak{g})$ is open in $\operatorname{Ad}(\mathfrak{g}/\mathfrak{z})$ and, because the latter is connected, we conclude that Π is surjective.

Now let $\Phi \in \operatorname{Ad} \mathfrak{g}$, and $\Pi(\Phi) = 1$, that is, $\Phi(X) - X \in \mathfrak{z}$, for all $X \in \mathfrak{g}$. Then $\Phi([X, Y]) = [\Phi(X), \Phi(Y)] = [X \bmod \mathfrak{z}, Y \bmod \mathfrak{z}] = [X, Y]$ shows that $\Phi = 1$ on $[\mathfrak{g}, \mathfrak{g}]$. Because $\Phi = 1$ on $\mathfrak{z}$ for every $\Phi \in \operatorname{Ad} \mathfrak{g}$, we get $\Phi = 1$ on $\mathfrak{g} = [\mathfrak{g}, \mathfrak{g}] + \mathfrak{z}$. That is, Φ is injective as well.

Because $\kappa_{\mathfrak{g}/\mathfrak{z}}$ is negative definite, see (3.6.4), Lemma 3.6.1 yields that $\operatorname{Ad}(\mathfrak{g}/\mathfrak{z})$ is compact. As Π is a continuous bijection: $\operatorname{Ad} \mathfrak{g} \to \operatorname{Ad}(\mathfrak{g}/\mathfrak{z})$, we get $\operatorname{Ad} \mathfrak{g}$ is compact as well, and we have proved (v) $\Rightarrow$ (ii).

If $\operatorname{Ad} \mathfrak{g}$ is contained in a compact subgroup of $\mathbf{GL}(\mathfrak{g})$, then there exists an $\operatorname{Ad} \mathfrak{g}$-invariant inner product β on $\mathfrak{g}$. We get (vi) from (ii) by differentiating $\beta(e^{t \operatorname{ad} X}(Y), e^{t \operatorname{ad} X}(Z)) = \beta(Y, Z)$ with respect to t at $t = 0$.

Now let β be as in (vi), and write $\mathfrak{g}'$ for the orthogonal complement of $\mathfrak{z}$ in $\mathfrak{g}$, with respect to β. If $X, Y \in \mathfrak{g}'$, then $\beta([X, Y], Z) = \beta(X, [Y, Z]) = 0$ for all $Z \in \mathfrak{z}$, or $[X, Y] \in \mathfrak{g}'$. That is, $\mathfrak{g}'$ is a Lie subalgebra of $\mathfrak{g}$, such that $\mathfrak{g} = \mathfrak{g}' \oplus \mathfrak{z}$. Also, because $\operatorname{Ad} \mathfrak{g}$ is contained in the orthogonal group with respect to β, we get, via (iii), and hence (v), that the Killing form is negative definite on $\mathfrak{g}'$. That is, we have proved (vi) $\Rightarrow$ (vii).

If (vii) holds, then, by Lemma 3.6.1, $\operatorname{Ad} \mathfrak{g}'$ is compact. On the other hand, $\mathfrak{z}(\mathfrak{g}') = \mathfrak{z}(\mathfrak{g}' + \mathfrak{z}) \cap \mathfrak{g}' = \mathfrak{z}(\mathfrak{g}) \cap \mathfrak{g}' = \mathfrak{z} \cap \mathfrak{g}' = 0$; so the Lie algebra $\operatorname{ad} \mathfrak{g}'$ of $\operatorname{Ad} \mathfrak{g}'$ is isomorphic to $\mathfrak{g}'$. If Z denotes any torus with Lie algebra equal to $\mathfrak{z}$, then $G = \operatorname{Ad} \mathfrak{g}' \times Z$ is a compact Lie group with Lie algebra isomorphic to $\mathfrak{g}$, and this proves that (vii) implies (i).

Finally, $[\mathfrak{g}, \mathfrak{g}] = [\mathfrak{g}' + \mathfrak{z}, \mathfrak{g}' + \mathfrak{z}] = [\mathfrak{g}', \mathfrak{g}'] \subset \mathfrak{g}'$. Together with $\mathfrak{g} = [\mathfrak{g}, \mathfrak{g}] + \mathfrak{z}$, which followed from (v), this yields $[\mathfrak{g}, \mathfrak{g}] = \mathfrak{g}'$. □

A Lie algebra is said to be *compact*, if any of the equivalent characterizations in Theorem 3.6.2 holds. Note that if $\mathfrak{z} \neq 0$, then obviously not **every** connected Lie group with Lie algebra equal to $\mathfrak{g}$ is compact. Also, $\operatorname{Aut} \mathfrak{g} = \operatorname{Aut} \mathfrak{g}' \times \mathbf{GL}(\mathfrak{z})$, $\operatorname{Der} \mathfrak{g} = \operatorname{Der} \mathfrak{g}' \times \mathbf{L}(\mathfrak{z}, \mathfrak{z}) = \operatorname{Ad} \mathfrak{g} \times \mathbf{L}(\mathfrak{z}, \mathfrak{z})$; so in this case the conclusion of Lemma 3.6.1 does not hold either. Finally, one can describe $\operatorname{Ad}\mathfrak{g}$ as the identity component of the compact algebraic group $\{\Phi \in \operatorname{Aut}\mathfrak{g} \mid \Phi|_{\mathfrak{z}} = \text{ identity on } \mathfrak{z}\} = \operatorname{Aut}(\mathfrak{g}/\mathfrak{z})$.

In Corollary 3.9.4 we shall see that if G is a connected compact Lie group, and the Lie algebra of G has zero center, then the fundamental group of G is finite. This can be applied to the adjoint group of a compact Lie algebra $\mathfrak{g}$. For any connected Lie group G with Lie algebra equal to $\mathfrak{g}/\mathfrak{z}$, the mapping Ad defines a covering: $G \to \operatorname{Ad}\mathfrak{g}$. Because the fiber is a subgroup of the fundamental group of $\operatorname{Ad}\mathfrak{g}$, it follows that G is compact.

Corollary 5.2.12 states that, for any Lie algebra $\mathfrak{g}$, the Killing form is nondegenerate if and only if $\mathfrak{g}$ is a so-called *semisimple* Lie algebra. Summarizing these remarks, we get:

(3.6.3) Corollary. *Let $\mathfrak{g}$ be a finite-dimensional Lie algebra over $\mathbf{R}$. Then the following conditions are equivalent:*

(i) $\mathfrak{g}$ has negative definite Killing form.
(ii) $\mathfrak{g}$ is compact and semisimple.
(iii) $\mathfrak{g}$ is compact and has zero center.
(iv) Every connected Lie group with Lie algebra isomorphic to $\mathfrak{g}$ is compact.

In any of these cases, $\mathfrak{g} = [\mathfrak{g}, \mathfrak{g}]$.

A *complex torus* is defined as a connected, compact and Abelian complex Lie group. That is, a Lie group isomorphic to $(\mathbf{C}^n, +)/\Delta$, where Δ is a discrete subgroup of $(\mathbf{C}^n, +)$ which contains a real basis of $\mathbf{C}^n \simeq \mathbf{R}^{2n}$. Note that as a real Lie group, $\mathbf{C}^n/\Delta$ is isomorphic to the standard $2n$-dimensional torus $(\mathbf{R}/\mathbf{Z})^{2n}$, cf. Corollary 1.12.4, but two Δ's only yield isomorphic complex Lie groups if they can be mapped to each other by an element of $\mathbf{SL}(n, \mathbf{C})$.

(3.6.4) Corollary. *A compact, connected, complex Lie group is a complex torus.*

Proof. We will regard the Lie algebra $\mathfrak{g}$ as a real Lie algebra, provided with a multiplication by i, which is a real linear mapping $J\colon \mathfrak{g} \to \mathfrak{g}$ such that $J^2 = -\mathrm{I}$ and $[J(X), Y] = J([X, Y]) = [X, J(Y)]$, for all $X, Y \in \mathfrak{g}$. The compactness implies that the Killing form κ of $\mathfrak{g}$ is negative semidefinite, and the kernel of κ is equal to the center $\mathfrak{z}$ of $\mathfrak{g}$, cf. Theorem 3.6.2.(vii).

Now $[J(X), [J(Y), Z]] = J([X, J([Y, Z])] = J \circ J([X, [Y, Z]] = -[X, [Y, Z]]$ shows that $\kappa(J(X), J(Y)) = -\kappa(X, Y)$. That is, κ is positive semidefinite as well, or $\kappa = 0$, or $\mathfrak{g} = \mathfrak{z}$ is Abelian. □

3.7 Maximal Tori

From now on, until Section 3.14, G will denote a connected, compact Lie group, with Lie algebra equal to $\mathfrak{g}$.

In this situation we know already several salient facts, see Proposition 3.1.3, Corollary 3.1.4, and Section 3.3. The concepts of root and root space, however, lead to considerable further clarification, as we want to show now.

A torus in G, that is, a connected, compact, Abelian subgroup of G, is said to be a *maximal torus* in G, if for any torus T' in G such that $T \subset T'$, we have $T = T'$. A subset S of G is said to be a *maximal Abelian subset* of G, if $S \subset \mathrm{Z}_G(S)$ and if, for any $S' \subset G$ such that $S \subset S' \subset \mathrm{Z}_G(S')$, we have $S = S'$. Clearly a torus in G is a maximal torus in G if it is a maximal Abelian subset of G, but there exist examples of maximal Abelian subsets that are no tori. For instance, the three rotations through the angle π, with axes equal to the three coordinate axes in $\mathbf{R}^3$, form a maximal Abelian subset of $\mathbf{SO}(3, \mathbf{R})$, as is not hard to verify.

We begin with some simple observations.

(3.7.1) Maximal Torus Theorem.

(i) *For any $X \in \mathfrak{g}^{\mathrm{reg}}$, $\mathfrak{t} := \mathfrak{g}_X$ is a maximal Abelian subspace of $\mathfrak{g}$. Furthermore, $T := G_X$ is a torus in G with Lie algebra equal to $\mathfrak{t}$, and T is a maximal Abelian subset of G. The same conclusions are valid if we replace X by $x \in G^{\mathrm{princ}}$.*

(ii) *All maximal Abelian subspaces $\mathfrak{t}$ of $\mathfrak{g}$, and maximal tori T in G, arise as $\mathfrak{g}_X$, and G_X, respectively, for some $X \in \mathfrak{g}^{\mathrm{reg}}$. Also, $\mathfrak{g}^{\mathrm{reg}} \cap \mathfrak{t} = \mathfrak{t}^{\mathrm{reg}}$, and it equals the complement of the root hyperplanes in $\mathfrak{t}$, for any maximal Abelian subspace $\mathfrak{t}$ of $\mathfrak{g}$.*

(iii) *The equation $T = \exp \mathfrak{t}$ defines a bijective correspondence between the maximal Abelian subspaces $\mathfrak{t}$ of $\mathfrak{g}$ and the maximal tori T in G. Every connected Abelian subgroup of G is contained in a maximal torus in G, and every Abelian subset of $\mathfrak{g}$ is contained in a maximal Abelian subspace of $\mathfrak{g}$.*

(iv) *All maximal tori in G are conjugate to each other, and $\mathrm{Ad}\, G$ acts transitively on the set of maximal Abelian subspaces of $\mathfrak{g}$. Each element of G is conjugate to an element of a given maximal torus, and $\mathrm{Ad}\, G(\mathfrak{t}) = \mathfrak{g}$ for any maximal Abelian subspace $\mathfrak{t}$ of $\mathfrak{g}$.*

Proof. (i) The observation that $X \in \mathfrak{g}_X$ implies that $\mathfrak{z}_{\mathfrak{g}}(\mathfrak{g}_X) \subset \mathfrak{g}_X$ and $\mathfrak{z}_{\mathfrak{g}}(G_X) \subset \mathrm{Z}_G(\mathfrak{g}_X) \subset G_X$. (Compare Lemma 3.2.3.(i).) This shows that if $\mathfrak{g}_X$, and G_X, is an Abelian subset of $\mathfrak{g}$, and G, respectively, then it is a maximal Abelian subset. That G_X is connected, and hence a torus with $T = \exp \mathfrak{t}$, follows from Theorem 3.3.1.(ii). In Corollary 3.3.2 also the centralizers of

$x \in G^{\text{princ}}$ have been identified as tori, actually equal to G_X for suitable $X \in \mathfrak{g}^{\text{reg}}$.

(iii) If S is a connected Abelian subgroup of G, then S^{cl}, its closure in G, is connected, Abelian, and now also compact, hence a torus in G. As tori become larger, their dimensions have to increase, after finitely many increases we therefore have to arrive at a maximal torus in G. This proves the last statement in (iii), the proof of the corresponding statement in $\mathfrak{g}$ starts by observing that the linear span of an Abelian subset of $\mathfrak{g}$ is an Abelian subspace of $\mathfrak{g}$.

Now let $\mathfrak{t}$ be a maximal Abelian subspace of $\mathfrak{g}$. Then $\exp \mathfrak{t}$ is a connected Abelian subgroup of G, hence contained in a maximal torus T in G. However, the Lie algebra of T is an Abelian subspace of $\mathfrak{g}$ containing $\mathfrak{t}$, so it is equal to $\mathfrak{t}$. The conclusion is that $T = \exp \mathfrak{t}$.

Now conversely suppose T is a maximal torus in G with Lie algebra $\mathfrak{t}$, and $\mathfrak{t}'$ a maximal Abelian subspace of $\mathfrak{g}$ containing $\mathfrak{t}$. Then the previous result gives that $\mathfrak{t}'$ is the Lie algebra of a maximal torus T', which obviously contains T and therefore is equal to T; hence $\mathfrak{t}' = \mathfrak{t}$. This shows that $\mathfrak{t}$ is maximal Abelian in $\mathfrak{g}$.

(ii) Let $\mathfrak{t}$ be a maximal Abelian subspace of $\mathfrak{g}$, and $X \in \mathfrak{t}$. Then:

$$\mathfrak{g}_X = \mathfrak{t} \oplus \sum_{\{\, \alpha \in R \mid \alpha(X) = 0 \,\}} (\mathfrak{g}_\alpha \oplus \mathfrak{g}_{-\alpha}) \cap \mathfrak{g}. \tag{3.7.1}$$

So, if $X \in \mathfrak{t} \setminus \bigcup_{\alpha \in R} \ker \alpha$, then $\mathfrak{g}_X = \mathfrak{t}$, which is Abelian; and this shows that $X \in \mathfrak{g}^{\text{reg}}$. Conversely, if $\alpha \in R$ and $\alpha(X) = 0$, then $\mathfrak{g}_X$ contains the space $\mathfrak{t} \oplus (\mathfrak{g}_\alpha \oplus \mathfrak{g}_{-\alpha}) \cap \mathfrak{g}$, which has larger dimension than $\mathfrak{t} = \mathfrak{g}_Y$ for $Y \in \mathfrak{t}^{\text{reg}}$. Because $\mathfrak{t}^{\text{reg}}$ is dense in $\mathfrak{t}$, this shows that X is not regular in $\mathfrak{g}$.

We have verified that $\mathfrak{t}^{\text{reg}} = \mathfrak{g}^{\text{reg}} \cap \mathfrak{t}$, but also, because this set is not void, that $\mathfrak{t} = \mathfrak{g}_X$ for some $X \in \mathfrak{g}^{\text{reg}}$. The corresponding statement for the maximal tori now follows from (iii).

(iv) This follows from the fact that there is only one principal orbit type, both in G and in $\mathfrak{g}$, cf. 3.1.(f) and 3.2.(f). The last statement follows from 3.1.(h) and 3.2.(h). □

The common dimension of the maximal tori in G, and of the maximal Abelian subspaces of $\mathfrak{g}$, is called the *rank* of G, and of $\mathfrak{g}$, respectively. It can also be defined as the dimension of the principal orbit type in the orbit space $(\mathbf{Ad}\, G)\backslash G$, and $(\mathrm{Ad}\, G)\backslash \mathfrak{g}$, respectively.

(3.7.2) Weyl's Covering Theorem. *Let T be a maximal torus in G, with the maximal Abelian subspace $\mathfrak{t}$ of $\mathfrak{g}$ as its Lie algebra. Then:*

(i) The Weyl group $W := \mathrm{N}(T)/T = \mathrm{N}(\mathfrak{t})/T = \mathrm{Ad}\, \mathrm{N}(\mathfrak{t})|_{\mathfrak{t}}$, is finite. The action of G by conjugation in G, and by the adjoint action in $\mathfrak{g}$, induces a G-equivariant real-analytic diffeomorphism:

$$G/T \times_W (G^{\text{reg}} \cap T) \xrightarrow{\sim} G^{\text{reg}}, \tag{3.7.2}$$

and

$$G/T \times_W \mathfrak{t}^{\mathrm{reg}} \xrightarrow{\sim} \mathfrak{g}^{\mathrm{reg}}, \tag{3.7.3}$$

respectively.

(ii) W *acts transitively on the set of connected components of* $G^{\mathrm{reg}} \cap T$*, and also on the set of Weyl chambers in* $\mathfrak{t}$.

(iii) $G^{\mathrm{reg}} \cap T = \{ x \in T \mid G_x \subset \mathrm{N}(T) \}$*, so* $G^{\mathrm{princ}} \cap T$ *is precisely the subset of* $G^{\mathrm{reg}} \cap T$ *on which* W *acts freely.*

Proof. (i) In 3.1.(d) we find the corresponding statement with G^{reg} replaced by the somewhat smaller open set G^{princ}. (Note that for (3.7.3) this question does not arise, because we have seen in Corollary 3.3.2 that $\mathfrak{g}^{\mathrm{princ}} = \mathfrak{g}^{\mathrm{reg}}$.)

So let us now prove (3.7.2). The set

$$U := G/T \times (G^{\mathrm{reg}} \cap T)$$

is precisely the open subset of $G/T \times T$ on which the mapping

$$\Gamma \colon (gT, t) \mapsto gtg^{-1} \colon G/T \times T \to G$$

has a bijective tangent mapping. Also, (ii) implies that Γ is surjective; because G^{reg} is conjugacy invariant, it follows that $\Gamma(U) = G^{\mathrm{reg}}$.

If $gtg^{-1} = hsh^{-1}$, for $g, h \in G$, and $s, t \in G^{\mathrm{reg}} \cap T$, then, writing $k = h^{-1}g$, we get $s = ktk^{-1}$. It follows that $G_s = kG_tk^{-1}$, and also, on taking Lie algebras, $\mathfrak{g}_s = \mathrm{Ad}\, k(\mathfrak{g}_t)$. Because $s, t \in T$ and T is Abelian, G_s and G_t both contain T; so $\mathfrak{g}_s$ and $\mathfrak{g}_t$ both contain $\mathfrak{t}$. Because $x \mapsto \dim \mathfrak{g}_x$ is constant on connected components of G^{reg}, and G^{reg} is connected, cf. 3.1.(g), we get $\dim \mathfrak{g}_s = \dim \mathfrak{t} = \dim \mathfrak{g}_t$; hence $\mathfrak{g}_s = \mathfrak{t} = \mathfrak{g}_t$, and therefore $\mathfrak{t} = \mathrm{Ad}\, k(\mathfrak{t})$. Thus $k \in \mathrm{N}(\mathfrak{t}) = \mathrm{N}(T)$, or $kT \in W$.

Because W acts analytically on both G/T (from the right) and on the open subset $G^{\mathrm{reg}} \cap T$ of T, the left hand side in (3.7.2) is a real-analytic manifold, "W-fold" covered by the set U. The mapping Γ induces a real-analytic, G-equivariant mapping from it to G^{reg}, and we have just shown that this mapping is bijective and has bijective tangent mappings. The inverse function theorem now implies that the inverse is real-analytic as well.

(ii) follows from the fact that both G^{reg} and $\mathfrak{g}^{\mathrm{reg}}$ are connected, cf. 3.1.(g) and 3.2.(g).

(iii) See the proof of (i). □

A generalization of (3.7.2) to arbitrary proper group actions can be found in Proposition 2.8.7.(ii). The basic fact that simplified the situation here is that G_x is connected for elements x of principal orbit type.

The diffeomorphisms (3.7.2) and (3.7.3) immediately lead to the **Weyl integration formula** in the group G, and the Lie algebra $\mathfrak{g}$, respectively, see Theorem 3.14.1.

3.8 Orbit Structure in the Lie Algebra

In 3.1.(h), we actually obtained that every element of G is conjugate to an element of C^{cl}, the closure of a connected component C of $G^{\mathrm{princ}} \cap T$ in T, and a similar statement holds in $\mathfrak{g}$. Let us first take a closer look at the Lie algebra, where the situation is simpler.

(3.8.1) Proposition. *The Weyl group W acts freely on $\mathfrak{t}^{\mathrm{reg}}$, and also freely and transitively on the set of Weyl chambers in $\mathfrak{t}$; consequently the number of Weyl chambers is equal to the number of elements of the Weyl group. For any Weyl chamber $\mathfrak{c}$ in $\mathfrak{t}$, the adjoint action induces a real-analytic G-equivariant diffeomorphism from $G/T \times \mathfrak{c}$ onto $\mathfrak{g}^{\mathrm{reg}}$. Finally, if $\mathfrak{c}^{\mathrm{cl}}$ denotes the closure of $\mathfrak{c}$ in $\mathfrak{t}$, then $\mathfrak{g} = \mathrm{Ad}\, G(\mathfrak{c}^{\mathrm{cl}})$.*

Proof. (For a completely different proof, see the beginning of the proof of Corollary 3.8.4.(ii).)

In the first sentence, the only new statement is that if $s \in W$ and $s(\mathfrak{c}) = \mathfrak{c}$, then $s = 1$; or that the group

$$\mathrm{N}(\mathfrak{c}) := \{ g \in G \mid \mathrm{Ad}\, g(\mathfrak{c}) = \mathfrak{c} \},$$

which appears in 3.2.(h), is equal to T. We now use the convexity of $\mathfrak{c}$, cf. (3.5.12). This yields that for every $X \in \mathfrak{c}$, the average

$$X^{av} := \#(\mathrm{N}(\mathfrak{c})/T)^{-1} \sum_{gT \in \mathrm{N}(\mathfrak{c})/T} \mathrm{Ad}\, g(X) \tag{3.8.1}$$

of the orbit of X under the action of the finite group $\mathrm{N}(\mathfrak{c})/T$ again belongs to $\mathfrak{c}$. Using the linearity of the mappings $\mathrm{Ad}\, g$, we see that X^{av} is a fixed point for $\mathrm{Ad}\, \mathrm{N}(\mathfrak{c})$. Because the action of $\mathrm{N}(T)/T$ on $\mathfrak{g}^{\mathrm{reg}} \cap \mathfrak{t}$ was free, and $\mathrm{N}(\mathfrak{c}) \subset \mathrm{N}(T)$, the desired conclusion that $\mathrm{N}(\mathfrak{c}) = T$ follows. □

The diffeomorphism: $G/T \times \mathfrak{c} \to \mathfrak{g}^{\mathrm{reg}}$ leads to an identification with $\mathfrak{c}$ of the regular (= principal) orbit type in the orbit space $(\mathrm{Ad}\, G)\backslash\mathfrak{g}$. Note that this is a typical "real" phenomenon: in the complexification $\mathfrak{h}$ of $\mathfrak{t}$, the complex root hyperplanes $\ker \alpha$ have real codimension equal to 2. This implies that their complement $\mathfrak{h} \setminus \bigcup_{\alpha \in R} \ker \alpha$ in $\mathfrak{h}$ is connected, and no analogue of Proposition 3.8.1 holds.

We now turn to the other orbit types. Because every $Y \in \mathfrak{g}$ is of the form $\mathrm{Ad}\, g(X)$ for some $g \in G$, and $X \in \mathfrak{c}^{\mathrm{cl}}$, we may restrict our attention to orbit types of $X \in \mathfrak{c}^{\mathrm{cl}}$.

According to 3.2.(d), we have a G-equivariant real-analytic diffeomorphism:

$$G/G_X \times_{\mathrm{N}(G_X)/G_X} (\mathfrak{g}_X^{\sim} \cap \mathfrak{z}(G_X)) \xrightarrow{\sim} \mathfrak{g}_X^{\sim}, \tag{3.8.2}$$

where $\mathfrak{g}_X^{\sim} \cap \mathfrak{z}(G_X)$ is an open subset of the Abelian subspace $\mathfrak{z}(G_X)$ of $\mathfrak{g}$. Because G_X is connected (Theorem 3.3.1.(ii)), we can also write $\mathfrak{z}(G_X) = \mathfrak{z}(\mathfrak{g}_X)$. In view of the formula (3.7.1) for $\mathfrak{g}_X$ we get:

$$\mathfrak{z}(\mathfrak{g}_X) = \bigcap_{\{\alpha\in R \mid \alpha(X)=0\}} \ker\alpha \cap \mathfrak{t}, \tag{3.8.3}$$

and

$$\mathfrak{g}_X^{\sim} \cap \mathfrak{z}(\mathfrak{g}_X) = \mathfrak{z}(\mathfrak{g}_X) \setminus \bigcup_{\{\beta\in R \mid \beta(X)\neq 0\}} \ker\beta. \tag{3.8.4}$$

In particular, $\mathfrak{g}_X^{\sim} \cap \mathfrak{z}(\mathfrak{g}_X)$ is the complement in $\mathfrak{z}(\mathfrak{g}_X)$ of finitely many hyperplanes, and its connected components are convex polyhedral cones, quite similar to the Weyl chambers in $\mathfrak{c}$.

The connected component $\mathfrak{f}$ of X in $\mathfrak{g}_X^{\sim} \cap \mathfrak{z}(\mathfrak{g}_X)$ is called the *face* of X in the closed convex polyhedron $\mathfrak{c}^{\mathrm{cl}}$. Clearly $\mathfrak{g}_X = \mathfrak{g}_Y$, $G_X = G_Y$ and $\mathfrak{g}_X^{\sim} = \mathfrak{g}_Y^{\sim}$ if X and Y belong to the same face $\mathfrak{f}$. Therefore we allow ourselves to write:

$$\mathfrak{g}_{\mathfrak{f}} = \mathfrak{g}_X, \quad G_{\mathfrak{f}} = G_X, \quad \mathfrak{g}_{\mathfrak{f}}^{\sim} = \mathfrak{g}_X^{\sim}, \quad \text{if } X \in \mathfrak{f}. \tag{3.8.5}$$

For the final result, we first need a lemma.

(3.8.2) Lemma. *If $X, Y \in \mathfrak{c}^{\mathrm{cl}}$ are in the same* $\mathrm{Ad}\, G$*-orbit, then* $X = Y$.

Proof. Write $Y = \mathrm{Ad}\, g(X)$, $g \in G$. We approximate X by $X' \in \mathfrak{c}$; then $Y' := \mathrm{Ad}\, g(X')$ is close to Y. Note that $[Y, \mathrm{Ad}\, g(X')] = \mathrm{Ad}\, g[\mathrm{Ad}\, g^{-1}(Y), X'] = \mathrm{Ad}\, g[X, X'] = 0$, or $Y' \in \mathfrak{g}_Y$.

Now $\mathfrak{t}$ is also a maximal Abelian subspace of the Lie algebra $\mathfrak{g}_Y$ of the connected, compact Lie group G_Y. Furthermore, if P denotes the choice of positive roots associated to $\mathfrak{c}$ as in (3.5.11), then $P_Y := \{\alpha \in P \mid \alpha(Y) = 0\}$ is a choice of positive roots for the root system of $\mathfrak{g}_Y$. An application of $\mathrm{Ad}\, G(\mathfrak{c}) = \mathfrak{g}^{\mathrm{reg}}$, with G, and $\mathfrak{g}$, replaced by G_Y, and $\mathfrak{g}_Y$, respectively, therefore yields the existence of an $h \in G_Y$, such that $Y'' := \mathrm{Ad}\, h^{-1}(Y') \in \mathfrak{t}$, and $i^{-1}\alpha(Y'') > 0$ for all $\alpha \in P_Y$.

Because $\mathrm{Ad}\, G_Y(Y) = \{Y\}$ and G_Y is compact, an application of the principle of uniform continuity of continuous functions on compacta yields the existence, for every neighborhood U of Y in $\mathfrak{g}$, of a neighborhood V of Y in $\mathfrak{g}$, such that $\mathrm{Ad}\, G_Y(V) \subset U$. We have $i^{-1}\beta(Y) > 0$ for all $\beta \in P \setminus P_Y$; by continuity we can find a neighborhood U of Y in $\mathfrak{g}$, such that $i^{-1}\beta > 0$ on $U \cap \mathfrak{t}$, for all $\beta \in P \setminus P_Y$. Taking X' sufficiently close to X, we get $Y' \in V$, hence $Y'' \in U \cap \mathfrak{t}$. But then $i^{-1}\alpha(Y'') > 0$ for all $\alpha \in P$, or $Y'' \in \mathfrak{c}(P) = \mathfrak{c}$, cf. (3.5.12). That is, $s = \mathrm{Ad}(h^{-1}g) = \mathrm{Ad}\, h^{-1} \circ \mathrm{Ad}\, g$ maps $X' \in \mathfrak{c}$ to $Y'' \in \mathfrak{c}$. Because $\mathfrak{g}_{X'} = \mathfrak{t} = \mathfrak{g}_{Y''}$, $s(\mathfrak{t}) = \mathfrak{t}$, and, because $s(X') = Y''$, also $s(\mathfrak{c}) = \mathfrak{c}$. That is, $h^{-1}g \in \mathrm{N}(\mathfrak{c}) = T$ (use Proposition 3.8.1), or $g \in hT \subset G_Y \cdot T = G_Y$. But then $g^{-1} \in G_Y$, or $X = \mathrm{Ad}\, g^{-1}(Y) = Y$. □

(3.8.3) Theorem. *Let $\mathfrak{c}$ be a Weyl chamber in the maximal Abelian subspace $\mathfrak{t}$ of $\mathfrak{g}$. Then:*

(i) The relation $\mathfrak{f} \cap \mathcal{T} \neq \emptyset$ sets up a bijective correspondence between the set of faces $\mathfrak{f}$ of $\mathfrak{c}^{\mathrm{cl}}$ and the set of connected components $\mathcal{T}$ of orbit types in $\mathfrak{g}$.

(ii) For each face $\mathfrak{f}$ of $\mathfrak{c}^{\mathrm{cl}}$, the adjoint action induces a real-analytic, G-equivariant diffeomorphism: $G/G_{\mathfrak{f}} \times \mathfrak{f} \xrightarrow{\sim} \mathcal{T}_{\mathfrak{f}}$. Here $\mathcal{T}_{\mathfrak{f}}$, the connected component of $\mathfrak{f}$ in $\mathfrak{g}_{\mathfrak{f}}^{\sim}$, is a locally closed, real-analytic submanifold of $\mathfrak{g}$, of codimension equal to:

$$(\dim \mathfrak{t} - \dim \mathfrak{f}) + 2 \sum_{\{\alpha \in P | \alpha(X)=0\}} \dim_{\mathbf{C}} \mathfrak{g}_\alpha,$$

which is > 3 if $\mathfrak{f} \neq \mathfrak{c}$.

(iii) The restriction to $\mathfrak{c}^{\mathrm{cl}}$ of the projection: $\mathfrak{g} \to (\mathrm{Ad}\, G)\backslash\mathfrak{g}$ is a homeomorphism from $\mathfrak{c}^{\mathrm{cl}}$ onto the orbit space $(\mathrm{Ad}\, G)\backslash\mathfrak{g}$.

Proof. (i) $\mathrm{Ad}\, G(\mathfrak{f})$ is equal to the connected component $\mathfrak{f}$ in $\mathfrak{g}_{\mathfrak{f}}^{\sim}$, which does not meet another face of $\mathfrak{c}^{\mathrm{cl}}$, cf. Lemma 3.8.2.

(ii) If $X \in \mathfrak{f}$, then the codimension of $\mathfrak{g}_{\mathfrak{f}}^{\sim} = \mathfrak{g}_X^{\sim}$ in $\mathfrak{g}$ is equal to $\dim \mathfrak{g}_X - \dim \mathfrak{z}(\mathfrak{g}_X)$, see 3.2.(d). Now apply (3.7.1) and (3.8.3). Note also that $\mathfrak{f}$ is an open subset of $\mathfrak{z}(\mathfrak{g}_X)$, so $\dim \mathfrak{f} = \dim \mathfrak{z}(\mathfrak{g}_X)$.

(iii) The mapping $\mathfrak{c}^{\mathrm{cl}} \to (\mathrm{Ad}\, G)\backslash\mathfrak{g}$ is continuous and proper, as the composition of the continuous and proper mappings $\mathfrak{c}^{\mathrm{cl}} \to \mathfrak{g}$ and $\mathfrak{g} \to (\mathrm{Ad}\, G)\backslash\mathfrak{g}$. For the properness of the first mapping, use that $\mathfrak{c}^{\mathrm{cl}}$ is closed in $\mathfrak{g}$, and for the properness of the second mapping, use the conic structure in the complement of the zero orbit. Because the mapping is injective in view of Lemma 3.8.2, its inverse exists, and it is continuous, because of the properness. □

Warning. The restriction to connected components of orbit types is essential in Theorem 3.8.3. Writing $R_X = \{\alpha \in R \mid \alpha(X) = 0\}$, we have $\mathfrak{g}_X^{\sim} = \mathfrak{g}_Y^{\sim}$ for $X, Y \in \mathfrak{c}^{\mathrm{cl}}$, if and only if $s^*(R_X) = R_Y$ for some $s \in W$, and this can happen for X, Y in different faces. For instance, for $\mathfrak{su}(3)$, of rank 2, there are three root lines, which are permuted by the Weyl group, so the walls of a given chamber belong to two different connected components of the same orbit type. For $\mathfrak{su}(2) \times \mathfrak{su}(2)$ the Weyl group does not permute the two root lines however, and there are four (connected) orbit types.

A codimension-one face of $\mathfrak{c}^{\mathrm{cl}}$ is called a *wall of the Weyl chamber* $\mathfrak{c}$. An element $X \in \mathfrak{c}^{\mathrm{cl}}$ lies in a wall if and only if the $\alpha \in P$ that vanish on X span a one-dimensional subspace of $i\mathfrak{t}^*$. In Theorem 3.10.2.(i), we shall see that a root β can only be a multiple of α, if $\beta = \pm\alpha$. So $X \in \mathfrak{c}^{\mathrm{cl}}$ lies in a wall if precisely one of the $\alpha \in P$ vanishes on X. These α are precisely the positive roots that are extremal points for the convex cone in $i\mathfrak{t}^*$ generated by P, that is, such that $\alpha = \sum_{\beta \in P} c_\beta \beta$, with $c_\beta \geq 0$ for all $\beta \in P$, implies that

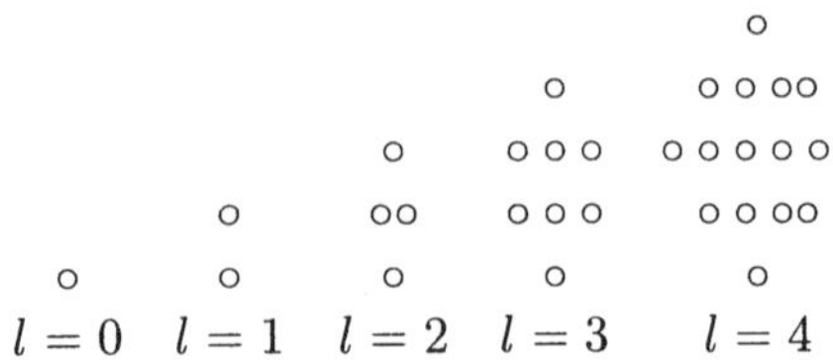

Fig. 3.8.1. Orbit type graphs

$c_\beta = 0$ for all $\beta \neq \alpha$. These α are called the *extremal roots* for the choice P of positive roots. Preceding Corollary 3.10.3, it is shown that the roots form a reduced root system in the sense of Section 5.5. There, before Theorem 5.5.9, a positive root is called a *simple root*, if it cannot be written as the sum of two positive roots. Lemma 5.5.10 shows that the set of simple roots is equal to the set of extremal roots; in the sequel we will only use the more customary term "simple root". The set of simple roots for P will be denoted by $S = S(P) = S(\mathfrak{c})$.

The relation $\alpha = 0$ on $\mathfrak{f}$ defines a bijection between S and the set of walls of $\mathfrak{c}$. It is also clear that every $\beta \in P$ can be written as $\beta = \sum_{\alpha \in S} c_\alpha \alpha$, with $c_\alpha \geq 0$ for all $\alpha \in S$. It follows in particular that:

$$\mathfrak{c} = \mathfrak{c}(P) = \mathfrak{c}(S) := \{\, X \in \mathfrak{t} \mid i^{-1}\alpha(X) > 0, \text{ for all } \alpha \in S \,\}. \tag{3.8.6}$$

Also, a face $\mathfrak{f}$ of $\mathfrak{c}^{\mathrm{cl}}$ is uniquely determined by the set:

$$S(\mathfrak{f}) := \{\, \alpha \in S \mid \alpha = 0 \text{ on } \mathfrak{f} \,\}, \tag{3.8.7}$$

which is a choice of simple roots for $\mathfrak{g}_\mathfrak{f}$.

From Theorem 5.5.9.(iv), we see that the simple roots are linearly independent. But this implies that on a suitable basis, the closure of the Weyl chamber is equal to a "corner": $X_1 \geq 0, \dots, X_l \geq 0$ in $\mathbf{R}^{\operatorname{rank}\mathfrak{g}}$. Here $l = \#(S)$ is called the *rank of the root system* R. Note that $l = \operatorname{rank}\mathfrak{g} - \dim \mathfrak{z} = \operatorname{rank}(\mathfrak{g}/\mathfrak{z})$, because $\mathfrak{z} = \bigcap_{\alpha \in R} \ker \alpha$, and every root is a linear combination of the simple ones. (Certainly the positive ones, but $R = P \cup -P$, cf. (3.5.10).)

It follows that the mapping: $\mathfrak{f} \mapsto S \setminus S(\mathfrak{f})$ is a bijection from the set of faces of $\mathfrak{c}^{\mathrm{cl}}$ onto the collection 2^S of all subsets of S, and that $\dim \mathfrak{t} - \dim \mathfrak{f} = \#(S \setminus S(\mathfrak{f}))$. In other words, this identifies the directed graph of orbit types, as introduced in the beginning of Section 2.8, with the lattice 2^S, with inclusion as the partial ordering in 2^S. So for the adjoint action of connected, compact Lie groups, this graph only depends on the number l. This is certainly not sufficient to reconstruct the Lie algebra, as we have seen with $\mathfrak{su}(3)$ and $(\mathfrak{su}(2))^2$.

Theorem 3.10.2.(i) also yields that $\dim_{\mathbb{C}} \mathfrak{g}_\alpha = 1$, for all $\alpha \in R$. So the codimension in Theorem 3.8.3.(ii) is actually equal to $\dim \mathfrak{t} - \dim \mathfrak{f} + 2\#\{\alpha \in P \mid \alpha = 0 \text{ on } \mathfrak{f}\}$; and this is precisely **equal to 3**, if $\mathfrak{f}$ is a wall. In any case, as a consequence of the estimate of the codimension of the singular orbit types, we get:

(3.8.4) Corollary.

(i) $\mathfrak{g}^{\mathrm{reg}}$ *is simply connected.*

(ii) G/H *is simply connected for every connected, compact subgroup* H *of* G *such that* $\operatorname{rank} H = \operatorname{rank} G$.

(iii) All adjoint orbits in $\mathfrak{g}$ *are simply connected.*

Proof. (i) 3.2.(d), combined with the estimate in Theorem 3.8.3.(ii), shows that $\mathfrak{g}^{\mathrm{sing}} = \mathfrak{g} \setminus \mathfrak{g}^{\mathrm{reg}}$ is stratified by submanifolds $\mathcal{T}$ of $\mathfrak{g}$, each of codimension ≥ 3 in $\mathfrak{g}$.

Any closed curve γ in $\mathfrak{g}^{\mathrm{reg}}$ can be contracted, within $\mathfrak{g}$, to a point. We may view γ as a continuous mapping: $\partial D \to \mathfrak{g}^{\mathrm{reg}}$, where D is the unit disc in the plane and ∂D its boundary circle; and then the contractability means that γ can be extended to a continuous mapping $\Gamma\colon D \to \mathfrak{g}$.

An elementary form of the transversality theorem states the following. Assume that X_0 is a compact subset of a submanifold X of Z, and that Y_0 is a compact subset of some manifold Y, and that $\dim X + \dim Y < \dim Z$. Then the C^1 mappings $f\colon Y \to Z$ such that $f(Y_0) \cap X_0 = \emptyset$, form a dense subset in the C^1 topology. (See for instance Guillemin and Pollack [1974].)

Suppose that we have already that $\Gamma'(D)$ is disjoint from the union $\mathcal{U}$ of all $\mathcal{T}$ of dimension $< k$, for some Γ' that is C^1 close to Γ. Recall that the $\mathcal{T}$ form a stratification; this implies that, for each $\mathcal{T}$, the set $\mathcal{T}^{\mathrm{cl}} \setminus \mathcal{T}$ is a union of strictly lower dimensional strata, cf. the definition preceding Theorem 2.7.3. So $\mathcal{U}$ is closed, and $\Gamma'(D) \subset V$, for some open subset V such that $V^{cl} \cap \mathcal{U} = \emptyset$. For each $\mathcal{T}$ of dimension k, $\mathcal{T}^{cl} \setminus \mathcal{T} \subset \mathcal{U}$; hence $V^{\mathrm{cl}} \cap \mathcal{T}$ is compact. Applying the transversality principle above, with X equal to the union of the k-dimensional strata, X_0 to its intersection with V^{cl}, Z to $\mathfrak{g} \setminus \mathcal{U}$, and Y_0 to D, we get by induction on k finally a mapping $\Gamma''\colon D \to \mathfrak{g}$, C^1 close to Γ, such that $\Gamma''(D) \cap \mathfrak{g}^{\mathrm{sing}} = \emptyset$, or $\Gamma''(D) \subset \mathfrak{g}^{\mathrm{reg}}$.

In other words, the curve $\gamma'' := \Gamma''|_D$ can be contracted, **within** $\mathfrak{g}^{\mathrm{reg}}$, to a point. Because γ'' is uniformly close to γ, it is homotopic, within the open subset $\mathfrak{g}^{\mathrm{reg}}$ of $\mathfrak{g}$, to γ; so γ can be contracted, within $\mathfrak{g}^{\mathrm{reg}}$, to a point as well. The conclusion is therefore that $\mathfrak{g}^{\mathrm{reg}}$ is simply connected.

(ii) We have the covering: $G/T \times \mathfrak{c} \to \mathfrak{g}^{\mathrm{reg}}$, with covering group $\mathrm{N}(\mathfrak{c})/T$, cf. 3.2.(h). Because $\mathfrak{g}^{\mathrm{reg}}$ is simply connected, this covering is a diffeomorphism (hence $\mathrm{N}(\mathfrak{c}) = T$). Lifting a closed curve in G/T to one in the product and then projecting the homotopy to a point in the product to the first coordinate, we get a contraction of the curve in G/T to a point. That is, G/T is simply connected as well.

The condition $\operatorname{rank} H = \operatorname{rank} G$ means that each maximal torus T in H

is also a maximal torus in G. In order to prove that G/H is simply connected, we use that the embedding $T \subset H$ induces a fibration: $\pi\colon G/T \to G/H$, with H/T as fiber. In order to apply the argument above to conclude that G/H is simply connected, we must be able to lift any closed curve γ in G/H to a closed curve δ in G/T, that is, one satisfying $\pi \circ \delta = \gamma$.

Using local trivializations of the bundle $\pi\colon G/T \to G/H$ over a sequence of coordinate patches along γ one can construct a continuous curve δ_0 in G/T such that $\pi \circ \delta_0 = \gamma$. (One may also replace γ by a smooth curve which is homotopic to it, and then take for δ_0 a horizontal curve over γ, with respect to some connection in the bundle.) Because H is connected, the fiber F over $\gamma(0) = \gamma(1)$ is connected as well. Note that $\delta_0(0)$ and $\delta_0(1)$ both lie in F. Let δ_1 be a curve in F, running from $\delta_0(1)$ back to $\delta_0(0)$. If δ denotes the curve "first δ_0 and then δ_1", then δ is a closed curve in G/T such that $\pi \circ \delta$ is homotopic to γ. Projecting a contraction to a point, in G/T, of δ, to G/H, we get a contraction to a point, in G/H, of γ.

(iii) For each $X \in \mathfrak{g}$, the adjoint action induces a diffeomorphism: $G/G_X \xrightarrow{\sim} \operatorname{Ad} G(X)$. Now $X \in \mathfrak{t}$, for a maximal Abelian subspace $\mathfrak{t}$ of $\mathfrak{g}$; if $T = \exp \mathfrak{t}$ is the corresponding maximal torus, we get $T \subset G_X$, so $\operatorname{rank} G_X = \operatorname{rank} G$. Because G_X is connected (cf. Theorem 3.3.1.(ii)), we can apply (ii). □

Remarks.

(a) In the proof above we just reconstructed the argument for the exactness at π_1(base) of the homotopy sequence:

$$\cdots \to \pi_1(\text{bundle}) \to \pi_1(\text{base}) \to \pi_0(\text{fiber})$$

in algebraic topology. (See Steenrod [1951], Section 17.11.)

(b) The road, of first proving that G_X is connected and then establishing that the adjoint orbit of X is simply connected, can also be traveled in the other direction. It can be shown that the restriction to $\operatorname{Ad} G(X)$, of the generic linear form on $\mathfrak{g}$, is a Morse function on $\operatorname{Ad} G(X)$, all of whose critical points have even indices. The existence of such a Morse function leads to very strong conclusions about the topology of $G/G_X = \operatorname{Ad} G(X)$, one of them being that G/G_X is simply connected. From the exactness at $\pi_0(G_X)$ of the homotopy sequence:

$$0 = \pi_1(G/G_X) \to \pi_0(G_X) \to \pi_0(G) = 0$$

it then follows that G_X is connected.

(c) The exactness at $\pi_1(G)$ of the homotopy sequence:

$$\cdots \to \pi_1(T) \to \pi_1(G) \to \pi_1(G/T) \to \cdots$$

for the bundle $G \to G/T$ with fiber T, is proved in a similar way as the one in (a). The simple connectedness of G/T now implies that the inclusion $T \subset G$

induces a surjective homomorphism: $\pi_1(T) \to \pi_1(G)$. Because $\exp\colon \mathfrak{t} \to T$ is the universal covering of T, we have:

$$\Lambda := \ker \exp \cap \mathfrak{t} \xrightarrow{\sim} \pi_1(T), \tag{3.8.8}$$

where the isomorphism is given by the mapping that assigns to $X \in \Lambda$ the homotopy class in T of the closed curve $t \mapsto \exp tX$, for $t \in [0,1]$. Composing this with the surjective homomorphism $\pi_1(T) \to \pi_1(G)$, we get the isomorphism:

$$\Lambda/\Lambda_0 \xrightarrow{\sim} \pi_1(G), \tag{3.8.9}$$

where Λ_0 is the additive subgroup of the lattice Λ, defined by:

$$\Lambda_0 := \{ X \in \Lambda \mid t \mapsto \exp tX, \text{ for } t \in [0,1], \text{ is contractible in } G \}. \tag{3.8.10}$$

In particular we recover that $\pi_1(G)$ is Abelian, a fact generally true for Lie groups (or even topological groups), cf. Proposition 1.13.2. Note that this in turn implies that the $\pi_1(G, x_0)$, for various base points x_0, are canonically isomorphic to each other, so we don't have to worry about base points here.

3.9 The Fundamental Group

We start with some results analogous to Corollary 3.8.4 and Proposition 3.8.1, for the action of conjugation in the group.

(3.9.1) Lemma.

(i) The set $G\backslash G^{\mathrm{reg}}$ is equal to the disjoint union of finitely many connected, locally closed, real-analytic submanifolds, the singular infinitesimal orbit types, each of codimension ≥ 3 in G.

(ii) The inclusion: $G^{\mathrm{reg}} \subset G$ induces an isomorphism: $\pi_1(G^{\mathrm{reg}}) \xrightarrow{\sim} \pi_1(G)$.

Proof. (i) All the properties of the singular set, except for the estimate of the codimension, have been summed up in 3.1.(g).

Let $x \in G$, because of Theorem 3.7.1.(iv) we may assume that $x \in T$, that is, we may write $x = \exp X$, with $X \in \mathfrak{t}$. Now $\mathrm{Ad}\, x = \mathrm{Ad} \exp X = \mathrm{e}^{\mathrm{ad}\, X}$, while $\mathrm{ad}\, X$ acts on $\mathfrak{g}_\alpha$ as multiplication by $\alpha(X)$, so $\mathrm{Ad}\, x$ acts on $\mathfrak{g}_\alpha$ as multiplication by $\mathrm{e}^{\alpha(X)}$. Hence:

$$\mathfrak{g}_x = \mathfrak{t} \oplus \sum_{\{ \alpha \in P \mid \alpha(X) \in 2\pi i \mathbf{Z} \}} (\mathfrak{g}_\alpha \oplus \mathfrak{g}_{-\alpha}) \cap \mathfrak{g}, \tag{3.9.1}$$

and

$$\mathfrak{z}(\mathfrak{g}_x) = \bigcap_{\{ \alpha \in P \mid \alpha(X) \in 2\pi i \mathbf{Z} \}} \ker \alpha \cap \mathfrak{t}. \tag{3.9.2}$$

So the codimension of the infinitesimal orbit type of x in G is equal to:

$$\dim \mathfrak{g}_x/\mathfrak{z}(\mathfrak{g}_x) = \text{ number of linearly independent roots } \alpha \\ \text{such that } \alpha(X) \in 2\pi i\mathbf{Z} + 2 \sum_{\{\alpha\in P \mid \alpha(X)\in 2\pi i\mathbf{Z}\}} \dim_{\mathbf{C}} \mathfrak{g}_\alpha .$$

This number is ≥ 3, unless it is equal to 0; in that case $\mathfrak{g}_x = \mathfrak{t}$, and x is regular.

(ii) The proof in Corollary 3.8.4 that $\mathfrak{g}^{\mathrm{reg}}$ is simply connected, can now be adapted to show that $\pi_1(G^{\mathrm{reg}}) = \pi_1(G)$. One replaces the disc D on which the contractions of loops were defined by a cylinder, that is, a circle $\times[0,1]$, on which now homotopies between loops are defined. □

(3.9.2) Lemma. *Let T be a maximal torus in G, with Lie algebra equal to $\mathfrak{t}$. Let B be a connected component of $G^{\mathrm{reg}} \cap T$. Define $\mathrm{N}(B) = \{ g \in G \mid gBg^{-1} = B \}$. Then:*

(i) $T \subset \mathrm{N}(B) \subset \mathrm{N}(T)$. Furthermore, $g \in \mathrm{N}(B)$ as soon as there exists some $t \in B$, such that $gtg^{-1} \in B$.

(ii) The action of conjugation induces a G-equivariant real-analytic diffeomorphism:

$$G/T \times_{\mathrm{N}(B)/T} B \xrightarrow{\sim} G^{\mathrm{reg}}.$$

(iii) The restriction to B^{cl} of the projection: $G \to (\mathbf{Ad}\, G)\backslash G$ induces a homeomorphism from the orbit space $(\mathrm{N}(B)/T)\backslash B^{\mathrm{cl}}$ of B^{cl} for the action of $\mathrm{N}(B)/T$ on it, onto the space of conjugacy classes.

Proof. (i) In the proof of Theorem 3.7.2.(i), we have already seen that $g \in \mathrm{N}(T)$ as soon as $s := gtg^{-1} \in T$ for some $t \in G^{\mathrm{reg}} \cap T$. If moreover $t, s \in B$, then gBg^{-1} is a connected component of $G^{\mathrm{reg}} \cap T$ that meets B, and therefore it is equal to B.

(ii) Combining Theorem 3.7.2.(i) with the result above, it follows that the described mapping is a diffeomorphism onto its image, which is an open and closed, nonvoid subset of G^{reg}. Because G^{reg} is connected, the image is equal to G^{reg}.

(iii) The mapping: $(\mathrm{N}(B)/T)\backslash B^{\mathrm{cl}} \to (\mathbf{Ad}\, G)\backslash G$ is continuous and surjective. The latter because the image is compact and contains the dense subset $(\mathbf{Ad}\, G)\backslash G^{\mathrm{reg}}$. In view of the compactness of the source space, it suffices therefore to prove that the mapping is injective, this will be done as in the proof of Lemma 3.8.2.

Suppose $x, y \in B^{\mathrm{cl}}$, $g \in G$, and $y = gxg^{-1}$. Approximate x by $x' \in B$, then $y' := gx'g^{-1}$ is close to y, and $x' \in G_x$ implies that $y' \in G_y$. Now (3.9.1) shows that T is a maximal torus in the connected, compact Lie group $H := (G_y)^\circ$. Moreover, the roots are precisely such that, if $'\mathfrak{t}^{\mathrm{reg}}$ denotes the regular set in $\mathfrak{t}$ with respect to H, then there is an open neighborhood U of 0 in $\mathfrak{t}$, such that $y \exp(U \cap' \mathfrak{t}^{\mathrm{reg}}) = (y \exp U) \cap G^{\mathrm{reg}}$. But this means that there exists an $h \in H$, such that $h^{-1}y^{-1}y'h \in y^{-1}B$, or, because $h \in G_y$, $h^{-1}gx'g^{-1}h = h^{-1}y'h \in B$. That is, $h^{-1}g \in \mathrm{N}(B)$, or, because $y = h^{-1}yh = (h^{-1}g)x(h^{-1}g)^{-1}$, y is in the $\mathrm{N}(B)$-orbit of x. □

The subgroup $\mathrm{N}(B)/T$ of the Weyl group is closely related to the fundamental group of G. In order to explain this, we turn to a connected component of $\exp^{-1}(G^{\mathrm{reg}}) \cap \mathfrak{t}$, it is called an *alcove* in $\mathfrak{t}$.

(3.9.3) Lemma. *Let B be a connected component of $G^{\mathrm{reg}} \cap T$. Then there exists a choice of positive roots P on $\mathfrak{t}$, with the following properties. On the alcove:*

(3.9.3) $$\mathfrak{b} = \{\, X \in \mathfrak{t} \mid 0 < (2\pi i)^{-1}\alpha(X) < 1, \text{ for all } \alpha \in P \,\},$$

the mapping $\exp|_{\mathfrak{b}}$ is a covering from $\mathfrak{b}$ onto B. If $X, Y \in \mathfrak{b}$, then $\exp X = \exp Y$ if and only if $X - Y \in \ker \exp \cap \mathfrak{z} = \Lambda \cap \mathfrak{z}$.

Proof. B contains a point, hence a connected component C of $G^{\mathrm{princ}} \cap T$, because at each point of B, the set T is a slice. In the proof of Proposition 3.1.3, we observed that $1 \in C^{\mathrm{cl}}$, so $1 \in (B)^{\mathrm{cl}}$ as well. This means that B contains elements of the form $x = \exp X$, with $X \in \mathfrak{t}$ as close to 0 as you like. The condition that x is regular means that $\alpha(X)$ does not belong to $2\pi i\mathbf{Z}$ for all roots α, so certainly $X \in \mathfrak{t}^{\mathrm{reg}}$. Let P be the choice of positive roots for the Weyl chamber to which X belongs. Define $\mathfrak{b}$ as in (3.9.3), this is clearly the connected component of X in $\exp^{-1}(G^{\mathrm{reg}}) \cap \mathfrak{t}$. The covering $\exp\colon \mathfrak{t} \to T$ maps $\mathfrak{b}$ onto a nonvoid, connected open subset of $G^{\mathrm{reg}} \cap T$ which meets B, so $\exp \mathfrak{b} \subset B$.

Now let $y \in B$. Take a curve γ in B, running from x to y. Then there is a unique curve δ in $\mathfrak{t}$, starting at X, such that $\gamma = \exp \circ \delta$. Now all boundary points of $\mathfrak{b}$ in $\mathfrak{t}$ are clearly mapped by exp to singular elements of G, so δ has to remain in $\mathfrak{b}$, in particular $y = \exp Y \in \exp \mathfrak{b}$, if we write $Y = \delta(1)$. This shows that $\exp \mathfrak{b} = B$. Furthermore, because $\exp\colon \mathfrak{t} \to T$ is a covering, we can find a neighborhood U of Y in $\mathfrak{t}$ such that U is disjoint from $Z + U$, whenever $Z \in \Lambda = \ker \exp$, and $Z \neq 0$. By shrinking U further, we may assume that $U \subset \mathfrak{b}$ and that U is convex. But then, for every $Z \in \Lambda$, either $Z + U$ is contained in $\mathfrak{b}$, or disjoint from it, because $\exp(Z + U) = \exp U \subset B \subset G^{\mathrm{reg}}$ shows that it cannot contain boundary points of $\mathfrak{b}$. This proves that $\exp|_{\mathfrak{b}}$ is a covering: $\mathfrak{b} \to B$.

Finally $Z \in \Lambda$ implies that $e^{\mathrm{ad}\, Z} = \mathrm{Ad}(\exp Z) = \mathrm{Ad}\, 1 = 1$, so $\alpha(Z) \in 2\pi i\mathbf{Z}$ **for all** $\alpha \in R$. Because $\exp X = \exp Y$, for $X, Y \in \mathfrak{b}$, implies that $X - Y \in \Lambda$, we see readily from the definition of $\mathfrak{b}$ that this can only happen if $\alpha(X - Y) = 0$ for all $\alpha \in R$, or $X - Y \in \mathfrak{z}$. □

(3.9.4) Corollary. *Let B be a connected component of $G^{\mathrm{reg}} \cap T$. Then we have an exact sequence:*

(3.9.4) $$0 \to \Lambda \cap \mathfrak{z} = \pi_1(\mathrm{Z}(G)^\circ) \to \pi_1(G) \to \mathrm{N}(B)/T \to 0,$$

where the arrow $\pi_1(G) \to \mathrm{N}(B)/T$ *comes from the covering:* $G/T \times B \to G^{\mathrm{reg}}$, *induced by the action of conjugation. The quotient* $\mathrm{N}(B)/T$ *is an Abelian subgroup of the (finite) Weyl group* $W = \mathrm{N}(T)/T$.

We conclude that $\pi_1(G)$ *is finite if and only if* $\mathfrak{z}$, *the center of* $\mathfrak{g}$, *is equal to zero. In particular,* $\pi_1(\mathrm{Ad}\,\mathfrak{g})$ *is finite, and the universal covering group of* $\mathrm{Ad}\,\mathfrak{g}$ *is compact.*

Proof. As already indicated in the Corollary, $G/T \times B \to G^{\mathrm{reg}}$ is a covering, with fiber $\mathrm{N}(B)/T$. In this case the exact π_1-sequence for coverings (compare with the proof of Corollary 3.8.4) takes the form:

$$0 \to \pi_1(G/T \times B) \to \pi_1(G^{\mathrm{reg}}) \to \mathrm{N}(B)/T \to 0.$$

Because G/T is simply connected (Corollary 3.8.4), $\pi_1(G/T \times B) = \pi_1(B)$. On the other hand, the convex polytope $\mathfrak{b}$ is simply connected, so $\exp\colon \mathfrak{b} \to B$ is the universal covering of B, with fiber equal to $\Lambda \cap \mathfrak{z}$, cf. Lemma 3.9.3. This leads to an isomorphism: $\Lambda \cap \mathfrak{z} \xrightarrow{\sim} \pi_1(B)$ as in (3.8.8). Finally, use Lemma 3.9.1.(ii).

Because $\pi_1(G)$ is Abelian, (see the remark after (3.8.10)), (3.9.4) is actually an exact sequence of Abelian groups. For the statements about the adjoint group of $\mathfrak{g}$, we refer to the remarks after Theorem 3.6.2. □

(3.9.5) Corollary.

(i) For every $x \in G^{\mathrm{reg}}$, *the inclusion of its conjugacy class* $\mathbf{Ad}\,G(x)$ *in* G *induces an injective mapping:* $G_x/(G_x)^\circ = \pi_1(\mathbf{Ad}\,G(x)) \to \pi_1(G)$.

(ii) If $x \in B$, *with* B *as in Corollary 3.9.4, then composing this injection with the mapping* $\pi_1(G) \to \mathrm{N}(B)/T$ *of (3.9.4), we get the inclusion* $G_x/T \to \mathrm{N}(B)/T$, *followed by the inversion in the latter group.*

(iii) If $\mathfrak{z} = 0$, *then there exists an* $x \in B$, *such that* $G_x = \mathrm{N}(B)$. *For such* x, *the inclusion* $\mathbf{Ad}\,G(x) \subset G$ *induces an isomorphism:* $\pi_1(\mathbf{Ad}\,G(x)) \xrightarrow{\sim} \pi_1(G)$.

(iv) G *is simply connected, if and only if* $\mathfrak{z} = 0$ *and* G *has no exceptional conjugacy classes. If this is the case, then* $G^{\mathrm{reg}} = G^{\mathrm{princ}}$ *is diffeomorphic to* $G/T \times \mathfrak{b}$, *where* $\mathfrak{b}$ *is the convex polytope defined in (3.9.3), which now is bounded.*

Proof. (i), (ii) Because of Lemma 3.9.2.(ii), we may assume from the outset that $x \in B$. Then $(G_x)^\circ = T$, and we have a covering $G/T \to G/G_x = \mathbf{Ad}\,G(x)$ with fiber G_x/T. Because G/T is simply connected, $\pi_1(\mathbf{Ad}\,G(x)) = G_x/T$ follows. The element $gT \in G_x/T$ here represents the homotopy class of the closed curve $\delta\colon t \mapsto \gamma(t)x\gamma(t)^{-1}$ in $\mathbf{Ad}\,G(x)$, where γ denotes any curve in G running from 1 to g. In terms of the covering: $G/T \times B \to G^{\mathrm{reg}}$, the curve δ lifts to the curve $t \mapsto (\gamma(t)T, x)$ in $G/T \times B$, running from $(1T, x)$ to (gT, x). Now the quotient $G/T \times_{\mathrm{N}(B)/T} B = G^{\mathrm{reg}}$ was obtained by dividing out the action of $y \in \mathrm{N}(B)$ on $G/T \times B$, whereby y acts as $\mathbf{Ad}\,y$ on B, and as right multiplication with y^{-1} on G/T. Therefore it is the covering transformation defined by $y = g^{-1}$ which maps $(1T, x)$ to (gT, x).

(iii) For any $g \in \mathrm{N}(B)$, the mapping $\exp\colon \mathfrak{b} \to B$, which now is a diffeomorphism, conjugates $\mathbf{Ad}\, g\colon B \to B$ with an affine linear transformation: $\mathfrak{b} \to \mathfrak{b}$. Using averaging as in the proof of Proposition 3.8.1, we arrive at the existence of a fixed point for the induced action of $\mathrm{N}(B)/T$ on $\mathfrak{b}$, or, going back to B, the existence of an $x \in B$ such that $\mathrm{N}(B) \subset G_x$. Because the other inclusion was part of Lemma 3.9.2.(i), the proof is complete.

(iv) follows immediately from (i)-(iii). □

Remarks. If $\mathfrak{z} \neq 0$, then the situation with (iii) is somewhat more complicated. In order to explain the situation, we introduce the connected Lie subgroup G' with Lie algebra equal to the derived Lie algebra $\mathfrak{g}' = [\mathfrak{g}, \mathfrak{g}]$, whose existence is guaranteed by Theorem 1.10.4. G' is called the *derived group* of the Lie group G.

From (i) ⇒ (vi) ⇒ (vii) in Theorem 3.6.2 (the proof of these implications was easy), it follows that $\mathfrak{g} = \mathfrak{g}' \oplus \mathfrak{z}$. Hence $\mathrm{ad}\,|_{\mathfrak{g}'}$ is an isomorphism of Lie algebras: $\mathfrak{g}' \xrightarrow{\sim} \mathrm{ad}\,\mathfrak{g}$; and therefore $\mathrm{Ad}\,|_{G'}$ is a Lie group covering: $G' \to \mathrm{Ad}\, G = Ad\mathfrak{g}$. Moreover, $\mathfrak{z}(\mathfrak{g}') = \mathfrak{z}(\mathfrak{g}' + \mathfrak{z}) \cap \mathfrak{g}' = \mathfrak{z} \cap \mathfrak{g}' = 0$. Hence also $\mathfrak{z}(\mathrm{ad}\,\mathfrak{g}) = 0$; so by Corollary 3.9.4, the fundamental group of the connected, compact group $\mathrm{Ad}\, G$ is finite. It follows that the fiber $G' \cap \mathrm{Z}(G)$ of the covering: $G' \to \mathrm{Ad}\, G$ is finite as well, in particular we may conclude: G' **is compact**.

The multiplication $(x, z) \mapsto xz$ defines a homomorphism of Lie groups $\mu\colon G' \times \mathrm{Z}(G)^\circ \to G$, with:

$$\ker \mu = \{\, (x, x^{-1}) \mid x \in G' \cap \mathrm{Z}(G)^\circ \,\} \simeq G' \cap \mathrm{Z}(G)^\circ = \mathrm{Z}(G') \cap \mathrm{Z}(G)^\circ,$$

which is finite. So μ is a covering, in particular it is surjective. Moreover, $G^{\mathrm{reg}} = (G')^{\mathrm{reg}} \cdot \mathrm{Z}(G)^\circ$, if T' is a maximal torus in G' and B' a connected component of $(G')^{\mathrm{reg}} \cap T'$, then $T = T' \cdot \mathrm{Z}(G)^\circ$ is a maximal torus in G and $B = B' \cdot \mathrm{Z}(G)^\circ$ is a connected component in $G^{\mathrm{reg}} \cap T$. We get an obvious injective homomorphism: $\mathrm{N}(B')/T' \to \mathrm{N}(B)/T$. Finally, if $x \in B'$, and $z \in \mathrm{Z}(G)^\circ$, then $G_{xz}/T = G'_x/T'$, which goes injectively into $\mathrm{N}(B')/T'$.

On the other hand the commutative diagram below with exact rows and columns leads to an exact sequence: $0 \to \mathrm{N}(B')/T' \to \mathrm{N}(B)/T \to \ker \mu \to 0$. So the difference between $\mathrm{N}(B)/T$ and $\mathrm{N}(B')/T'$ (that is, the maximum at which G_x/T can get for $x \in B$), is measured by $\ker \mu$. Because it is easy (starting with a G' with a nontrivial center) to construct examples with $\ker \mu \neq 1$, we see that in general Corollary 3.9.5.(iii) does not hold.

$$
\begin{array}{ccccccccc}
 & & & & 0 & & & & \\
 & & & & \downarrow & & & & \\
 & & 0 & \longrightarrow & \ker\mu & & & & \\
 & & \downarrow & & \downarrow & & & & \\
0 & \longrightarrow & \pi_1(\mathrm{Z}(G)^\circ) & \longrightarrow & \pi_1(G) & \longrightarrow & \mathrm{N}(B)/T & \longrightarrow & 0 \\
 & & \downarrow & & \downarrow & & \downarrow & & \\
0 & \longrightarrow & \pi_1(\mathrm{Z}(G)^\circ) & \longrightarrow & \pi_1(G' \times \mathrm{Z}(G)^\circ) & \longrightarrow & \pi_1(G') = \mathrm{N}(B')/T' & \longrightarrow & 0 \\
 & & \downarrow & & \downarrow & & & & \\
 & & 0 & & 0 & & & &
\end{array}
$$

It follows from the considerations above that G **has no exceptional orbits if and only if** G' **is simply connected**, but it is not clear how useful this criterion is in a given situation.

Now let us take a somewhat closer look at the affine linear transformations in $\mathfrak{t}$ that appeared in the proof of Corollary 3.9.5.(iii).

The lattice $\Lambda = \ker \exp \cap \mathfrak{t}$ acts by translations $+Y\colon X \mapsto X+Y$, for $Y \in \Lambda$. Together with the linear transformations s of the Weyl group $W = \mathrm{N}(T)/T$, these generate a group of affine linear transformations in $\mathfrak{t}$, called the *affine Weyl group* $\Lambda \cdot W$ of $\mathfrak{t}$, with respect to the lattice Λ. The mapping $(Y, s) \mapsto (+Y) \circ s$ defines a bijection: $\Lambda \times W \xrightarrow{\sim} \Lambda \cdot W$, which pulls back the composition in $\Lambda \cdot W$ to the product:

$$(Y, s) \cdot (Y', s') = (Y + s(Y'), ss'), \tag{3.9.5}$$

because $s \circ (+Y) = (+s(Y)) \circ s$; note that Λ is W-invariant.

In general, if C is a group with subgroups A and B, then the product mapping $(a, b) \to ab$ defines a bijective mapping from $A \times B$ onto C, if and only if $A \cdot B = C$ and $A \cap B = 1$. The group C is said to be the *semidirect product* of A and B if moreover A is a normal subgroup of C; and then $B \xrightarrow{\sim} C/A$ is an isomorphism of groups. So here $\Lambda \cdot W$ is equal to the semidirect product of Λ and W, with Λ as the normal subgroup.

The covering map $(gT, X) \mapsto \exp \operatorname{Ad} g(X)\colon G/T \times \mathfrak{b} \to G^{\mathrm{reg}}$ yields the isomorphism:

$$\pi_1(G) \xrightarrow{\sim} \mathrm{N}_{\Lambda \cdot W}(\mathfrak{b}) := \{\, \phi \in \Lambda \cdot W \mid \phi(\mathfrak{b}) = \mathfrak{b} \,\}, \tag{3.9.6}$$

if we also observe that:

$$\mathrm{N}(B)/T = \mathrm{N}_{\Lambda \cdot W}(\mathfrak{b})/(\Lambda \cap \mathfrak{z}), \tag{3.9.7}$$

see Lemma 3.9.3. This is a quite straightforward variation of Corollary 3.9.4.

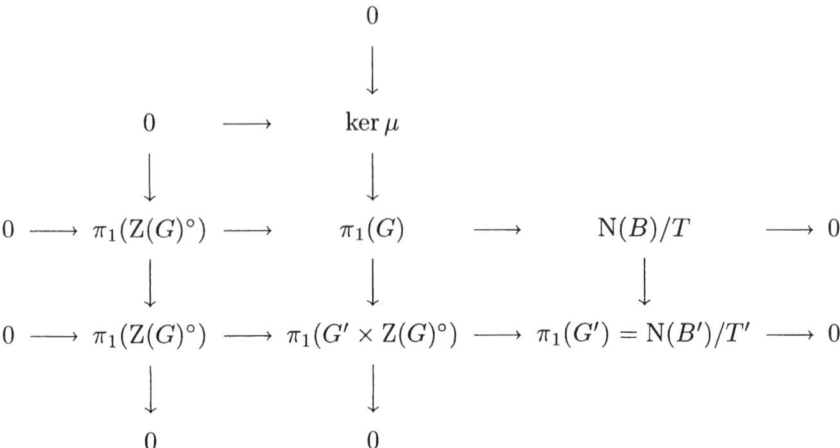

If $\phi \in \Lambda \cdot W$, then $\phi(0) \in \Lambda$. If moreover $\phi(\mathfrak{b}) = \mathfrak{b}$, then $\phi(\mathfrak{b}^{\mathrm{cl}}) = \mathfrak{b}^{\mathrm{cl}}$; hence $\phi(0) \in \mathfrak{b}^{\mathrm{cl}}$. If conversely $Y \in \Lambda \cap \mathfrak{b}^{\mathrm{cl}}$, then $-Y$ maps $\mathfrak{b}$ to a connected component $\mathfrak{b}'$ of $\exp^{-1}(G^{\mathrm{reg}}) \cap \mathfrak{t}$ that contains 0 in its closure. Near 0 only the root hyperplanes appear as a boundary, so $\mathfrak{b}'$ coincides near 0 with a Weyl chamber. Because the Weyl group acts freely and transitively on the set of Weyl chambers (Proposition 3.8.1), there is a unique $s \in W$ such that $s(\mathfrak{b}) = \mathfrak{b}'$; near 0, but then this holds globally, because $s(\mathfrak{b})$ is a connected component of $\exp^{-1}(G^{\mathrm{reg}}) \cap \mathfrak{t}$. So $\phi = (+Y) \circ s$ maps $\mathfrak{b}$ onto itself, and it is the unique element of $\mathrm{N}_{\Lambda \cdot W}(\mathfrak{b})$ that maps 0 to Y. This shows that $\phi \mapsto \phi(0)$ defines a bijection from $\mathrm{N}_{\Lambda \cdot W}(\mathfrak{b})$ onto $\mathfrak{b}^{\mathrm{cl}} \cap \Lambda$.

Now let $\phi = (+Y) \circ s \in \mathrm{N}_{\Lambda \cdot W}(\mathfrak{b})$, with $Y \in \Lambda$, and $s \in W$. We can write $s = \mathrm{Ad}\, g|_{\mathfrak{t}}$, for some $g \in \mathrm{N}(T)$. Choose any curve γ in G, running from 1 to g. Next take any $X \in \mathfrak{b}$. Then $\phi(X) \in \mathfrak{b}$; and, because $\mathfrak{b}$ is convex, $t\phi(X) + (1-t)X \in \mathfrak{b}$, for all $t \in [0,1]$. That is, the curve:

$$\delta_X : t \mapsto \gamma(t)^{-1} \exp(t\phi(X) + (1-t)X)\gamma(t) \colon [0,1] \to G \tag{3.9.8}$$

runs entirely in G^{reg}, if $X \in \mathfrak{b}$. Moreover, $\delta_X(0) = \exp X$, $\delta_X(1) = g^{-1} \exp \phi(X) g = g^{-1} \exp(s(X) + Y) g = g^{-1} \exp s(X) g = \exp X$, so δ_X is closed. Running through the isomorphism in (3.9.6), we see that the homotopy class of δ_X is the element of $\pi_1(G)$ that corresponds to $\phi \in \mathrm{N}_{\Lambda \cdot W}(\mathfrak{b})$. If X is a fixed point of ϕ, that is, $g \in G_{\exp X}$, then $\delta_X(t) = \gamma(t)^{-1} \exp X \gamma(t)$; and we recover the closed curve in the conjugacy class of $x = \exp X$, as discussed in the proof of Corollary 3.9.5.(ii).

However, as closed curves in G, the δ_X are defined for all $X \in \mathfrak{t}$, and are homotopic to each other. In particular we may consider $\delta_0 \colon t \mapsto \gamma(t)^{-1}(\exp tY)\gamma(t)^{-1}$. Now the fact that $\exp Y = 1$ implies that $\delta_{X,s} \colon t \mapsto \gamma(st)^{-1}(\exp tY)\gamma(st)$, with t running from 0 to 1, is a closed curve, for every $s \in [0,1]$. Letting s drop from 1 to 0, we get a homotopy from δ_X to the curve $t \mapsto \exp tY$, with t running from 0 to 1. In this way we have obtained an explicit set of representatives in $\pi_1(T)$, mapped bijectively onto $\pi_1(G)$ under the natural projection $\pi_1(T) \to \pi_1(G)$, cf. (3.8.9).

We now summarize this alternative formulation of Corollary 3.9.5 in:

(3.9.6) Corollary. *Let G be a connected, compact Lie group with Lie algebra $\mathfrak{g}$. Let $\mathfrak{t}$ be a maximal Abelian subspace of $\mathfrak{g}$, W the Weyl group of $\mathfrak{t}$, and $\Lambda \cdot W$ the affine Weyl group of $\mathfrak{t}$ with respect to the lattice $\Lambda = \ker \exp \cap \mathfrak{t}$. Let $\mathfrak{b}$ be the alcove in $\mathfrak{t}$ defined in (3.9.3), write $\mathrm{N}_{\Lambda \cdot W}(\mathfrak{b}) = \{\phi \in \Lambda \cdot W \mid \phi(\mathfrak{b}) = \mathfrak{b}\}$. Then we have:*

(i) $\phi \mapsto \phi(0)$ defines a bijection from $\mathrm{N}_{\Lambda \cdot W}(\mathfrak{b})$ onto the set $\mathfrak{b}^{\mathrm{cl}} \cap \Lambda$.

(ii) Composing this bijection with the mapping that assigns to any $Y \in \Lambda$ the homotopy class in $\pi_1(G)$ of the closed curve $t \mapsto \exp tY$, with t running from 0 to 1, we get an isomorphism of groups: $\mathrm{N}_{\Lambda \cdot W}(\mathfrak{b}) \xrightarrow{\sim} \pi_1(G)$.

(iii) The isomorphism in (ii) is equal to the one induced by the covering:

$$(gT, X) \mapsto \exp \mathrm{Ad}\, g(X) \colon G/T \times \mathfrak{b} \to G^{\mathrm{reg}},$$

with covering group $\mathrm{N}_{\Lambda\cdot W}(\mathfrak{b})$.

(iv) *If G is simply connected, then the mapping: $X \mapsto \mathbf{Ad}\, G(\exp X)$ defines a homeomorphism from $\mathfrak{b}^{\mathrm{cl}}$ onto $(\mathbf{Ad}\, G)\backslash G$, the space of conjugacy classes in G.*

Proof. Only the last statement needs a further comment. Combining Lemma 3.9.2.(iii) with the triviality of $\mathrm{N}(B)/T$, we get that B^{cl} is homeomorphic to $(\mathbf{Ad}\, G)\backslash G$, so it remains to show that the exponential mapping is injective on $\mathfrak{b}^{\mathrm{cl}}$. Let $X, Y \in \mathfrak{b}^{\mathrm{cl}}$, and $\exp X = \exp Y$, that is, $Z := Y - X \in \Lambda$. Then, for any root α, we have $|(2\pi i)^{-1}\alpha(X) - (2\pi i)^{-1}\alpha(Y)| \leq 1$, so $|(2\pi i)^{-1}\alpha(tZ)| < 1$ for $0 \leq t < 1$. Perturbing Z slightly to a regular element Z', we get that tZ' belongs to an alcove $\mathfrak{b}'$ if $0 < t < 1 - \delta$, where δ is arbitrarily small. That is, $Z \in (\mathfrak{b}')^{\mathrm{cl}} \cap \Lambda$, $0 \in (\mathfrak{b}')^{\mathrm{cl}}$. So $\pi_1(G) = 0$, combined with (i), (ii), yields that $Z = 0$. □

In Proposition 3.11.1, the subgroup Λ_0 of the lattice Λ, defined in (3.8.10), satisfying $\pi_1(G) = \Lambda/\Lambda_0$, will be computed in terms of the root structure. Even for the formulation, some more information about the Weyl group is needed; this is the reason why we did not do it here.

If $\mathfrak{g}$ is simple, then the closure of the alcove is a simplex, see the remark following (5.6.3). In general, it is a Cartesian product of simplices, one for each simple summand, and a line segment for each one-dimensional summand of the center.

3.10 The Weyl Group as a Reflection Group

Crossing a wall of a Weyl chamber $\mathfrak{c}$, corresponding to a simple root α, one enters an adjacent Weyl chamber $\mathfrak{c}'$. According to Proposition 3.8.1, there is exactly one $s \in W$, such that $s(\mathfrak{c}) = \mathfrak{c}'$. In order to describe the action of this Weyl group element, we need some more insight in the structure of the root spaces $\mathfrak{g}_\alpha$.

(3.10.1) Lemma. *Let G be a connected, compact Lie group of rank equal to one. Then G is isomorphic to the circle, to $\mathbf{SO}(3, \mathbf{R})$, or to $\mathbf{SU}(2)$.*

Proof. $\operatorname{rank} G = 1$ means that a maximal Abelian subspace $\mathfrak{t}$ of the Lie algebra $\mathfrak{g}$ of G is isomorphic to the real line. If $\mathfrak{g}$ is Abelian, then $\mathfrak{g} = \mathfrak{t}$, and we conclude that G is a circle; so we may assume from now on that $\mathfrak{g}$ is not Abelian.

Then $\mathfrak{t} \setminus \mathfrak{t}^{\mathrm{reg}} = 0$, so $\mathfrak{t}^{\mathrm{reg}}$ has two Weyl chambers, denoted by $\mathbf{R}_{>0}$, and $\mathbf{R}_{<0}$, respectively. In Proposition 3.8.1, we have seen that $\operatorname{Ad} G(\mathbf{R}_{>0}) = \mathfrak{g}^{\mathrm{reg}}$. Here we can replace $\mathfrak{t}$ by any line through the origin in $\mathfrak{g}$. So the conclusion is that a given regular $\operatorname{Ad} G$-orbit in $\mathfrak{g}$ meets every half-line emanating from the origin in $\mathfrak{g}$; hence it must be equal to a sphere around the origin, with respect to any $\mathrm{Ad}\, G$-invariant inner product in $\mathfrak{g}$. Using multiplications with positive scalars, we see that $\mathfrak{g}^{\mathrm{reg}} = \mathfrak{g} \setminus \{0\}$; and this set is fibered by the $\mathrm{Ad}\, G$-orbits, the concentric spheres around the origin.

A maximal torus T in G has a maximal Abelian subspace $\mathfrak{t}$ as its Lie algebra, so it is a circle group. G/T is diffeomorphic to a regular adjoint orbit, and hence to a sphere. We know already that $\dim G/T = \dim \mathfrak{g} - \dim \mathfrak{t}$ is even, cf. (3.5.8), and because we assumed that it is positive, $\dim G/T \geq 2$.

The Weyl group permutes the two connected components, the half lines in $\mathfrak{t}^{\mathrm{reg}}$ (Theorem 3.7.2.(ii)), so there exists $g \in \mathrm{N}(T)$ such that $\mathrm{Ad}\, g$ acts as -1 on $\mathfrak{t}$. Because G is connected, there exists a smooth curve $\gamma\colon [0,1] \to G$, such that $\gamma(t) = 1$, $\gamma(t) = g$. The map $\Gamma\colon (t,s) \mapsto \gamma(s) t \gamma(s)^{-1}\colon T \times [0,1] \to G$ therefore is a homotopy of loops in G, starting, for $s = 0$, with the circle $t \mapsto t\colon T \to G$, and ending, for $s = 1$, with $t \mapsto t^{-1}\colon T \to G$, which is the same circle run through backwards.

Let π denote the canonical projection: $G \to G/T$. Then $\Sigma = \pi \circ \Gamma$ is a smooth mapping: $T \times [0,1] \to G/T$, such that $\Sigma(t,0) = \Sigma(t,1) =$ the base point $b = 1T$, for all $t \in T$. (This means that Σ can be viewed as a continuous mapping from the 2-sphere into G/T.)

Now assume that $\dim G/T > 2$. Then Sard's theorem (cf. Guillemin and Pollack [1974]) implies that $\Sigma(T \times [0,1])$ has measure zero in G/T, so in particular there exists a point $p \in G/T \setminus \Sigma(T \times [0,1])$. Now the sphere G/T minus the point p is diffeomorphic to a vector space, so we can contract Σ to its base point b. That is, there exists a smooth mapping $\Phi\colon T \times [0,1] \times [0,1] \to G/T$, such that $\Phi(t,s,0) = \Sigma(t,s)$, and $\Phi(t,0,r) = \Phi(t,1,r) = \Phi(t,s,1) = b$, for all $(t,s,r) \in T \times [0,1] \times [0,1]$.

The mapping Φ can be "lifted" to a smooth mapping $\Omega\colon T \times [0,1] \times [0,1] \to G$, such that $\Phi = \pi \circ \Omega$, and such that $\Omega(t,s,0) = \Gamma(t,s)$ for all $(t,s) \in T \times [0,1]$. One way of doing this, is to take a smooth connection in the bundle $G \to G/T$, and to define $r \mapsto \Omega(t,s,r)$ as the horizontal curve in G over $r \mapsto \Phi(t,s,r)$, starting for $r = 0$ at $\Gamma(t,s)$.

The "base point conditions" now mean that all the points $\Omega(t,0,r)$, $\Omega(t,1,r)$, $\Omega(t,s,1)$ lie in T, for all $(t,s,r) \in T \times [0,1] \times [0,1]$. This implies that **within** T, the identity curve $t \mapsto \Omega(t,0,0) = \Gamma(t,0) = t$ is homotopic to $t \mapsto \Omega(t,0,1)$ (let r run from 0 to 1), and then to $t \mapsto \Omega(t,1,1)$ (let s run from 0 to 1), and finally to $t \mapsto \Omega(t,1,0) = \Gamma(t,1) = t^{-1}$ (let r go back from 1 to 0). Clearly this is impossible (using the covering $\exp\colon \mathfrak{t} \to T$, we see that $\pi_1(T)$ is isomorphic to $\mathbf{Z}$, and we would have arrived at $1 = -1$ in $\mathbf{Z}$); and the conclusion is that $\dim G/T = 2$, or $\dim G = 3$.

Now identify $\mathfrak{g}$ with $\mathbf{R}^3$ by means of an orthonormal basis with respect to the $\mathrm{Ad}\, G$-invariant inner product on $\mathfrak{g}$. Then Ad is a homomorphism of Lie groups: $G \to \mathbf{SO}(3,\mathbf{R})$. Furthermore, $\mathfrak{z} = 0$, because otherwise $\mathfrak{t} = \mathfrak{z}$, and $\mathfrak{g}$ would be Abelian. This makes $\mathrm{T}_1 \mathrm{Ad} = \mathrm{ad}$ injective, and because $\dim \mathfrak{g} = 3 = \dim \mathfrak{so}(3,\mathbf{R})$, it is bijective. Since G is connected, it follows that

$\mathrm{Ad}\colon G \to \mathbf{SO}(3,\mathbf{R})$ is a covering of Lie groups, cf. the definition preceding Corollary 1.10.5. So G is isomorphic to a quotient of the universal covering $\mathbf{SU}(2)$ of $\mathbf{SO}(3,\mathbf{R})$, by a subgroup of the center $\{-1,1\}$ of $\mathbf{SU}(2)$. See 1.2.a,b, and the discussion following Lemma 1.13.3. The only possibilities therefore are that $G = \mathbf{SU}(2)$ or $G = \mathbf{SO}(3,\mathbf{R}) = \mathrm{Ad}\,\mathbf{SU}(2)$. □

Remark. The proof actually discussed the exactness at $\pi_1(T)$ of the homotopy sequence $\cdots \to \pi_2(G/T) \to \pi_1(T) \to \pi_1(G) \to \cdots$, and observed that $\pi_2(S) = 0$, if S is a sphere of dimension > 2.

(3.10.2) Theorem. *Let G be a connected, compact Lie group, $\mathfrak{t}$ a maximal Abelian subspace of the Lie algebra $\mathfrak{g}$ of G, and α a root of $\mathfrak{t}$. Then:*

(i) $\dim_{\mathbf{C}} \mathfrak{g}_\alpha = 1$, *and* $\mathfrak{g}_{k\alpha} = 0$, *if* $k \in \mathbf{C}$ *and* $k \neq -1, 0, 1$.

(ii) $[\,\mathfrak{g}_\alpha, \mathfrak{g}_{-\alpha}\,] \subset \mathfrak{h} = \mathfrak{g}_0$, *and* $\mathfrak{g}^{(\alpha)} :=$ *the real part of* $\mathfrak{g}_\alpha \oplus \mathfrak{g}_{-\alpha} \oplus [\,\mathfrak{g}_\alpha, \mathfrak{g}_{-\alpha}\,]$ *is a 3-dimensional Lie subalgebra of* $\mathfrak{g}$, *isomorphic to* $\mathfrak{so}(3,\mathbf{R}) = \mathfrak{su}(2)$.

(iii) $G^{(\alpha)} := \exp \mathfrak{g}^{(\alpha)}$ *is a compact Lie subgroup of* G, *isomorphic to either* $\mathbf{SO}(3,\mathbf{R})$ *or* $\mathbf{SU}(2)$.

(iv) If $\alpha^\vee$ *denotes the unique element of* $[\,\mathfrak{g}_\alpha, \mathfrak{g}_{-\alpha}\,]$ *such that* $\alpha(\alpha^\vee) = 2$, *then* $i\alpha^\vee \in \mathfrak{t}$ *and* $\exp 2\pi i\alpha^\vee = 1$, *while* $\beta(\alpha^\vee) \in \mathbf{Z}$, *for all* $\beta \in R$.

(v) There exists $g \in G^{(\alpha)}$ *such that* $\mathrm{Ad}\, g$ *leaves the root hyperplane* $\ker\alpha \cap \mathfrak{t}$ *pointwise fixed, and maps* $i\alpha^\vee$ *to* $-i\alpha^\vee$.

Proof. (i) Choose $X \in \ker\alpha \cap \mathfrak{t}$ such that $\beta(X) \neq 0$ for all $\beta \in R$ that are not a complex (hence real) multiple of α. Then G_X° is a connected, compact Lie subgroup of G, with Lie algebra equal to:

$$\mathfrak{g}_X = \mathfrak{t} \oplus \sum_{k\neq 0} (\mathfrak{g}_{k\alpha} \oplus \mathfrak{g}_{-k\alpha}) \cap \mathfrak{g}, \tag{3.10.1}$$

cf. (3.7.1). (Actually, $G_X^\circ = G_X$, see Theorem 3.3.1.(ii), but we won't need this here.) So $\mathfrak{z}(\mathfrak{g}_X) = \ker\alpha \cap \mathfrak{t}$. The Lie algebra of the connected, compact Lie group $G' := G_X^\circ / \mathrm{Z}(G_X^\circ)$ is equal to $\mathfrak{g}_X/\mathfrak{z}(\mathfrak{g}_X)$, which has $\mathfrak{t}/(\ker\alpha \cap \mathfrak{t})$ as maximal Abelian subspace. So $\mathrm{rank}\, G' = 1$, and from Lemma 3.10.1 we conclude that $\mathfrak{g}_X/\mathfrak{z}(\mathfrak{g}_X)$ is isomorphic to $\mathfrak{so}(3,\mathbf{R})$. The 3-dimensionality implies that $\dim \sum_{k\neq 0} (\mathfrak{g}_{k\alpha} \oplus \mathfrak{g}_{-k\alpha}) \cap \mathfrak{g} = 2$, and this proves (i).

(ii) From (1.2.12) we see that Lie bracket of the two root spaces in $\mathfrak{so}(3,\mathbf{R})$ is equal to the Cartan subalgebra. In view of the surjective Lie algebra homomorphism: $\mathfrak{g}_X \to \mathfrak{g}_X/\mathfrak{z}(\mathfrak{g}_X) = \mathfrak{so}(3,\mathbf{R})$, this implies that $[\,\mathfrak{g}_\alpha, \mathfrak{g}_{-\alpha}\,]$ is nonzero, and contained in $\mathfrak{h}$. In turn this yields that $[\,\mathfrak{g}_\alpha, \mathfrak{g}_{-\alpha}\,] + \mathfrak{g}_\alpha + \mathfrak{g}_{-\alpha}$ is a Lie algebra. We have $[\,\mathfrak{g}_\alpha, \mathfrak{g}_{-\alpha}\,]^{\mathrm{conj}} = [\,\mathfrak{g}_{-\alpha}, \mathfrak{g}_\alpha\,] = [\,\mathfrak{g}_\alpha, \mathfrak{g}_{-\alpha}\,]$, so the real part of $[\,\mathfrak{g}_\alpha, \mathfrak{g}_{-\alpha}\,]$ is a real 1-dimensional linear subspace of $\mathfrak{t}$, a fact that we shall need in (iv). The mapping $Y \mapsto Y + \mathfrak{z}(\mathfrak{g}_X)$ now defines an isomorphism: $\mathfrak{g}^{(\alpha)} \xrightarrow{\sim} \mathfrak{g}_X/\mathfrak{z}(\mathfrak{g}_X) = \mathfrak{so}(3,\mathbf{R})$.

(iii) The Lie group $G^{(\alpha)}$ exists according to Theorem 1.10.3. The adjoint representation $\mathrm{Ad}\colon G^{(\alpha)} \to \mathrm{Ad}\, G^{(\alpha)} = \mathrm{Ad}\, \mathfrak{g}^{(\alpha)} = \mathbf{SO}(3,\mathbf{R})$ is a covering, and again the conclusion is that $G^{(\alpha)}$ is isomorphic to $\mathbf{SO}(3,\mathbf{R})$ or to $\mathbf{SU}(2)$. In particular, this implies that $G^{(\alpha)}$ is compact; the surjectivity of the exponential mapping, cf. Corollary 3.1.4, then shows that $G^{(\alpha)} = \exp \mathfrak{g}^{(\alpha)}$.

(iv) Considering $x := \exp \pi i \alpha^\vee$ as an element of the Lie group $G^{(\alpha)}$, we get $\operatorname{Ad} x = \mathrm{e}^{\operatorname{ad}(\pi i \alpha^\vee)} = 1$ on $\mathfrak{g}^{(\alpha)}$. But then $x = 1$ if $G^{(\alpha)} = \mathbf{SO}(3,\mathbf{R})$, whereas $x \in \{-1, 1\}$ if $G^{(\alpha)} = \mathbf{SU}(2)$. In either case, $\exp 2\pi i \alpha^\vee = x^2 = 1$, in $G^{(\alpha)} \subset G$.

Furthermore, $1 = \operatorname{Ad} \exp 2\pi i \alpha^\vee = \mathrm{e}^{\operatorname{ad}(2\pi i \alpha^\vee)}$ acts as multiplication by $\beta(2\pi i \alpha^\vee) = 2\pi i \beta(\alpha^\vee)$ on $\mathfrak{g}_\beta$, hence $\beta(\alpha^\vee) \in \mathbf{Z}$, for every $\beta \in R$.

(v) Clearly there exists a rotation in the 3-dimensional space $\mathfrak{g}^{(\alpha)}$ that maps $i\alpha^\vee$ to $-i\alpha^\vee$. Because $\operatorname{Ad} G^{(\alpha)} = \mathbf{SO}(3,\mathbf{R})$, it follows that there exists $g \in G^{(\alpha)}$ such that $\operatorname{Ad} g(i\alpha^\vee) = -i\alpha^\vee$. (Actually, these are precisely the elements g that represent the nontrivial Weyl group element for the maximal Abelian subspace $[\,\mathfrak{g}_\alpha, \mathfrak{g}_{-\alpha}\,] \cap \mathfrak{g}$ of $\mathfrak{g}^{(\alpha)}$.) On the other hand, the action of $\operatorname{Ad} G^{(\alpha)}$ on $\mathfrak{g}$ leaves $\ker \alpha \cap \mathfrak{t} = \mathfrak{z}_{\mathfrak{g}}(\mathfrak{g}^{(\alpha)})$ pointwise fixed. □

Remark. The conclusion $[\,\mathfrak{g}_\alpha, \mathfrak{g}_{-\alpha}\,] \subset \mathfrak{g}_0$ also follows from:

(3.10.2) $$[\,\mathfrak{g}_\alpha, \mathfrak{g}_\beta\,] \subset \mathfrak{g}_{\alpha+\beta}, \quad \text{for any } \alpha, \beta \in R.$$

(3.10.2) follows immediately from the Jacobi identity, and the argument works for any complex Lie algebra, cf. (4.11.5).

The complex linear mapping:

(3.10.3) $$s_\alpha \colon X \mapsto X - \alpha(X)\alpha^\vee \colon \mathfrak{h} \to \mathfrak{h}$$

leaves $\ker \alpha$ pointwise fixed and maps $\alpha^\vee$ to $-\alpha^\vee$, so it is equal to the complex linear extension of $\operatorname{Ad} g|_{\mathfrak{t}}$, with g as in Theorem 3.10.2.(v). Because $\operatorname{Ad} g(\mathfrak{t}) = \mathfrak{t}$, we get $g \in \mathrm{N}(\mathfrak{t}) = \mathrm{N}(T)$, or: **$s_\alpha|_{\mathfrak{t}}$ belongs to the Weyl group $W = W(\mathfrak{g}, \mathfrak{t})$.**

We also note that $\alpha^\vee$ is orthogonal to $\ker \alpha$, with respect to every $\operatorname{Ad} \mathfrak{g}$-invariant bilinear form β on $\mathfrak{g}_{\mathbf{C}}$. Indeed, if $X \in \mathfrak{g}_\alpha$, $Y \in \mathfrak{g}_{-\alpha}$, and $H \in \mathfrak{h}$, then $\beta([\,X,Y\,], H) = -\beta(Y, [\,X, H\,]) = \beta(Y, \alpha(H)X) = 0$, whenever $\alpha(H) = 0$. For this reason, s_α is referred to as the *orthogonal reflection in the root hyperplane* $\ker \alpha$. Note also that $(s_\alpha)^2 = 1$.

The reflection $s_\alpha = s_\alpha^{-1}$ is the restriction to $\mathfrak{h}$ of an automorphism Φ of the complex Lie algebra $\mathfrak{g}_{\mathbf{C}}$. In general, if $\Phi \in \operatorname{Aut} \mathfrak{g}_{\mathbf{C}}$, and $\Phi(\mathfrak{h}) = \mathfrak{h}$, then we have for any $X \in \mathfrak{h}$, $Y \in \mathfrak{g}_\beta$, that $[\,X, \Phi(Y)\,] = \Phi([\,\Phi^{-1}(X), Y\,]) = \Phi(\beta(\Phi^{-1}(X))Y) = \beta(\Phi^{-1}(X))\Phi(Y)$. That is,

(3.10.4) $$\Phi(\mathfrak{g}_\beta) = \mathfrak{g}_{\Phi\beta},$$

if we write, as usual, $(\Phi^{-1})^*(\beta) = \Phi\beta$. In particular,

(3.10.5) $$s_\alpha\beta \in R, \quad \text{whenever } \alpha, \beta \in R.$$

That is, R forms a *reduced root system* in $V = i\mathfrak{t}^* \cap \mathfrak{z}^0$, in the sense of Section 5.5, with the $\alpha^\vee$ as the corresponding coroots. Conversely, Theorem 5.7.4

and Theorem 5.7.7 together imply that every reduced reduced root system is equal to the root system of a compact semisimple Lie algebra. Moreover, using Theorem 5.7.1, one can prove that two such Lie algebras are isomorphic if and only if their root systems are isomorphic. So the classification of the reduced root systems in Section 5.6 can also be viewed as a classification of the compact semisimple Lie algebras.

(3.10.3) Corollary. *The Weyl group* $W = \operatorname{Ad} \mathrm{N}(\mathfrak{t})|_{\mathfrak{t}} = \mathrm{N}(T)/T$ *is equal to the group* W_R *generated by the orthogonal reflections in the root hyperplanes. That is, it can be identified with the Weyl group of the root system* R*, and of the dual root system* $R^\vee$ *of the coroots, respectively.*

3.11 The Stiefel Diagram

In this section, we will determine the additive subgroup Λ_0 of the lattice $\Lambda = \ker \exp \cap \mathfrak{t}$, which was defined in (3.8.10), and had the property that $\pi_1(G) = \Lambda/\Lambda_0$.

Recall that the set $\exp^{-1}(G^{\mathrm{reg}}) \cap \mathfrak{t}$ is equal to the complement in $\mathfrak{t}$ of the *affine root hyperplanes*:

$$H(\alpha, k) := \{\, X \in \mathfrak{t} \mid \alpha(X) = 2\pi i k \,\}, \quad \text{for } \alpha \in R \text{ and } k \in \mathbf{Z}. \tag{3.11.1}$$

The system of affine root hyperplanes is also called the *Stiefel diagram* in $\mathfrak{t}$, with respect to the root system R. The alcove $\mathfrak{b}$, defined in (3.9.3), which played such an important part in the description of $\pi_1(G)$ in Corollary 3.9.4–6, was one of the connected components of $\exp^{-1}(G^{\mathrm{reg}}) \cap \mathfrak{t}$; we now use the word alcove for any connected component of this set.

If $Y \in H(\alpha, k)$, then we have:

$$\begin{aligned} s_\alpha(X - Y) + Y &= X - Y - \alpha(X - Y)\alpha^\vee + Y = X - \alpha(X)\alpha^\vee + \alpha(Y)\alpha^\vee \\ &= s_\alpha(X) + 2\pi i k \alpha^\vee. \end{aligned}$$

This shows that:

$$(+Y) \circ s_\alpha \circ (-Y) = (+2\pi i k \alpha^\vee) \circ s_\alpha \tag{3.11.2}$$

actually does not depend on the choice of $Y \in H(\alpha, k)$. It leaves Y, and therefore every point of $H(\alpha, k)$, fixed, and clearly interchanges the two connected components (half spaces) of $\mathfrak{t} \setminus H(\alpha, k)$, and is called the *orthogonal reflection in the affine root hyperplane* $H(\alpha, k)$.

Now we define the *coroot lattice* as:

$$\begin{aligned} \Lambda_{R^\vee} := {} & \text{the subgroup of } (\mathfrak{t}, +) \text{ generated} \\ & \text{by the vectors } 2\pi i \alpha^\vee, \text{ with } \alpha \in R. \end{aligned} \tag{3.11.3}$$

Because also the set of dual roots is invariant under W, it is clear that the group W'_R generated by the reflections in the affine root hyperplanes is contained in the group $\Lambda_{R^\vee} \cdot W$ generated by the translations over elements of the coroot lattice, and the Weyl group W. Also, from Theorem 3.10.2.(iv), we see that $2\pi i\alpha^\vee \in \Lambda$ for each $\alpha \in R$; and this implies: $\Lambda_{R^\vee} \subset \Lambda$.

(3.11.1) Proposition.

(i) *The coroot lattice $\Lambda_{R^\vee}$ is equal to the set Λ_0 of $X \in \Lambda$ for which the closed curve $\gamma_X : t \mapsto \exp tX$, with $t \in [0,1]$, is contractible in G. The mapping that assigns to each $X \in \Lambda$ the homotopy class in $\pi_1(G)$ of γ_X, induces an isomorphism: $\Lambda/\Lambda_{R^\vee} \xrightarrow{\sim} \pi_1(G)$.*

(ii) *The group W'_R is equal to $\Lambda_{R^\vee} \cdot W$, and acts freely and transitively on the set of alcoves. The group $\Lambda \cdot W$ is equal to the semidirect product of the normal subgroup $\Lambda_{R^\vee} \cdot W$ with $\mathrm{N}_{\Lambda \cdot W}(\mathfrak{b})$, where $\mathrm{N}_{\Lambda \cdot W}(\mathfrak{b})$ is as in Corollary 3.9.6.*

Proof. In the proof of Theorem 3.10.2.(iv), we actually found that $\exp \pi i\alpha^\vee = 1$ if $G^{(\alpha)} = \mathbf{SO}(3,\mathbf{R})$. So in this case $t \mapsto \exp t2\pi i\alpha^\vee$, for $t \in [0,1]$, is actually a closed curve in an $\mathbf{SO}(3,\mathbf{R})$, run through twice. Because $\mathbf{SO}(3,\mathbf{R})$ has a 2-fold simply connected covering (viz. $\mathbf{SU}(2)$), this curve is contractible in $G^{(\alpha)} \subset G$. In the other case $G^{(\alpha)} = \mathbf{SU}(2)$ itself is simply connected and again the curve is contractible in G. That is, we have improved the estimate $\Lambda_{R^\vee} \subset \Lambda$ to $\Lambda_{R^\vee} \subset \Lambda_0$.

Translations over $Y \in \Lambda$ leave $\exp^{-1}(G^{\mathrm{reg}}) \cap \mathfrak{t}$ invariant and the same is true with elements of W (because G^{reg} is invariant under conjugations). It follows that $\theta(\mathfrak{b})$ is an alcove if $\mathfrak{b}$ is an alcove and $\theta \in \Lambda \cdot W$. In other words, $\Lambda \cdot W$ maps alcoves to alcoves; the same holds for the subgroup W'_R of $\Lambda \cdot W$.

Next we observe that W'_R acts transitively on the set of alcoves. For the proof, suppose that $\mathfrak{b}$ and $\mathfrak{b}'$ are alcoves, and choose $X \in \mathfrak{b}$ and $Y \in \mathfrak{b}'$. By perturbing one of these points, if necessary, we can arrange that the straight line segment between X and Y meets only one affine root hyperplane at a time, thereby crossing transversally the boundary between two adjacent alcoves, which locally coincides with an affine root hyperplane. Now define by induction over j the elements $\theta_j \in W'_R$ by: $\theta_0 = 1$, and, at the j-th crossing: $\theta_j = \theta_{j-1}$, followed by the reflection in the affine root hyperplane at which we have arrived. By induction it follows that θ_j maps the initial alcove onto the alcove which will be entered after the j-th crossing, so $\theta_k(\mathfrak{b}) = \mathfrak{b}'$, if k is the number of crossings.

It immediately follows that the product mapping: $W'_R \times \mathrm{N}_{\Lambda \cdot W}(\mathfrak{b}) \to \Lambda \cdot W$ is surjective. Indeed, if $\mu \in \Lambda \cdot W$, then, for the alcove $\mathfrak{b}$ of (3.9.3), $\mu(\mathfrak{b})$ is an alcove, so there exists a $\theta \in W'_R$, such that $\mu(\mathfrak{b}) = \theta(\mathfrak{b})$, or $\phi(\mathfrak{b}) = \mathfrak{b}$, or $\phi \in \mathrm{N}_{\Lambda \cdot W}(\mathfrak{b})$, if we write $\phi = \theta^{-1} \circ \mu$. But this means that $\mu = \theta \circ \phi$, with $\theta \in W'_R$, $\phi \in \mathrm{N}_{\Lambda \cdot W}(\mathfrak{b})$.

On the other hand, if $\phi \in \Lambda_0 \cdot W \cap \mathrm{N}_{\Lambda \cdot W}(\mathfrak{b})$, then $\phi(0) \in \Lambda_0$ and

$\phi \in \mathrm{N}_{\Lambda \cdot W}(\mathfrak{b})$, and it follows from Corollary 3.9.6.(i) that $\phi = 1$. This is the same as saying that the group $\Lambda_0 \cdot W$ acts freely on the set of alcoves. Because $W_R' \subset \Lambda_0 \cdot W$ (and the latter is a group), we conclude that $W_R' = \Lambda_0 \cdot W$. In particular, because $W_R' \subset \Lambda_{R^\vee} \cdot W \subset \Lambda_0 \cdot W$, we also get $\Lambda_{R^\vee} \cdot W = \Lambda_0 \cdot W$, and it follows (by evaluating at 0) that $\Lambda_{R^\vee} = \Lambda_0$.

For the final statement we have still to check that $\Lambda_{R^\vee} \cdot W$ is normal in $\Lambda \cdot W$. This amounts to showing that $(+Y) \circ s \circ (-Y) = (Y - s(Y)) \circ s \in \Lambda_{R^\vee} \cdot W$, or $Y - s(Y) \in \Lambda_{R^\vee}$, if $Y \in \Lambda$, and $s \in W$. For $s = s_\alpha$, with $\alpha \in R$, this follows because $\alpha(Y) \in 2\pi i \cdot \mathbf{Z}$ whenever $\exp Y = 1$ ($\Rightarrow e^{\mathrm{ad}\, Y} = \mathrm{Ad} \exp Y = 1$). Using Corollary 3.10.3 we then obtain by induction for all $s \in W$, in view of $Y - s \cdot s_\alpha(Y) = (Y - s_\alpha(Y)) + \big(s_\alpha(Y) - s(s_\alpha(Y))\big)$. □

Remarks.

(a) The connectedness of G implies that the center $\mathrm{Z}(G)$ of G is equal to $\mathrm{Z}_G(\mathfrak{g}) = \ker \mathrm{Ad}$, so $\exp^{-1}(\mathrm{Z}(G)) \cap \mathfrak{t}$ is equal to:

$$\Lambda_Z := \{\, X \in \mathfrak{t} \mid \alpha(X) \in 2\pi i \mathbf{Z}, \text{ for all } \alpha \in R\,\}. \tag{3.11.4}$$

In general we have the inclusions:

$$\Lambda_{R^\vee} \subset \Lambda \subset \Lambda_Z. \tag{3.11.5}$$

One of these inclusions can only be an equality if $\mathfrak{z} = 0$, because Λ is discrete and spans $\mathfrak{t}$, whereas on the one hand the linear span of $\Lambda_{R^\vee}$ is equal to $\mathfrak{g}' \cap \mathfrak{t}$ (which has zero intersection with $\mathfrak{z}$), and on the other hand $\mathfrak{z} \subset \Lambda_Z$.

However, from Proposition 3.11.1 we may read off that:

$$\pi_1(\mathrm{Ad}\, \mathfrak{g}) = \Lambda_Z / (\Lambda_{R^\vee} + \mathfrak{z}), \tag{3.11.6}$$

and this is finite, confirming the last statement in Corollary 3.9.4 in a somewhat different way.

(b) If $\mathfrak{z} = 0$, then Λ_Z is called the *central lattice* in $\mathfrak{t}$. The condition that $\mathfrak{z} = 0$ means that $\mathrm{Ad}\colon G \to \mathrm{Ad}\, \mathfrak{g}$ is a covering, and, if $(\mathrm{Ad}\, \mathfrak{g})^\sim$ denotes the universal covering of $\mathrm{Ad}\, \mathfrak{g}$, we have a sequence of Lie group coverings:

$$(\mathrm{Ad}\, \mathfrak{g})^\sim \to G \to \mathrm{Ad}\, \mathfrak{g}, \tag{3.11.7}$$

The covering groups are $(\mathrm{Ad}\, \mathfrak{g})^\sim / G = \Lambda / \Lambda_{R^\vee}$, and $G / \mathrm{Ad}\, \mathfrak{g} = \Lambda_Z / \Lambda$, so in this sense the sequence (3.11.7) is dual to (3.11.5). The extreme case $\Lambda_{R^\vee} = \Lambda$ occurs if G is simply connected, or has center isomorphic to the group in (3.11.6); it actually suffices to require that it has the same number of elements. The other extreme $\Lambda = \Lambda_Z$ occurs precisely when G is isomorphic to $\mathrm{Ad}\, \mathfrak{g}$, or has trivial center.

(c) Every positive root is a sum of simple roots, cf. Theorem 5.5.9.(i). Because $R = P \cup -P$, it follows that every root is a linear combination of simple roots with integer coefficients. It follows that the central lattice is also given by:

$$\text{(3.11.8)} \qquad \Lambda_Z = \{\, X \in \mathfrak{t} \mid \alpha(X) \in 2\pi i\mathbf{Z}, \text{ for all } \alpha \in S \,\}.$$

The corresponding dual statement is that every coroot is a linear combination with integer coefficients of the $\alpha^\vee$, for $\alpha \in S$. Because the simple roots, and therefore also the simple coroots, are linearly independent, cf. Theorem 5.5.9.(iv), it follows that $(c_\alpha)_{\alpha\in S} \mapsto \sum_{\alpha\in S} c_\alpha 2\pi i\alpha^\vee$ is a linear isomorphism: $\mathbf{I}\colon \mathbf{R}^S \to \mathfrak{t}\cap\mathfrak{g}'$, such that:

$$\text{(3.11.9)} \qquad \Lambda_{R^\vee} = \left\{\, \textstyle\sum_{\alpha\in S} n_\alpha 2\pi i\alpha^\vee \mid n_\alpha \in \mathbf{Z} \,\right\} = \mathbf{I}(\mathbf{Z}^S).$$

It follows that:

$$\mathbf{I}^{-1}(\Lambda_Z) = \left\{\, (c_\beta)_{\beta\in S} \mid \textstyle\sum_{\beta\in S} \alpha(\beta^\vee) c_\beta \in \mathbf{Z}, \text{ for all } \alpha \in S \,\right\} = C^{-1}(\mathbf{Z}^S),$$

where the *Cartan matrix* C is defined by:

$$\text{(3.11.10)} \qquad C_{\alpha,\beta} = \alpha(\beta^\vee), \quad \text{with } \alpha, \beta \in S.$$

In passing, we mention that the root system, and therefore the whole Lie algebra structure, is determined by the Cartan matrix, cf. Corollary 5.5.12.

We view here C as a linear mapping: $\mathbf{R}^S \to \mathbf{R}^S$, which is invertible because the simple roots, and coroots, respectively, are linearly independent. Also, $C(\mathbf{Z}^S) \subset \mathbf{Z}^S$, because $\alpha(\beta^\vee) \in \mathbf{Z}$ for all $\alpha, \beta \in S$, cf. Theorem 3.10.2.(iv). In particular, if $\mathfrak{z} = 0$, then $\Lambda_Z/\Lambda_{R^\vee} \cong C^{-1}(\mathbf{Z}^S)/\mathbf{Z}^S \cong \mathbf{Z}^S/C(\mathbf{Z}^S)$. On the other hand, C induces a covering of tori $C\colon \mathbf{R}^S/\mathbf{Z}^S \to \mathbf{R}^S/\mathbf{Z}^S$, and comparing volume, we find that the fiber, which is isomorphic to $\mathbf{Z}^S/C(\mathbf{Z}^S)$, has $\det C$ many elements. In view of (3.11.6), we can therefore conclude:

$$\text{(3.11.11)} \qquad \#(\pi_1(\operatorname{Ad} G)) = \#\big(\Lambda_Z/(\Lambda_{R^\vee} + \mathfrak{z})\big) = \det(\alpha(\beta^\vee))_{\alpha,\beta\in S}.$$

For the computation of $\Lambda_Z/\Lambda_{R^\vee}$ for the simple Lie algebras see Proposition 5.6.4.

3.12 Unitary Groups

The next level in detail of our understanding of compact Lie groups is obtained by analyzing and using the properties of (restricted) root systems, see the discussion preceding Corollary 3.10.3. Because this does not belong so much to the central theme of this chapter, we do not treat this here, and refer the reader to Section 5.5 instead. However, we do not want to leave this subject without taking a look at the simplest series of connected, compact groups, the special unitary groups (see Section 1.2.b for the definition), in order to illustrate the theory.

If $x \in G := \mathbf{SU}(n)$, it has an eigenvalue, which has absolute value equal to 1, and a corresponding eigenvector e_1. Because x leaves the hermitian inner product in $\mathbf{C}^n$ invariant, it also leaves invariant the orthogonal complement of e_1 in $\mathbf{C}^n$, which is a complex linear subspace of complex dimension $n-1$. Repeating the procedure, one finds a (complex) basis $e_1, \dots, e_n$ of eigenvectors, with corresponding eigenvalues $\sigma_1, \dots, \sigma_n$, and such that $< e_p, e_q >= 0$ whenever $p \neq q$. One can also arrange that the basis is orthonormal, that is, in addition $< e_p, e_p >= 1$, for all p; and then the linear transformation y that sends the standard basis to this one has determinant equal to 1. That is, $y^{-1}xy$ has a diagonal matrix for some $y \in \mathbf{SU}(n)$.

If $x \in \mathbf{SU}(n)$ is diagonal and all eigenvalues are different from each other, then a simple argument (or computation) shows that $y \in \mathbf{SU}(n)$ commutes with x if and only if y is also diagonal. In other words, $G_x = T := \{ y \in G \mid y \text{ is diagonal} \}$. This group is isomorphic (as a Lie group) to C^{n-1}, where C is the unit circle in the complex plane $\mathbf{C}$; we can freely select for instance the first $n-1$ eigenvalues in C, but then the last one is uniquely determined by the condition that $\det y = 1$. On the other hand it is also easily verified that $\dim G_x > n-1$ as soon as two eigenvalues of x agree. Conclusions: G is connected, T is a maximal torus in G, each element of G is conjugate to an element of T, and

$$G^{\text{reg}} = G^{\text{princ}} = \{ x \in G \mid \text{ all eigenvalues of } x \text{ have multiplicity one} \}.$$

According to Corollary 3.9.5.(iv), $G^{\text{reg}} = G^{\text{princ}}$ implies that $\mathbf{SU}(n)$ **is simply connected**, a fact that we will verify below again using the criterion of Corollary 3.9.6, and Proposition 3.11.1.(i), respectively. In the meantime, the reader is invited to give a more elementary proof.

The Lie algebra of T is the maximal Abelian subspace $\mathfrak{t}$ of $\mathfrak{g} = \mathfrak{su}(n)$, consisting of all diagonal matrices X with purely imaginary entries $i\theta_1, \dots, i\theta_n$, such that $\sum_j \theta_j = 0$. In this case the complexification $\mathfrak{g}_{\mathbf{C}}$ of $\mathfrak{g}$ is readily available in the form of $\mathfrak{sl}(n, \mathbf{C})$, the space of all complex $n \times n$-matrices with trace equal to 0. If we denote by $E_{p,q}$ the matrix which has a 1 in the p-th row and the q-th column, and zeros elsewhere, then the root spaces are the complex 1-dimensional spaces $\mathbf{C} \cdot E_{p,q}$, $p \neq q$; and the corresponding root form $\alpha_{p,q}$ assigns to the X above the number $i\theta_p - i\theta_q$. So the root hyperplanes correspond to $\theta_p = \theta_q$, and we get confirmed that their complement, consisting of the X for which all eigenvalues are different, is the complement of the root hyperplanes.

The Weyl chambers correspond to the various orderings that can occur between the $\theta_j \in \mathbf{R}$. It will be convenient to take the set of $X \in \mathfrak{t}$ such that $\theta_1 > \theta_2 > \dots > \theta_n$ as the positive Weyl chamber; we shall denote it by $\mathfrak{t}^+$, cf. (3.5.14). The corresponding choice P of positive roots consists of the $\alpha_{p,q}$ with $p < q$. Because the Weyl group acts freely and transitively on the set of Weyl chambers, that is, the set of orderings between the numbers $1, \dots, n$, we get that it is isomorphic to the **group S_n of permutations of n**

elements. The walls of the Weyl chamber $\mathfrak{t}^+$ correspond to the occurrence of precisely one equality sign $\theta_p = \theta_{p+1}$ instead of the inequality $\theta_p > \theta_{p+1}$. The simple roots are the $\alpha_{p,p+1}$, where $p = 1, \dots, n-1$; clearly these are linearly independent, compare this with Theorem 5.5.9.(iv). The reflections in the simple root hyperplanes correspond to the "nearest neighbor switches" $p \leftrightarrow p+1$ in the permutation group S_n; and in this context Corollary 3.10.3 recovers the well-known fact that every permutation in n elements can be obtained by successively carrying out at most $\frac{1}{2}n(n-1)$ nearest neighbor switches. The faces of $(\mathfrak{t}^+)^{\mathrm{cl}}$ correspond to the choices of intervals in the indices p, where the $<$-sign has not been changed into $=$.

The alcove $\mathfrak{b}$ of (3.9.3) is the part of $\mathfrak{t}^+$ satisfying in addition $\theta_1 - \theta_n < 2\pi$. Because $\sum_j \theta_j = 0$, and $(\theta_1, \dots, \theta_n) \mapsto (\theta_1 - \theta_2, \dots, \theta_1 - \theta_n)$ is a linear isomorphism from this hyperplane onto $\mathbf{R}^{n-1}$, we see that $\mathfrak{b}$ is affinely isomorphic to the set: $\{\, \mu \in \mathbf{R}^{n-1} \mid 0 < \mu_{n-1} < \dots < \mu_1 < 1 \,\}$, whose closure is the standard $(n-1)$-dimensional simplex. Because Λ consists of the θ such that $\theta_j \in 2\pi\mathbf{Z}$, for all j, we also see that $\mathfrak{b}^{\mathrm{cl}} \cap \Lambda$ only consists of $\theta = 0$; so Corollary 3.9.6 yields again that $\mathbf{SU}(n)$ is simply connected. Note that $\mathfrak{b}^{\mathrm{cl}} \cap \Lambda_Z$ consists of the θ of the form $\theta_j = 2\pi\frac{n-p}{n}$, for $1 \le j \le p$, and $\theta_k = -2\pi\frac{p}{n}$, for $p+1 \le k \le n$, one for each $p \in \{1, \dots, n-1\}$. Under exp, these elements correspond precisely to the center $Z = \{\, \mathrm{e}^{2\pi i p/n} \mid p \in \mathbf{Z} \,\} = \mathbf{Z}/(n\mathbf{Z})$ of $\mathbf{SU}(n)$, confirming that, according to Corollary 3.9.6, they also are in bijective correspondence with $\pi_1(\mathrm{Ad}\,\mathbf{SU}(n))$.

The dual roots, cf. Theorem 3.10.2.(iv), are given by $(\alpha_{p,q})^\vee = E_{p,p} - E_{q,q}$, for $p \neq q$. In the picture below for $n = 3$, we have sketched the affine root hyperplanes, the positive Weyl chamber $\mathfrak{t}^+$, and the alcove in it that has the origin as a boundary point (called $\mathfrak{b}$ in (3.9.3)). We have also indicated the coroot lattice $\Lambda_{R^\vee}$ and the central lattice Λ_Z. Of course the metric in the picture is determined by the Killing form, so that the dual roots are orthogonal to the root hyperplanes. Because it is easily verified that $i\theta_p \in 2\pi i\mathbf{Z}$ for all p, if and only if $X \in \sum \mathbf{Z} \cdot 2\pi i \cdot (E_{p,p} - E_{q,q})$, we get that $\Lambda := \ker \exp \cap \mathfrak{t} = \Lambda_{R^\vee}$; and this shows again that $\mathbf{SU}(n)$ is simply connected.

We have seen above that $\mathbf{SU}(n)$ acts transitively (and freely) on the space of oriented orthonormal bases ("frames") $e_1, \dots, e_n$ in $\mathbf{C}^n$. To such a basis we can assign the sequence $L_k = \sum_{j=1}^k \mathbf{C} \cdot e_j$, for $k = 1, \dots, n-1$, of complex linear subspaces of $\mathbf{C}^n$. Any sequence L_k, for $k = 1, \dots, n-1$, of complex linear subspaces of $\mathbf{C}^n$ such that $L_k \subset L_{k+1}$ and $\dim L_k = k$, for all k, is called a *flag* in $\mathbf{C}^n$. The flags in $\mathbf{C}^n$ together form a complex projective algebraic variety, called the *flag variety* F of $\mathbf{C}^n$. $\mathbf{SU}(n)$ acts transitively on F, and the stabilizer of the flag of the standard basis is equal to our maximal torus T, as is easily verified. So G/T, which we had identified with the conjugacy classes of regular elements in G, or with the orbits under the adjoint representation of regular elements in $\mathfrak{g}$, can also be identified with the flag variety F.

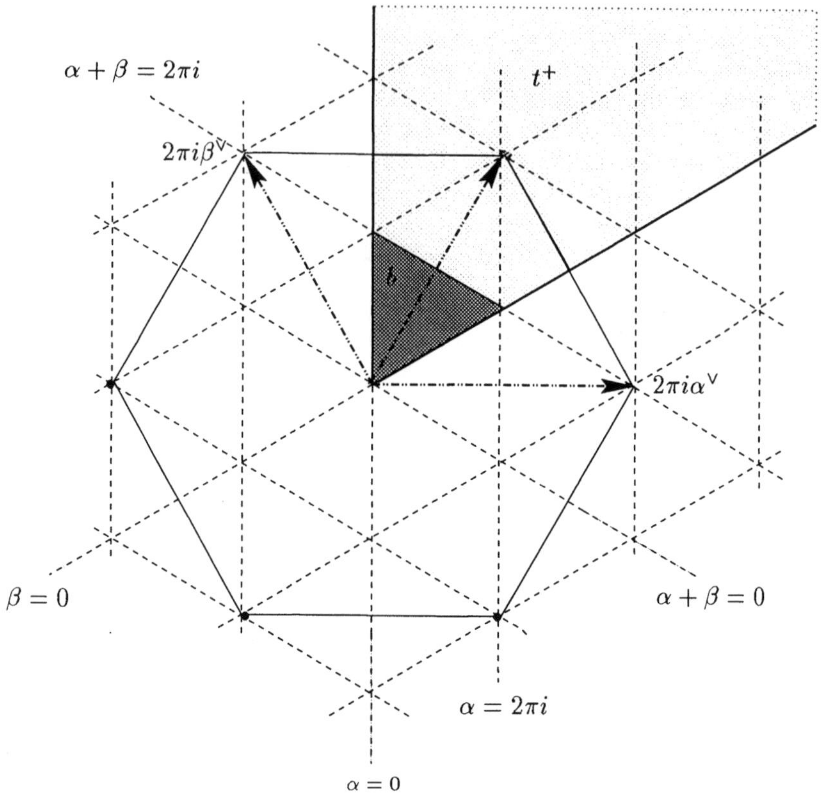

Fig. 3.12.1. n = 3, $\alpha=\alpha_{1,2}$, $\beta=\alpha_{2,3}$

The point is now that on F also the much bigger complex group $G_{\mathbf{C}} = \mathbf{SL}(n, \mathbf{C})$ acts transitively. The stabilizer of the standard flag is equal to the group B of the upper triangular matrices (with the diagonal included) in $\mathbf{SL}(n, \mathbf{C})$. The action is complex-analytic (even complex algebraic) so the complex structure on $F = G/T$ can also be obtained from the identification $F = G_{\mathbf{C}}/B$.

Similarly the other adjoint orbits G/G_X, with X in a face of $(\mathfrak{t}^+)^{\mathrm{cl}}$, can be identified with spaces of *partial flags*, the increasing sequences of complex linear subspaces $L_1 \subset L_2 \subset \ldots \subset L_p$, with prescribed dimensions $n_1 < n_2 < \ldots < n_p$. The partial flag varieties are complex projective algebraic varieties too, acted upon transitively by $\mathbf{SL}(n, \mathbf{C})$. For $p = 1$, we get the familiar *Grassmann varieties* $\mathrm{G}(n, k, \mathbf{C})$, the space of k-dimensional complex linear subspaces of $\mathbf{C}^n$; they correspond to the 1-dimensional faces of $(\mathfrak{t}^+)^{\mathrm{cl}}$.

This fact, that the adjoint orbits can be provided with the structure of a complex projective algebraic variety, with a complex Lie group acting transitively on it, turns out to be true for all connected compact Lie groups G, cf. Lemma 4.12.6; and these spaces almost classify the homogeneous compact complex-analytic spaces, cf. Proposition 11.6.4 and the remarks following it.

3.13 Integration

This section contains a general discussion of invariant densities, especially on homogeneous spaces. Although in the next section this is used only in a relatively simple special case, the general discussion will be useful for future reference. The reader who already is familiar with it, could pass directly to Section 3.14.

A *density* (that is, an unoriented volume form) θ on a manifold M is an assignment, to each local coordinate chart κ (from an open subset V_κ of M onto an open subset U_κ of $\mathbf{R}^m$, with $m = \dim M$), of a function θ_k on U_κ, with the substitution rule:

(3.13.1) $$\theta_\kappa(z) = \theta_{\kappa'}(\kappa' \circ \kappa^{-1}(z)) \cdot |\det D(\kappa' \circ \kappa^{-1})(z)|,$$

if κ' is another coordinate chart, and $z \in U_\kappa \cap \kappa(V_{\kappa'})$.

One can also express this by saying that θ is a section of the line bundle $D(M)$ over M (the density bundle), whose fiber over $x \in M$ is equal to the (1-dimensional) space of all functions θ_x on $\bigwedge^m \mathrm{T}_x M$, such that $\theta_x(cv) = |c|\theta_x(v)$, for all $c \in \mathbf{R}$ and $v \in \bigwedge^m \mathrm{T}_x M$.

θ is said to be C^k, and positive, if each θ_k is C^k, and everywhere positive, respectively. Using C^k partitions of unity, we see that every (paracompact, Hausdorff) C^k manifold M admits positive densities of class C^k, here $0 \leq k \leq \infty$.

If f is a C^k function with compact support on M then, using a C^k partition of unity, we can write f as a finite sum:

(3.13.2) $$f = \sum f_\kappa, \quad \text{with } f_\kappa \in \mathrm{C}_c^k(M) \text{ and support of } f_\kappa \subset V_\kappa.$$

Now assume that θ is continuous. The *integral of f against the density θ* is then defined as:

(3.13.3) $$\int_M f \cdot \theta = \sum \int_{U_\kappa} f_\kappa(\kappa^{-1}(y))\theta_\kappa(y)\, dy.$$

Using the formula for substitutions of variables in an integral,

(3.13.4) $$\int_U f(x)\, dx = \int_V f(\Phi(y)) \cdot |\det D\Phi(y)|\, dy,$$

where Φ is a C^1 diffeomorphism from an open subset V of $\mathbf{R}^m$ to another open subset U of $\mathbf{R}^m$, and f a continuous function on U, with compact support contained in U, one sees that (3.13.1) guarantees that the right hand side in (3.13.3) is independent of the way we have written f as in (3.13.2).

The standard theory of Lebesgue integration now consists of defining the space $\mathrm{L}^1(M,\theta)$ of *Lebesgue integrable functions on M, with respect to the density* θ, as the completion of $\mathrm{C}_c(M)$ with respect to the integral norm $f \mapsto \int |f| \cdot \theta$. Similarly, the space $\mathrm{L}^p(M,\theta)$ is defined as the completion of $\mathrm{C}_c(M)$ with respect to the norm $f \mapsto (\int |f|^p \cdot \theta)^{1/p}$. This definition allows to prove many results first only in the space $\mathrm{C}_c(M)$, and then extending them to these completions by continuity. (The more tricky part of Lebesgue theory then of course is to identify the elements of the completion with ordinary functions on M.)

The familiar formula for substitutions in an integral now immediately gets the following generalization to manifolds. If Φ is a diffeomorphism from a manifold N onto M, and N is provided with the standard positive density τ, then

$$\int_M f \cdot \theta = \int_N \Phi^*(f \cdot \theta) = \int_N f \circ \Phi \cdot \Phi^* \theta = \int_N (f \circ \Phi) \cdot J_\Phi \cdot \tau,$$

where the *absolute Jacobian* J_Φ is defined as the positive function $(\Phi^*\theta)/\tau$. In turn, $J_\Phi(y) = |\det(L^{-1} \circ \mathrm{T}_y \Phi)|$, if L is an auxiliary linear mapping: $\mathrm{T}_y N \to \mathrm{T}_x M$, with $x = \Phi(y)$, such that $L^* \theta_x = \tau_y$. By the principle above, $f \in \mathrm{L}^1(M,\theta)$ if and only if $(f \circ \Phi) \cdot J_\Phi \in \mathrm{L}^1(N,\tau)$; and if this is the case, then the integrals agree.

If θ is positive, then $\mu_\theta \colon f \mapsto \int f \cdot \theta$ is a *positive Radon measure* on M, that is, a linear form μ on $\mathrm{C}_c(M)$, the space of continuous functions with compact support on M, such that moreover $\mu(f) > 0$, whenever $f \in \mathrm{C}_c(M)$, $f \geq 0$, $f \neq 0$.

Every nowhere zero, continuous volume form Ω on M (that is, differential form of degree equal to the dimension of M) gives rise to a positive, continuous density $\theta = |\,\Omega\,|$ on M, by means of the definition

$$\text{(3.13.5)} \qquad \kappa^*(\theta_\kappa dx_1 \wedge \cdots \wedge dx_m) = \Omega, \quad \text{on } V_\kappa,$$

where we only allow local coordinate charts κ for which the resulting functions θ_k are positive. Note that, given the continuous, nowhere zero volume form Ω, there exists an atlas $\mathcal{A}$ of such coordinate charts κ, and that:

$$\text{(3.13.6)} \qquad \det D(\kappa' \circ \kappa^{-1})(z) > 0, \text{ if } z \in U_\kappa \cap \kappa(V_{\kappa'}), \quad \text{for } \kappa, \kappa' \in \mathcal{A}.$$

The choice of such an atlas is said to be an *orientation* of the manifold M. The manifold M is called *orientable* if it has such an atlas, and this is equivalent to the existence of a nowhere vanishing continuous volume form on M.

If a group G acts on M by means of diffeomorphisms, then one has an induced action on $C_c(M)$ defined by $(g, f) \mapsto (g^{-1})^*(f)$, where we use the standard notation:

$$(\Phi^*(f))(x) = f(\Phi(x)) \tag{3.13.7}$$

for the *pull back* of a function f by a mapping Φ. (Actually this is a special case of the pull back of differential forms used above.) The adjoint of Φ^*, acting on measures, is called the *push forward* by Φ, and will be denoted by Φ_*. Then $(g, \mu) \mapsto g_*\mu$ defines an action of G on the space of measures on M. All such induced actions of G on spaces of functions, and their dual spaces of distributions, are always by means of continuous linear transformations in these function, and distribution, spaces, respectively, that is, these are representations in the sense of Chapter 4. Now the measure μ on M is said to be *G-invariant* if $g^*\mu = \mu$ for all $g \in G$, and similarly the density θ is said to be invariant if μ_θ is G-invariant. It is also easy to verify that, for a volume form Ω on M, the density $\theta = |\,\Omega\,|$ is G-invariant, if and only if $g^*\Omega = \pm\Omega$, for all $g \in G$.

Let G be a Lie group acting transitively (and smoothly) on a manifold M. According to Section 1.11, there is a G-equivariant diffeomorphism from the quotient space G/H onto M, for a suitable closed Lie subgroup H of G; here the action of G on G/H is by means of multiplications from the left. If we denote the Lie algebra of G, and H, by $\mathfrak{g}$, and $\mathfrak{h}$, respectively, then the tangent space of G/H at the base point $b = 1H$ is identified with $\mathfrak{g}/\mathfrak{h}$, in such a way that the tangent mapping, at b, of the canonical projection $\Pi\colon G \to G/H$ is equal to the canonical projection $\pi\colon \mathfrak{g} \to \mathfrak{g}/\mathfrak{h}$. Here $\Pi\colon G \to G/H$ is a (real-analytic) principal fiber bundle, with structure group H, acting on G by means of multiplications from the right.

The stabilizer group of the base point $1H$ is equal to H, it acts on $\mathrm{T}_b(G/H) = \mathfrak{g}/\mathfrak{h}$ via $(h, X + \mathfrak{h}) \mapsto \operatorname{Ad} h(X) + \mathfrak{h}$, this is obtained by differentiating:

$$hxH = hxh^{-1}H, \quad \text{with } h \in H, x \in G, \tag{3.13.8}$$

with respect to x at $x = 1$, in the direction of $X \in \mathfrak{g}$.

It follows that $h^{-1} \in H$ acts on the space of volume forms Ω_b on $\mathfrak{g}/\mathfrak{h}$ (the antisymmetric m-linear forms on $\mathfrak{g}/\mathfrak{h}$, with $m = \dim \mathfrak{g}/\mathfrak{h}$) via multiplication by:

$$\det(\operatorname{Ad} h)_{\mathfrak{g}/\mathfrak{h}} = \bigl(\det(\operatorname{Ad} h)|_{\mathfrak{g}}\bigr)/\bigl(\det(\operatorname{Ad} h)|_{\mathfrak{h}}\bigr). \tag{3.13.9}$$

The volume form Ω_b on $\mathfrak{g}/\mathfrak{h} = \mathrm{T}_b(G/H)$ extends to a *G-invariant volume form* θ on G/H (that is, a differential form on G/H of degree $m = \dim G/H$), if and only if Ω_b remains fixed under the action of H. If this is the case, then Ω is uniquely determined by Ω_b, and is real-analytic. Clearly, Ω is nonzero everywhere, and zero everywhere, if and only if $\Omega_b \neq 0$, and $\Omega_b = 0$, respectively. In turn, the existence of a nonzero G-invariant volume form Ω on G/H is equivalent to the condition that:

$$\det(\operatorname{Ad} h)_{\mathfrak{g}/\mathfrak{h}} = 1, \quad \text{for all } h \in H. \tag{3.13.10}$$

The existence of a G-invariant density θ, and measure μ_θ, respectively, on G/H is equivalent to the slightly weaker condition:

$$|\det(\operatorname{Ad} h)_{\mathfrak{g}/\mathfrak{h}}| = 1, \quad \text{for every } h \in H. \tag{3.13.11}$$

Note that (3.13.11) is equivalent to (3.13.10), if and only if G/H is orientable, or $\det(\operatorname{Ad} h)_{\mathfrak{g}/\mathfrak{h}} > 0$ for all $h \in H$; and this in turn is the case if H is connected. Using convolutions as in (4.4.1), one can show that any G-invariant distribution u on G/H is of the form μ_θ for a smooth G-invariant density θ on G/H, which in turn is locally equal to $|\Omega|$ for a locally G-invariant volume form Ω on G/H. This implies that u is uniquely determined up to a constant factor. If G/H is orientable, then $u = \theta_{|\Omega|}$ for a G-invariant volume form Ω on G/H, and we are back in the situation we started out with.

In general the function $h \mapsto |\det(\operatorname{Ad} h)_{\mathfrak{g}/\mathfrak{h}}| : H \to \mathbf{R}_{>0}$ is a continuous homomorphism from H to the multiplicative group $\mathbf{R}_{>0}$. So if H **is compact**, the image is a compact subgroup of $\mathbf{R}_{>0}$, which can only be $\{1\}$, and the conclusion is that G/H carries a G-invariant positive density.

A special case occurs when $H = \{1\}$, that is, $G/H = G$, viewed as a G-homogeneous space via the action by left multiplications. In this case any nonzero volume form Ω_1 on $\mathfrak{g}$ gives rise to a unique left-invariant volume form Ω on G; we get a corresponding left-invariant density, measure, and orientation, respectively on G. The left-invariant measure is also called the (left-) *Haar measure* on G, the corresponding density is usually denoted by dx. If G is compact, then the unique Haar measure such that:

$$\int_G 1\, dx = 1, \tag{3.13.12}$$

is the one that is used in the procedure of averaging over G, and is called the *normalized Haar measure on G*.

If Ω is a left-invariant volume form on G then, for each $g \in G$, we have:

$$\begin{aligned} \mathrm{L}(x)^*\,\mathrm{R}(g)^*\Omega &= \big(\mathrm{R}(g)\circ \mathrm{L}(x)\big)^*\Omega = \big(\mathrm{L}(x)\circ \mathrm{R}(g)\big)^*\Omega = \mathrm{R}(g)^*\,\mathrm{L}(x)^*\Omega \\ &= \mathrm{R}(g)^*\Omega, \end{aligned}$$

that is $\mathrm{R}(g)^*\Omega$ is left-invariant as well. It follows that it is a constant multiple of Ω. Evaluating $\mathrm{R}(g)^*\Omega = \mathrm{R}(g)^*\,\mathrm{L}(g^{-1})^*\Omega = (\mathbf{Ad}\, g^{-1})^*\Omega$ at 1, we get:

$$\mathrm{R}(g)^*\Omega = \det \operatorname{Ad} g^{-1}\cdot \Omega, \quad \text{for } g \in G. \tag{3.13.13}$$

In particular, there exists a *bi-invariant* (that is, a left-invariant and right-invariant) volume form on G, if and only if $\det \operatorname{Ad} g = 1$ for all $g \in G$; that is, the adjoint representation maps into the special linear group of G. For the existence of a bi-invariant density, and measure, respectively on G, the condition reads:

$$|\det \operatorname{Ad} g| = 1, \quad \text{for all } g \in G; \tag{3.13.14}$$

in this case the group G is said to be *unimodular*. As before, every compact Lie group is unimodular. This implies also that:

$$\int_G f(x^{-1})\,dx = \int_G f(x)\,dx, \quad \text{for } f \in \mathrm{C}_c(G). \tag{3.13.15}$$

Now let $\pi\colon M \to B = G\backslash M$ be a principal fiber bundle with the Lie group G as structure group, cf. (1.11.18). For any positive, continuous density dg on G, and any $f \in \mathrm{C}_c(M)$, we get a new continuous function $\int_G f\colon x \mapsto \int_G f(g^{-1}x)\,dg$, obtained from f by "*integration over the fiber*". Clearly $\int_G f$ is G-invariant for every $f \in \mathrm{C}_c(M)$ if and only if dg is right-invariant. If this is the case, $\int_G f$ can be regarded as a function on B, this actually defines a continuous linear mapping $\int_G\colon \mathrm{C}_c(M) \to \mathrm{C}_c(B)$.

For any positive, continuous density db on B, the product densities of these two on the local trivializations piece together to a positive, continuous density dx on M, such that, for all $f \in \mathrm{C}_c(M)$:

$$\int_M f(x)\,dx = \int_{G\backslash M}\left(\int_G f(gx)\,dg\right) d(Gx) = \int_B\left(\int_G f\right)(b)\,db. \tag{3.13.16}$$

The proof is by observing that, on the domain $G \times U$ of a retrivialization $\delta\colon (g, b) \mapsto (g\mathcal{X}(b), b)$, we have

$$\begin{aligned}\int_{G\times U} f &= \int_U\left(\int_G f(g,b)\,dg\right) db = \int_U\left(\int_G f(g\mathcal{X}(b),b)\,dg\right) db\\ &= \int_U\left(\int_G (\delta^* f)(g,b)\,dg\right) db = \int_{G\times U} \delta^* f.\end{aligned}$$

Next, applying in the integrations over G the substitution of variables $g = h^{-1}g'$, and using the analogue of (3.13.13) for right-invariant densities on G, we obtain that the density dx on M satisfies:

$$\int_M f(hx)\,dx = |\det \operatorname{Ad} h^{-1}| \int_M f(x)\,dx, \quad \text{for } h \in G. \tag{3.13.17}$$

So this density dx on M is G-invariant if and only if G is unimodular. If conversely the positive, continuous density dx on M satisfies (3.13.17) and dg is a right-invariant density on G, then there is a unique positive, continuous density db on B such that (3.13.16) holds. In this situation we write:

$$dx = dg\,db, \quad \text{and} \quad db = \frac{dx}{dg} \tag{3.13.18}$$

Remarks.

(a) There exists a G-invariant positive, continuous density dx on M and a positive, continuous density db on B such that (3.13.16) hold if and only if the mappings $\mathcal{X}: U \to G$, which appear in the retrivializations, can be chosen such that they only take values in the kernel of the homomorphism $g \mapsto |\det \mathrm{Ad}\, g| : G \to \mathbf{R}_{>0}$.

(b) If we start out with a left-invariant density on G, then we end up with the formula:

$$\int_M f(x)\, dx = \int_{G\backslash M} \left(\int_G f(g^{-1}x)\, dg \right) d(Gx), \tag{3.13.16'}$$

and the density dx on M again satisfies (3.13.17).

(c) In the case that $\pi: M \to B$ is a covering, which is a special case of a principal fiber bundle, the group G is discrete (0-dimensional), and it is customary to use the the counting measure on G. Then G is unimodular, so dx is G-invariant, and (3.13.16) takes the form:

$$\int_G f(x)\, dx = \int_B (\sum_{g \in G} f(gx))\, d(Gx). \tag{3.13.16''}$$

(d) In the principal fibration $G \to G/H$, two actions are considered on G: the action of G by means of left multiplications, and the action $h \mapsto \mathrm{R}(h)^{-1}$ of H by means of right multiplications. Taking the left-invariant density on H, (3.13,16',17) read:

$$\int_G f(x)\, dx = \int_{G/H} \left(\int_H f(xh)\, dh \right) d(xH), \tag{3.13.19}$$

and

$$\int_G f(xh)\, dx = |\det(\mathrm{Ad}\, h)|_{\mathfrak{h}}| \int_G f(x)\, dx, \quad \text{for } h \in H, \tag{3.13.20}$$

for any densities dx, and $d(xH)$, respectively, for which (3.13.19) holds. It is clear now that dx is left-invariant (and hence (3.13.20) holds for all $h \in G$), if and only if $d(xH)$ is G-invariant, which in turn is equivalent to (3.13.10). One gets all the invariances one could wish for, if both H and G are unimodular.

3.14 The Weyl Integration Theorem

If G is a compact Lie group, and H is a compact subgroup, then the invariant positive densities on them, which always exist, are both left- and right-invariant, and the quotient density on G/H is G-invariant as well. See Section 3.13.

(3.14.1) Theorem [Weyl's integration formula]. *Let G be a connected, compact Lie group with Lie algebra $\mathfrak{g}$, T a maximal torus in G with Lie algebra $\mathfrak{t}$, $W = \mathrm{N}(T)/T$ the corresponding Weyl group. Let dx, and dt, be invariant positive densities on G, and T, respectively, and provide G/T with the quotient density $d(xT) = dx/dt$. On $\mathfrak{g}$, and $\mathfrak{t}$, we take the constant densities, equal to $(dx)_1$, and $(dt)_1$, respectively. Then:*

(i) The mapping that assigns to $f \in \mathrm{C}(G)$ the function

$$F\colon (xT, t) \mapsto f(xtx^{-1})|\det(\mathrm{Ad}\, t - \mathrm{I})_{\mathfrak{g}/\mathfrak{t}}| \quad \text{on } G/T \times T,$$

extends to a topological isomorphism from $\mathrm{L}^1(G)$ onto $\mathrm{L}^1(G/T \times T)^W$, the space of W-invariant functions on $G/T \times T$. Here $sT \in W$, with $s \in \mathrm{N}(T)$, acts on (xT, t) by sending it to $(xs^{-1}T, sts^{-1})$. Moreover, if $f \in \mathrm{L}^1(G)$, then:

$$\begin{aligned} &\int_G f(x)\, dx \\ &= \#(W)^{-1} \int_T \left(\int_{G/T} f(xtx^{-1})\, d(xT) \right) |\det(\mathrm{I} - \mathrm{Ad}\, t)_{\mathfrak{g}/\mathfrak{t}}|\, dt. \end{aligned} \tag{3.14.1}$$

(ii) The mapping that assigns to $\phi \in \mathrm{C}_c(\mathfrak{g})$ the function

$$\Phi\colon (gT, X) \mapsto \phi(\mathrm{Ad}\, g(X))|\det(\mathrm{ad}\, X)_{\mathfrak{g}/\mathfrak{t}}| \quad \text{on } G/T \times \mathfrak{t}$$

extends to a topological isomorphism: $\mathrm{L}^1(\mathfrak{g}) \xrightarrow{\sim} \mathrm{L}^1(G/T \times \mathfrak{t})^W$, where now $sT \in W$ acts on (gT, X) by sending it to $(gs^{-1}T, \mathrm{Ad}\, s(X))$. Moreover, if $\phi \in \mathrm{L}^1(\mathfrak{g})$, then:

$$\begin{aligned} &\int_{\mathfrak{g}} \phi(X)\, dX \\ &= \#(W)^{-1} \int_{\mathfrak{t}} \left(\int_{G/T} \phi(\mathrm{Ad}\, g(X))\, d(gT) \right) |\det(\mathrm{ad}\, X)_{\mathfrak{g}/\mathfrak{t}}|\, dX. \end{aligned} \tag{3.14.2}$$

Proof. (i) Consider the mapping $\Gamma\colon (xT, t) \mapsto xtx^{-1}\colon G/T \times T \to G$. The transformations $\mathrm{L}(x) \times \mathrm{L}(t)$, and $\mathrm{R}(t)^{-1} \circ \mathbf{Ad}\, x^{-1}$, in $G/T \times T$, and G, respectively, preserve the given densities in these spaces. We have:

$$\begin{aligned} &\mathrm{R}(t)^{-1} \circ \mathbf{Ad}\, x^{-1} \circ \Gamma \circ (\mathrm{L}(x) \times \mathrm{L}(t))(yT, s) \\ &= x^{-1}((xy)(ts)(xy)^{-1})xt^{-1} = ytsy^{-1}t^{-1}. \end{aligned}$$

The tangent mapping of this at $(1T, 1)$ is equal to:

$$A\colon (Y + \mathfrak{t}, S) \mapsto (\mathrm{I} - \mathrm{Ad}\, t)(Y) + S\colon \mathfrak{g}/\mathfrak{t} \times \mathfrak{t} \to \mathfrak{g}.$$

Because of the choice of the density on G/T as the quotient one, the mapping $L\colon Y + S \mapsto (Y + \mathfrak{t}, S)\colon \mathfrak{g} \to \mathfrak{g}/\mathfrak{t} \times \mathfrak{t}$ is density preserving. Here $Y \in \mathfrak{q}$, $S \in \mathfrak{t}$, and $\mathfrak{q} = [\mathfrak{t}, \mathfrak{g}]$ is the complementary subspace to $\mathfrak{t}$ in $\mathfrak{g}$ as in (3.5.8), it is $\mathrm{Ad}\, t$-invariant. Because $L \circ A\colon (Y + \mathfrak{t}, S) \mapsto ((\mathrm{I} - \mathrm{Ad}\, t)(Y) + \mathfrak{t}, S)$, we see that $J_\Gamma(xT, t) = |\det(\mathrm{I} - \mathrm{Ad}\, t)_{\mathfrak{g}/\mathfrak{t}}|$.

Now Γ defines a covering: $G/T \times (G^{\mathrm{reg}} \cap T) \to G^{\mathrm{reg}}$, with covering group W, cf. Theorem 3.7.2.(i). The set $G \setminus G^{\mathrm{reg}}$, being equal to a union of finitely many locally closed, connected real-analytic submanifolds $\mathcal{T}$ of positive codimension (even ≥ 3), has zero Lebesgue measure; for instance, one may use that each $\mathcal{T}$ is equal to a countable union of compact subsets of $\mathcal{T}$, of which it is easy to prove that these have zero measure. In view of the covering $\exp\colon \mathfrak{t} \to T$, and the description in the proof of Lemma 3.9.1.(i) of $\exp^{-1}(G^{\mathrm{reg}}) \cap \mathfrak{t}$ as the complement of the affine root hyperplanes (3.11.1), the proof that $T' := T \setminus G^{\mathrm{reg}}$ has Lebesgue measure zero in T is even more elementary. Using also (3.13.16), and the fact that W leaves J_Γ invariant because it acts on $\mathfrak{g}$ as automorphisms preserving $\mathfrak{t}$, we now get, for any $f \in \mathrm{C}(G)$:

$$\begin{aligned}
&\int_{G/T \times T} f(xtx^{-1}) J_\Gamma(t)\, dt\, d(xT) = \int_{G/T \times T'} f(xtx^{-1}) J_\Gamma(t)\, dt\, d(xT) \\
&= \int_{G/T \times T'} \sum_{w \in W} f((xw^{-1})(wtw^{-1})(xw^{-1})^{-1}) J_\Gamma(wtw^{-1})\, dt\, d(xT) \\
&= \#(W) \int_{G/T \times T'} f(xtx^{-1}) J_\Gamma(t)\, dt\, d(xT) = \#(W) \int_{G'} f(x)\, dx \\
&= \#(W) \int_G f(x)\, dx.
\end{aligned}$$

The crucial identity is of course the fourth one, where we applied the formula for substitutions in an integral to the diffeomorphism: $G/T \times_W T' \xrightarrow{\sim} G^{\mathrm{reg}}$ induced by Γ.

In passing we have also seen that the function F, appearing in the integral over $G/T \times T$, is W-invariant, and by continuity we see that $f \mapsto \#(W)^{-1} F$ extends to an isometric embedding of $\mathrm{L}^1(G)$ into $\mathrm{L}^1(G/T \times T)^W$.

Suppose now conversely that $F \in \mathrm{L}^1(G/T \times T)^W$. Then the restriction of F/J_Γ to $G/T \times T'$ is equal to the pull back of a measurable function f_1 on $G/T \times_W T'$, the quotient space for the proper and free action of W on $G/T \times T'$. In turn, f_1 is equal to the pull back of a measurable function f on G^{reg} under the diffeomorphism: $G/T \times_W T' \xrightarrow{\sim} G^{\mathrm{reg}}$. Reading backwards, we see from (3.14.1), approximating $|f|$ from below by integrable functions, that the integral of $|f|$ over G is bounded above by $\#(W)^{-1}$ times the integral of $|F|$ over $G/T \times T$; here we did not discriminate between the domains G^{reg} and G, respectively $G/T \times T'$ and $G/T \times T$, because the complements have measure zero. The conclusion is that $f \in \mathrm{L}^1(G)$, and that F is the function assigned to f as in the theorem.

(ii) The proof goes exactly along the same lines as as for (i), with Γ replaced by $\Omega\colon (gT, X) \mapsto \operatorname{Ad} g(X)\colon G/T \times \mathfrak{t} \to \mathfrak{g}$. This time we determine the Jacobian by investigating the tangent mapping of:

$$(hT, Y) \mapsto \operatorname{Ad} g^{-1} \circ \operatorname{Ad}(gh)(X+Y)$$

at $(1T, 0)$. This is equal to $(U+\mathfrak{t}, Z) \mapsto [U, X] + Z : \mathfrak{g}/\mathfrak{t} \times \mathfrak{t} \to \mathfrak{g}$, and therefore $J_\Omega(gT, X) = |\det(\operatorname{ad} X)_{\mathfrak{g}/\mathfrak{t}}|$. □

Observing that the determinant of a real linear mapping is equal to the determinant of its complex linear extension, we see from the root space decomposition (3.5.4), (3.5.7), that the Jacobian which appears in (3.14.2) can be expressed in terms of the roots. For this purpose, we introduce, for any choice of positive roots P, the function:

(3.14.3) $$\varpi = \varpi_P\colon X \mapsto \prod_{\alpha \in P} \alpha(X)\colon \mathfrak{t} \to i^{\#(P)}\mathbf{R}.$$

Then, using (3.5.10) and the facts that $-\alpha(X) = \overline{\alpha(X)}$ (notice that $\alpha(X)$ is purely imaginary) and $\dim_{\mathbf{C}} \mathfrak{g}_\alpha = 1$ for every $\alpha \in R$ (Theorem 3.10.2.(i)), we get:

(3.14.4) $$\det(\operatorname{ad} X)_{\mathfrak{g}/\mathfrak{t}} = \varpi(X)\overline{\varpi(X)}, \quad \text{for } X \in \mathfrak{t}.$$

Using the covering $\exp\colon \mathfrak{t} \to T$, we easily can give an explicit formula for the Jacobian factor appearing in (3.14.1). Writing $t \in T$ as $t = \exp X$, with $X \in \mathfrak{t}$, we recall that $\operatorname{Ad} t = \operatorname{Ad} \exp X = e^{\operatorname{ad} X}$ acts on $\mathfrak{g}_\alpha$ as multiplication by the scalar:

(3.14.5) $$t^\alpha := e^{\alpha(X)}.$$

The fact that t^α lies on the unit circle in $\mathbf{C}$ corresponds to $t^{-\alpha} = (t^\alpha)^{\mathrm{conj}}$, so now the Jacobian in (3.14.1) is given by:

(3.14.6) $$\det(\mathrm{I} - \operatorname{Ad} t)_{\mathfrak{g}/\mathfrak{t}} = \delta(t)\overline{\delta(t)},$$

where we have used the function

(3.14.7) $$\delta := \delta_P : t \mapsto \prod_{\alpha \in P} (1 - t^{-\alpha})\colon T \to \mathbf{C}.$$

Note that in both cases the determinants themselves already are positive, so that we actually did not need the absolute value signs in (3.14.1), (3.14.2).

Under the assignment of Theorem 3.14.1, the conjugacy invariant functions f on G, and the Ad-invariant functions ϕ on $\mathfrak{g}$, correspond bijectively to functions F, and Φ, respectively that do not depend on the first variable xT and that are Weyl group invariant as a function of the second variable. Because all these spaces are closed linear subspaces of the corresponding function spaces, Theorem 3.14.1 now leads immediately to:

(3.14.2) Corollary. *Set* $c := \int_{G/T} d(xT) = \int_G dx / \int_T dt$, $c' = (c/\#(W))^{1/2}$. *Then:*

(i) The assignment $f \mapsto (f|_T) | \delta |^2$ *defines a topological linear isomorphism from* $\mathrm{L}^1(G)^{\mathrm{Ad}\, G}$, *the space of Lebesgue integrable class functions on* G, *onto the space* $\mathrm{L}^1(T)^W$ *of Lebesgue integrable functions on* T *that are Weyl group invariant. For* $f \in \mathrm{L}^1(G)^{\mathrm{Ad}\, G}$, *we have the integral formula:*

$$\int_G f(x)\, dx = \#(W)^{-1} c \int_T f(t) \,|\, \delta(t) \,|^2\, dt. \tag{3.14.8}$$

Secondly, the assignment $f \mapsto c'\delta(f|_T)$ *defines a (unitary) linear isomorphism from the Hilbert space* $\mathrm{L}^2(G)^{\mathrm{Ad}\, G}$ *of square integrable class functions on* G *onto the Hilbert space* $\mathrm{L}^2(T)^W$ *of Weyl group invariant, square integrable functions on* T.

(ii) The assignment $\phi \mapsto (\phi|_{\mathfrak{t}}) | \varpi |^2$ *defines a topological isomorphism:* $\mathrm{L}^1(\mathfrak{g})^{\mathrm{Ad}\, G} \xrightarrow{\sim} \mathrm{L}^1(\mathfrak{t})^W$; *moreover, if* $\phi \in \mathrm{L}^1(\mathfrak{g})^{\mathrm{Ad}\, G}$, *then:*

$$\int_{\mathfrak{g}} \phi(X)\, dX = \#(W)^{-1} c \int_{\mathfrak{t}} \phi(H) \,|\, \varpi(H) \,|^2\, dH. \tag{3.14.9}$$

The assignment $\phi \mapsto c'\varpi(\phi|_{\mathfrak{t}})$ *defines a unitary transformation from* $\mathrm{L}^2(\mathfrak{g})^{\mathrm{Ad}\, G}$ *onto* $\mathrm{L}^2(\mathfrak{t})^W$.

In a more metric approach, one would start with a left- and right-invariant Riemannian structure β on G and take the densities on G, T, G/T that assign the value 1 to any orthonormal basis in a tangent space. These Riemannian structures are, via the mapping $\beta \mapsto B = \beta_1$, in bijective correspondence with the $\mathrm{Ad}\, G$-invariant inner products B on $\mathfrak{g}$. (If $\mathfrak{z} = 0$, we could take B equal to minus the Killing form, cf. Section 3.6. Also, if $\mathfrak{g}$ is simple, then this is the only choice up to a positive factor, but in general there is more freedom.)

Because the eigenspaces of orthogonal linear transformations are orthogonal to each other, it follows that the decomposition $\mathrm{T}_x\, G = \mathrm{T}_x(\mathbf{Ad}\, G(x)) \oplus \mathrm{T}_x(G_x)$ of Proposition 3.1.1 is an orthogonal one, for every $x \in G$. In particular, for $x \in G^{\mathrm{reg}} \cap T$, the space $\mathrm{T}_x(\mathbf{Ad}\, G(x))$ is the orthogonal complement of $\mathrm{T}_x(T)$ in $\mathrm{T}_x\, G$. If $x \in G^{\mathrm{princ}} \cap T$, and B_r is a ball of sufficiently small radius r around x in T, then the action of conjugation defines a diffeomorphism from $G/T \times B_r$ onto the r-neighborhood U_r of $\mathbf{Ad}\, G(x)$ in G. Because the ($\dim G/T$-dimensional) volume of $\mathbf{Ad}\, G(x)$ can be defined as the limit, for $r \to 0$, of $\mathrm{vol}_G(U_r)/\mathrm{vol}_T(B_r)$, we get, applying (3.14.8) to f equal to the characteristic function of U_r, that:

$$|\, \delta(t) \,|^2 = \text{the volume of the conjugacy class of } t. \tag{3.14.10}$$

If the conjugacy class of t is exceptional, then we have to divide in the left hand side by $\#(G_x/T)$. At this point, it might be interesting to note that, at any critical point X in the alcove $\mathfrak{b}$, cf. (3.9.3), the Hessian of the function: $f\colon X \mapsto |\,\delta(\exp X)\,|^2$ is equal to

$$(Y, Z) \mapsto |\,\delta(\exp X)\,|^2 \sum_{\alpha\in P} (1 - \cos\alpha(X)/i)^{-1}\alpha(Y)\alpha(Z);$$

so it is negative definite if $\mathfrak{z} = 0$. By elementary Morse theory, using that $\mathfrak{b}$ is contractible as a convex polytope, we get that f has a unique critical point X_{max} in $\mathfrak{b}$; this is the point where f attains its absolute maximum. As a consequence of the uniqueness statement, X_{max} is a fixed point for any transformation: $\mathfrak{b} \to \mathfrak{b}$ that leaves f invariant. Therefore the conjugacy class in G of the element $x_{\mathrm{max}} = \exp X_{\mathrm{max}}$, characterized by the properties that it is regular and that its covering in $G^{\sim}$ has maximal Euclidean volume, is one of the regular conjugacy classes for which the fundamental group is canonically isomorphic to the fundamental group of G, cf. Corollary 3.9.5.(iii) and its proof. For a picture of the level curves of f for $G = \mathbf{SU}(3)$, see Fig.3.14.1.

Fig. 3.14.1. Level curves of the volume of conjugacy classes

Similarly the factor $|\,\varpi(H)\,|^2$ has the interpretation of the ($\dim G/T$-dimensional) volume of the adjoint orbit through $X \in \mathfrak{g}^{\mathrm{reg}} \cap \mathfrak{t}$. The orbits, both in the group and in the Lie algebra, literally shrink around the lower dimensional singular orbits, when approaching such a singular orbit.

Note also that the zeros of δ, and ϖ, which occur precisely at the singular set in G, and $\mathfrak{g}$, respectively, cause the restriction: $f \mapsto f|_T$ and $\phi \mapsto \phi|_{\mathfrak{t}}$, to be **not** continuous from $\mathrm{L}^1(G)^{\mathrm{Ad}\, G}$ to $\mathrm{L}^1(T)^W$, and from $\mathrm{L}^1(\mathfrak{g})^{\mathrm{Ad}\, G}$ to $\mathrm{L}^1(\mathfrak{t})^W$, respectively. This in contrast with the fact that these restrictions **are** topological isomorphisms between the corresponding spaces of continuous functions, because the orbit spaces $W\backslash T$, and $W\backslash\mathfrak{t}$, are homeomorphic to $(\mathbf{Ad}\, G)\backslash G$, and $(\mathrm{Ad}\, G)\backslash\mathfrak{g}$, respectively. See the slightly stronger Lemma 3.9.2.(iii), and Theorem 3.8.3.(iii), respectively.

The factor $c = \mathrm{vol}(G/T)$, which appears in (3.14.8) and in (3.14.9), can be determined explicitly once we can evaluate the integrals in the left and right hand side for a suitable invariant and integrable function f, and ϕ, respectively. One could for instance take, in (3.14.9), ϕ equal to the characteristic function of the unit ball in $\mathfrak{g}$; then one is left with the computation of the integral of the polynomial $\varpi(H)\overline{\varpi(H)}$ over the unit ball in $\mathfrak{t}$.

Instead, we take $\phi(X) = \mathrm{e}^{-<X,X>/2}$; in the notation we use the inner product B to identify $\mathfrak{g}$, and $\mathfrak{t}$, respectively, with its dual, and we also write $\mu(X) = <X, \mu>$, for a linear form μ. Then, as is well-known, the left hand side of (3.14.9) is equal to $(2\pi)^{\dim \mathfrak{g}/2}$. On the other hand, for any **polynomial** f, the function

$$F_t\colon X \mapsto (4\pi t)^{-\dim \mathfrak{t}/2} \int_{\mathfrak{t}} \mathrm{e}^{-<X-Y,X-Y>/4t}\, f(Y)\, dY$$

is a polynomial of degree $\leq \deg f$, depending smoothly on $t > 0$. This function F_t satisfies the differential equation:

$$\frac{\partial F_t}{\partial t} = \Delta F_t, \quad \text{with the boundary condition} \quad \lim_{t\to 0, t>0} F_t = f.$$

Here Δ denotes the Laplace operator with respect to the given inner product, $\Delta = \sum_j \frac{\partial^2}{\partial Y_j^2}$ on any orthonormal basis. Now Δ leaves the finite-dimensional space of polynomials of degree $\leq \deg f$ invariant, and because it decreases degrees, actually acts on this space as a nilpotent operator. It follows that: $F_t = \sum_{k\geq 0}(k!)^{-1}(t\Delta)^k f$, and therefore, taking $t = \frac{1}{2}$, we get:

$$\int_{\mathfrak{t}} \mathrm{e}^{-<Y,Y>/2}\, f(Y)\, dY = (2\pi)^{\dim \mathfrak{t}/2} \sum_{k\geq 0} \frac{1}{k!}((\frac{1}{2}\Delta)^k f)(0),$$

where both sums are finite. In our case, $f : Y \mapsto \prod_{\alpha\in R} \alpha(Y)$ is a homogeneous polynomial of degree $2p$, if we write $p = \#(P) = \frac{1}{2}(\dim \mathfrak{g} - \dim \mathfrak{t})$. This implies that only the term with $k = p$ gives a nonzero contribution. We therefore arrive at:

(3.14.11) $$(2\pi)^p \#(W) = cd, \quad \text{where } p = \#(P),$$

and

(3.14.12) $$d = (p!\, 2^p)^{-1} \sum_{\sigma} < \alpha_{\sigma(1)}, \alpha_{\sigma(2)} > \ldots < \alpha_{\sigma(2p-1)}, \alpha_{\sigma(2p)} > .$$

Here $j \mapsto \alpha_j$ is an enumeration of R, and the sum is over all permutations of $\{1, \dots, 2p\}$.

This formula is explicit, but quite cumbersome in practical computations. A simpler formula can be given by using the fact that ϖ is W-anti-invariant, actually:

$$\varpi(Y) = \frac{1}{p!} \sum_{s \in W} \det s\big((s\rho)(Y)\big)^p, \quad \text{with} \quad \rho = \frac{1}{2} \sum_{\alpha \in P} \alpha.$$

This follows from:

$$\prod_{\alpha \in P} \left(e^{\alpha(X)/2} - e^{-\alpha(X)/2} \right) = \sum_{s \in W} (\det s)\, e^{s \cdot \rho(X)},$$

cf. (4.9.23), by inserting $X = tY$ and comparing the coefficients of t^p in the Taylor expansion. Furthermore, each W-anti-invariant polynomial has ϖ as a factor, because anti-invariance of f under s_α implies that $f = 0$ on $\ker \alpha$, for each $\alpha \in P$.

Because Δ commutes with W, we have that $\Delta\varpi$ is W-anti-invariant, and because $\deg \Delta\varpi < \deg \varpi$, the conclusion is that $\Delta\varpi = 0$. Because

$$\frac{1}{2}\Delta(f^2) = \sum_j \Big(\frac{\partial f}{\partial Y_j}\Big)^2 + f\Delta f$$

for any function f, and because Δ commutes with all constant coefficient linear partial differential operators, the conclusion is that:

$$\begin{aligned}
\frac{1}{p!}\Big(\frac{1}{2}\Delta\Big)^p(\varpi^2) &= \frac{1}{p!} \sum_j \Big(\frac{\partial^p \varpi}{\partial Y_{j(1)} \cdots \partial Y_{j(p)}} \Big)^2 \\
&= \frac{1}{p!} \sum_j \left(\sum_{s \in W} \operatorname{sgn} s \frac{1}{p!} \frac{\partial^p (s\rho)}{\partial Y_{j(1)} \cdots \partial Y_{j(p)}} \right)^2 \\
&= \frac{1}{p!} \sum_j \Big(\sum_{s \in W} \operatorname{sgn} s (s\rho)_{j(1)} \cdots (s\rho)_{j(p)} \Big)^2 \\
&= \frac{1}{p!} \sum_j \sum_{s,s'} \operatorname{sgn} s \operatorname{sgn} s' (s\rho)_{j(1)} \cdots (s\rho)_{j(p)} (s'\rho)_{j(1)} \cdots (s'\rho)_{j(p)} \\
&= \frac{1}{p!} \sum_{s,s'} \operatorname{sgn} s \operatorname{sgn} s' < s\rho, s'\rho >^p = \frac{1}{p!} \sum_{s,s'} \operatorname{sgn}(s^{-1}s') < \rho, s^{-1}s'\rho >^p \\
&= \frac{1}{p!} \#(W) \sum_s \operatorname{sgn} s \big(s\rho(B^{-1}\rho)\big)^p = \#(W) \varpi(B^{-1}\rho),
\end{aligned}$$

where $B^{-1}\rho$ is the element of $\mathfrak{h}$ such that $\mu(B^{-1}\rho) = <\rho, \mu>$ for all $\mu \in \mathfrak{h}^*$. Also, j ranges over all mappings from $\{1, \dots, p\}$ to $\{1, \dots, \dim \mathfrak{t}\}$, and $(s\rho)_k$ equals the k-th coordinate of $s\rho$ with respect to the chosen orthonormal basis in $\mathfrak{t}$.

We have now arrived at:

$$\mathrm{vol}(G/T) = \frac{(2\pi)^{\#(P)}}{(-1)^p} \prod_{\alpha \in P} < \rho, \alpha > . \tag{3.14.13}$$

If one takes, for $\mathfrak{z} = 0$, $B = -\kappa$, then also the factors -1 in (3.14.13) get nicely absorbed in the product.

3.15 Nonconnected Groups

We conclude this chapter with some remarks about the case that the compact Lie group G may have several connected components.

We start with the adjoint action of G on its Lie algebra $\mathfrak{g}$. This time $\mathrm{Ad}\, G$, still a compact subgroup of $\mathrm{Aut}\,\mathfrak{g}$, need not be equal to $\mathrm{Ad}\,\mathfrak{g}$, the adjoint group of $\mathfrak{g}$. We can only say that $(\mathrm{Ad}\, G)^\circ = \mathrm{Ad}\,\mathfrak{g}$, and that the component group $\mathrm{Ad}\, G/\,\mathrm{Ad}\,\mathfrak{g}$ is finite.

Fix a maximal Abelian subspace $\mathfrak{t}$ of $\mathfrak{g}$ and a Weyl chamber $\mathfrak{c}$ in $\mathfrak{t}$, cf. the discussion preceding Lemma 3.5.2. If $\Phi \in \mathrm{Aut}\,\mathfrak{g}$, then, for any $X \in \mathfrak{c}$, we have $\Phi(X) \in \mathfrak{g}^{\mathrm{reg}}$, and according to Proposition 3.8.1 there exists $\Psi \in \mathrm{Ad}\,\mathfrak{g}$, such that $\Theta(X) \in \mathfrak{c}$, if we write $\Theta = \Psi^{-1} \circ \Phi \in \mathrm{Aut}\,\mathfrak{g}$. But then, observing that $\mathfrak{g}_X = \mathfrak{t} = \mathfrak{g}_{\Theta(X)}$, we get $\Theta(\mathfrak{t}) = \mathfrak{t}$, and $\Theta(\mathfrak{t}^{\mathrm{reg}}) = \Theta(\mathfrak{t}^{\mathrm{reg}})$, hence $\Theta(\mathfrak{c}) = \mathfrak{c}$, because Θ maps $X \in \mathfrak{c}$ to a point of the same connected component of $\mathfrak{t}^{\mathrm{reg}}$. So, introduce the notation:

$$\mathrm{N}_{\mathrm{Aut}\,\mathfrak{g}}(\mathfrak{c}) := \{\Phi \in \mathrm{Aut}\,\mathfrak{g} \mid \Phi(\mathfrak{c}) = \mathfrak{c}\}. \tag{3.15.1}$$

We have proved that the product induces a surjective mapping: $\mathrm{Ad}\,\mathfrak{g} \times \mathrm{N}_{\mathrm{Aut}\,\mathfrak{g}}(\mathfrak{c}) \to \mathrm{Aut}\,\mathfrak{g}$.

If $\Theta \in \mathrm{Ad}\,\mathfrak{g}$, and $\Theta(\mathfrak{c}) = \mathfrak{c}$, then Proposition 3.8.1 also yields that $\Theta \in \mathrm{e}^{\mathrm{ad}\,\mathfrak{t}}$. We have obtained:

(3.15.1) Proposition.

(i) *For every* $\Phi \in \mathrm{Aut}\,\mathfrak{g}$, *there exist* $\Psi \in \mathrm{Ad}\,\mathfrak{g}$ *and* $\Theta \in \mathrm{N}_{\mathrm{Aut}\,\mathfrak{g}}(\mathfrak{c})$, *such that* $\Phi = \Psi \circ \Theta$. *If* $\Psi' \in \mathrm{Ad}\,\mathfrak{g}$ *and* $\Theta' \in \mathrm{N}_{\mathrm{Aut}\,\mathfrak{g}}(\mathfrak{c})$, *with* $\Phi' = \Psi' \circ \Theta'$, *then* $\Psi' \in \Psi\,\mathrm{e}^{\mathrm{ad}\,\mathfrak{t}}$, *and* $\Theta' \in \mathrm{e}^{\mathrm{ad}\,\mathfrak{t}}\,\Theta$.

(ii) *For every* $x \in G$, *there exist* $g \in G^\circ$, $h \in \mathrm{N}_G(\mathfrak{c})$, *with:*

$$\mathrm{N}_G(\mathfrak{c}) := \{h \in G \mid \mathrm{Ad}\,h(\mathfrak{c}) = \mathfrak{c}\}, \tag{3.15.2}$$

such that $x = gh$. *If* $g' \in G^\circ$, $h' \in \mathrm{N}_G(\mathfrak{c})$, *and* $x = g'h'$, *then* $g' \in gT$, $h' \in Th = hT$, *where* $T = \exp\mathfrak{t}$ *is the maximal torus in* G *with Lie algebra equal to* $\mathfrak{t}$. *Finally,* $\mathrm{N}_G(\mathfrak{c})^\circ = T = \mathrm{N}_G(\mathfrak{c}) \cap G^\circ$, *and the injections* $\mathrm{N}_G(\mathfrak{c}) \to G$ *and* $T \to G^\circ$, *yield an isomorphism:*

$$\mathrm{N}_G(\mathfrak{c})/T \to G/G^\circ. \tag{3.15.3}$$

Remark. If $\mathfrak{z} = 0$, then we can take $G = \operatorname{Aut}\mathfrak{g}$ (with Lie algebra isomorphic to $\mathfrak{g}$ via ad), so $\mathrm{N}_{\operatorname{Aut}\mathfrak{g}}(\mathfrak{c})^\circ = \mathrm{e}^{\operatorname{ad}\mathfrak{t}}$, and we get an isomorphism:

$$\mathrm{N}_{\operatorname{Aut}\mathfrak{g}}(\mathfrak{c})/\,\mathrm{e}^{\operatorname{ad}\mathfrak{t}} \xrightarrow{\sim} \operatorname{Aut}\mathfrak{g}/(\operatorname{Aut}\mathfrak{g})^\circ \simeq \operatorname{Aut}\mathfrak{g}/\operatorname{Ad}\mathfrak{g}. \tag{3.15.4}$$

(Note that if $\mathfrak{z} \neq 0$, then $\mathrm{N}_{\operatorname{Aut}\mathfrak{g}}(\mathfrak{c})$ gets a nontrivial contribution from $\mathbf{GL}(\mathfrak{z})$.)

Every $\Phi \in \operatorname{Aut}\mathfrak{g}$ such that $\Phi(\mathfrak{t}) = \mathfrak{t}$ induces an automorphism of the root system $R = R(\mathfrak{g}, \mathfrak{t})$. If $\Phi(\mathfrak{c}) = \mathfrak{c}$, then Φ also leaves invariant the set $S = S(\mathfrak{c})$ of simple roots, corresponding to the Weyl chamber $\mathfrak{c}$. The set of simple roots S is provided with a combinatorial structure, called the *Dynkin diagram*, which encodes the Cartan matrix (3.11.10), and from which the whole root structure can be reconstructed. See the definition preceding Theorem 5.5.15. A permutation Φ of S is called an *automorphism of the Dynkin diagram* if $(\Phi(\alpha))(\Phi(\beta)^\vee) = \alpha(\beta^\vee)$, for all $\alpha, \beta \in S$; we will denote the group of these automorphisms by $\operatorname{Aut}(S)$. These are in bijective correspondence with the automorphisms of the root system that leave S invariant, see Corollary 5.5.11. It is proved in Corollary 5.6.2, that every automorphism of the Dynkin diagram actually is induced by some $\Phi \in \mathrm{N}_{\operatorname{Aut}\mathfrak{g}}(\mathfrak{c})$, and that $\mathrm{Z}_{\operatorname{Aut}\mathfrak{g}}(\mathfrak{t}) = \mathrm{e}^{\operatorname{ad}\mathfrak{t}}$. These statements lead to another isomorphism:

$$\mathrm{N}_{\operatorname{Aut}\mathfrak{g}}(\mathfrak{c})/\,\mathrm{e}^{\operatorname{ad}\mathfrak{t}} \xrightarrow{\sim} \operatorname{Aut}(S). \tag{3.15.5}$$

Combined with (3.15.4), we therefore get that **the component group of the automorphism group of $\mathfrak{g}$ is isomorphic to the automorphism group of the Dynkin diagram**, if $\mathfrak{z} = 0$.

(3.15.2) Proposition.

(i) The regular set $\mathfrak{g}^{\mathrm{reg}}$ in $\mathfrak{g}$ for the adjoint action of G is the same as for $\operatorname{Ad}G^\circ = \operatorname{Ad}\mathfrak{g}$. Recall that every $X \in \mathfrak{g}^{\mathrm{reg}}$ can be brought into $\mathfrak{c}$ by means of $\operatorname{Ad}\mathfrak{g}$.

(ii) If $X \in \mathfrak{c}$, then X is of principal orbit type for the adjoint action of G on $\mathfrak{g}$, if and only if $\mathrm{N}_G(\mathfrak{c})/\,\mathrm{Z}_G(\mathfrak{t})$ acts freely on X, that is, $G_X = \mathrm{Z}_G(\mathfrak{t})$.

(iii) There exist $X \in \mathfrak{c}$, such that $G_X = \mathrm{N}_G(\mathfrak{c})$.

Proof. (i) is obvious, because the infinitesimal action of G is equal to the infinitesimal action of G°.

(ii) Clearly $X \in \mathfrak{c}$ implies $G_X \subset \mathrm{N}_G(\mathfrak{c})$. The finite group $\mathrm{N}_G(\mathfrak{c})/\,\mathrm{Z}_G(\mathfrak{t})$ (indeed, $T \subset \mathrm{Z}_G(\mathfrak{t})$ and $\mathrm{N}_G(\mathfrak{c})/T$ is finite) acts on $\mathfrak{c}$. The elements of the principal orbit type for a finite group action are characterized by the condition that their stabilizer group $G_X/\,\mathrm{Z}_G(\mathfrak{t})$ must have the minimal number of elements. Counting connected components of G_X, we see that this corresponds exactly to the condition that X is of principal orbit type for the adjoint action of G on $\mathfrak{g}$. Furthermore, these minimal stabilizer groups $G_X/\,\mathrm{Z}_G(\mathfrak{t})$, by continuity, must be constant as a function of X in the principal orbit type. Because the latter is an open subset of $\mathfrak{c}$, hence of $\mathfrak{t}$, the conclusion is that $G_X = \mathrm{Z}_G(\mathfrak{t})$ if and only if X is of principal orbit type.

(iii) Fixed points for the linear action of the finite group $\mathrm{N}_G(\mathfrak{c})/\mathrm{Z}_G(\mathfrak{t})$ on the convex set $\mathfrak{c}$ are obtained by averaging any orbit in $\mathfrak{c}$, as in the proof of Proposition 3.8.1. □

The "largest" group to which Proposition 3.15.2 applies is the group A of automorphisms of $\mathfrak{g}$ that act as the identity on the center $\mathfrak{z}$ of $\mathfrak{g}$. We have $\mathrm{Ad}\,\mathfrak{g} \subset \mathrm{Ad}\,G \subset A$ if G is a compact Lie group with Lie algebra equal to $\mathfrak{g}$. It follows from Lemma 3.6.1 that $\mathrm{ad}\,\mathfrak{g} = A^\circ$, so $\mathrm{Ad}\,G$ is an open subgroup of A. Also, A is compact.

The action of conjugation by G on G° can be given a similar description. If C' denotes the connected component of $G^{\mathrm{reg}} \cap T$ that near 1 is equal to $\exp(\mathfrak{c} \cap V)$, with V a small neighborhood of 0 in $\mathfrak{t}$, then $\mathrm{N}_G(C') = \mathrm{N}_G(\mathfrak{c})$ acts on C', and the elements in C' which are of principal orbit type for the action of G coincide with the elements of principal orbit type for the action of $\mathrm{N}_G(\mathfrak{c})$. Note that for all elements of C', the Lie algebra of the centralizer is equal to $\mathfrak{t}$; and that we have a G-equivariant analytic diffeomorphism: $G/\mathrm{Z}_G(T) \times_{\mathrm{N}(T)/\mathrm{Z}(T)} C' \xrightarrow{\sim} G^{\mathrm{reg}} \cap G^\circ$.

Let us now turn to the action of conjugation of G on the other connected components of G. The idea is to bring the description of 3.1.(h) in a more detailed form, by means of the structure theory of the adjoint representation, as it has been developed up till now.

(3.15.3) Lemma. *For each $X \in \mathfrak{g}$, $x \in G_X$, there exists $z \in G^{\mathrm{princ}} \cap G_X \cap G_x$, arbitrarily close to x.*

Proof. Let $x \in G_X$. Then $X \in \mathfrak{g}_x$, hence, for each $t \in \mathbf{R}$, we have $\exp tX \in G_x$. For sufficiently small t, the element $y := (\exp tX)x$ is in the slice at x, hence $\mathfrak{g}_y \subset \mathfrak{g}_x$. On $\mathfrak{g}_x$, we get $\mathrm{Ad}\,y = (\mathrm{Ad}\exp tX)\,\mathrm{Ad}\,x = \mathrm{e}^{t\,\mathrm{ad}\,X}$, so for sufficiently small $t \neq 0$, we have:

$$\mathfrak{g}_y = \ker(\mathrm{Ad}\,y - \mathrm{I}) \cap \mathfrak{g}_x = \ker(\mathrm{e}^{t\,\mathrm{ad}\,X} - \mathrm{I}) \cap \mathfrak{g}_x = \ker \mathrm{ad}\,X \cap \mathfrak{g}_x = \mathfrak{g}_X \cap \mathfrak{g}_x.$$

Now, near y, the set G_y is a slice at y, so all nearby elements in G are conjugate to nearby points in G_y. Because G^{princ} is dense and conjugacy-invariant, there exist $z \in G^{\mathrm{princ}} \cap y(G_y)^\circ$, arbitrarily close to y. Because $\mathfrak{g}_y = \mathfrak{g}_X \cap \mathfrak{g}_x$, $y \in G_X \cap G_x$, we have $y(G_y)^\circ \subset G_X \cap G_x$, and because y could be chosen arbitrarily close to x, the lemma is proved. □

(3.15.4) Lemma.

(a) For each $u \in G$, $\mathfrak{g}_u$ contains regular elements.

(b) If σ is an automorphism of G and G^σ denotes the set of fixed points of σ in G, then $G^{\mathrm{reg}} \cap G^\sigma$ is dense in G^σ.

Proof.

(a) Take $X \in \mathfrak{c} \subset \mathfrak{g}^{\mathrm{reg}}$, such that $G_X = \mathrm{N}_G(\mathfrak{c})$, see Proposition 3.15.2.(iii). According to (3.15.3), the connected component C of u in G contains some $x \in \mathrm{N}_G(\mathfrak{c})$. Lemma 3.15.3 yields the existence of $z \in G^{\mathrm{princ}} \cap G_X \cap C$. That is, $\mathfrak{g}_z$ contains the regular element X. Because there is only one principal orbit type in C we get that $\mathfrak{g}_u$ contains $\mathrm{Ad}\, g(\mathfrak{g}_z)$ for some $g \in G$; $\mathrm{Ad}\, g(X) \in \mathfrak{g}_u \cap \mathfrak{g}^{\mathrm{reg}}$.

(b) For each automorphism Φ of G, $\phi := \mathrm{T}_1 \Phi$ is an automorphism of $\mathfrak{g}$ which leaves $\mathfrak{g}' := [\mathfrak{g}, g]$ and $\mathfrak{z} := \ker \mathrm{ad}$ invariant, as well as the lattice $\ker \exp \cap \mathfrak{z}$ in $\mathfrak{z}$. It follows that the elements of the Lie algebra of $\mathrm{Aut}(G)$ act on $\mathfrak{g}$ as derivations which vanish on $\mathfrak{z}$. Using Lemma 3.6.1 we get that ad is an isomorphism from $\mathfrak{g}'$ onto the Lie algebra of $\mathrm{Aut}(G)$. For any $x \in G^\sigma$, let $A := \{\Phi \in \mathrm{Aut}(G) \mid \Phi(x) = x\}$. This is a Lie subgroup of $\mathrm{Aut}(\mathfrak{g})$, with Lie algebra equal to $\mathfrak{a} := \{\mathrm{ad}\, X \mid X \in \mathfrak{g}_x\}$. Also, $\sigma \in A$.

Applying (a) with G and u replaced by A and σ, respectively, we get $X \in \mathfrak{g}_x \cap \mathfrak{g}'$ such that the centralizer of $\mathrm{ad}\, X$ in $\mathfrak{g}_x \cap \mathfrak{g}'$ is Abelian and $\mathrm{ad}\, \sigma(X) = \sigma \circ \mathrm{ad}\, X \circ \sigma^{-1} = \mathrm{ad}\, X$. Because $\mathrm{ad}\,|_{\mathfrak{g}'}$ is an isomorphism, this means that $\mathfrak{g}_X \cap \mathfrak{g}_x$ is Abelian and $\sigma(X) = X$. Let $t \in \mathbf{R}$. Then $y := x \cdot \exp(t\, X) \in G^\sigma$. Because $\mathfrak{g}_y = \mathfrak{g}_x \cap \mathfrak{g}_X$, or $y \in G^{\mathrm{reg}}$, if t is nonzero and sufficiently small, the proof is complete. □

(3.15) Corollary. *Let σ be an automorphism of a simply connected compact Lie group G. Then the fixed point set G^σ of σ in G is connected.*

Proof. In view of Lemma 3.15.4.(b), it suffices to prove that each $x \in G^{\mathrm{reg}} \cap G^\sigma$ belongs to $(G^\sigma)^\circ$, the identity component of G^σ. Because x is regular, $\mathfrak{t} := \mathfrak{g}_x := \ker(\mathrm{Ad}\, x - \mathrm{I})$ is a maximal Abelian subspace of $\mathfrak{g}$. It follows from Corollary 3.9.5.(iv) and Corollary 3.3.2 that $T := \exp \mathfrak{t} = G_x$, so $x \in T$. Recall that in Section 3.11 the alcoves in $\mathfrak{t}$ were defined as the connected components of $\mathfrak{t} \cap \exp^{-1}(G^{\mathrm{reg}})$.

From Proposition 3.11.1 we see that $\Lambda := \ker \exp|_{\mathfrak{t}}$ is equal to the coroot lattice $\Lambda_{R^\vee}$, and that $\Lambda \cdot W$ acts freely and transitively on the set of alcoves. Moreover, the closure $\mathfrak{b}^{\mathrm{cl}}$ of each alcove $\mathfrak{b}$ contains exactly one element of Λ, cf. Corollary 3.9.6.(i),(ii).

Let us denote $\mathrm{T}_1\, \sigma$ with the same letter σ. Then σ leaves $\mathfrak{t}$, Λ and the union of the alcoves invariant, in particular it permutes the alcoves. Choose $X \in \mathfrak{t} \cap \exp^{-1}(x)$, the regularity of x implies that $X \in \mathfrak{b}$ for some alcove $\mathfrak{b}$. Because $\Lambda \cdot W$ acts transitively on the set of alcoves, and W leaves the set of alcoves $\mathfrak{b}$ with $0 \in \mathfrak{b}^{\mathrm{cl}}$ invariant, we can arrange that $0 \in \mathfrak{b}^{\mathrm{cl}}$. On the other hand $\sigma(x) = x$ is equivalent to $Y := \sigma(X) - X \in \Lambda$, which implies that $\sigma(\mathfrak{b}) = Y + \mathfrak{b}$. Now $0 \in \mathfrak{b}^{\mathrm{cl}}$ implies that $0 = \sigma(0) \in \sigma(\mathfrak{b}^{\mathrm{cl}}) = (\sigma(\mathfrak{b}))^{\mathrm{cl}}$, and also $Y = Y + 0 \in Y + \mathfrak{b}^{cl} = (Y + \mathfrak{b})^{\mathrm{cl}} = (\sigma(\mathfrak{b}))^{\mathrm{cl}}$. The uniqueness of $\Lambda \cap \mathfrak{b}^{\mathrm{cl}}$ therefore yields that $Y = 0$, or $\sigma(X) = X$. But then $\exp(t\, X) \in G^\sigma$ for all $t \in \mathbf{R}$, letting t run from 0 to 1 we see that $x = \exp X$ belongs to the identity component of g^σ. □

As another application, we shall describe an analogue of the Weyl covering (3.7.2). Let $x \in G^{\mathrm{princ}}$, choose $X \in \mathfrak{g}_x \cap \mathfrak{g}^{\mathrm{reg}}$. Then $\mathfrak{t} := \mathfrak{g}_X$ is a maximal Abelian subspace of $\mathfrak{g}$, and X belongs to a Weyl chamber $\mathfrak{c}$ of $\mathfrak{t}$. Furthermore $\mathrm{Ad}\, x(X) = X$ implies that $\mathrm{Ad}\, x(\mathfrak{t}) = \mathfrak{t}$, hence $\mathrm{Ad}\, x(\mathfrak{c}) = \mathfrak{c}$.

Write $S := (G_x)^{\circ}$, this is a subtorus of T, with Lie algebra equal to $\mathfrak{s} := \mathfrak{g}_x$. Because $x \in \mathrm{N}_G(\mathfrak{c})$, the map $\mathrm{Ad}\, x$ leaves $\mathfrak{t}$ invariant, and $\ker(\mathrm{Ad}\, x - \mathrm{I}) = \mathfrak{s}$; so $\mathfrak{t}' := (\mathrm{Ad}\, x - \mathrm{I})(\mathfrak{t})$ is a complementary subspace to $\mathfrak{s}$ in $\mathfrak{t}$, on which $\mathrm{Ad}\, x - \mathrm{I}$ is an invertible linear transformation. Note that $\mathrm{Ad}\, x|_{\mathfrak{t}}$ belongs to the finite group $\mathrm{N}_G(\mathfrak{c})/\,\mathrm{Z}_G(\mathfrak{t})$; therefore it has finite order, and all eigenvalues of $\mathrm{Ad}\, x|_{\mathfrak{t}}$ are roots of unity.

(3.10.4) shows that $(\mathrm{Ad}\, x)^*$ maps roots to roots (and permutes P, and S, respectively), more precisely,

$$\mathrm{Ad}\, x(\mathfrak{g}_\beta) = \mathfrak{g}_{\mathrm{Ad}\, x\beta}. \tag{3.15.6}$$

The orbits in R, under the cyclic group generated by the action of $\mathrm{Ad}\, x$ in R, are cycles, of the form $\mathrm{Ad}\, x^k\alpha$, all different from each other for $k \in \{0, \dots, m-1\}$, while $\mathrm{Ad}\, x^m\alpha = \alpha$. (The orders $m = \#(\sigma)$ of the various cycles σ may differ from each other.) $\mathrm{Ad}\, x^m$ acts on $\mathfrak{g}_\alpha$ as multiplication by a number $c \in \mathbf{C}$ on the unit circle. On a basis $Y, \mathrm{Ad}\, x(Y), \dots, \mathrm{Ad}\, x^{m-1}(Y)$, with $Y \in \mathfrak{g}_\alpha \setminus \{0\}$, the mapping $\mathrm{Ad}\, x$ has the matrix:

$$\begin{pmatrix} 0 & 0 & \cdot & \cdot & 0 & c \\ 1 & 0 & \cdot & \cdot & 0 & 0 \\ \cdot & \cdot & \cdot & \cdot & \cdot & \cdot \\ 0 & 0 & \cdot & \cdot & 1 & 0 \end{pmatrix}.$$

So $\det(\mathrm{I} - \mathrm{Ad}\, x) = 1 - c$. Also note that $\mathrm{Ad}\, x^m$ acts as multiplication by the scalar $c = c_\sigma(x)$ on the whole space

$$\mathfrak{g}_\sigma := \sum_{\alpha \in \sigma} \mathfrak{g}_\alpha. \tag{3.15.7}$$

We get the same conclusion if x is replaced by $u \in xT$, the connected component of x in $\mathrm{N}_G(\mathfrak{c})$. Writing $u = xt$, with $t \in T$,

$$u^m = x^m(x^{1-m}tx^{m-1})(x^{2-m}tx^{m-2}) \cdots (x^{-1}tx)t,$$

so the action of this element on $\mathfrak{g}_\alpha$ is equal to multiplication by $c_\sigma(x)t^{\Sigma(\sigma)}$, where $\Sigma(\sigma) = \sum_{j=0}^{m-1} \mathrm{Ad}\, x^j\alpha$, which is the sum over all elements in the $\mathrm{Ad}\, x$-cycle of α in P. That is,

$$c_\sigma(x.t) = c_\sigma(x)t^{\Sigma(\sigma)}, \quad \text{with } t \in T, \text{ and } \Sigma(\sigma) = \sum_{\alpha \in \sigma} \alpha. \tag{3.15.8}$$

We have also used the notation (3.14.5) here.

Recall from Section 3.1.h, that all conjugacy classes in the connected component of x in G already meet the subset xS of xT. Writing $u = xs$, $s \in S$, we get: $u \in G^{\mathrm{reg}} \Leftrightarrow \mathfrak{g}_u = \mathfrak{s} = \mathfrak{g}_x \Leftrightarrow c_\sigma(u) = c_\sigma(x)s^{\Sigma(\sigma)} \neq 1$, for each $\mathrm{Ad}\, x$-cycle σ in P. More precisely, $\mathfrak{g}_u$ is equal to the direct sum of $\mathfrak{s}$ and, for each cycle $\sigma \in P$ satisfying $c_\sigma(x)s^{\Sigma(\sigma)} = 1$, a real 2-dimensional linear subspace of $\sum_{\alpha\in\sigma}(\mathfrak{g}_\alpha + \mathfrak{g}_{-\alpha}) \cap \mathfrak{g} = \mathrm{Re}\, \mathfrak{g}_\sigma$. Furthermore, $\mathfrak{z}(\mathfrak{g}_u) = \bigcap_{\{\sigma | c_\sigma(u)=1\}} \ker \Sigma(\sigma) \cap \mathfrak{s}$. Note that $\mathfrak{s} = \mathfrak{g}_x$ implies that $\alpha = \beta$ on $\mathfrak{s}$, for all $\alpha, \beta \in \sigma$; hence $\Sigma(\sigma) = \#(\sigma)\alpha$ on $\mathfrak{s}$, for each $\alpha \in \sigma$. Also, because $\mathfrak{s}$ contains regular elements (Lemma 3.15.4), $\ker \Sigma(\sigma) \cap \mathfrak{s} = \ker \alpha \cap \mathfrak{s} \neq \mathfrak{s}$. So we see that here again **the codimension** ($= \dim \mathfrak{g}_u/\mathfrak{z}(G_u)$) **of each singular infinitesimal orbit type is** ≥ 3.

The observations above imply that $G^{\mathrm{reg}} \cap xS$ is an open subset of xS, the complement of which in xS is equal to the union of finitely many "translated subtori" of codimension 1 in xS. The analogue of (3.7.2) is the G-equivariant diffeomorphism:

$$G/S \times_{\mathrm{N}(xS)/S} (G^{\mathrm{reg}} \cap xS) \xrightarrow{\sim} {}^{x}G^{\mathrm{reg}}, \tag{3.15.9}$$

where we have written $\mathrm{N}(xS) = \{ g \in G \mid gxSg^{-1} = xS \}$, and ${}^{x}G^{\mathrm{reg}}$ is the union of the conjugacy classes that meet the connected component of x in G.

Taking invariant measures dg, and ds, on G, and S, respectively, and the corresponding quotient measure on G/S, cf. Section 3.13, we obtain the following analogue of the Weyl integration formula (3.14.1):

$$\begin{aligned} \int_G f(g)\, dg &= \sum \#(\mathrm{N}(x.S)/S)^{-1} |\det(\mathrm{I} - \mathrm{Ad}\, x)_{\mathfrak{t}/\mathfrak{s}}| \\ &\times \int_S \left(\int_{G/S} f(gxsg^{-1})\, d(gG_x) \right) \delta(xs)\overline{\delta(xs)}\, ds. \end{aligned} \tag{3.15.10}$$

Here the summation is over the conjugacy classes in the component group G/G°. For each such conjugacy class we have chosen a representative $x \in G^{\mathrm{princ}}$, $\mathfrak{s} = \mathfrak{g}_x$, $S = \exp \mathfrak{s} = (G_x)^\circ$. Furthermore, $\mathfrak{t}$ is a maximal Abelian subspace of $\mathfrak{g}$, containing $\mathfrak{s}$. One can make a fixed choice of $\mathfrak{t}$ and a positive Weyl chamber $\mathfrak{c}$ in $\mathfrak{t}$, thereby taking all representatives $x \in G^{\mathrm{princ}} \cap \mathrm{N}_G(\mathfrak{c})$. However, S (even its dimension) usually will strongly depend on the choice of the conjugacy class in G/G°. Finally we have written

$$\delta(u) = \det(\mathrm{I} - \mathrm{Ad}\, u^{-1})_{\mathfrak{n}} = \prod_\sigma (1 - c_\sigma(u)^{-1}), \quad \text{for } u \in U, \tag{3.15.11}$$

where $\mathfrak{n} = \sum_{\alpha\in P} \mathfrak{g}_\alpha$. Further σ runs over the $\mathrm{Ad}\, u$-cycles in P, and $c_\sigma(u) = c_\sigma(x)t^{\Sigma(\sigma)} = c_\sigma(x)t^{\#(e)\alpha}$ is the eigenvalue of $\mathrm{Ad}\, u^{\#(e)}$ on $\mathfrak{g}_\alpha$, for $\alpha \in \sigma$ and $u = xt$, $t \in T$.

For conjugacy invariant functions, (3.15.10) leads to integrals over the translated tori xS, one for each conjugacy class in G/G°, where also the dimension of $S = (G_x)^\circ$ may vary for the different connected components. It is more convenient however to regard (3.15.10) in this case as an integral over the **group** $U := \mathrm{N}_G(\mathfrak{c})$, which is a Lie group with Abelian Lie algebra (equal to $\mathfrak{t}$), and which meets every connected component of G, cf. Proposition 3.15.1.

For this purpose, we will replace the action of G by the action by conjugation of G° on the various connected components of G. In this way one obtains a version of (3.15.10) in which the sum is taken over the connected components of G. In each of the connected components of G, one chooses a representative u, of principal orbit type and normalizing $\mathfrak{c}$. The groups $\mathrm{N}(uS)$ have to be replaced by $\mathrm{N}(uS)\cap G^\circ$, and the group G/S, which appears in the integration over the conjugacy classes, by G°/S. (Warning: if we have normalized the measure on G such that $\int_G dg = 1$, then $\int_{G^\circ} dg = \#(G/G^\circ)^{-1}$.)

If $g \in G^\circ$, $gug^{-1} = us$, $s \in S$, then $\mathrm{Ad}\, u \circ \mathrm{Ad}\, g^{-1}(X) = \mathrm{Ad}\, g^{-1} \circ \mathrm{Ad}\, u \circ \mathrm{Ad}\, s(X) = \mathrm{Ad}\, g^{-1}(X)$, for each $X \in \mathfrak{s} = \mathfrak{g}_u$; or $\mathrm{Ad}\, g^{-1}$ leaves $\mathfrak{s}$ invariant. Because $\mathfrak{s}$ contains regular elements, cf. Lemma 3.15.4, $\mathrm{Ad}\, g^{-1}(X) \in \mathfrak{t}$, for some $X \in \mathfrak{t} \cap \mathfrak{g}^{\mathrm{reg}}$; and taking centralizers, we get that g^{-1} belongs to the normalizer of $\mathfrak{t}$ in G°, or $gT \in W$, the Weyl group of $\mathfrak{t}$. Moreover, $\mathrm{Ad}\, g|_{\mathfrak{t}}$ commutes with $\mathrm{Ad}\, u|_{\mathfrak{t}}$.

Now suppose conversely that $g \in G^\circ$, $\mathrm{Ad}\, g(\mathfrak{t}) = \mathfrak{t}$, and $\mathrm{Ad}\, g|_{\mathfrak{t}}$ commutes with $\mathrm{Ad}\, u|_{\mathfrak{t}}$; write $y = gug^{-1}$. Then $\mathrm{Ad}\, y = \mathrm{Ad}\, u$ on $\mathfrak{t}$; thus $\mathrm{Ad}\, y(X) = X$, for all $X \in \mathfrak{s}$. Taking $X \in \mathfrak{s} \cap \mathfrak{g}^{\mathrm{reg}}$, we get $y \in G_X = \mathrm{N}_G(\mathfrak{c})$, so $u^{-1}y \in G^\circ \cap \mathrm{N}_G(\mathfrak{c}) = T$, or $gug^{-1} = y = ut$ for some $t \in T$. Replacing g by τg, $\tau \in T$, makes that y gets replaced by $\tau u \tau^{-1} t = u(u^{-1}\tau u)\tau^{-1} t$. Writing $\tau = \exp U$, with $U \in \mathfrak{t}$, we obtain $(u^{-1}\tau u)\tau^{-1} = \exp(\mathrm{Ad}\, u^{-1} - \mathrm{I})(U)$. Because $\mathfrak{t} = \mathfrak{s} + \mathfrak{s}'$, where $\mathfrak{s}' = (\mathrm{Ad}\, u^{-1} - \mathrm{I})(\mathfrak{t})$, the mapping $V \mapsto (\exp V)S$ is surjective: $\mathfrak{s}' \to T/S$, and it follows that we can choose $\tau \in T$ such that $huh^{-1} \in uS$, if we write $h = \tau g$. Note that $\mathrm{Ad}\, h(\mathfrak{s}) = \mathfrak{s}$, and this implies that also $huSh^{-1} = uS$, or $h \in \mathrm{N}(uS)$.

We have therefore proved that the mapping $g \mapsto gT$ is surjective from $\mathrm{N}(uS) \cap G^\circ$ onto the subgroup:

(3.15.12) $$W(uT) := \{\, s \in W \mid s \text{ commutes with } \mathrm{Ad}\, u|_{\mathfrak{t}} \,\}.$$

The notation expresses that the right hand side does not change if u is replaced by ut, for $t \in T$, so it only depends on the connected component uT of the group U. Because the kernel is equal to $\mathrm{N}(uS) \cap T$, this yields the isomorphism:

(3.15.13) $$(\mathrm{N}(uS) \cap G^\circ)/(\mathrm{N}(uS) \cap T) \xrightarrow{\sim} W(uT).$$

The integral formula with G replaced by $U = \mathrm{N}_G(\mathfrak{c})$, based on the action by conjugation of T on the various connected components of U, takes the form:

(3.15.14) $$\begin{aligned} &\int_U \phi(u)\, du \\ &= \sum_{uT} n(u)^{-1} |\det(\mathrm{I} - \mathrm{Ad}\, u)_{\mathfrak{t}/\mathfrak{s}}| \int_S \int_{T/S} \phi(tust^{-1})\, d(tT_u)\, ds. \end{aligned}$$

Here the summation is over $U/U^\circ = U/T \cong G/G^\circ$, so the same as for G, while u is a representative of principal orbit type in each connected component of

U, and $n(u) = \#((\mathrm{N}(uS) \cap T)/S)$, whereas $S = (G_u)^\circ$. Combined with (3.15.11,12), this yields that:

$$\begin{aligned}\#\big((\mathrm{N}(uS) \cap G^\circ)/S\big) &= \#\big((\mathrm{N}(uS) \cap G^\circ)/(\mathrm{N}(uS) \cap T)\big)\#\big((\mathrm{N}(uS) \cap T)/S\big) \\ &= \#\big(W(uT)\big)n(u).\end{aligned}$$

We have proved:

(3.15.6) Proposition. *For any conjugacy invariant function f on G, define the conjugacy invariant function ϕ on $U = \mathrm{N}_G(\mathfrak{c})$ by*

$$\phi(u) = \#(W(uT))^{-1} f(u)\delta(u)\overline{\delta(u)}.$$

Then $f \in \mathrm{L}^1(G)$ if and only if $\phi \in \mathrm{L}^1(U)$, and $\int_G f(x)\,dx = \int_U \phi(u)\,du$ in this case.

Examples. In Section 3.4, we considered the normalizer $\mathrm{N}(T)$ of a maximal torus T in a connected, compact Lie group as an interesting example of an nonconnected compact Lie group. Because the Lie algebra $\mathfrak{t}$ is Abelian, it consists entirely of regular elements, so here Lemma 3.15.4 for instance does not yield anything new.

Another type of examples is provided by the automorphism groups of compact Lie algebras $\mathfrak{g}$ with zero center. As discussed after Proposition 3.15.1, the component group here can be identified with the automorphims group of the Dynkin diagram. For instance, for $\mathfrak{g} = \mathfrak{su}(n)$, there is an automorphism which, for each $1 \leq p \leq n-1$, interchanges the simple root $\alpha_{p+1,p}$ with $\alpha_{n-p+1,n-p}$ (cf. Section 3.12). It is easy to see that the fixed point set in $\mathfrak{t}$ of this automorphism contains regular elements.

3.16 Exercises

(3.1) Exercise. Let G be a connected, compact Lie group with Lie algebra $\mathfrak{g}$. (i) Prove that the mapping $x \mapsto x^n$ of G into itself is surjective, for every $n > 1$. (ii) Show that there exists a compact set $C \subset \mathfrak{g}$ such that $\exp C = G$. (iii) Let $\Phi\colon G \to H$ be a surjective homomorphism of Lie groups and suppose H is commutative. Show that the restriction of Φ to a maximal torus in G is also surjective.

(3.2) Exercise. Show that the subgroup of diagonal matrices with entries $\epsilon_j = \pm 1$ satisfying $\epsilon_1 \cdots \epsilon_n = 1$, is a maximal Abelian subgroup of $\mathbf{SO}(n, \mathbf{R})$ but not a torus.

(3.3) Exercise. Prove that the product of a finite number of Lie algebras is a compact Lie algebra if and only if every factor is compact. Show that a subalgebra of a compact Lie algebra is a compact Lie algebra.

(3.4) Exercise. Let $\mathfrak{g}$ be a compact Lie algebra. Suppose that $\mathfrak{t}$ and $\mathfrak{t}'$ are two maximal Abelian subspaces of $\mathfrak{g}$, that $\mathfrak{a}$ is a subset of $\mathfrak{t}$, and that $\Phi \in \mathrm{Aut}\,\mathfrak{g}$ satisfies $\Phi(\mathfrak{a}) \subset \mathfrak{t}'$. Then there exists $\Psi \in \mathrm{Ad}\,\mathfrak{g}$ such that $\Phi \circ \Psi$ maps $\mathfrak{t}$ onto $\mathfrak{t}'$ and coincides with Φ on $\mathfrak{a}$.

(3.5) Exercise. Let G be a connected Lie group with a Lie algebra $\mathfrak{g}$ and suppose that the closure of $\mathrm{Ad}\,G$ in $\mathrm{Aut}\,\mathfrak{g}$ is compact. Show that the quotient of G by its center is compact, and that consequently $\mathrm{Ad}\,G$ is compact.
Hint: Use that there exists an $\mathrm{Ad}\,G$-invariant inner product on $\mathfrak{g}$.

(3.6) Exercise. (i) Let $\Phi\colon G \to G'$ be a homomorphism of connected, compact Lie groups. Then if $T \subset G$ is a maximal torus, so is $\Phi(T) \subset G'$.
Hint: Let T' be a maximal torus in G' with generating element x'. Any $x \in \Phi^{-1}(\{x'\})$ is contained in a maximal torus in G.
(ii) Suppose $\ker \Phi \subset \mathrm{Z}(G)$. Show that every maximal torus in G is the preimage under Φ of a maximal torus in G'.

(3.7) Exercise. Let G be a connected, compact Lie group. (i) Prove that $\mathrm{Z}(G)$ is the intersection of all maximal tori in G. (ii) Let $x \in G$. Show that $(G_x)^\circ$ is the union of all maximal tori of G that contain x. (iii) Let T be a torus in G. Prove that the centralizer of T is the union of all maximal tori in G that contain T.

(3.8) Exercise. Let G be a connected, compact Lie group, and G' a connected closed subgroup of G, with Lie algebras $\mathfrak{g}$, and $\mathfrak{g}'$, respectively. (i) Prove that every maximal torus T' in G' is of the form $G' \cap T$ with T a maximal torus in G. Let W, and $W' = \mathrm{N}_{G'}(T')/T'$, be the corresponding Weyl groups of G, and G', respectively. Show that $W' \simeq F/F'$, where F is the subgroup of W that leaves $G' \cap T$ stable as a set, and F' is the normal subgroup of F that fixes $G' \cap T$ elementwise. (ii) Let $\mathfrak{t}$, and $\mathfrak{t}'$, be the Lie algebra of T, and T'. Show that every root of $\mathfrak{t}'$ is the restriction to $\mathfrak{t}'$ of at least one root of $\mathfrak{t}$.
Hint: The complexification $\mathfrak{g}'_{\mathbf{C}}$ is stable under $\mathrm{Ad}\,t$ for all $t \in T'$, and $\mathfrak{g}'_{\mathbf{C}}$ is the direct sum of $\mathfrak{t}'_{\mathbf{C}}$ and the $\mathfrak{g}_{\alpha'}$, where α' runs through the set of restrictions to $\mathfrak{t}'$ of the roots of $\mathfrak{t}$, and $\mathfrak{g}_{\alpha'}$ denotes the sum of the $\mathfrak{g}_\alpha$ for the roots α whose restriction to $\mathfrak{t}'$ is α'.

(3.9) Exercise. Let G be a connected, compact Lie group and let T be a maximal torus in G. A homomorphism $\xi\colon T \to \mathbf{C} \setminus \{0\}$ is a global root of T if $\{\, Y \in \mathfrak{g}_{\mathbf{C}} \mid \mathrm{Ad}\,x(Y) = \xi(x) Y, \text{ for all } x \in T \,\}$ is nonzero. (i) Show that a global root ξ of T takes values in $\mathrm{U}(1)$. (ii) Prove that the map $\xi \mapsto \mathrm{T}_1 \xi \in$

$\mathbf{L}(\mathfrak{t}, i\mathbf{R})$ is a bijection between the set of global roots of T and the set R of roots of $\mathfrak{t}$. (iii) We write ξ_α for the global root of T corresponding to $\alpha \in R$, and $T_\alpha = \ker \xi_\alpha$. Show that the Lie algebra of G_x, for $x \in T$, is given by:

$$\mathfrak{g}_x = \ker(\mathrm{Ad}\, x - \mathrm{I}) = \mathfrak{t} \oplus \sum_{\{\alpha \in R \mid x \in T_\alpha\}} (\mathfrak{g}_\alpha \oplus \mathfrak{g}_{-\alpha}) \cap \mathfrak{g}.$$

(iv) Verify $\mathrm{Z}(G) = \bigcap_{\alpha \in R} T_\alpha$. (v) Let S be a closed subgroup of T and define $R(S)$ to be the system of roots associated with the group $\mathrm{Z}_G(S)$ and its maximal torus T. Prove that $R(S) = \{\alpha \in R \mid \alpha(S) = \{1\}\}$. Show that $\mathrm{Z}(\mathrm{Z}_G(S)^\circ) = \bigcap_{\alpha \in R(S)} T_\alpha$.

(3.10) Exercise. Let G be a semisimple, connected, compact Lie group $\neq \{0\}$ with Lie algebra $\mathfrak{g}$, and let T be a maximal torus in G with Lie algebra $\mathfrak{t}$. Suppose Φ is an automorphism of G with tangent mapping $\phi \in \mathrm{Aut}\, \mathfrak{g}$ at 1 satisfying $\Phi(T) = T$ and $\alpha \circ \phi = \alpha$ for every root α of $\mathfrak{t}$. Let $\phi_{\mathbf{C}} \in \mathrm{Aut}\, \mathfrak{g}_{\mathbf{C}}$ denote the complexification of $\phi \in \mathrm{Aut}\, \mathfrak{g}$. (i) Prove that for every α belonging to the set of roots R there exists $a_\alpha \in \mathbf{U}(1)$ such that $\phi_{\mathbf{C}}(Y) = a_\alpha Y$, for all $Y \in \mathfrak{g}_\alpha$, and $a_\alpha a_{-\alpha} = 1$, $a_\alpha a_\beta = a_{\alpha+\beta}$ if $\alpha + \beta \in R$. Show there exists $X \in \mathfrak{t}$ such that $\mathrm{e}^{\alpha(X)} = a_\alpha$ for each $\alpha \in R$. (ii) Deduce from (i) that the group $\mathrm{Aut}\, G / \mathbf{Ad}\, G$ is finite.

(3.11) Exercise. Let G be a connected, compact Lie group, let T be a maximal torus in G, and let W be the corresponding Weyl group. (i) Prove that the canonical injection of T into G induces a homeomorphism of the orbit space $W \backslash T$ onto $(\mathbf{Ad}\, G) \backslash G$, the space of conjugacy classes in G. (ii) Show that there exists a canonical isomorphism of complex vector spaces $\mathrm{C}((\mathbf{Ad}\, G) \backslash G) \to \mathrm{C}(T)^W$, between the space of continuous conjugacy invariant functions on G and the space of continuous W-invariant functions on T. (iii) Let H be closed subgroup of G containing T. Denote by W_H the subgroup $\mathrm{N}_H(T)/T$ of W. Prove that the group of components H/H° of H is isomorphic with the quotient group W_H / W_{H°.

(3.12) Exercise. (i) Let G be a connected Lie group with a compact Lie algebra. Prove that G is isomorphic to $(G' \times T \times V)/F$, where G' is a compact, simply connected Lie group having semisimple Lie algebra, T is a torus, V a vector space of finite dimension, and F is a finite normal subgroup. Show that G contains a maximal compact subgroup K and that K is connected, and that $\mathrm{Z}(G)$ contains a closed vector subgroup W such that G is the direct product $K \times W$. (ii) Let G be a connected, compact Lie group. Prove that G is isomorphic to $(G_1 \times \cdots \times G_d \times T)/F$, where the G_i are compact, simply connected Lie groups having semisimple Lie algebras without proper ideals, T is a torus, and F is a finite normal subgroup.

(3.13) Exercise. Let G be a connected, compact Lie group, let $\mathfrak{t}$ be a maximal Abelian subspace in the Lie algebra $\mathfrak{g}$ of G, and let R be the corresponding set of roots. Prove that the set of coroots $\{\alpha^\vee \mid \alpha \in R\}$ is a root system in the dual space of $\mathfrak{t}$.

(3.14) Exercise. Let G be a connected, compact Lie group and let $\mathfrak{t}$ be a maximal Abelian subspace in the Lie algebra $\mathfrak{g}$ of G. Furthermore Λ denotes $\ker \exp \cap \mathfrak{t}$ and $\Lambda_{R^\vee}$ the coroot lattice. (i) Prove that for any alcove $\mathfrak{b}$ the set $\Lambda \cap \mathfrak{b}^{\mathrm{cl}}$ is a system of representatives for $\Lambda/\Lambda_{R^\vee} \simeq \pi_1(G)$. In particular, the order of $\pi_1(G)$ is the number of elements in $\Lambda \cap \mathfrak{b}^{\mathrm{cl}}$. (ii) Suppose the center of $\mathfrak{g}$ is 0 and let Λ_Z be the central lattice in $\mathfrak{t}$. Prove that there is precisely one simply connected, compact Lie group $\widetilde{G}$ covering G with $\Lambda(\widetilde{G}) = \Lambda_{R^\vee}$, $\mathrm{Z}(\widetilde{G}) = \Lambda_Z/\Lambda_{R^\vee}$ and $\Lambda(\widetilde{G}/\mathrm{Z}(\widetilde{G})) = \Lambda_Z$.

(3.15) Exercise. Let $G = \mathbf{SU}(2)$. Verify that:

$$\bar{\imath} = \begin{pmatrix} i & 0 \\ 0 & -i \end{pmatrix}, \quad \bar{\jmath} = \begin{pmatrix} 0 & -1 \\ 1 & 0 \end{pmatrix}, \quad \bar{k} = \begin{pmatrix} 0 & i \\ i & 0 \end{pmatrix}$$

form a basis for the Lie algebra $\mathfrak{g}$, and that a basis for $\mathfrak{g}_{\mathbf{C}}$ is given by H, $X = -\frac{1}{2}(\bar{\jmath} + i\bar{k})$, and $Y = \frac{1}{2}(\bar{\jmath} - i\bar{k})$ as in Exercise 2.1. Let $T = \{\operatorname{diag}(z, \bar{z}) \mid z \in \mathbf{U}(1)\}$. Deduce from the commutation rules in this exercise that the set $\{\pm\alpha\}$ of roots of $\mathfrak{t}$ is given by $\alpha(itH) = 2it$, and that $\xi_\alpha(\exp itH) = \mathrm{e}^{2it}$ for all $t \in \mathbf{R}$. Show that $|\delta(\exp itH)|^2 = 4\sin^2 t$, and that for $f \in \mathrm{L}^1(G)^{\operatorname{Ad} G}$:

$$\int_G f(x)\,dx = \frac{2}{\pi}\int_0^\pi f\left(\begin{pmatrix} \mathrm{e}^{it} & 0 \\ 0 & \mathrm{e}^{-it} \end{pmatrix}\right)\sin^2 t\,dt.$$

Prove that $f_n \colon T \to \mathbf{C}$ given by $f_n(\exp itH) = \frac{\sin(n+1)t}{\sin t}$ is the restriction to T of a continuous conjugacy invariant function $f_n \colon G \to \mathbf{C}$, and show $\int_G f_n(x)\,dx = 1$ for $n = 0$, and $\int_G f_n(x)\,dx = 0$ for $n \in \mathbf{Z}_{>0}$. Prove $\int_G f_m(x)\overline{f_n(x)}\,dx = \delta_{mn}$ for $m, n \in \mathbf{Z}_{>0}$.

3.17 Notes

Most of the results of this chapter are, in some form or another, due to H. Weyl [1925/26], especially his Chapters III, IV on the general theory. (Chapter I deals with $\mathbf{SL}(n, \mathbf{C})$ and $\mathbf{SU}(n)$, and Chapter II with the complex symplectic and orthogonal groups, respectively, and their unitary restrictions.)

To start with, the Weyl covering theorem, cf. (3.7.2), can be found in loc. cit., Kap.IV, §1. The argument there is based on the fact that G^{reg} is connected, which is the same one behind our proofs (cf. Corollary 2.8.3). Apparently Weyl considered the connectedness of the complement of finitely many manifolds of codimension ≥ 2 as evident (and we basically agree). The cited section also contains an intriguing space of "flats", which must be the space $G/\mathrm{N}(T)$ of maximal tori in G. The projection onto it, about which Weyl talks, could be the projection: $G/T \times_{\mathrm{N}(T)/T} (G^{\mathrm{reg}} \cap T) \to G/\mathrm{N}(T)$ induced by projection onto the first factor, here the space on the left hand side is diffeomorphic to G^{reg}.

The covering theorem of course implies that each element of G is conjugate to an element of T (hence $\exp \mathfrak{g} = G$), and that all maximal tori are conjugate. A completely different proof of the latter has been given by A. Weil [1935]. His idea of studying fixed points, for the action of an element $g \in G$ on the homogeneous space G/T, has led to many variations. Hunt [1956] gave a proof of the theorem that all maximal Abelian subspaces of $\mathfrak{g}$ are conjugate to each other, by studying the critical points of the restriction to adjoint orbits of linear forms on $\mathfrak{g}$. Remark (b) after Corollary 3.8.4, which we owe to V. W. Guillemin, is based on the observation that these functions can also be used to study the topology of the adjoint orbits. Still another proof of the conjugacy of maximal tori has been given by Hopf [1940/41] and [1942/43]. The first paper contains, in Nr.23, an elegant proof of the theorem on connectedness of centralizers of tori (Theorem 3.3.1.(iii)). The second one contains the statement that G/T is simply connected, in a footnote.

In Kap.IV, §2, Weyl proceeds to prove that $\mathfrak{g}^{\mathrm{reg}}$ is simply connected, cf. Corollary 3.8.4. Also here, Weyl's proof is extremely short. He then essentially derives Corollary 3.9.4, although his estimate for the number of elements of $\pi_1(G)$ is somewhat coarse. The identification of $\pi_1(G)$ with the lattice points in the closure of the alcove $\mathfrak{b}$, cf. Corollary 3.9.6, has been found by É. Cartan [1927-II], where also the affine Weyl group appears. Proposition 3.11.1 is due to Stiefel [1941/42], p.361. We found the expositions on $\pi_1(G)$ in Helgason [1962], Ch.VII, §6 and in Loos [1969-II], Ch.V very helpful. For a very different proof that $\pi_1(\mathrm{Ad}\,\mathfrak{g})$ is finite for any compact Lie algebra $\mathfrak{g}$, see Bourbaki [1982], §1, No.4, Th.1.

Immediately after this, in Kap.IV, §2, loc. cit., Weyl determines the Jacobian (3.14.6) in the integration formula (3.14.1). (This was obtained at about the same time also by É. Cartan [1925], No.13.) For the generalities about integration on manifolds with group actions, as in Section 3.13, see also Helgason loc. cit., Ch.X, §1.

Weyl's monumental paper concludes with the determination of the characters and dimensions of the irreducible representations of G, based on the L^2-isomorphism: $f \mapsto c'\delta f|_T$ of Corollary 3.14.2. We will present this in Theorems 4.9.1 and 4.9.2. Finally, (3.14.11,12) is due to Freudenthal and de Vries [1969], Section 37, whereas (3.14.13) has been found by Harish-Chandra [1975]. Another efficient formula was obtained by Steinberg, see the Appendix in Harder [1971]. See also Macdonald [1980].

In his derivation of the characters, a central role is played by the group generated by the orthogonal reflections in the root hyperplanes, called (S) by Weyl. At the end of the paper he concludes that it is equal to $\mathrm{N}(T)/T$, but with a very indirect proof; and he expresses the wish for a direct proof. This is provided by Cartan [1927-III], Chap.I, No.8, where he proves Theorem 3.10.2.(v), knowing (i), (ii) already for a long time from his investigations on the root structure. A proof of Theorem 5.5.9.(ii) which is based on the extremality of the simple roots, is given in Cartan [1928]. This also contains a general proof of the fact that $\mathfrak{b}^{\mathrm{cl}}$ is a simplex if $\mathfrak{g}$ is a simple Lie algebra, a fact which was verified in the previous Annali paper by going through the classification.

The characteristic equation (for the eigenvalues of $\mathrm{ad}\, X$) was introduced by Lie (cf. Lie [1888], p.590) in his search for 2-dimensional Lie subalgebras. It was developed by Killing [1888-90] and Cartan [1894], as the main tool in their classification of the simple complex Lie algebras. We used only the most elementary part of this tool here, but in Section 5.4 we shall return to it in a more systematic way.

Because of severe criticism of, among others, Lie [1893], §142, Killing's contribution to the classification has been underrated; especially after the superior treatment by É. Cartan, although the latter makes it clear enough that many of the basic ideas came from Killing.

In Killing [1888], the coefficients of the characteristic polynomial for $\mathrm{ad}\, X$ are introduced as functions which are $\mathrm{Aut}\, \mathfrak{g}$-invariant. Killing then concentrates on the linear form $X \mapsto \mathrm{tr}(\mathrm{ad}\, X)$, but does not pay any special attention to the quadratic one, related to the Killing form $X \mapsto \mathrm{tr}((\mathrm{ad}\, X)^2)$ (cf. Section 3.6 and (5.1.14)). In contrast, the Killing form is fundamental in the criteria of Cartan [1894] for solvability, and semisimplicity, respectively, of Lie algebras. Perhaps as a historical compensation, the name "Killing form" became customary in the 1950's, cf. Borel et Serre [1953], Séminaire Sophus Lie [1954/55], Jacobson [1962], p.69.

Lemma 3.6.1, stating that every derivation of a semisimple Lie algebra is inner, is due to Zassenhaus [1939], with exactly the same proof. We learned about compact Lie algebras from Helgason [1962], Ch.II, §§5 and 6, see also Bourbaki [1982], §1, No.3, Prop.1. Jacobson, loc. cit., defines a Lie algebra to be compact if its Killing form is negative definite, compare Corollary 3.6.3.

The proof which we gave of Lemma 3.10.1 originates with Samelson [1940/41], for the application to Theorem 3.10.2, see also Hopf [1942/43]. The theorem that the Weyl group acts freely and transitively on the set of Weyl chambers (Proposition 3.8.1) is usually derived from the properties of the Weyl group as a reflection group, cf. Bourbaki [1968], Chap.5, §3, No.2, Th.1.(iii).

Concerning Theorem 3.8.3.(iii) and Lemma 3.9.2.(iii), in the older literature, say before 1945, "fundamental domains" are restricted to the set of regular elements. Serre [1954/55] gives a simple proof of the fact that $W \backslash T$ is homeomorphic to the space of conjugacy classes $(\mathbf{Ad}\, G) \backslash G$. A complex companion of this is the theorem of Springer and Steinberg [1970], p.195, that in a complex, reductive, connected, algebraic group, the semisimple conjugacy classes are in natural correspondence with $T_{\mathbf{C}}/W$, where $T_{\mathbf{C}}$ is a complex algebraic torus. See also Proposition 4.10.2. Note that in this case, there is no analogue of Theorem 3.8.3.(iii), or Corollary 3.9.6.(iv), because the regular set in $T_{\mathbf{C}}$ is connected. Returning to the compact case, it is elementary that $W\backslash\mathfrak{t}$ is homeomorphic to $\mathfrak{c}^{\mathrm{cl}}$, cf. Bourbaki [1968], Chap.V, §3, No.3. So Theorem 3.8.3.(iii) can also be obtained by combining $W\backslash\mathfrak{t} \xrightarrow{\sim} (\mathrm{Ad}\, G)\backslash\mathfrak{g}$ with $\mathfrak{c}^{\mathrm{cl}} \xrightarrow{\sim} W\backslash\mathfrak{t}$.

Corollary 3.15.5 is due to Borel [1961], Théorème 3.4. He remarks that a similar proof yields that the fixed point set of any semisimple automorphism of a simply connected complex semisimple Lie group is also connected. See also Steinberg [1968], in which Corollary 9.9 is a converse, the complex analogue of our Corollary 3.9.5.(iii).

About nonconnected compact Lie groups, the complex version of Lemma 3.15.3 goes back to Gantmakher [1939]. It is based on an analysis of the action, by conjugation, of the identity component of the automorphism group on the other components. It also contains the principle that, in noncompact semisimple Lie groups, the semisimple elements are the ones for which the centralizers are slices. The isomorphism between the component group of Aut $\mathfrak{g}$ and the automorphism group of the Dynkin diagram we found in Freudenthal and de Vries [1969], 33.9. Other facts about nonconnected, compact Lie groups can be found in de Siebenthal [1956]. The description following Lemma 3.15.4 was suggested to us by T. A. Springer.

References for Chapter Three

Borel, A. (1961) Sous groupes commutatifs et torsion des groupes de Lie compacts connexes. Tôhoku Math. J. **13** (1961) 216-240 = Œuvres Collected Papers, Vol. II, pp. 139-163. Springer-Verlag, Berlin Heidelberg New York Tokyo 1983

Borel, A., Serre, J-P. (1953) Sur certains sous-groupes des groupes de Lie compacts. Comment. Math. Helv. **27** (1953) 128-139 = Œuvres Collected Papers, Vol. I, pp. 217-228. Springer-Verlag, Berlin Heidelberg New York Tokyo 1983

Bourbaki, N. (1968) Éléments de Mathématique. Groupes et Algèbres de Lie, Chap. 4, 5, 6. Hermann, Paris

Bourbaki, N. (1982) Éléments de Mathématique. Groupes et Algèbres de Lie, Chap. 9. Masson, Paris

Cartan, É. (1894) Sur la structure des groupes de transformations finis et continus. Thèse, Paris, 1894 = Œuvres Complètes, Partie I, pp. 137-287. Gauthier-Villars, Paris 1952

Cartan, É. (1925) Les tenseurs irréductibles et les groupes linéaires simples et semi-simples. Bull. Sc. Math. **49** (1925) 130-152 = Œuvres Complètes, Partie I, pp. 531-553. Gauthier-Villars, Paris 1952

Cartan, É. (1927-II) La géométrie des groupes simples. Ann. Mat. Pura Appl. **4** (1927) 209-256 = Œuvres Complètes, Partie I, pp. 793-840. Gauthier-Villars, Paris 1952

Cartan, É. (1927-III) Sur certaines formes riemanniennes remarquables des géométries à groupe fondamental simple. Ann. Sci. École Norm. Sup. **44** (1927) 345-467 = Œuvres Complètes, Partie I, pp. 867-989. Gauthier-Villars, Paris 1952

Cartan, É. (1928) Complément au mémoire: Sur la géométrie des groupes simples. Ann. Mat. Pura Appl. **5** (1928) 253-260 = Œuvres Complètes, Partie I, pp. 1003-1010. Gauthier-Villars, Paris 1952

Freudenthal, H., de Vries, H. (1969) Linear Lie Groups. Academic Press, New York London

Gantmakher, F. (1939) Canonical representation of automorphisms of a complex semi-simple Lie group. Mat. Sbornik **5** (47) (1939) 101-146 (English, Russian summary: Math. Rev. **1** (1940) 163; Jahrb. ü. d. Fortschr. d. Math. **65** (1939) 1131)

Guillemin, V., Pollack, A. (1974) Differential Topology. Prentice-Hall, Englewood Cliffs

Harder, G. (1971) A Gauss-Bonnet formula for discrete arithmetically defined groups. Ann. Sci. École Norm. Sup. **4** (1971) 409-445

Harish-Chandra (1975) Harmonic analysis on real reductive groups I. J. of Functional Analysis. **19** (1975) 104-204 = Collected Papers, Vol. IV, pp. 102-202. Springer-Verlag, New York Berlin Heidelberg Tokyo 1984

Helgason, S. (1962) Differential Geometry and Symmetric Spaces. Academic Press, New York London

Hopf, H. (1940/41) Über den Rang geschlossener Liescher Gruppen. Comment. Math. Helv. **13** (1940/41) 119-143 = Selecta, pp. 152-174. Springer-Verlag, Berlin Göttingen Heidelberg New York 1964

Hopf, H. (1942/43) Maximale Toroide und singuläre Elemente in geschlossenen Lieschen Gruppen. Comment. Math. Helv. **15** (1942/43) 59-70

Hunt, G.A. (1956) A theorem of Élie Cartan. Proc. Amer. Math. Soc. **7** (1956) 307-308

Jacobson, N. (1962) Lie Algebras. Interscience Publishers, New York London Sydney

Killing, W. (1888-1890) Die Zusammensetzung der stetigen endlichen Transformationsgruppen, I, II, III, IV. Math. Ann. **31** (1888) 252-290; **33** (1889) 1-48; **34** (1889) 57-122; **36** (1890) 161-189

Loos, O. (1969-II) Symmetric Spaces, II: Compact Spaces and Classification. W.A. Benjamin, New York Amsterdam

Macdonald, I.G. (1980) The volume of a compact Lie group. Invent. Math. **56** (1980) 93-95

Samelson, H. (1940/41) Über die Sphären, die als Gruppenräume auftreten. Comment. Math. Helv. **13** (1940/41) 144-155

Serre, J-P. (1954/55) Tores maximaux des groupes de Lie compacts. In: Séminaire Sophus Lie. Exposé No. 23. Secrétariat Mathématique, Paris

de Siebenthal, J. (1956) Sur les groupes de Lie compacts non-connexes. Comment. Math. Helv. **31** (1956) 41-89

Springer, T.A., Steinberg, R. (1970) Conjugacy classes. In: Seminar on Algebraic Groups and Related Finite Groups, Lecture Notes in Math. **131**, pp. 167-266. Springer-Verlag, Berlin Heidelberg New York

Steenrod, N. (1951) The Topology of Fibre Bundles. Princeton Math. Series, No.14. Princeton University Press, Princeton

Steinberg, R. (1968) Endomorphisms of Linear Algebraic Groups. Mem. Amer. Math. Soc. **80**. Amer. Math. Soc., Providence

Stiefel, E. (1941/42) Über eine Beziehung zwischen geschlossenen Lie'schen Gruppen und diskontinuierlichen Bewegungsgruppen euklidischer Räume und ihre Anwendung auf die Aufzählung der einfachen Lie'schen Gruppen. Comment. Math. Helv. **14** (1941/42) 350-380

Weil, A. (1935) Démonstration topologique d'un théorème fondamental de Cartan. C.R. Acad. Sci. Paris Sér. I. Math. **200** (1935) 518-520 = Collected Papers, Vol. I, pp. 109-111. Springer-Verlag, New York Heidelberg Berlin 1979

Weyl, H. (1925/26) Theorie der Darstellung kontinuierlicher halb-einfacher Gruppen durch lineare Transformationen, I, II, III, Nachtrag. Math. Z. **23** (1925) 271-309; **24** (1926) 328-376, 377-395, 789-791 = Gesammelte Abh., Band II, pp. 543-647. Springer-Verlag, Berlin Heidelberg New York 1968

Zassenhaus, H. (1939) Über Liesche Ringe mit Primzahlcharakteristik. Abh. Math. Sem. Univ. Hamburg **13** (1939) 1-100

Chapter 4

Representations of Compact Groups

4.0 Introduction

In this introduction, we give the basic definitions of representation theory, followed by a summary of the main results for compact (Lie) groups.

Let G be a Lie group, and V a finite-dimensional vector space. A *representation of G in V* is defined as a continuous homomorphism $\pi\colon G \to \mathbf{GL}(V)$. Because every continuous homomorphism between Lie groups is real-analytic (Corollary 1.10.9), this definition agrees with Definition 1.10.1. Another way of saying this is that π defines a real-analytic action of G on V (Section 1.11), by means of linear transformations. The adjoint representation (1.1.12) is a representation of G in its own Lie algebra $\mathfrak{g}$, this example we already have met numerous times.

If M is a manifold, and $A\colon (g,x) \mapsto A(g)(x) : G \times M \to M$ is a C^k action of G on M (with $0 \leq k \leq \omega$), then, for each $f \in \mathrm{C}^k(M)$, the function $A(g)(f)\colon x \mapsto f(gx^{-1})$ is another C^k function on M. The mapping $A(g)\colon f \mapsto A(g)(f)$ is a linear mapping from $\mathrm{C}^k(M)$ onto itself. Furthermore, the mapping $A\colon (g,f) \mapsto A(g)(f)$ is continuous: $G \times \mathrm{C}^k(M) \to \mathrm{C}^k(M)$.

Because such induced actions on function spaces form a central theme in the theory of representations, and because, at least for compact groups, the differentiable structure will not be used for quite some time, the definition of a representation has been generalized in the following way. A *topological group* is a group G that at the same time is a Hausdorff topological space, in such a way that the multiplication: $(g,h) \mapsto gh$, and the inversion: $g \mapsto g^{-1}$, is continuous: $G \times G \to G$, and $G \to G$, respectively. If V is a topological vector space (which usually will be locally convex and complete), then a *representation of the topological group G in the topological vector space V* is defined as a homomorphism $\pi\colon G \to \mathbf{GL}(V)$, such that the mapping: $(g,v) \mapsto \pi(g)(v)$ is continuous: $G \times V \to V$. We will also write $V = V_\pi$.

An example is provided by the action of G on $\mathrm{C}(G)$ (or $\mathrm{C_c}(G)$, the space of compactly supported continuous functions on G), induced by the action of G on itself by left and right multiplications respectively. These are called the *left* and *right regular representation* of G, denoted by L and R respectively. (Note that the action of G on itself by right multiplication is given by $(g, x) \mapsto xg^{-1}$, so the induced action of the function space is given by $(g, f) \mapsto (x \mapsto f(xg).)$ One has also the representation of $G \times G$ on $\mathrm{C}(G)$, induced by the left-right action of $G \times G$ on G.

A representation π of G in a complete, locally convex, topological vector space V is called *irreducible* if there are no $\pi(G)$-invariant, closed linear subspaces U of V, other than $U = 0$ or $U = V$. (A subspace U is said to be *$\pi(G)$-invariant* if $\pi(g)(U) \subset U$, for every $g \in G$.) The representation is said to be *completely reducible* if, for every $\pi(G)$-invariant, closed linear subspace U of V, there is another $\pi(G)$-invariant, closed linear subspace U' of V, such that $V = U \oplus U'$. Note that all linear subspaces of V are automatically closed, if V is finite-dimensional.

The representations $\sigma\colon G \to \mathbf{GL}(U)$ and $\tau\colon G \to \mathbf{GL}(V)$ respectively are said to be *equivalent* if there is a topological linear isomorphism L from U onto V, such that $L \circ \sigma(g) = \tau(g) \circ L$, for all $g \in G$. The equivalence class of the representation π will be denoted by $[\pi]$. The set of equivalence classes of irreducible representations of G is called the *dual* $\widehat{G}$ of G.

If the representation $\pi\colon G \to \mathbf{GL}(V)$ is finite-dimensional, then:

$$\chi_\pi : g \mapsto \operatorname{tr} \pi(g)$$

is a continuous, conjugacy-invariant function (class function) on G, called the *character* of the representation. Note that $\chi_\pi(1) = \dim V$, called the *dimension* d_π of the representation π. Clearly, $\chi_\sigma = \chi_\tau$ if σ, τ are equivalent finite-dimensional representations of G.

We are now ready to formulate our goals regarding the representation theory of a **compact** topological group G. We will restrict ourselves here to **complex** representations, that is, homomorphisms from G to the group of complex linear transformations of a complex vector space V, because for those the formulation is somewhat simpler than for the real ones. (For some remarks on the latter, see Section 4.8.)

(i) G carries a unique left invariant measure: $f \mapsto \int_G f(g)\, dg$, such that $\int_G dg = 1$. It is automatically right invariant, and is also called *averaging over G*.

(ii) Every irreducible representation π of G is finite-dimensional. The **Peter-Weyl theorem** states that the characters of the irreducible representations of G form a countable orthonormal basis of the Hilbert space of square-integrable, complex-valued, conjugacy-invariant functions on G. In particular, inequivalent irreducible representations have different characters (actually orthogonal to each other with respect to the L^2-inner product in the function space).

(iii) Let σ be a representation of G in the complete, locally convex, topological vector space U, and let π be an irreducible representation of G. The

π-*isotypical subspace* U_π of U is defined as the sum of all d_π-dimensional, $\sigma(G)$-invariant linear subspaces V of U, such that $g \mapsto \pi(g)|_V$ is equivalent to π. Then:

$$E_\pi := \mathrm{d}_\pi \int_G \chi_\pi(g^{-1})\sigma(g)\, dg$$

is a continuous linear projection from U onto U_π. The U_π, for $[\pi] \in \widehat{G}$, are closed linear subspaces of U, and their sum:

$$U^{\mathrm{fin}} = \sum_{[\pi]\in\widehat{G}} U_\pi$$

is direct. It is an immediate consequence of the Peter-Weyl theorem that U^{fin} is dense in U. The space U^{fin} is also equal to the space of *G-finite vectors* in U, that is, the $u \in U$ such that the $\sigma(g)u$, for $g \in G$, span a finite-dimensional linear subspace of U. Finally, if U_π is finite-dimensional, then it can actually be written as a direct sum of copies of π; the number, $\dim U_\pi / \mathrm{d}_\pi$, of these is called the *multiplicity* $[\sigma\colon \pi]$ of $[\pi]$ in σ.

(iv) For the right regular representation R in $\mathrm{C}(G)$ (or in $\mathrm{L}^2(G)$), the multiplicity of $[\pi] \in \widehat{G}$ in R is equal to d_π. The π-isotypical subspace M_π is also equal to the $\pi^\vee$-isotypical subspace for the left regular representation L of G in $\mathrm{C}(G)$. Here $\pi^\vee : g \mapsto \pi(g^{-1})^* : G \to \mathbf{GL}(V_\pi^*)$ is the *contragredient* or *dual* representation of π. The space M_π is irreducible for the left-right action of $G \times G$ on $\mathrm{C}(G)$. By (iii), the direct sum $M = \oplus_{[\pi]\in\widehat{G}} M_\pi$, which is orthogonal with respect to the L^2-inner product, is dense in $\mathrm{C}(G)$. The space M of functions of finite type is also called the *space of matrix coefficients*. If H is a closed subgroup of G, then the $\pi^\vee$-isotypical subspace of $\mathrm{C}(G/H) = \mathrm{C}(G)^H$ is the space M_π^H of $\mathrm{R}(H)$-fixed elements in M_π; so the multiplicity of $\pi^\vee$ in $\mathrm{C}(G/H)$ is $\leq \mathrm{d}_\pi$. The decomposition:

$$\mathrm{C}(G/H) = \left(\oplus_{[\pi]\in\widehat{G}} M_\pi^H\right)^{\mathrm{cl}}$$

is called the *Fourier decomposition* of the homogeneous G-space G/H; the elements of the M_π^H here play the role of the harmonic oscillations in the classical Fourier decomposition of the functions on the circle, that is, the periodic functions on $\mathbf{R}$.

(v) The compact group G is a Lie group if and only if it has no arbitrarily small subgroups. In Corollary 14.6.2 it will even be shown that if this is the case, then G has the structure of a real affine algebraic set, with the matrix coefficients of the finite-dimensional real representations as the real-valued polynomial functions on G. With this structure, the multiplication, and inversion, is a polynomial mapping: $G \times G \to G$, and $G \to G$, respectively, making G into a *real affine algebraic group*. It has a natural *complexification* $G_{\mathbf{C}}$, and these complexifications of the compact

Lie groups are precisely the so-called reductive complex affine algebraic groups. See Section 14.6.

(vi) For a connected, compact Lie group G, the conjugacy-invariant functions are determined by their restrictions to a maximal torus T in G. The **Weyl character formula** is an explicit formula for the restrictions to T of the characters of the irreducible representations π of G; and an explicit formula for the dimensions $\mathrm{d}_\pi = \chi_\pi(1)$ follows. The irreducible representations themselves will be constructed by means of a complex structure on G/T. Finally, we will make some remarks on the characters of nonconnected compact Lie groups.

(vii) Another application of the Peter-Weyl theorem is the theorem that compact portions of C^k-actions of a compact Lie group G can be equivariantly embedded in a finite-dimensional linear action (representation) of G. This is also the first step in the proof of the **Mostow embedding theorem** of Section 2.

4.1 Schur's Lemma

Recall that a mapping ϕ is said to *intertwine* the actions A and B of G if $\phi \circ A(g) = B(g) \circ \phi$, for all $g \in G$, cf. (2.2.3). If σ, and τ, are representations in vector spaces U, and V, over a field k, then the space of k-linear mappings $\phi\colon U \to V$ that intertwine σ with τ, will be denoted by $\mathrm{I}_k(\sigma,\tau)$, or $\mathrm{I}(\sigma,\tau)$ if there is no danger of confusion.

Note that $\mathrm{I}(\sigma,\tau)$ is a k-linear subspace of $\mathbf{L}(U,V)$. Moreover, $\mathrm{I}(\sigma,\sigma)$ is a subalgebra of $\mathbf{L}(U,U)$, with composition of linear mappings as the multiplication. It contains the identity mapping in U. Also, if $A \in \mathrm{I}(\sigma,\sigma)$ is invertible in $\mathbf{L}(U,U)$, then $A^{-1} \in \mathrm{I}(\sigma,\sigma)$. We recall that an algebra $\mathcal{A}$ over k with unit is said to be a *division algebra*, if every nonzero element of $\mathcal{A}$ has an inverse.

(4.1.1) Schur's lemma. *Let G be any group. Let σ and τ be an irreducible representation of G in the finite-dimensional vector space U and V respectively. Then:*

(i) Every nonzero $L \in \mathrm{I}(\sigma,\tau)$ is an isomorphism: $U \to V$, making σ equivalent to τ.

(ii) $\mathrm{I}(\sigma,\sigma)$ is a division algebra. If k is algebraically closed, then $\mathrm{I}(\sigma,\sigma) = \{c\mathrm{I} \mid c \in k\}$.

(iii) If $\mathrm{I}(\sigma,\tau)$ is nonzero, then it is a one-dimensional left, and right, module over $\mathrm{I}(\tau,\tau)$, and $\mathrm{I}(\sigma,\sigma)$, respectively.

Proof. (i) From $L \circ \sigma(g) = \tau(g) \circ L$, we see that $\ker L$ is $\sigma(g)$-invariant; and $L(U)$ is $\tau(g)$-invariant, for each $g \in G$. The irreducibility of both representations, combined with $L \neq 0$, yields that $\ker L = 0$ and $L(U) = V$, so L is a linear isomorphism from U onto V.

(ii) The first statement follows immediately from (i). For the second statement, we use that if $A \in \mathrm{I}(\sigma, \sigma)$, $c \in k$, then $A - c\,\mathrm{I} \in \mathrm{I}(\sigma, \sigma)$. If k is algebraically closed, then there exists $c \in k$ such that $\det(A - c\,\mathrm{I}) = 0$, or $A - c\,\mathrm{I}$ is not invertible, which in turn implies that $A - c\,\mathrm{I} = 0$.

(iii) is obvious in view of (i). □

In the sequel of Chapter 4, we shall assume that $k = \mathbf{C}$, the algebraically closed field of the complex numbers. (The only exception is Section 4.8.)

(4.1.2) Corollary. *If G is Abelian, then every finite-dimensional irreducible representation π of G is one-dimensional. For every $x \in G$, the operator $\pi(x)$ is equal to multiplication by the nonzero complex number $\chi_\pi(x)$, and χ_π is a continuous homomorphism from G to the multiplicative group $\mathbf{C}^\times = \{ c \in \mathbf{C} \mid c \neq 0 \}$.*

Proof. For every $x \in G$, the operator $\pi(x)$ commutes with every $\pi(y)$, for $y \in G$; so by Schur's lemma it is equal to a scalar multiplication. But then every one-dimensional linear subspace is $\pi(G)$-invariant; so the irreducible representation π is one-dimensional. □

Example. If $G = \mathbf{R}/\mathbf{Z}$, the circle, then the continuous homomorphisms: $G \to \mathbf{C}^\times$ are the functions $x \mapsto \mathrm{e}^{2\pi i n x}$, with $n \in \mathbf{Z}$. Slightly more generally, if G is a torus with Lie algebra $\mathfrak{g}$, and lattice $\Lambda = \ker \exp$, then the continuous homomorphisms: $G \to \mathbf{C}^\times$ are the functions $x \mapsto x^\mu$, where μ runs over the linear forms: $\mathfrak{g} \to i\mathbf{R}$, such that $\mu(\Lambda) \subset 2\pi i \mathbf{Z}$. Here we use the notation $x^\mu = \mathrm{e}^{\mu(X)}$, if $x = \exp X$ with $X \in \mathfrak{g}$, as in (3.14.5).

A very convenient tool in the theory of representations will be the use of invariant Hermitian inner products. In Corollary 4.1.3 below we shall see that for irreducible finite-dimensional representations, such an inner product is uniquely determined up to a positive scalar factor. For the convenience of the reader, and for later reference, we begin with a review of the relevant basic definitions.

A mapping L from a complex vector space V to another complex vector space U is called *complex antilinear* if L is a real linear mapping (considering V and U as vector spaces over $\mathbf{R}$), such that $L(cv) = \bar{c}L(v)$, for all $c \in \mathbf{C}$ and $v \in V$. A *sesquilinear form* on V is a function: $(v, v') \mapsto < v, v' >: V \times V \to \mathbf{C}$, such that $v \mapsto < v, v' >$ is complex linear: $V \to \mathbf{C}$, for each $v' \in V$, and $v' \mapsto < v, v' >$ is complex antilinear: $V \to \mathbf{C}$, for each $v \in V$. It is called a *Hermitian inner product* in V if in addition $< v, v > > 0$, for all $v \in V$, $v \neq 0$. In this case $v \mapsto |v| := \sqrt{< v, v >}$ is a norm in V; and the topology of V will be the one for which the sets $\{ v' \in V \mid |v - v'| < r \}$, for $r > 0$, form a basis of neighborhoods of the point $v \in V$. If V is complete with respect to this norm, then V is called a *Hilbert space*, with this Hermitian inner product. Note that V is automatically complete if V is finite-dimensional.

The Hermitian inner product in the Hilbert space V induces an antilinear mapping A from V to the space V' of continuous complex-linear forms on V, defined by:

$$A(w)(v) = < v, w > = \overline{< w, v >}, \quad \text{for } v, w \in V. \tag{4.1.1}$$

This mapping A is injective, because:

$$A(v) = 0 \Rightarrow < v, v > = A(v)(v) = 0 \Rightarrow v = 0;$$

and the Riesz representation theorem says that A is surjective as well. Furthermore, the antilinear mapping $A^{-1}\colon V' \to V = (V')'$ (providing V' with the weak topology of pointwise convergence) comes from a Hermitian inner product on V', whereas the associated norm is equal to the operator norm in V'. Because V' is complete with respect to the latter norm, V' is a Hilbert space with respect to the Hermitian inner product A^{-1}, this is called the *dual Hilbert space* of V. See Treves [1967] for these and other basic facts about Hilbert spaces.

If $L \in \mathbf{L}(V, U)$, a continuous linear mapping from a complete, locally convex topological vector space V to another one U, then the *transposed* mapping $L'\colon U' \to V'$ is defined by $L'(\mu)(v) = \mu(L(v))$, for all $\mu \in U'$, $v \in V$. The transposed mapping L' is continuous: $U' \to V'$, with respect to the weak topology in U' and V' of pointwise convergence of the continuous linear forms. If V and U are Hilbert spaces with Hermitian inner products A_V and A_U respectively, then the *Hilbert space adjoint* $L^*\colon U \to V$ of L is defined as $L^* = A_V^{-1} \circ L' \circ A_U$, or:

$$< L(v), u >_U = < v, L^*(u) >_V, \quad \text{for } v \in V, u \in U. \tag{4.1.2}$$

It is straightforward that L^* has the same operator norm as L, so in particular L^* is continuous as well.

If π is a representation of G in the complete, locally convex topological vector space V, then the *contragredient representation* $\pi^\vee$ of G in $V' := \mathbf{L}(V, \mathbf{C})$ is defined by:

$$\pi^\vee(x) := \pi(x^{-1})' \in \mathbf{GL}(V'), \quad \text{for } x \in G. \tag{4.1.3}$$

If V is a Hilbert space, then the representation π is said to be *unitary* if each $\pi(x)$, for $x \in G$, is a unitary transformation in V, that is:

$$< \pi(x)(v), \pi(x)(w) > = < v, w >, \quad \text{for all } v, w \in V. \tag{4.1.4}$$

In the terminology above, this means that $\pi(x^{-1}) = \pi(x)^{-1} = \pi(x)^* = A^{-1} \circ \pi(x)' \circ A$, for each $x \in G$; or A intertwines π with $\pi^\vee$. It follows that π and $\pi^\vee$ are equivalent as real representations. However, because A is complex **anti**linear, this does not imply that π and $\pi^\vee$ are equivalent as complex representations, and often they are not, cf. Section 4.8. Also note that $\pi^\vee$ is unitary with respect to the Hermitian inner product A^{-1} in V'.

(4.1.3) Corollary. *Let π be an irreducible representation of an arbitrary group G in the finite-dimensional vector space V.*

(i) Let A be a $\pi(G)$-invariant, nondegenerate sesquilinear (complex bilinear) form on V. Then any $\pi(G)$-invariant sesquilinear (complex bilinear respectively) form B on V is a complex multiple of A. If A, B are Hermitian inner products, then the factor is a positive real number.

(ii) Every equivalence between two irreducible, finite-dimensional, unitary representations can be made unitary by multiplying it with a suitable positive factor.

Proof. (i) Viewing A and B as complex antilinear mappings: $V \to V^*$ which intertwine π with $\pi^\vee$, the mapping $A^{-1} \circ B$ is complex linear: $V \to V$, and intertwines π with itself. So Schur's lemma 4.1.1 yields that $A^{-1} \circ B = c\mathrm{I}$, or $B = cA$, for some $c \in \mathbf{C}$. A similar, slightly simpler, argument works for the complex bilinear forms.

(ii) Applying (i) to $B := L' \circ A_U \circ L$, where $L \in \mathbf{L}(V, U)$ intertwines π with a unitary representation in U, we get $B = cA$, for some $c > 0$. But this means that $c^{-1/2} L$ is unitary: $V \to U$. □

4.2 Averaging

From now on in this chapter, unless explicitly stated otherwise, G is a compact topological group

As a first consequence, G carries a unique left invariant measure: $f \mapsto \int_G f(x)\, dx$: $\mathrm{C}(G) \to \mathbf{C}$, such that $\int_G dx = 1$. This measure is automatically right invariant as well, and satisfies $\int_G f(x^{-1})\, dx = \int_G f(x)\, dx$, for $f \in \mathrm{C}(G)$. It will be called *averaging* over G. For a proof, see Section 10.3. If G is a compact **Lie** group, then averaging is defined by a bi-invariant density on G, which automatically is real-analytic, cf. (4.10.10-15).

The process of averaging can be extended to continuous functions f on G with values in a complete, locally convex, topological vector space V. Using the uniform continuity of f on the compact space G, one can find, for each continuous seminorm ν on V and each $\epsilon > 0$, a finite partition of unity with nonnegative continuous functions h_j on G, such that $\nu(f(x) - f(y)) < \epsilon$, if $x, y \in G$ are contained in the support of the same h_j. Choosing $x_j \in \operatorname{supp} h_j$, we see that the "*Riemann sums*":

$$\sum_j \int_G h_j(x)\, dx\, f(x_j)$$

in V form a Cauchy net, converging to an element of V, which by definition is the integral $\int_G f(x)\, dx$ of f over G. Clearly, for every continuous linear form μ on V, we have $\mu \circ f \in \mathrm{C}(G)$, and:

$$\mu\Big(\int_G f(x)\,dx\Big) = \int_G \mu(f(x))\,dx, \quad \text{for } \mu \in V'. \tag{4.2.1}$$

Of course, if V is finite-dimensional, then these definitions are much more elementary, and (4.2.1) then just expresses that the averaging is defined coordinate-wise.

If π is a representation of G in the complete, locally convex, topological vector space V, then one can define, for every $f \in \mathrm{C}(G)$, the linear mapping $\pi(f)\colon V \to V$ by:

$$\pi(f)(v) := \int_G f(x)\,\pi(x)(v)\,dx, \quad \text{for } v \in V. \tag{4.2.2}$$

A particular case occurs for $f = \mathbf{1}$, the function on G which is constant equal to 1. In this case, the corresponding operator:

$$\mathrm{av}(\pi)\colon v \mapsto \int_G \pi(x)(v)\,dx\colon V \to V \tag{4.2.3}$$

is called the *average of the representation* π.

For each $v \in V$, the approximating sums:

$$\sum_j \int_G h_j(x)\,dx\, f(x_j)\pi(x_j)(v)$$

form a bounded subset of V. So, if V is a barreled space, we can use the Banach-Steinhaus theorem to conclude that $\pi(f)$ is continuous: $V \to V$. See Bourbaki [1964], Ch.III, §3, No.6. In decreasing degree of generality (but increasing degree of familiarity), this conclusion holds if V is a Fréchet space, Banach space, or a Hilbert space.

(4.2.1) Proposition [Averaging principle]. *Let G be a compact group, and π a representation of G in the complete, locally convex, topological vector space V. Then $\mathrm{av}(\pi)$ is a linear projection from V onto the space $V^{\pi(G)} := \{v \in V \mid \pi(x)(v) = v\}$ of fixed points for the action π of G on V. If π is a unitary representation of G in the Hilbert space V, then $\mathrm{av}(\pi)$ is equal to the orthogonal projection from V onto $V^{\pi(G)}$.*

Proof. For any $v \in V$, $g \in G$, we have:

$$\begin{aligned}\pi(g)\circ \mathrm{av}(\pi)(v) &= \pi(g)\Big(\int_G \pi(x)(v)\,dx\Big) = \int_G \pi(g)\circ\pi(x)(v)\,dx \\ &= \int_G \pi(gx)(v)\,dx = \int_G \pi(y)(v)\,dy = \mathrm{av}(\pi)(v),\end{aligned}$$

where we have used the continuity of $\pi(g)$ in the second identity, and the left invariance of averaging in the fourth one. So $\mathrm{av}(\pi)(V) \subset V^{\pi(G)}$.

On the other hand, if $v \in V^{\pi(G)}$, then we get:

$$\mathrm{av}(\pi)(v) = \int_G \pi(x)(v)\,dx = \int_G v\,dx = v,$$

using that $\int_G dx = 1$; so $\mathrm{av}(\pi)$ acts on $V^{\pi(G)}$ as the identity.

If π is unitary, then:

$$\begin{aligned}< \mathrm{av}(\pi)(v), w > &= < \int_G \pi(x)(v)\,dx, w > = \int_G < \pi(x)(v), w > \,dx \\ &= \int_G < v, \pi(x)^{-1}(w) > \,dx = \int_G < v, \pi(x^{-1})(w) > \,dx \\ &= \int_G < v, \pi(x)(w) > \,dx = < v, \int_G \pi(x)(w)\,dx > \\ &= < v, \mathrm{av}(\pi)(w) >,\end{aligned}$$

for all $v, w \in V$. That is, $\mathrm{av}(\pi)$ is self-adjoint, which proves the last statement of the averaging principle. □

The averaging principle, in a real version, has been used in Bochner's linearization theorem (Theorem 2.2.1), and also in the construction of invariant Riemannian structures for proper group actions in Proposition 2.5.2. The latter is related to the following application to representation theory.

If π is a unitary representation of G in a Hilbert space V, then the fact that $\pi(x)^* = \pi(x^{-1}) \in \pi(G)$ yields that the orthogonal complement $U^\perp$ of any $\pi(G)$-invariant closed linear subspace U of V is $\pi(G)$-invariant. Because $V = U \oplus U^\perp$, it follows that such a representation is *completely reducible*, according to the definition in Section 4.0.

(4.2.2) Corollary. *Let π be a representation of the compact group G in the Hilbert space V. Then there exists a Hermitian inner product on V, defining the same topology, for which the representation π is unitary. It follows that the representation π is completely reducible.*

In particular, every representation π of a compact group G in a finite-dimensional vector space V is completely reducible; V can be written as the direct sum of $\pi(G)$-invariant subspaces V_j, such that $\pi|_{V_j} : x \mapsto \pi(x)|_{V_j}$ is irreducible for each j.

Proof. By the Banach-Steinhaus theorem (cf. Bourbaki, loc. cit.), there is a constant $C > 0$ such that $< \pi(x)(v), \pi(x)(v) > \leq C < v, v >$, for all $x \in G$, $v \in V$; and from this in turn we read off: $< v, v > \leq C < \pi(x)(v), \pi(x)(v) >$, for all $x \in G$, $v \in V$. Defining

$$<< v, v' >> = \int_G < \pi(x)(v), \pi(x)(v') > \,dx,$$

we get that $C^{-1} < v, v > \le << v, v >> \le C < v, v >$, for all $v \in V$; so the double brackets define a Hermitian inner product on V, whose norm is equivalent to the original one. Furthermore, the new inner product, being the average of the original one, is $\pi(G)$-invariant. That is, the representation π is unitary with respect to the new inner product.

The last statement follows because every finite-dimensional vector space V can be provided with a Hermitian inner product; in the finite-dimensional case it then suffices to observe that the average of a positive definite Hermitian form is positive definite. The splitting of V into irreducible invariant subspaces goes by induction on the dimension of V, starting with a $\pi(G)$-invariant linear subspace V_1 of V of minimal positive dimension, and applying the induction hypothesis to the orthogonal complement of V_1 in V. □

If σ, and τ, is a representation of G in V, and U, respectively, then the *exterior tensor product representation* $\sigma^\vee \otimes^{\mathrm{e}} \tau$ of $\sigma^\vee$ with τ is defined as the representation of $G \times G$ in $\mathbf{L}(V, U)$, given by:

(4.2.4) $$(\sigma^\vee \otimes^{\mathrm{e}} \tau)(x, y)(L) := \tau(y) \circ L \circ \sigma(x^{-1}), \quad \text{for } x, y \in G, L \in \mathbf{L}(V, U).$$

The notation is reminiscent of the concept of the *tensor product* $\mu \otimes u \in \mathbf{L}(V, U)$, defined for $\mu \in V'$, $u \in U$ by:

(4.2.5) $$(\mu \otimes u)(v) := \mu(v)u, \quad \text{for } v \in V.$$

These are just the elements of $\mathbf{L}(V, U)$ of rank (= dimension of the image space) equal to 1. The linear span of these tensor products is equal to the space of $L \in \mathbf{L}(V, U)$ of finite rank. In particular, if V and U are finite-dimensional, then they span all of $\mathbf{L}(V, U)$; and it is customary to write:

(4.2.6) $$\mathbf{L}(V, U) = V^* \otimes U,$$

the tensor product of the spaces V^* and U.

In contrast with the exterior tensor product representation, the *tensor product representation* $\sigma^\vee \otimes \tau$ of $\sigma^\vee$ with τ is defined as the representation of G in $\mathbf{L}(V, U)$, given by:

(4.2.7) $$(\sigma^\vee \otimes \tau)(x)(L) := \tau(x) \circ L \circ \sigma(x^{-1}), \quad \text{for } x \in G, L \in \mathbf{L}(V, W).$$

In other words, $\sigma^\vee \otimes \tau = (\sigma^\vee \otimes^{\mathrm{e}} \tau) \circ \Delta$, the composition of the exterior tensor product with Δ, where Δ denotes the homomorphism: $G \to G \times G$ that sends $x \in G$ to the diagonal element $(x, x) \in G \times G$.

If V, U are Hilbert spaces, then one defines the *Hilbert-Schmidt inner product* of the finite rank operators $L, M \in \mathbf{L}(V, U)$ by:

(4.2.8) $$< L, M >_{\mathrm{HS}} := \mathrm{tr}(M^* \circ L).$$

Here $M^* \colon U \to V$ is the Hilbert space adjoint of M. This norm is stronger than the operator norm in $\mathbf{L}(V, U)$, so the completion of the space of finite rank operators with respect to the Hilbert-Schmidt norm can be identified with a linear subspace $\mathbf{L}(V,U)_{\mathrm{HS}}$, whose elements are called the *Hilbert-Schmidt operators* in $\mathbf{L}(V,U)$. Because:

$$\begin{aligned}< (\epsilon'\otimes u')^*\circ(\epsilon\otimes u)(v), v'> &=< (\epsilon\otimes u)(v), (\epsilon'\otimes u')(v')>\\ &=< \epsilon(v)u, \epsilon'(v')u'>= \epsilon(v)\overline{\epsilon'(v')}< u,u'>\\ &= \epsilon(v) < A^{-1}\epsilon', v'>< u,u'>,\end{aligned}$$

we get:

$$(\epsilon'\otimes u')^*\circ(\epsilon\otimes u) =< u,u'> (\epsilon\otimes A^{-1}\epsilon'),$$

so:

$$<\epsilon\otimes u, \epsilon'\otimes u'>_{\mathrm{HS}}=< u,u'> \epsilon(A^{-1}\epsilon') =< u,u'>< \epsilon,\epsilon'> .$$

In turn we read off from this that if σ and τ is a unitary representation of G in V and U respectively, then $\sigma^\vee\otimes^{\mathrm{e}}\tau$ is a unitary representation of $G\times G$ in $\mathbf{L}(V,U)_{\mathrm{HS}}$.

(4.2.3) Corollary. *Let σ, and τ, be a unitary representation of the compact group G in the finite-dimensional Hilbert space V, and U, respectively. Then the orthogonal projection of $\mathbf{L}(V,U)_{\mathrm{HS}}$ onto $\mathrm{I}(\sigma,\tau)\cap\mathbf{L}(V,U)_{\mathrm{HS}}$, the linear subspace of operators that intertwine σ with τ, is equal to the restriction of $\mathrm{av}(\sigma^\vee\otimes\tau)$ to $\mathbf{L}(V,U)_{\mathrm{HS}}$.*

Proof. $T\in\mathbf{L}(V,U)$ intertwines σ with τ if and only if it is fixed under the representation $\sigma^\vee\otimes\tau$ of G in $\mathbf{L}(V,U)$. Now apply the averaging principle 4.2.1. For the orthogonality, observe that we just have verified that $\sigma^\vee\otimes\tau$ is unitary in $\mathbf{L}(V,U)_{\mathrm{HS}}$. □

4.3 Matrix Coefficients and Characters

Let π be a representation of the topological group G in the finite-dimensional complex vector space V. Because π is a continuous mapping from G to the complex vector space $\mathbf{L}(V,V)$ of complex linear mappings: $V\to V$, it leads in a natural way to continuous complex-valued functions on G by taking $\mu\circ\pi$, where μ is a complex linear form on $\mathbf{L}(V,V)$.

Dropping the adjective "complex" in the sequel, there is a canonical nondegenerate symmetric bilinear form τ on $\mathbf{L}(V,V)$, defined by

(4.3.1) $$\tau(L,M)=\mathrm{tr}(L\circ M),\quad \text{for } L,M\in\mathbf{L}(V,V),$$

called the *trace form* on $\mathbf{L}(V,V)$. (Note that the Killing form of a Lie algebra $\mathfrak{g}$, introduced in Section 3.6, is equal to the pull back of the trace form on $\mathbf{L}(\mathfrak{g},\mathfrak{g})$, under the mapping $\mathrm{ad}\colon\mathfrak{g}\to\mathbf{L}(\mathfrak{g},\mathfrak{g})$.)

The nondegeneracy of the trace form implies that every linear form μ on $\mathbf{L}(V,V)$ is of the form $M \mapsto \tau(M,L)$, for a unique $L \in \mathbf{L}(V,V)$; so our function $\mu \circ \pi$ on G can be written as:

$$m_{\pi,L} \colon x \mapsto \operatorname{tr}(\pi(x) \circ L). \tag{4.3.2}$$

A special case is obtained by taking $L = \mathrm{I}$, when we get:

$$\chi_\pi := x \mapsto \operatorname{tr} \pi(x), \tag{4.3.3}$$

called the *character* of the representation π. Noting that:

$$\begin{aligned} \chi_\pi(gxg^{-1}) &= \operatorname{tr} \pi(gxg^{-1}) = \operatorname{tr}(\pi(g) \circ \pi(x) \circ \pi(g)^{-1}) \\ &= \operatorname{tr}(\pi(x) \circ \pi(g)^{-1} \circ \pi(g)) = \operatorname{tr} \pi(x) = \chi_\pi(x), \end{aligned}$$

for all $g, x \in G$, we see that **characters are conjugacy-invariant functions on** G. Also, providing V with a Hermitian inner product for which π is unitary, cf. Corollary 4.2.2, we have:

$$\overline{\operatorname{tr} \pi(x)} = \operatorname{tr}(\pi(x)^*) = \operatorname{tr} \pi(x^{-1}) = \operatorname{tr} \pi^\vee(x);$$

hence:

$$\overline{\chi_\pi(x)} = \chi_\pi(x^{-1}) = \chi_{\pi^\vee}(x), \quad \text{for } x \in G. \tag{4.3.4}$$

The linear subspace:

$$\mathrm{M}_\pi := \{\, m_{\pi,L} \in \mathrm{C}(G) \mid L \in \mathbf{L}(V,V) \,\} \tag{4.3.5}$$

of the space $\mathrm{C}(G)$ of complex-valued continuous functions on G is called the *space of matrix coefficients* of the representation π. The name comes from the fact that any linear form μ on $\mathbf{L}(V,V)$ is a linear combination of the matrix coefficients $L \mapsto L^j_k$, with respect to a basis $e_1, \dots, e_n$ in V; so the $f \in \mathrm{M}_\pi$ are the linear combinations of the functions $x \mapsto \pi(x)^j_k$, where $j, k \in \{1, \dots, n\}$. The definition (4.3.5) can be considered as a coordinate free description of the space of matrix coefficients.

For every $\epsilon \in V^*$, $e \in V$, $M \in \mathbf{L}(V,V)$, we have:

$$\operatorname{tr}(M \circ (\epsilon \otimes e)) = \epsilon(M(e)). \tag{4.3.6}$$

If $e_1, \dots, e_n$ is a basis of V, then the *dual basis* $\epsilon^1, \dots, \epsilon^n \in V^*$ is defined by:

$$\epsilon^j(e_k) = \begin{cases} 0, & \text{for } k \neq j; \\ 1, & \text{for } k = j. \end{cases} \tag{4.3.7}$$

It follows that the matrix coefficients of $M \in \mathbf{L}(V,V)$ with respect to the basis $e_1, \dots, e_n$ are given by $M^j_k = \epsilon^j(M(e_k)) = \operatorname{tr}(M \circ (\epsilon^j \otimes e_k))$; or $x \mapsto \pi(x)^j_k$ is equal to the function $f_{\pi,L}$, with $L = \epsilon^j \otimes e_k$.

Remark. In Section 4.6, it will be shown that, for a compact group G, the space $\mathrm{M}(G)$ of matrix coefficients, that is, the union of the M_π, where π ranges over the finite-dimensional representations of G, is dense in $\mathrm{C}(G)$. This theorem of Peter-Weyl then has far-reaching consequences for the general representations of G. For many noncompact Lie groups G, there exist not sufficiently many finite-dimensional representations of G to get a similar density statement. It then turns out to be fruitful to allow infinite-dimensional representations π of G; in this case one still can define $m_{\pi,L}$ as in (4.3.2), for operators $L \in \mathbf{L}(V,V)$ of finite rank, that is, of finite dimension for the image space $L(V)$. For such operators, the trace is defined as the trace of $L|_{L(V)}$, which is a linear mapping from the finite-dimensional vector space $L(V)$ to itself.

By taking suitable completions, one can also make continuous extensions of the trace to certain classes of operators L of infinite rank. For instance, if V is a Hilbert space, then the operators $L \in \mathbf{L}(V,V)$ that are of trace class, that is, for which $L^* \circ L \in \mathbf{L}(V,V)_{\mathrm{HS}}$, will do. Unfortunately, the character cannot be defined in this way, because the identity in V is not of trace class if V is an infinite-dimensional Hilbert space. For certain classes of noncompact Lie groups, the character can however be defined as a generalized function (distribution) on G, cf. Harish-Chandra [1967]. For the character of certain infinite-dimensional representations of compact groups G, see (4.6.7).

Recall the **left and the right regular representation** of G on $\mathrm{C}(G)$, denoted by L and R^* respectively, and defined by, for $f \in \mathrm{C}(G)$, $g, h, x \in G$:

$$\text{(4.3.8)} \qquad \big(\mathrm{L}(g)f\big)(x) = f(g^{-1}x), \quad \big(\mathrm{R}^*(h)f\big)(x) = f(xh).$$

The notation is adapted to the notation $\mathrm{L}(g)$ and $\mathrm{R}(h)$ for left multiplication by g and right multiplication by h respectively, cf. (1.3.1,2). Denoting, for any mapping Φ, the *pull back of the function f under Φ* by:

$$\Phi^*(f)\colon x \mapsto f(\Phi(x)),$$

we have $(\Phi \circ \Omega)^* = \Omega^* \circ \Phi^*$. Because $\mathrm{R}\colon h \mapsto \mathrm{R}(h)$ is an anti-homomorphism, $\mathrm{R}^*\colon h \mapsto \mathrm{R}(h)^*$ is a homomorphism. Similarly, the mapping: $g \mapsto \mathrm{L}(g^{-1})^*$ is a homomorphism; it is natural to denote this then simply by L.

These two representations commute, so they can be combined to give a representation of $G \times G$ on $\mathrm{C}(G)$, defined by:

$$\text{(4.3.9)} \qquad ((g,h),f) \mapsto \big(\mathrm{L}(g) \circ \mathrm{R}^*(h)\big)(f)\colon (G \times G) \times \mathrm{C}(G) \to \mathrm{C}(G),$$

called the *regular representation* $\mathrm{L} \times \mathrm{R}^*$ of $G \times G$ in $\mathrm{C}(G)$.

After all these definitions, we are now ready to formulate some facts about the matrix coefficients.

(4.3.1) Lemma. *Let π be a representation of G in the finite-dimensional vector space V. Then the mapping $m_\pi\colon A \mapsto m_{\pi,A}$ is a linear mapping from $\mathbf{L}(V,V)$ into $\mathrm{C}(G)$; it intertwines the representation $\pi^\vee \otimes^{\mathrm{e}} \pi$ of $G \times G$ in $\mathbf{L}(V,V)$, with the regular representation of $G \times G$ in $\mathrm{C}(G)$. In particular, the space $\mathrm{M}_\pi := m_\pi(\mathbf{L}(V,V))$ of π-matrix coefficients is a subspace of $\mathrm{C}(G)$, which is invariant under both the left and the right action of G on $\mathrm{C}(G)$. Furthermore, $\dim \mathrm{M}_\pi \le \mathrm{d}_\pi{}^2$.*

Proof. We have:

$$\begin{aligned} m_{\pi,(\pi^\vee\otimes^{\mathrm{e}}\pi)(g,h)(A)}(x) &= \operatorname{tr}\bigl(\pi(x)\circ\pi(h)\circ A\circ\pi(g^{-1})\bigr) \\ &= \operatorname{tr}\bigl(\pi(g^{-1})\circ\pi(x)\circ\pi(h)\circ A\bigr) = \operatorname{tr}\bigl(\pi(g^{-1}xh)\circ A\bigr) = m_{\pi,A}(g^{-1}xh). \end{aligned}$$

□

(4.3.2) Lemma.

(i) Let V and U be finite-dimensional vector spaces, $A \in \mathbf{L}(V,V)$, $B \in L(U,U)$. Then the trace of the linear mapping:

$$B \odot A\colon C \mapsto B\circ C\circ A\colon \mathbf{L}(V,U) \to \mathbf{L}(V,U) \tag{4.3.10}$$

is equal to $(\operatorname{tr} A)(\operatorname{tr} B)$.

(ii) If σ and τ is a representation of G in V and U respectively, then:

$$\chi_{\sigma^\vee\otimes^{\mathrm{e}}\tau}(g,h) = \chi_\sigma(g^{-1})\chi_\tau(h), \quad \textit{for } (g,h)\in G\times G. \tag{4.3.11}$$

Proof. (i) Let e_j and f_k denote a basis in V and U respectively, with dual basis ϵ^j and ϕ^k in V^* and U^* respectively. Then we have:

$$\begin{aligned} \operatorname{tr}(B\odot A) &= \sum_{j,k}(e_j\otimes\phi^k)\bigl((\epsilon^j\circ A)\otimes(B\circ f_k)\bigr) = \sum_{j,k}\epsilon^j(Ae_j)\phi^k(Bf_k) \\ &= (\operatorname{tr} A)(\operatorname{tr} B). \end{aligned}$$

(ii) Applying this to $A = \sigma(g^{-1})$, $B = \tau(h)$, we obtain that (4.3.11) follows. □

Remark. If f, and g, is a function defined on the set X, and Y, respectively, then the function $f\otimes g\colon (x,y)\mapsto f(x)g(y)$, defined on $X\times Y$, is called the *tensor product of the functions* f and g. Note that $fg = (f\otimes g)\circ\Delta$, if $X = Y$. Using (4.3.4), we can write (4.3.11) as:

$$\chi_{\sigma^\vee\otimes^{\mathrm{e}}\tau} = \chi_{\sigma^\vee}\otimes\chi_\tau, \quad \text{hence } \chi_{\sigma^\vee\otimes\tau} = \chi_{\sigma^\vee}\chi_\tau. \tag{4.3.12}$$

An important Hilbert space for the representation theory of G will be the space $\mathrm{L}^2(G)$, the completion of $\mathrm{C}(G)$ with respect to the Hermitian inner product:

(4.3.13) $$< f,g > = \int_G f(x)\overline{g(x)}\,dx, \quad \text{for } f,g \in \mathrm{C}(G).$$

Due to the left and right invariance of averaging, both the left and the right regular representation of G, and therefore also the regular representation of $G \times G$, are unitary representations in $\mathrm{L}^2(G)$.

Remark. If G is only locally compact, then one may replace averaging over G by any left, and right, Haar measure, respectively, cf. Section 10.3. Defining $\mathrm{L}^2(G)$ as the completion of $\mathrm{C_c}(G)$, the space of compactly supported continuous functions on G, with respect to (4.3.1), then one still gets that the left, and right, respectively, regular representation defines a unitary representation of G in $\mathrm{L}^2(G)$. This is one of the starting points of the generalization of the representation theory to noncompact groups.

(4.3.3) Lemma. *Let σ and τ be a unitary representation of G in the finite-dimensional Hilbert space V and U respectively. Then we have:*

(4.3.14) $$< m_{\tau,B}, m_{\sigma,A} > = \mathrm{tr}_{\mathbf{L}(V,U)}\big(\mathrm{av}(\sigma^\vee \otimes \tau) \circ (B \odot A^*)\big),$$

for each $A \in \mathbf{L}(V,V)$, $B \in \mathbf{L}(U,U)$.

Proof. We have:

$$\begin{aligned}
\mathrm{tr}\big(\tau(x)\circ B\big)\overline{\mathrm{tr}\big(\sigma(x)\circ A\big)} &= \mathrm{tr}\big(\tau(x)\circ B\big)\,\mathrm{tr}\big((\sigma(x)\circ A)^*\big)\\
&= \mathrm{tr}\big(\tau(x)\circ B\big)\,\mathrm{tr}\big(A^*\circ\sigma(x)^*\big) = \mathrm{tr}\big((\tau(x)\circ B)\odot(A^*\circ\sigma(x^{-1}))\big)\\
&= \mathrm{tr}\big((\tau(x)\odot\sigma(x^{-1}))\circ(B\odot A^*)\big) = \mathrm{tr}\big((\sigma^\vee\otimes\tau)(x)\circ(B\odot A^*)\big).
\end{aligned}$$

Here we used, in the third identity, Lemma 4.3.2.(i). And also the identity $\sigma(x)^* = \sigma(x^{-1})$, because σ is unitary. Integration over x yields (4.3.14). □

Let us temporarily define the dual $\widehat{G}$ of G as the set of equivalence classes of **finite-dimensional** irreducible representations of G. Later, in Corollary 4.4.3, we shall see that every irreducible representation of a compact group is finite-dimensional.

(4.3.4) Theorem [Orthogonality relations]. *Let σ, and τ, be an irreducible unitary representation of G in the finite-dimensional Hilbert space V, and U, respectively. Then we have the following.*

(i) If σ is not equivalent to τ, then M_σ is orthogonal to M_τ in $\mathrm{L}^2(G)$.

(ii) $< m_{\sigma,A}, m_{\sigma,B} > = \mathrm{d}_\sigma^{-1} < A, B >_{\mathrm{HS}}$, for $A, B \in \mathbf{L}(V,V)$.

(iii) $m_\sigma : A \mapsto m_{\sigma,A}$ is an equivalence between the representation $\sigma^\vee \otimes^{\mathrm{e}} \sigma$ of $G \times G$ in $\mathbf{L}(V,V)$, and the regular representation of $G \times G$ in M_σ. In particular, $\dim \mathrm{M}_\sigma = \mathrm{d}_\sigma^{\,2}$.

(iv) The characters χ_π, for $[\pi] \in \widehat{G}$, form an orthonormal system in $\mathrm{L}^2(G)$.

(v) *If the topology of* G *has a countable basis, in particular if* G *is a compact Lie group, then* $\widehat{G}$ *is countable.*

Proof. (i) If σ is not equivalent to τ, then Schur's lemma 4.1.1 implies that $\mathrm{I}(\sigma, \tau) = 0$, so $\mathrm{av}(\sigma^\vee \otimes \tau) = 0$, because of Corollary 4.2.3. Now (i) follows in view of (4.3.14).

(ii) Schur's lemma, together with Corollary 4.2.3, also yields that $\mathrm{av}(\sigma^\vee \otimes \sigma)$ is equal to the orthogonal projection onto $\{ c\mathrm{I} \mid c \in \mathbf{C} \}$. Now the linear form $\epsilon\colon A \mapsto {\mathrm{d}_\sigma}^{-1} \operatorname{tr} A$ on $\mathbf{L}(V, V)$, takes the value 1 on I; so

$$\operatorname{tr}_{\mathbf{L}(V,V)}\big(\mathrm{av}(\sigma^\vee \otimes \sigma) \circ (A \odot B^*)\big) = \epsilon\big((A \odot B^*)(\mathrm{I})\big) = {\mathrm{d}_\sigma}^{-1} \operatorname{tr}(A \circ B^*).$$

(iii) m_σ is surjective: $\mathbf{L}(V, V) \to \mathrm{M}_\sigma$ by definition, and injective in view of (ii). We have seen in Lemma 4.3.1 that it intertwines $\sigma^\vee \otimes^{\mathrm{e}} \sigma$ with the regular representation of $G \times G$.

(iv) The orthogonality of χ_σ and χ_τ, if $[\sigma] \neq [\tau]$, follows from (i); whereas (ii) with $A = B = \mathrm{I}$ yields that $< \chi_\sigma, \chi_\sigma > = {\mathrm{d}_\sigma}^{-1} \operatorname{tr} \mathrm{I} = 1$.

(v) If the topology of G has a countable basis, that is, G is metrizable, then $\mathrm{C}(G)$, and hence $\mathrm{L}^2(G)$, has a countable dense subset C, see Dieudonné [1968], (7.4.4). Therefore any orthonormal system f_j in $\mathrm{L}^2(G)$ can be at most countable, because the open balls around the f_j with radius $2^{-1/2}$ are disjoint, so the mapping which assign to f_j an element of C at distance $< 2^{-1/2}$ is injective. □

The orthonormality of the characters of irreducible finite-dimensional representations leads to the following additional information about the decomposition of arbitrary finite-dimensional representations into irreducible ones, as described in Corollary 4.2.2.

(4.3.5) Corollary. *Let* σ *be a representation of* G *in the finite-dimensional vector space* U*. Then we have the following.*

(i) *Let* U *be equal to the direct sum of* $\sigma(G)$*-invariant irreducible subspaces* V_j*. For each* $[\pi] \in \widehat{G}$*, the number* $[\sigma : \pi]$ *of indices* j *such that* $\sigma|_{V_j} \in [\pi]$*, is equal to* $< \chi_\sigma, \chi_\pi >$*. This number therefore is independent of the particular way in which* σ *is decomposed into irreducibles, and is called the* multiplicity *of* π *in* σ*.*

(ii) *If* τ *is another finite-dimensional representation of* G*, then:*

$$< \chi_\sigma, \chi_\tau > = \sum_{[\pi] \in \widehat{G}} [\sigma : \pi][\tau : \pi],$$

a finite sum of positive integers, only with contributions from irreducible representations that occur both in σ *and in* τ*.*

(iii) σ *is equivalent to* τ *if and only if* $\chi_\sigma = \chi_\tau$*.*

(iv) σ *is irreducible if and only if* $< \chi_\sigma, \chi_\sigma > = 1$*.*

Proof. We have $\chi_\sigma = \sum_{[\pi]\in\widehat{G}}[\sigma : \pi]\chi_\pi$ (a finite sum). Using the orthonormality of the χ_π, for $[\pi] \in \widehat{G}$, we now obtain the assertions (i) and (ii). Note also that $\chi_\sigma = \chi_\tau$ implies that $[\sigma : \pi] = [\tau : \pi]$, for all $[\pi] \in \widehat{G}$; hence σ is equivalent to τ. Finally,

$$< \chi_\sigma, \chi_\sigma > = \sum_{[\pi]\in\widehat{G}} [\sigma : \pi]^2;$$

which is equal to 1 if and only if $[\sigma : \pi] = 1$, for only one $[\pi] \in \widehat{G}$, and $= 0$, for the others. □

(4.3.6) Corollary. *Let π be an irreducible representation of G in the finite-dimensional vector space V. Then:*

(i) The exterior tensor product representation $\pi^\vee \otimes^e \pi$ of $G \times G$ in $\mathbf{L}(V, V)$ is irreducible, as well as the regular representation of $G \times G$ in the space of matrix coefficients M_π. Any equivalence $m\colon \mathbf{L}(V, V) \to \mathrm{M}_\pi$ is a scalar multiple of m_π.

(ii) The restriction to M_π of the left and the right regular representation of G is equal to a direct sum of d_π copies of $\pi^\vee$ and of π respectively. Also, $\dim \mathrm{M}_\pi = d_\pi{}^2$.

Proof. (i) By (4.3.12), $\chi_{\pi^\vee\otimes^e\pi} = \chi_{\pi^\vee} \otimes \chi_\pi$, the product with its complex conjugate is equal to $|\chi_\pi|^2 \otimes |\chi_\pi|^2$. Integrating this over $G \times G$ we get the result $< \chi_\pi, \chi_\pi >< \chi_\pi, \chi_\pi > = 1$, proving the irreducibility of $\pi^\vee \otimes^e \pi$, using Corollary 4.3.5.(iv). In Theorem 4.3.4.(iii) we have seen that m_π is an equivalence: $\mathbf{L}(V, V) \to \mathrm{M}_\pi$; and Schur's lemma 4.1.1 contains the statement that any other equivalence is a scalar multiple of m_π.

(ii) $\mathrm{L}(g)|_{\mathrm{M}_\pi} = (\mathrm{L} \times \mathrm{R}^*)(g, 1)|_{\mathrm{M}_\pi}$, so its trace is equal to $\chi_{\pi^\vee\otimes^e\pi}(g, 1) = \chi_{\pi^\vee}(g)\chi_\pi(1) = \chi_{\pi^\vee}(g)\,\mathrm{d}_\pi$. Similarly, the character of $h \mapsto \mathrm{R}^*(h)|_{\mathrm{M}_\pi}$ is equal to $d_{\pi^\vee}\chi_\pi = \mathrm{d}_\pi\,\chi_\pi$. □

Remark. The proof of (i) actually shows that the exterior tensor product, of any two irreducible finite-dimensional representations $\sigma^\vee$, τ of G, is an irreducible representation of $G \times G$.

4.4 *G*-types

Let σ be a representation of G in the complete, locally convex, topological vector space U. The element $u \in U$ is called a *$\sigma(G)$-finite vector* if the linear subspace S of U, spanned by the $\sigma(x)(u)$, for $x \in G$, is finite-dimensional. The $\sigma(G)$-finite vectors in U form a linear subspace of U, which will be denoted by $U^{\mathrm{fin}} = U^{\mathrm{fin}_\sigma}$.

Now S is the direct sum of irreducible $\sigma(G)$-invariant linear subspaces (see Corollary 4.2.2), so U^{fin} can also be described as the algebraic sum (not direct) of all the irreducible, finite-dimensional, $\sigma(G)$-invariant linear subspaces of U. For any $[\pi] \in \widehat{G}$, the sum of all the $\sigma(G)$-invariant finite-dimensional linear subspaces V of U, such that $x \mapsto \sigma(x)|_V$ is equivalent to π, is called the *π-isotypical subspace* U_π of U; it is contained in U^{fin}. The U_π, with $[\pi] \in \widehat{G}$, are called the *G-types* in U (for the representation σ).

(4.4.1) Lemma. *Let $f \in \mathrm{C}(G)$, and $\sigma(f)$ as in (4.2.2). Then:*

(i) $\sigma(f)(V) \subset V$, for every $\sigma(G)$-invariant closed linear subspace V of U.

(ii) If f is conjugacy-invariant, then $\sigma(f)$ commutes with $\sigma(x)$, for every $x \in G$, and with $\sigma(\phi)$, for $\phi \in \mathrm{C}(G)$.

(iii) If f is conjugacy-invariant, and V is an irreducible, finite-dimensional, $\sigma(G)$-invariant linear subspace of U, then $\sigma(f)$ acts on V as multiplication by the scalar ${\mathrm{d}_\pi}^{-1} < f, \chi_{\pi^\vee} >$. Here π denotes the representation $x \mapsto \sigma(x)|_V$.

Proof. (i) follows because for each $v \in V$ the approximating Riemann sums for $\int_G f(x)\,\sigma(x)(v)\,dx$ are finite linear combinations of elements $\sigma(x)(v)$, with $x \in G$, and therefore belong to V.

(ii) We have:

$$\begin{aligned}
\sigma(x) \circ \sigma(f) \circ \sigma(x)^{-1}(u) &= \sigma(x)\Big(\int_G f(y)\,\sigma(y) \circ \sigma(x)^{-1}(u)\,dy\Big) \\
&= \int_G f(y)\,\sigma(x) \circ \sigma(y) \circ \sigma(x)^{-1}(u)\,dy = \int_G f(y)\,\sigma(xyx^{-1})(u)\,dy \\
&= \int_G f(x^{-1}zx)\,\sigma(z)(u)\,dz = \int_G f(z)\,\sigma(z)(u)\,dz = \sigma(f)(u).
\end{aligned}$$

(iii) $\sigma(f)|_V$ interwines the irreducible, finite-dimensional representation π with itself; so by Schur's lemma 4.1.1, it is equal to multiplication by a scalar $c \in \mathbf{C}$. Now:

$$\begin{aligned}
c\,\mathrm{d}_\pi &= \mathrm{tr}\big(\sigma(f)|_V\big) = \mathrm{tr}\Big(\int_G f(x)\,(\sigma(x)|_V)\,dx\Big) = \int_G f(x)\chi_\pi(x)\,dx = < f, \overline{\chi_\pi} > \\
&= < f, \chi_{\pi^\vee} >,
\end{aligned}$$

cf. (4.3.4). □

(4.4.2) Proposition. *Let σ be a representation of G in the complete, locally convex, topological vector space U. Then, for each $[\pi] \in \widehat{G}$, the mapping $E_\pi = \mathrm{d}_\pi\,\sigma(\chi_{\pi^\vee})$ is a linear projection from U onto U_π; and for each $u \in U_\pi$, the linear span of the $\sigma(x)(u)$, with $x \in G$, has dimension $\leq {d_\pi}^2$. Furthermore, $E_\pi|_{U_\tau} = 0$, if $[\tau] \in \widehat{G}$ and $[\tau] \neq [\pi]$; and U^{fin} is equal to the direct sum of the U_π, for $[\pi] \in \widehat{G}$. Finally, if U is a Hilbert space and σ is unitary, then E_π is an orthogonal projection.*

Proof. From Lemma 4.4.1.(iii), we see that E_π acts as the identity on U_π, and is equal to 0 on U_τ, if $[\tau] \in \widehat{G}$, $[\tau] \neq [\pi]$. This show that the sum $U^{\mathrm{fin}} = \oplus_{[\pi]\in\widehat{G}} U_\pi$ is direct; indeed, if $u \in U_\pi$ also belongs to the sum of the U_τ, with $[\tau] \in \widehat{G}$ such that $[\tau] \neq [\pi]$, then $u = E_\pi(u) = 0$.

Now let $u = E_\pi(v)$, and denote by S the closure of the linear subspace of U, spanned by the $\sigma(x)(u)$, for $x \in G$. For $\mu \in S'$, we set $f_\mu \colon x \mapsto \mu(\sigma(x)(u)) \in \mathrm{C}(G)$. Clearly, $\mu \mapsto f_\mu$ is an injective linear mapping: $S' \to \mathrm{C}(G)$.

On the other hand, writing $t = \mathrm{d}_\pi\, \chi_\pi$, we have:

$$\mu\big(\sigma(x)(u)\big) = \mu\Big(\sigma(x)\Big(\int_G t(y^{-1})\, \sigma(y)(v)\, dy\Big)\Big) = \int_G t(y^{-1})\, \mu\big(\sigma(xy)(v)\big)\, dy$$
$$= \int_G t(z^{-1}x)\, \mu\big(\sigma(z)(v)\big)\, dz.$$

This shows that f_μ is contained in the closure of the linear subspace of $\mathrm{C}(G)$, spanned by the $\mathrm{L}(z)(\chi_\pi)$, for $z \in G$. However, $\chi_\pi \in \mathrm{M}_\pi$, which is $\mathrm{L}(G)$-invariant and has dimension $\leq {d_\pi}^2$ (Lemma 4.3.1). So $f_\mu \in \mathrm{M}_\pi$, for all $\mu \in S'$; and therefore $\dim S' \leq {d_\pi}^2$. Because U is locally convex, S is locally convex; and the Hahn-Banach theorem (cf. Bourbaki [1966], Ch.II, §3, No.2) implies that the common kernel of the $\mu \in S'$ is equal to 0, which in turn shows that $\dim S = \dim S' \leq {d_\pi}^2$.

But this means that $u \in U^{\mathrm{fin}}$; and we have the decomposition $u = \sum_{[\tau]\in\widehat{G}} u_\tau$ (a finite sum), with $u_\tau \in U_\tau$. For any $\mu \in U'$, $[\tau] \in \widehat{G}$, the function $f_{\mu,\tau} \colon x \mapsto \mu\big(\sigma(x)(u_\tau)\big)$ belongs to M_τ, whereas $f_\mu = \sum_{[\tau]\in\widehat{G}} f_{\mu,\tau} \in \mathrm{M}_\pi$. In view of Theorem 4.3.4.(i), this can only happen if $f_{\mu,\tau} = 0$, for all $[\tau] \neq [\pi]$. Again using the Hahn-Banach theorem, it follows that $u_\tau = 0$, for all $[\tau] \neq [\pi]$; or $u \in U_\pi$.

Finally, the selfadjointness of E_π, if σ is unitary, is a consequence of the following lemma, and the fact that $\chi_{\tau^\vee} = \overline{\chi_\tau}$, cf. (4.3.4). □

(4.4.3) Lemma. *Let σ be a unitary representation of G in the Hilbert space U, let $f \in \mathrm{C}(G)$. Then the Hilbert space adjoint of $\sigma(f)$ is equal to $\sigma(f)^* = \sigma(f^*)$, if we write $f^*(x) = \overline{f(x^{-1})}$, $x \in G$.*

Proof. We have:

$$< \sigma(f)(v), u > = < \int_G f(x)\, \sigma(x)(v)\, dx, u > = \int_G f(x) < \sigma(x)(v), u > dx$$
$$= \int_G f(x) < v, \sigma(x)^{-1}(u) > dx = \int_G < v, \overline{f(x)}\sigma(x^{-1})(u) > dx$$
$$= \int_G < v, f^*(x)\, \sigma(x)(u) > dx = < v, \int_G f^*(x)\, \sigma(x)(u)\, dx >$$
$$= < v, \sigma(f^*)(u) > .$$

□

In Proposition 4.4.4 below we shall discuss the various ways in which the G-type U_π for the general representation σ can be further decomposed as a direct sum of copies of π. Because the only tool in the proof is Schur's lemma, this already could have been presented in Section 4.1.

(4.4.4) Proposition. *Let σ be a representation of an arbitrary group G in an arbitrary complex vector space U. Let π be an irreducible representation of G in the finite-dimensional vector space V, and write $\mathrm{I}(\pi, \sigma)$ for the linear subspace of $\mathbf{L}(V, U)$ consisting of the $A \in \mathbf{L}(V, U)$ that intertwine π with σ. Then:*

(i) If A_j, with $j \in J$, form a linearly independent system in $\mathrm{I}(\pi, \sigma)$, then the $V_j := A_j(V)$, for $j \in J$, form a system of irreducible $\sigma(G)$-invariant subspaces of U_π, and the sum $\sum_{j\in J} V_j$ is direct. If U_π is completely reducible, then $U_\pi = \sum_{j\in J} V_j$ if and only if the A_j, for $j \in J$, form a basis of $\mathrm{I}(\pi, \sigma)$, in the algebraic sense.

(ii) The number $[\sigma : \pi]$ of any maximal system of irreducible $\sigma(G)$-invariant subspaces of U_π that form a direct sum, is equal to $\dim \mathrm{I}(\pi, \sigma)$. Moreover, assuming that U_π is finite-dimensional and completely reducible, we have:

$$\begin{aligned}\dim \mathrm{I}(\sigma^\vee, \pi^\vee) = \dim \mathrm{I}(\pi, \sigma) = [\sigma : \pi] = [\sigma^\vee : \pi^\vee] = \dim \mathrm{I}(\pi^\vee, \sigma^\vee)\\ = \dim \mathrm{I}(\sigma, \pi).\end{aligned}$$

(iii) If V, and U_π, carries a $\pi(G)$-invariant, and $\sigma(G)$-invariant, respectively, Hermitian inner product, then the $(V_j)_{j\in J}$ form a (maximal) orthogonal system of copies of π, if and only if the $(A_j)_{j\in J}$ form a (maximal) orthogonal system in $\mathrm{I}(V, U)$ with respect to the Hermitian inner product $(L, M) \mapsto \mathrm{tr}(M^* \circ L)$.

Proof. (i) Suppose $(A_j)_{j\in J}$ is a system of nonzero elements in $\mathrm{I}(\pi, \sigma)$ such that the sum $S := \sum_{j\in J} A_j(V)$ is direct. Write π_j for the linear projection from S onto $A_j(V)$, which is equal to 0 on the A_j, for $k \neq j$. By Schur's lemma 4.1.1.(ii), each A_j is a linear isomorphism from V onto $A_j(V)$. Now let $A \in \mathrm{I}(\pi, \sigma)$ be such that $T := A(V) \cap S$ is not equal to 0. Then $A(V)$ is $\sigma(G)$-invariant, and irreducible because it is equivalent to V. Now T is a nonzero, $\sigma(G)$-invariant linear subspace of $A(V)$; hence $T = A(V)$, or $A(V) \subset S$. Write $B_j := A_j^{-1} \circ \pi_j \circ A \colon V \to V$. Then $B_j \in \mathrm{I}(\pi, \pi)$, so by Schur's lemma, $B_j = c_j \mathrm{I}$, for some $c_j \in \mathbf{C}$. But this means that, for each $v \in V$:

$$A(v) = \sum_{j\in J} \pi_j \circ A(v) = \sum_{j\in J} A_j \circ B_j(v) = \sum_{j\in J} A_j(c_j v) = \sum_{j\in J} c_j A_j(v),$$

or $A = \sum_{j\in J} A_j$. That is, if A is linearly independent of the A_j, with $j \in J$, then the sum $\sum_{j\in J} A_j(V) + A(V)$ is direct.

Now suppose that U_π is completely reducible, and that the A_j form a basis of $\mathrm{I}(\pi, \sigma)$. If $u \in U_\pi \setminus S$, then R, the space spanned by the $\sigma(x)(u)$, for $x \in G$, is a finite-dimensional $\sigma(G)$-invariant subspace of U_π, which is a direct sum of copies of π. Not all of these copies can be contained in S, because then $u \in S$. Let $A(V)$, for some $A \in \mathrm{I}(\pi, \sigma)$, be a copy of π that is not contained in S. By the argument above this yields that $A(V) \cap S = 0$; or A is linearly independent of the A_j, a contradiction. The conclusion therefore is that $U_\pi = \sum_{j \in J} A_j(V)$.

(ii) The first statement follows from (i); and we have the second and fourth identity in the second statement.

The first and the last identity follow from the observation that $A \mapsto A^*$ is an isomorphism: $\mathrm{I}(\sigma, \tau) \to \mathrm{I}(\tau^\vee, \sigma^\vee)$, for any pair σ, τ of finite-dimensional representations of G. By taking orthogonal complements, an invariant splitting leads to a corresponding invariant splitting for the contragredient representation, leading to the third identity.

(iii) For each $A, B \subset \mathrm{I}(\pi, \sigma)$, we have $B^* \circ A \in \mathrm{I}(\pi, \pi)$; so by Schur's lemma, $B^* \circ A = c\,\mathrm{I}$, for some $c \in \mathbf{C}$. Now $\mathrm{tr}\, B^* \circ A = \mathrm{d}_\pi\, c$; so if $\mathrm{tr}\, B^* \circ A = 0$, then $c = 0$. In turn, $B^* \circ A = 0$ means that

$$< A(v), B(v') > = < B^* \circ A(v), v' > = 0,$$

for all $v, v' \in V$; or $A(V)$ is orthogonal to $B(V)$. □

Remarks.

(a) It follows from (i) that for each $v \in V$, $v \neq 0$, the mapping $A \mapsto A(v)$ is an injective linear mapping: $\mathrm{I}(\pi, \sigma) \to U_\pi$. The image Z has the remarkable property that it intersects each copy of π in a one-dimensional linear subspace. If U_π is completely reducible, then this defines a bijective correspondence between the set of direct sum decompositions of M_π into copies of π, and the set of direct sum decompositions of $\mathrm{I}(\pi, \sigma)$, and of Z, respectively, into one-dimensional linear subspaces. In the situation (iii), $A \mapsto A(v)$ is an isometric embedding: $\mathrm{I}(\pi, \sigma)_{\mathrm{HS}} \to Z$ if and only if $< v, v > = \mathrm{d}_\pi$.

(b) The maximality in the last assertion of Proposition 4.4.4 means that $\sum_{j \in J} V_j$ is dense in U_π, and $\sum_{j \in J} \mathbf{C} \cdot L_j$ is dense in $\mathrm{I}(\pi, \sigma)_{\mathrm{HS}}$. If U_π is a completely reducible, infinite-dimensional Hilbert space, then $\mathrm{I}(\pi, \sigma)_{\mathrm{HS}}$ is infinite-dimensional as well, and complete, because V is finite-dimensional. In this case, $\mathrm{I}(\pi, \sigma) \neq \sum_{j \in J} \mathbf{C} \cdot A_j$; and therefore also $\sum_{j \in J} V_j \neq U_\pi$. This is based on the well-known fact that if H is an infinite-dimensional Hilbert space, and e_j a maximal orthonormal system in H, then one can always find an $h \in H$ that is not a finite linear combination of the e_j.

If $f, g \in \mathrm{C}(G)$, then one defines the *convolution* $f * g \in \mathrm{C}(G)$ by:

$$(f * g)(x) = \int_G f(y)\, g(y^{-1}x)\, dy = \int_G f(xz^{-1})\, g(z)\, dz, \tag{4.4.1}$$

the "average of the $f(y)\,g(z)$, with $x = yz$".

With the convolution as the product, $\mathrm{C}(G)$ is an associative algebra, as is not hard to verify. Also, the convolution is commutative, if and only if G is commutative.

The convolution can also be written as:

$$f * g = \mathrm{L}(f)(g). \tag{4.4.2}$$

On the other hand,

$$\begin{aligned}\big(\mathrm{R}^*(f)(g)\big)(x) &= \int_G f(y)\,g(xy)\,dy = \int_G f(x^{-1}z)\,g(z)dz\\ &= \int_G f^\vee(z^{-1}x)\,g(z)\,dz\end{aligned}$$

shows that:

$$\mathrm{R}^*(f)(g) = g * f^\vee, \tag{4.4.3}$$

if we write:

$$f^\vee(x) = f(x^{-1}), \quad \text{for } x \in G. \tag{4.4.4}$$

If σ is a representation of G in the complete, locally convex, topological vector space U, then:

$$\sigma(x)\circ\sigma(f)\circ\sigma(y^{-1}) = \sigma\big(\mathrm{L}(x)\circ\mathrm{R}^*(y)f\big), \quad \text{for } x, y \in G,\ f \in \mathrm{C}(G), \tag{4.4.5}$$

and, corresponding to (4.4.2), one gets:

$$\sigma(f)\circ\sigma(g) = \sigma(f * g), \quad \text{for } f, g \in \mathrm{C}(G). \tag{4.4.6}$$

This turns U into a $(\mathrm{C}(G), *)$-*module*, that is, for each $u \in U$, $f \in \mathrm{C}(G)$, the element $\sigma(f)(u)$ is viewed as a product $fu \in U$ of $u \in U$ with the element f of the algebra $(\mathrm{C}(G), *)$. The mapping $(f, u) \mapsto fu$ is complex bilinear: $\mathrm{C}(G) \times U \to U$, and (4.4.6) expresses that it also satisfies the associative law $f(gu) = (f * g)u$, with $f, g \in \mathrm{C}(G)$, $u \in U$.

(4.4.5) Theorem. *Let $[\pi] \in \widehat{G}$. Then:*

(i) For the right regular representation in $\mathrm{C}(G)$, and $\mathrm{L}^2(G)$, respectively, the π-isotypical subspace is equal to M_π. The projection $E_\pi\colon \mathrm{L}^2(G) \to \mathrm{M}_\pi$ is given by:

$$f \mapsto \mathrm{d}_\pi\, f * \chi_\pi = \mathrm{d}_\pi\, \chi_\pi * f, \tag{4.4.7}$$

and this projection is $\mathrm{L}^2(G)$-orthogonal. The space M_π is equal to the linear span of the $\mathrm{R}^*(x)(\chi_\pi) = \mathrm{L}(x)(\chi_\pi)$, for $x \in G$. Finally,

$$\chi_\pi * \chi_\pi = {\mathrm{d}_\pi}^{-1}\chi_\pi; \quad \chi_\pi * \chi_\tau = 0, \quad \text{if } [\tau] \in \widehat{G},\ [\tau] \neq [\pi].$$

(ii) M_π *is also equal to the* $\pi^\vee$*-isotypical subspace for the left regular representation, and to the* $\pi^\vee \otimes^{\mathrm{e}} \pi$*-isotypical subspace for the regular representation* $\mathrm{L} \times \mathrm{R}^*$ *of* $G \times G$*. The mapping* $A \mapsto \mathrm{d}_\pi\, m_{\pi,A}$ *is an anti-isomorphism from the algebra* $\mathbf{L}(V_\pi, V_\pi)$ *onto the algebra* M_π *with convolution as the product.*

(iii) The direct sum $\mathrm{M}(G)$ *of the* M_π*, for* $[\pi] \in \widehat{G}$*, called the space of matrix coefficients of* G*, is equal to: (a) the space of* $\mathrm{R}^*(G)$*-finite vectors; (b) the space of* $\mathrm{L}(G)$*-finite vectors; (c) the space of* $(\mathrm{L} \times \mathrm{R}^*)(G \times G)$*-finite vectors. With convolution as the product, it is an associative algebra over* $\mathbf{C}$*; and as such it is a direct sum of the irreducible subalgebras* M_π*, for* $[\pi] \in \widehat{G}$*.*

(iv) $\mathrm{M}(G)$ *is equal to the union of the* M_σ*, where* σ *runs over all finite-dimensional representations of* G*. With respect to the "ordinary" product of pointwise multiplication, it is a commutative algebra over* $\mathbf{C}$*, with unit element.*

Proof. We have already seen in Corollary 4.3.4.(ii) that M_π is contained in the π-isotypical subspace for the right regular representation, and in the $\pi^\vee$-isotypical subspace for the left regular representation.

The formula (4.4.7) for E_π follows by combining the definition in Proposition 4.4.2, with $\sigma = \mathrm{R}^*$, with (4.4.3). The fact that $\chi_\pi \in \mathrm{M}_\pi$, and that M_π is $\mathrm{R}^*(G)$-invariant, then yields that E_π maps into M_π, that is, the π-isotypical subspace for R^* is contained in M_π. A similar argument yields that M_π is also equal to the $\pi^\vee$-isotypical subspace for L. The equations for the convolutions of the characters follow from $E_\pi(\chi_\pi) = \chi_\pi$, and $E_\pi(\chi_\tau) = 0$.

In order to prove the second statement in (ii), we write:

$$\begin{aligned}
(m_{\pi,A} * m_{\pi,B})(x) &= \int_G \mathrm{tr}\big(\pi(y) \circ A\big) \cdot \mathrm{tr}\big(\pi(y^{-1}) \circ \pi(x) \circ B\big)\, dy \\
&= \int_G \mathrm{tr}\big(\pi(y) \circ A\big) \cdot \mathrm{tr}\big(\pi(x) \circ B \circ \pi(y^{-1})\big)\, dy \\
&= \int_G \mathrm{tr}\big(\big(\pi(y) \circ A\big) \odot \big(\pi(x) \circ B \circ \pi(y^{-1})\big)\big)\, dy \\
&= \int_G \mathrm{tr}\big(\big(\pi(y) \odot \pi(y^{-1})\big) \circ \big(A \odot \big(\pi(x) \circ B\big)\big)\big)\, dy \\
&= \mathrm{tr}\big(\mathrm{av}\big(\pi^\vee \otimes \pi\big) \circ \big(A \odot \big(\pi(x) \circ B\big)\big)\big) \\
&= \mathrm{tr}\big(C \mapsto {\mathrm{d}_\pi}^{-1}\, \mathrm{tr}\big(A \circ C \circ \pi(x) \circ B\big)\, \mathrm{I}\big) \\
&= {\mathrm{d}_\pi}^{-1}\, \mathrm{tr}\big(C \mapsto \mathrm{tr}\big(C \circ \pi(x) \circ B \circ A\big)\, \mathrm{I}\big) = {\mathrm{d}_\pi}^{-1}\, \mathrm{tr}\big(\pi(x) \circ B \circ A\big) \\
&= {\mathrm{d}_\pi}^{-1} m_{\pi, B \circ A}.
\end{aligned}$$

Here we used Lemma 4.3.2.(i) in the third identity.

The first statement in (iv) follows from the fact that $\mathrm{M}_\sigma + \mathrm{M}_\tau = M_{\sigma \oplus \tau}$, where $\sigma \oplus \tau$, the *direct sum* of the representations σ, and τ, in V, and U, respectively, is defined as the representation:

$$x \mapsto \big((v, u) \mapsto (\sigma(x)(v), \tau(x)(u))\big) \quad \text{in } V \times U.$$

In order to prove that $\mathrm{C}(G)^{\mathrm{fin}}$ is an algebra, it suffices to prove that $\chi_\sigma \chi_\tau \in \mathrm{C}(G)^{\mathrm{fin}}$, if $[\sigma], [\tau] \in \widehat{G}$, in view of the fact that the M_π are equal to the linear span of the translates of χ_π. However, $\chi_\sigma \chi_\tau$ is equal to the character of $\sigma \otimes \tau$, cf. (4.3.11) with $g = h$.

Finally, $\sum_{[\pi] \in \widehat{G}} \mathrm{M}_\pi$ is contained in the space of $(\mathrm{L} \times \mathrm{R}^*)(G \times G)$-finite vectors, which is contained in the space of $\mathrm{R}^*(G)$-finite vectors $= \sum_{[\pi] \in \widehat{G}} \mathrm{M}_\pi$. We get that the space of $(\mathrm{L} \times \mathrm{R}^*)(G \times G)$-finite vectors is equal to the direct sum of the M_π. These are irreducible and of the mutually inequivalent types $\pi^\vee \otimes^{\mathrm{e}} \pi$, as we have seen in Corollary 4.3.6.(i). □

Remark. The multiplicities of the irreducible finite-dimensional representations of $G \times G$ in $\mathrm{L}^2(G)$ are all ≤ 1. Note that for instance the $\sigma^\vee \otimes^{\mathrm{e}} \tau$, for $[\sigma], [\tau] \in \widehat{G}$, $[\sigma] \neq [\tau]$, don't occur in $\mathrm{L}^2(G)$. (See also the remark after Corollary 4.3.6.)

4.5 Finite Groups

In the special case that the group G is finite, the space $\mathrm{C}(G)$ is equal to the space $\mathbf{C}^G$ of all functions: $G \to \mathbf{C}$; this is a vector space of finite dimension equal to $\#(G)$, the number of elements of G. For any representation π of G in a vector space V, and any $v \in V$, the subspace spanned by the $\pi(x)(v)$, for $x \in G$, has dimension $\leq \#(G)$. So $V^{\mathrm{fin}} = V$; and $\dim V \leq \#(G)$, if the representation π is irreducible.

The mapping $[\pi] \mapsto \chi_\pi$ is a bijection from $\widehat{G}$ onto an orthonormal basis of the space of conjugacy-invariant functions on G, i.e. the space of functions on the set of conjugacy classes in G. So, **the number of elements of $\widehat{G}$ is equal to the number of conjugacy classes in G.**

Because $\mathbf{C}^G = \mathrm{M}(G)$ is equal to the direct sum of the M_π, for $[\pi] \in \widehat{G}$, and $\dim \mathrm{M}_\pi = \mathrm{d}_\pi{}^2$, cf. Corollary 4.3.6.(ii), we get:

(4.5.1)
$$\#(G) = \sum_{[\pi] \in \widehat{G}} \mathrm{d}_\pi{}^2.$$

Note that $\mathbf{C}^G$, with convolution, is a finite-dimensional associative algebra; it is a direct sum of the irreducible subalgebras M_π, for $[\pi] \in \widehat{G}$, which are (anti-)isomorphic to the matrix algebras $\mathbf{L}(V_\pi, V_\pi)$, cf. Theorem 4.4.5.(ii). This is a special case of a theorem of Molien [1893], which says that every simple finite-dimensional associative algebra is a matrix algebra.

Using for instance (4.5.1), we get that the following conditions (a)–(e) are equivalent:

(a) G is Abelian.
(b) The number of conjugacy classes in G is equal to the number of elements of G.
(c) $\#(\widehat{G}) = \#(G)$.
(d) $\mathrm{d}_\pi = 1$, for all $[\pi] \in \widehat{G}$.
(e) The convolution in $\mathbf{C}^G$ is Abelian.

This provides a converse to Corollary 4.1.2 in the case of finite groups. Also note that the common kernel of the χ_π, for $[\pi] \in \widehat{G}$, is equal to $\{1\}$, if G is Abelian.

If G' denotes the subgroup of G generated by the elements $xyx^{-1}y^{-1}$, the commutator of $x, y \in G$, then G' is the smallest normal subgroup H of G such that G/H is Abelian. The homomorphisms $\chi\colon G \to \mathbf{C}^\times$ (that is, the characters of the one-dimensional representations of G) are precisely the $\mu \circ \pi$, where μ ranges over the characters of G/G', and π denotes the projection: $G \to G/G'$. So one has $\#(G)/\#(G')$ many of these, and their common kernel is equal to G'.

In general it can be a hard problem to determine the characters of the irreducible representations of a finite group explicitly, see for instance Curtis and Reiner [1981/87].

4.6 The Peter-Weyl Theorem

For G equal to the circle $\mathbf{R}/\mathbf{Z}$, the space $\mathrm{M}(G)$ of matrix coefficients consists of the finite linear combinations of the functions $f_n\colon x \mapsto \mathrm{e}^{2\pi i n x}$, with $n \in \mathbf{Z}$, also called the *trigonometric polynomials*. (See the Example after Corollary 4.1.2.) It was realized long ago, that, for many applications, it is crucial to know that this system is *complete*, for instance, in the sense that every continuous function can be approximated uniformly with trigonometric polynomials. Using the orthonormality of the f_n, it follows then relatively easily that, for any $f \in \mathrm{L}^2(\mathbf{R}/\mathbf{Z})$, we have the *Fourier series* $f = \sum_n < f, f_n > f_n$, with convergence with respect to the L^2-norm. These facts will now be generalized to arbitrary compact groups G.

(4.6.1) Theorem. *Let $f \in \mathrm{C}(G)$. For every $\epsilon > 0$, there exists a $g \in \mathrm{M}(G)$ such that $|f(x) - g(x)| \leq \epsilon$, for all $x \in G$.*

Proof. As f is continuous and G is compact, there is a neighborhood U of 1 in G such that $|f(x) - f(y)| \leq \frac{1}{2}\epsilon$, whenever $xy^{-1} \in U$.

We can find an auxiliary function $\phi \in \mathrm{C}(G)$ such that: (a) $\phi(x) = 0$, if $x \in G \setminus U$; (b) $\phi(x) \geq 0$, for all $x \in G$; (c) $\phi(x) = \phi(x^{-1})$, for all $x \in G$; and finally (d) $\int_G \phi(x)\, dx = 1$. This is done by first taking a continuous function satisfying (a), (b), and $\phi(1) > 0$. Then replace it by $\phi\phi^\vee$, so that now it satisfies (a), (b), (c), and $\phi(1) > 0$. Finally, multiply it by $(\int_G \phi(x)\,dx)^{-1}$, in order to also get (d).

It follows that, for every $x \in G$:

$$|f(x) - (\phi * f)(x)| = |f(x) \int_G \phi(xy^{-1})\,dy - \int_G \phi(xy^{-1})\, f(y)\,dy|$$
$$\leq \int_G \phi(xy^{-1})|f(x) - f(y)|\,dy \leq \frac{1}{2}\epsilon \int_G dx \leq \frac{1}{2}\epsilon.$$

The mapping $\mathrm{L}(\phi)\colon g \mapsto \phi * g$ is an integral operator with continuous kernel $K\colon (x,y) \mapsto \phi(xy^{-1})\colon G \times G \to \mathbf{R}$. Using the Cauchy-Schwarz inequality, one sees that such an operator maps the unit ball in $\mathrm{L}^2(G)$ onto a uniformly bounded and equicontinuous subset of $\mathrm{C}(G)$, which is relatively compact in $\mathrm{C}(G)$, hence in $\mathrm{L}^2(G)$, because of the Ascoli theorem.

Moreover,

$$\overline{K(y,x)} = \phi(yx^{-1}) = \phi((xy^{-1})^{-1}) = \phi(xy^{-1}) = K(x,y),$$

showing that $\mathrm{L}(\phi)$ is self-adjoint as an operator in $\mathrm{L}^2(G)$. (We could also have used Lemma 4.4.3 here.) According to the classical theory of compact self-adjoint operators in Hilbert spaces, the orthogonal complement of $\ker \mathrm{L}(\phi)$ in $\mathrm{L}^2(G)$ has a countable orthonormal basis of eigenvectors f_j, with $j = 1, 2, \dots$, associated with eigenvalues $l_j \neq 0$, whereas $\lim_{j\to\infty} l_j = 0$. (See for instance Walter [1986], Kap.V, §28, Satz VII, for a simple proof.) In particular, we can write:

$$f = f_0 + \sum_{j>0} < f, f_j > f_j, \quad \text{with } f_0 \in \ker \mathrm{L}(\phi),$$

and with convergence in $\mathrm{L}^2(G)$. Applying $\mathrm{L}(\phi)$, we get:

$$\phi * f = \sum_{j>0} < f, f_j > l_j\, f_j,$$

with uniform convergence in $\mathrm{C}(G)$. Note that $f_j = l_j^{-1} \phi * f_j \in \mathrm{C}(G)$.

The result above implies that, for each $l \neq 0$, the eigenspace $V(l)$ of $\mathrm{L}(\phi)$ is finite-dimensional. Also, because $\mathrm{R}^*(x)$ commutes with $\mathrm{L}(\phi)$, the space $V(\mathrm{L})$ is $\mathrm{R}^*(x)$-invariant, for each $x \in G$. This implies that all f_j are $\mathrm{R}^*(G)$-finite vectors, that is, they belong to $\mathrm{M}(G)$, see Theorem 4.4.5. We get the existence of some $h \in \mathrm{M}(G)$, such that:

$$|(\phi * f)(x) - h(x)| \leq \frac{1}{2}\epsilon, \quad \text{for all } x \in G.$$

Combining this with the estimate above for the unform distance from f to $\phi * f$, we obtain the theorem. □

Applications. As a first result, we have that:

$$\#(\widehat{G}) < \infty \Rightarrow \dim \mathrm{M}(G) < \infty \Rightarrow \dim \mathrm{C}(G) < \infty \Rightarrow \#(G) < \infty,$$

so these four conditions are equivalent. See also Section 4.5.

Secondly, the following conditions (a)–(d) are equivalent:

(a) G is Abelian.
(b) $\mathrm{d}_\pi = 1$, for all $[\pi] \in \widehat{G}$.
(c) All $f \in \mathrm{C}(G)$ are conjugacy-invariant.
(d) The convolution in $\mathrm{C}(G)$ is commutative.

One may prove for instance (a) $\Rightarrow$ (b) $\Rightarrow$ (c) $\Rightarrow$ (d), and (c) $\Rightarrow$ (a), (d) $\Rightarrow$ (b). Theorem 4.6.1 is used as follows: if (b) holds, then $\mathrm{M}(G)$ is spanned by characters, so consists of conjugacy-invariant functions. The density of $\mathrm{M}(G)$ in $\mathrm{C}(G)$ then implies (c). In this way we obtain a converse to Corollary 4.1.2.

Thirdly, for any topological group G, one defines the *derived group* G' as the closure in G of the subgroup generated by the commutators $xyx^{-1}y^{-1}$, for $x, y \in G$. That is, G' is the smallest **closed** normal subgroup H of G such that G/H is commutative. If G is compact, then an application of Theorem 4.6.1 to G/G' yields that G' is equal to the common kernel of the continuous homomorphisms: $G \to \mathbf{C}^\times$, cf. the end of Section 4.5.

If G is a connected, compact Lie group, then G' coincides with the derived Lie group of G, which was introduced in the remark after Corollary 3.9.5. Note that G/G' is isomorphic to the torus $\mathrm{Z}(G)^\circ/(G' \cap \mathrm{Z}(G)^\circ)$ in this case.

(4.6.2) Corollary. *The regular representation* $\mathrm{L} \times \mathrm{R}^*$ *of* $G \times G$ *in* $\mathrm{L}^2(G)$ *is equal to the Hilbert space direct sum of the mutually orthogonal,* $(\mathrm{L} \otimes \mathrm{R}^*)$*-irreducible linear subspaces* M_π*, with* $[\pi] \in \widehat{G}$*. (For characterizations of the space* M_π *of matrix coefficients of* π*, and of the orthogonal projection* E_π *onto* M_π*, see Theorem 4.4.5.) That is, for every* $f \in \mathrm{L}^2(G)$*, we have:*

$$f = \sum_{[\pi] \in \widehat{G}} \mathrm{d}_\pi\, \chi_\pi * f, \quad \text{with convergence in } \mathrm{L}^2(G), \tag{4.6.1}$$

and we have the formula of Parseval-Plancherel:

$$< f, f > = \sum_{[\pi] \in \widehat{G}} \mathrm{d}_\pi < \pi(f), \pi(f) >_{\mathrm{HS}} . \tag{4.6.2}$$

Finally, if $f \in \mathrm{L}^2(G)$ *is conjugacy-invariant, then:*

$$f = \sum_{[\pi] \in \widehat{G}} < f, \chi_\pi > \chi_\pi, \quad \text{converging in } \mathrm{L}^2(G), \tag{4.6.3}$$

and

$$< f, f > = \sum_{[\pi] \in \widehat{G}} | < f, \chi_\pi > |^2. \tag{4.6.4}$$

Proof. Because $\mathrm{M}(G)$ is dense in $\mathrm{C}(G)$, with respect to the topology of uniform convergence (Theorem 4.6.1), which is stronger than the topology of L^2-convergence, and because $\mathrm{C}(G)$ is dense in $\mathrm{L}^2(G)$, we get that $\mathrm{M}(G)$ is dense in the Hilbert space $\mathrm{L}^2(G)$. In turn, the direct sum $\mathrm{M}(G) = \oplus_{[\pi]\in\widehat{G}} \mathrm{M}_\pi$ is orthogonal, cf. Theorem 4.3.4.(i). It then is standard Hilbert space theory, that

$$\sum_{[\pi]\in\widehat{G}} < E_\pi(f), E_\pi(f) > = < f, f >$$

is finite, and that $\sum_{[\pi]\in\widehat{G}} E_\pi(f) = f$, with convergence in $\mathrm{L}^2(G)$.

In order to prove (4.6.2), we start, using Lemma 4.4.3 and (4.4.6), with:

$$\begin{aligned} < \pi(f), \pi(f) >_{\mathrm{HS}} &= \mathrm{tr}\big(\pi(f) \circ \pi(f)^*\big) = \mathrm{tr}\big(\pi(f) \circ \pi(f^*)\big) = \mathrm{tr}\, \pi(f * f^*) \\ &= \mathrm{tr} \int_G (f * f^*)(x)\, \pi(x)\, dx = \int_G (f * f^*)(x)\, \chi_\pi(x)\, dx \\ &= \int_G \int_G f(y)\, f^*(y^{-1}x)\, \chi_\pi(x)\, dy\, dx \\ &= \int_G \int_G f(y)\, \overline{f(x^{-1}y)}\, \chi_\pi(x)\, dx\, dy = < \chi_\pi * \overline{f}, \overline{f} > . \end{aligned}$$

Hence:

$$\mathrm{d}_\pi < \pi(f), \pi(f) >_{\mathrm{HS}} = < E_\pi(\overline{f}), \overline{f} > .$$

Using (4.6.1), the sum of this over the $[\pi] \in \widehat{G}$ is equal to $< \overline{f}, \overline{f} >$.

If f is conjugacy-invariant, then we use (4.4.3) and Lemma 4.4.1.(iii), applied to M_π, to obtain:

$$E_\pi(f) = \mathrm{d}_\pi\, \chi_\pi * f = \mathrm{d}_\pi\, \mathrm{R}^*(f^\vee)(\chi_\pi) = < f^\vee, \chi_{\pi^\vee} > \chi_\pi = < f, \chi_\pi > \chi_\pi .$$

This proves (4.6.3); and (4.6.4) is a direct consequence. □

Remarks.

(a) From the proof of (4.6.2), we see that:

(4.6.5) $$< E_\pi(f), E_\pi(f) > = \mathrm{d}_\pi \cdot < \pi(\overline{f}), \pi(\overline{f}) >_{\mathrm{HS}}, \quad \text{for } f \in \mathrm{L}^2(G).$$

(b) In a convergent sum of nonnegative real numbers, like (4.6.2) or (4.6.4), only countably many terms can be nonzero. This means that for any $f \in \mathrm{L}^2(G)$, only countably many of the $E_\pi(f) = \mathrm{d}_\pi\, f * \chi_\pi$, with $[\pi] \in \widehat{G}$, can be nonzero; and if f is conjugacy-invariant, only countably many of the $< f, \chi_\pi >$. In view of Theorem 4.3.4.(v), this remark is only relevant if the topology of G has no countable basis, for instance, for the Cartesian product of uncountably many copies of the group $\{1, -1\}$. (Because of the Tychonoff theorem, this product of compact spaces is compact.)

(c) An amusing formulation of (4.6.1) can be given as follows. For any Radon measure μ on G, that is, a continuous linear form on $\mathrm{C}(G)$, and any $f \in \mathrm{C}(G)$, the formula:

$$(\mu * f)(x) = \mu((\mathrm{R}^*(x)f)^\vee)$$

defines a continuous function $\mu * f$ on $\mathrm{C}(G)$. The linear operator $\mu *: f \mapsto \mu * f$ in $\mathrm{C}(G)$ is continuous, with operator norm equal to the operator norm of μ. If δ, the *Dirac measure* at $1 \in G$, denotes the measure: $f \mapsto f(1)$, then $\delta * f = f$; so $\delta *$ is the identity in $\mathrm{C}(G)$. If θ is another Radon measure on G, then one may define the Radon measure $\mu * \theta$ by $(\mu * \theta)(f) = \mu((\theta * f^\vee)^\vee)$; this definition is chosen such that $(a\, dx) * (b\, dx) = (a * b)\, dx$, if $a, b \in \mathrm{C}(G)$. With convolution as the product, the space $\mathrm{Meas}(G)$ of Radon measures on G is an associative algebra, with unit element equal to δ, in contrast with $(\mathrm{C}(G), *)$, which has no unit as soon as G is infinite ($\Leftrightarrow$ not discrete). Now we provide $\mathrm{Meas}(G)$ with the topology w of *convergence in* $\mathrm{L}^2(G)$ of the $\mu * f$, for $f \in \mathrm{C}(G)$. This topology is rather weak, but it still separates points in $\mathrm{Meas}(G)$, because $\mu * f = 0$, for all $f \in \mathrm{C}(G)$, implies that $\mu = 0$. With this topology w, we can now phrase (4.6.1) as:

$$\delta = \sum_{[\pi] \in \widehat{G}} \mathrm{d}_\pi\, \chi_\pi, \quad \text{with convergence in } (\mathrm{Meas}(G), w). \tag{4.6.6}$$

If G is a Lie group, then w is stronger than the weak topology in the space of distributions on G, so then the identity (4.6.6) holds in the sense of distributions. Formula (4.6.6) is sometimes also referred to as the *Plancherel formula* for G.

For any compact group G, one could take $\mathrm{M}(G)$ as the space of "test functions" on G, so that the space $\mathrm{M}(G)^*$ of complex linear forms: $\mathrm{M}(G) \to \mathbf{C}$ would become a space of "distributions", or "generalized functions" on G. The Peter-Weyl theorem guarantees that the mapping: $\mu \mapsto \mu|_{\mathrm{M}(G)}$ is injective: $\mathrm{Meas}(G) \to \mathrm{M}(G)^*$; so measures, and therefore also integrable functions on G can be identified with elements of $\mathrm{M}(G)^*$. Note that the mapping $u \mapsto (u|_{\mathrm{M}_\pi})_{[\pi] \in \widehat{G}}$ is a linear isomorphism from $\mathrm{M}(G)^*$ onto the Cartesian product of all the $(\mathrm{M}_\pi)^* \cong \mathrm{M}_{\pi^\vee}$. In it, $\mathrm{M}(G)$ can be recognized as the subspace of the elements having only finitely many components nonzero.

If σ is a representation of G in an infinite-dimensional space V, such that, for each $[\pi] \in \widehat{G}$, the π-isotypical subpace V_π is finite-dimensional, then it is natural to define the *distributional character* $\chi_\sigma \in \mathrm{M}(G)^*$ of σ by:

$$\chi_\sigma(\phi) = \mathrm{tr}\, \sigma(\phi), \quad \text{for } \phi \in \mathrm{M}(G). \tag{4.6.7}$$

This in view of the fact that $\sigma(\phi)$ is a finite rank operator, as it maps V into V_π if $\phi \in \mathrm{M}_\pi$. Compare the remark after (4.3.7). A closer examination of each case at hand then should lead to continuity of χ_σ with respect to topologies on $\mathrm{M}(G)$ that are weaker than the algebraic direct sum topology, so that χ_σ can be identified with an element of a correspondingly smaller distribution space.

Writing $< u, \phi > = u(\phi^\vee)$, for $u \in \mathrm{M}(G)^*$, $\phi \in \mathrm{M}(G)$, we have the following generalization of Corollary 4.3.5.(i):

(4.6.8) $$[\sigma : \pi] = < \chi_\sigma, \chi_\pi >, \quad \text{for } [\pi] \in \widehat{G}.$$

This is based on $[\sigma : \pi] = \mathrm{d}_\pi{}^{-1} \operatorname{tr} E_\pi = \operatorname{tr} \sigma(\chi_{\pi^\vee})$, cf. Proposition 4.4.2.

For the next corollary of (4.6.1), recall the definition of the space of $\sigma(G)$-finite vectors for a representation σ, above Lemma 4.4.1, and its properties discussed in Proposition 4.4.2.

(4.6.3) Corollary. *Let σ be a representation of G in the complete, locally convex topological vector space U. Then:*

(i) The space U^{fin} of $\sigma(G)$-finite vectors is dense in U.
(ii) If U^{fin} is finite-dimensional, then $U = U^{\text{fin}}$.
(iii) If σ is irreducible, then U is finite-dimensional.

Proof. (i) In view of the Hahn-Banach theorem it is sufficient to show that $\mu \in U'$ equals 0, if μ annihilates U^{fin}. From the proof of Proposition 4.4.2, we pick up that, for any $v \in U$,

$$\mu\big(\sigma(x)(E_\pi(v))\big) = \mathrm{d}_\pi (f_\mu * \chi_\pi)(x), \quad \text{where } f_\mu(y) := \mu\big(\sigma(y)(v)\big).$$

Now $E_\pi(v) \in U^{\text{fin}}$, hence $\sigma(x)(E_\pi(v)) \in U^{\text{fin}}$, which therefore is annihilated by μ. The conclusion is that $\mathrm{d}_\pi\, f_\mu * \chi_\pi = 0$, for every $[\pi] \in \widehat{G}$. Applying (4.6.1), we obtain that $f_\mu = 0$. Evaluation at $y = 1$ yields $\mu(v) = 0$.

(ii) Every finite-dimensional linear subspace of a topological vector space is a closed subspace, because it is complete.

(iii) If σ is irreducible, then U^{fin} can contain only one finite-dimensional irreducible subspace, so it is equal to that one. Therefore W^{fin} is finite-dimensional, now apply (ii). □

(4.6.4) Lemma. *For every neighborhood Ω of 1 in G, there exists a finite-dimensional representation τ of G, such that $\ker \tau \subset \Omega$.*

Proof. Let $x \in G$, $x \neq 1$. There exists $f \in \mathrm{C}(G)$ such that $f(x) \neq f(1)$, so by Theorem 4.6.1, there exists $m \in \mathrm{M}(G)$ such that $m(x) \neq m(1)$. From Theorem 4.4.5.(iv) we get that $m \in \mathrm{M}_\sigma$, for some finite-dimensional representation σ of G; this in turn implies that $\sigma(x) \neq \sigma(1) = \mathrm{I}$. That is, the $G \setminus \ker \sigma$, where σ runs over the finite-dimensional representations of G, form an open covering of $G \setminus \{1\}$, and therefore in particular of the compact set $G \setminus \Omega$. The conclusion is that there is a finite set F of finite-dimensional representations of G, such that $G \setminus \Omega \subset \cup_{\sigma \in F} G \setminus \ker \sigma$. Taking for τ the direct sum of the $\sigma \in F$, we get $\ker \tau = \cap_{\sigma \in F} \ker \sigma \subset \Omega$. □

For each representation π of G in a finite-dimensional vector space V, the set $\pi(G)$ is a compact subgroup of $\mathbf{GL}(V)$, and therefore a Lie subgroup of $\mathbf{GL}(V)$, cf. Corollary 1.10.7. The representation π induces a bijective continuous mapping from the compact space $G/\ker \pi$ onto $\pi(G)$, which therefore has a continuous inverse. Because it is also an isomorphism of groups, it follows that, as a topological group, $G/\ker\pi$ is isomorphic to the compact Lie subgroup $\pi(G)$ of $\mathbf{GL}(V)$. If π, τ are such representations, then $(\ker\pi)\cap(\ker\tau)$ is equal to the kernel of the direct sum of π and τ. This shows that the $\ker\pi$ form a filterbase, and Lemma 4.6.4 just expresses that this filterbase is finer than the filter of neighborhoods of 1 in G. That is, G, **as a topological group, is isomorphic to the projective limit of the compact linear Lie groups** $G/\ker\pi$**, where** π **runs over all finite-dimensional representations of** G. (See Bourbaki [1965], Ch.I, §4, No.4 for the definition of projective limits.)

(4.6.5) Corollary. *For a compact group G, the following conditions (i)–(iii) are equivalent:*

(i) There exists a neighborhood Ω of 1 in G that contains no closed normal subgroups of G, other than $\{1\}$.
(ii) G is isomorphic to a closed subgroup of $\mathbf{GL}(V)$, for some vector space V of finite dimension.
(iii) G is a Lie group.

Proof. If (i) holds then, noting that the kernel of any continuous homomorphism is a closed normal subgroup, we read off from Lemma 4.5.4 that there exists a continuous homomorphism $\sigma\colon G\to\mathbf{GL}(V)$, such that V is finite-dimensional and $\ker\sigma=\{1\}$. But then $\sigma(G)$ is a compact subgroup of $\mathbf{GL}(V)$, and σ has a continuous inverse; so σ is an isomorphism of topological groups from G onto the closed subgroup $\sigma(G)$ of $\mathbf{GL}(V)$, and this proves (ii).

The assertion (ii) $\Rightarrow$ (iii) follows by applying von Neumann's theorem, which says that a closed subgroup of $\mathbf{GL}(V)$, for V finite-dimensional, is a Lie subgroup, cf. Corollary 1.10.7.

Now (iii) $\Rightarrow$ (i). Let G be a Lie group with Lie algebra equal to $\mathfrak{g}$. Let $X\mapsto|X|$ be any norm in $\mathfrak{g}$. According to Proposition 1.3.4, there exists an $r>0$, such that the exponential mapping is a diffeomorphism from $B_r:=\{\,X\in\mathfrak{g}\mid |X|<r\,\}$ onto an open neighborhood U_r of 1 in G. For $X\in\mathfrak{g}$ such that $0<|X|<\frac{1}{2}r$, let $n=[\frac{1}{2}r/|X|]$, the largest integer $\leq\frac{1}{2}r/|X|$. Then

$$\frac{1}{2}r/|X|<n+1<\frac{1}{2}r/|X|+\frac{1}{2}r/|X|=r/|X|,$$

so $(n+1)X\in B_r\setminus B_{\frac{1}{2}r}$. Writing $x=\exp X$, we have $x^{n+1}=\exp(n+1)X$; and using that exp is bijective: $B_r\to U_r$, we get that for each $x\in U_{\frac{1}{2}r}$ there exists $n\in\mathrm{N}$, such that x^{n+1} is not contained in $U_{\frac{1}{2}r}$. In other words, $U_{\frac{1}{2}r}$ does not contain any subgroup of G other than $\{1\}$. □

This result is a solution of *Hilbert's fifth problem*, asking for a characterization of Lie groups in topological terms, in the case of compact groups. Corollary 14.6.2 implies that a compact Lie group G actually is an affine algebraic $\mathbf{R}$-group, where the algebra A of polynomial functions on G is equal to the space of matrix coefficients of finite-dimensional **real** representations of G. The group operations of multiplication and inversion are polynomial mappings, and each finite-dimensional representation of G is polynomial too.

A compact Lie group G has a natural complexification $G_{\mathbf{C}}$, which is an affine algebraic $\mathbf{C}$-group. Each continuous homomorphism of compact Lie groups : $G \to H$ extends to a unique-complex analytic, even $\mathbf{C}$-algebraic, homomorphism : $G_{\mathbf{C}} \to H_{\mathbf{C}}$ and the groups $G_{\mathbf{C}}$ thus obtained are precisely the so-called reductive linear complex Lie groups, see Corollary 14.6.7.

We conclude this section with an application of the Peter-Weyl theorem to actions of compact Lie groups in manifolds. A local version can be found in Theorem 2.2.1.

(4.6.6) Theorem. *Let G be a compact Lie group acting in a C^k fashion on the C^k manifold M, with $1 \le k \le \omega$. For every compact subset K of M, there exists a representation π of G in a finite-dimensional vector space V and a C^k mapping $\epsilon\colon M \to V$ such that: (i) ϵ intertwines the action of G on M with π, and: (ii) $\epsilon|_\Omega$ is an embedding: $\Omega \to V$, for some open neighborhood Ω of K in M.*

Proof. There exists a finite-dimensional vector space E such that the set:

$$\mathcal{I} = \{\, \phi \in \mathrm{C}^k(M,E) \bigm| \phi|_\Omega \text{ is an embedding } : \Omega \to E, \\ \text{for some open neighborhood } \Omega \text{ of } K \text{ in } M \,\}$$

is nonvoid; moreover, $\mathcal{I}$ is an open subset of $\mathrm{C}^k(M,E)$, even for the C^1 topology. See for instance Bourbaki [1982], Chap.9, §9, No.1, Prop.4 and 3.

The formula $\big(\mathrm{L}(g)(\phi)\big)(x) = \phi(g^{-1}\cdot x)$, for $x \in M$, $g \in G$, defines a representation L of G in $\mathrm{C}^k(M,E)$. The space of $\mathrm{L}(G)$-finite vectors is dense in $\mathrm{C}^k(M,E)$, cf. Corollary 4.6.3.(i), so it meets the nonvoid open subset $\mathcal{I}$. That is, there exists $\phi \in \mathcal{I}$, such that the $\mathrm{L}(g)(\phi)$, for $g \in G$, span a finite-dimensional linear subspace F of $\mathrm{C}^k(M,E)$.

With $V := \mathbf{L}(F,E)$, define the representation π of G in V, by $\pi(g)(A) := A \circ \mathrm{L}(g^{-1})$, for $g \in G$, $A \in \mathbf{L}(F,E)$. Secondly, define the mapping $\mathrm{ev}\colon M \to \mathbf{L}(F,E)$, by $\mathrm{ev}(x)(f) := f(x)$, if $x \in M$, $f \in F$ ("evaluation"). Then:

$$\mathrm{ev}(g\cdot x)(f) = f(g\cdot x) = \big(\mathrm{L}(g^{-1})(f)\big)(x) = \mathrm{ev}(x)\circ \mathrm{L}(g^{-1})(f),$$

for all $g \in G$, $x \in M$, $f \in F$; and this shows that ev intertwines the action of G on M with π.

The proof is completed by observing that the composition of ev with the linear mapping:

$$A \mapsto A(\phi)\colon\ \mathbf{L}(F,E) \to E$$

is equal to ϕ, which is a C^k embedding: $\Omega \to E$. This implies that ev is a C^k embedding: $\Omega \to V$, for some open neighborhood Ω of K in M. □

Remarks.

(a) Concerning the first paragraph of the proof, one has the much stronger theorem that, if M is equal to the union of countably many compact subsets, then there exists a C^k embedding from M into $\mathbf{R}^p$, for $p = 2 \dim M$. For $k \leq \infty$, this is the Whitney embedding theorem, and in the real-analytic case it is the, still deeper, embedding theorem of Grauert [1958]. The result that we used here is comparatively easy to prove. It could be tempting to prove a global embedding theorem, starting with the fact that the set of C^k embeddings $\phi\colon M \to E$ is nonvoid, and open in $C^k(M, E)$ with respect to the Whitney C^1 topology. Unfortunately, $C^k(M, E)$ is not a topological vector space with respect to the Whitney topology, so the rest of the argument does not go through.

(b) In the proof of Theorem 4.6.6, we only used that the action of G in M defines a continuous homomorphism from G to the space $\mathrm{Diff}^k(M)$ of C^k diffeomorphisms of M, instead of the stronger assumption that the action is C^k. Because the representation π is real-analytic: $G \to \mathbf{GL}(V)$, it follows that the action is C^k on some open neighborhood of K; and as this is true for every compact subset K of M, the conclusion is that necessarily the action of G on M is C^k. This observation also follows from Theorem 11.5.1, the theorem of Bochner and Montgomery. That result is much more general, but it has a much longer proof.

It follows from Theorem 4.6.6 that every C^k action of a compact Lie group G on a compact manifold M is C^k equivalent to the restriction of π to a $\pi(G)$-invariant, compact, C^k submanifold of V, where π is a suitable representation of G in a finite-dimensional vector space V. Applying this to G-homogeneous spaces, we obtain:

(4.6.7) Corollary. *Let G be a compact Lie group, H a closed subgroup of G. Then there exists a representation π of G in a finite-dimensional vector space V, and an element $v \in V$, such that H is equal to the stabilizer of v in V.*

Proof. The set G/H has the structure of a compact, real-analytic manifold, such that the left action of G on G/H is real-analytic, see Corollary 1.11.3. According to Theorem 4.6.6, there exists a representation π of G in a finite-dimensional vector space V, and an injective mapping $\epsilon\colon G/H \to V$, such that $\epsilon(gxH) = \pi(g)(\epsilon(xH))$, for all $g, x \in G$. Putting $x = 1$, $v = \epsilon(1H)$, we get, for any $g \in G$, that $\pi(g)(v) = \epsilon(gH)$. This is equal to $v = \epsilon(1H)$ if and only if $\epsilon(gH) = \epsilon(1H)$, or $gH = 1H$; or $g \in H$, in view of the injectivity of ϵ. □

4.7 Induced Representations

Let H be a closed subgroup of the compact group G, and let σ be a representation of H in the finite-dimensional vector space U. The action $(h, f) \mapsto \sigma(h) \circ f$ of H on the space $\mathrm{C}(G, U)$ of continuous functions $f\colon G \to U$, will also be denoted by σ. On the other hand, each $g \in G$ acts on $f \in \mathrm{C}(G, U)$ by sending it to $\mathrm{L}(g)(f)\colon x \mapsto f(g^{-1}x)$, and $\mathrm{R}^*(g)(f)\colon x \mapsto f(xg)$, respectively, generalizing (4.1.8) to vector-valued functions. The linear transformations $\sigma(h)$, $\mathrm{L}(g)$, $\mathrm{R}^*(g')$ in $\mathrm{C}(G, U)$, for $h \in H$, $g, g' \in G$, all commute with each other. So:

$$\sigma \circ \mathrm{R}^*\colon h \mapsto \sigma(h) \circ \mathrm{R}^*(h) \tag{4.7.1}$$

is a representation of H in $\mathrm{C}(G, U)$; and its space of fixed elements,

$$\begin{aligned} &\mathrm{C}(G, U)^{\sigma\circ\mathrm{R}^*} \\ &= \{\, f \in \mathrm{C}(G, U) \mid f(xh) = \sigma(h)^{-1}\big(f(x)\big), \text{ for all } x \in G, h \in H \,\} \end{aligned} \tag{4.7.2}$$

is invariant under $\mathrm{L}(G)$.

The homomorphism:

$$y \mapsto \mathrm{L}(y)|_{\mathrm{C}(G,U)^{\sigma\circ\mathrm{R}^*}} : G \to \mathbf{GL}(\mathrm{C}(G, U)^{\sigma\circ\mathrm{R}^*}) \tag{4.7.3}$$

is called the *induced representation* of G on $\mathrm{C}(G, U)^{\sigma\circ\mathrm{R}^*}$, *defined by the representation* σ of H, and will be denoted by $\mathrm{Ind}_H^G \sigma$.

If G is a Lie group, then the homogeneous vector bundle $G \times_H U$ over G/H, with fiber U, is defined in Section 2.4 as the orbit space for the action $(h, (x, u)) \mapsto (xh^{-1}, \sigma(h)(u))$ of H on $G \times U$. A section of $G \times_H U \to G/H$ is a mapping which assigns to each xH, for $x \in G$, an H-orbit $\{\, (xh^{-1}, \sigma(h)(u)) \mid h \in H \,\}$; this amounts to the same as assigning to each $x \in G$ an element $u = f(x) \in U$ in such a way that $f(xh^{-1}) = \sigma(h)\big(f(x)\big)$, for all $h \in H$. That is, the space $\mathrm{C}(G, U)^{\sigma\circ\mathrm{R}^*}$ can be identified with *the space of continuous sections of the homogeneous vector bundle* $G \times_H U$ *over* G/H, with fiber U, and defined by the representation σ of H.

(4.7.1) Theorem [Frobenius reciprocity]. *Let H be a closed subgroup of the compact group G. Let σ and π be a representation of H, and G, in the finite-dimensional vector space U, and V, respectively. Then:*

(i) The mapping $\delta\colon A \mapsto \big(v \mapsto A(v)(1)\big)$ defines a linear isomorphism from $\mathrm{I}(\pi, \mathrm{Ind}_H^G \sigma)$ onto $\mathrm{I}(\pi|_H, \sigma)$.

(ii) If σ and π are irreducible, then we have:

$$[\mathrm{Ind}_H^G \sigma : \pi] = [\pi|_H : \sigma].$$

Proof. (i) If $A \in \mathrm{I}(\pi, \mathrm{Ind}_H^G \sigma)$, then, for any $x \in G$, $v \in V$:

$$\delta(A)\big(\pi(x)(v)\big) = A\big(\pi(x)(v)\big)(1) = \mathrm{L}(x)(A(v))(1) = A(v)(x^{-1}).$$

So, if $\delta(A) = 0$, then $A(v) = 0$, for all $v \in V$; or $A = 0$, which shows that δ is injective. Furthermore, if $v \in V$, $x \in H$, then:

$$\delta(A)\big(\pi(x)(v)\big) = A(v)(x^{-1}) = \sigma(x)\big(A(v)(1)\big) = \sigma(x)\big(\delta(A)(v)\big),$$

which shows that $\delta(A) \in \mathrm{I}(\pi|_H, \sigma)$.

Now let $B \in \mathrm{I}(\pi|_H, \sigma)$. Define, for each $v \in V$, $x \in G$,

$$A(v)(x) = B\big(\pi(x^{-1})(v)\big).$$

Then we have, for each $h \in H$:

$$\begin{aligned} A(v)(xh) &= B\big(\pi((xh)^{-1})(v)\big) = B\big(\pi(h^{-1}) \circ \pi(x^{-1})(v)\big) \\ &= \sigma(h^{-1})\big(B(\pi(x^{-1})(v))\big) = \sigma(h^{-1})\big(A(v)(x)\big), \end{aligned}$$

or: $A(v) \in \mathrm{C}(G, U)^{\sigma \circ \mathrm{R}^*}$.

Furthermore, if $y \in G$, then:

$$\begin{aligned} A\big(\pi(y)(v)\big)(x) &= B\big(\pi(x^{-1}) \circ \pi(y)(v)\big) = B\big(\pi((y^{-1}x)^{-1})(v)\big) = A(v)(y^{-1}x) \\ &= \big(\mathrm{L}(y)(A(v))\big)(x), \end{aligned}$$

for all $x \in G$; or $A(\pi(y)(v)) = \mathrm{L}(y)(A(v))$, for all $y \in G$, $v \in V$; and this means that:

$$A \in \mathrm{I}(\pi, \mathrm{Ind}_H^G \sigma).$$

Finally, the equality $\delta(A)(v) = A(v)(1) = B(v)$, for all $v \in V$, shows that $\delta(A) = B$, and this proves the surjectivity of δ.

(ii) From Proposition 4.4.4.(ii) we read off that:

$$\dim \mathrm{I}(\pi, \mathrm{Ind}_H^G \sigma) = [\mathrm{Ind}_H^G \sigma : \pi],$$

if π is irreducible; whereas the identity between the third and the last quantity in loc. cit. shows that $[\pi|_H : \sigma] = \dim \mathrm{I}(\pi|_H, \sigma)$, if σ is irreducible. □

Remarks.

(a) Usually only the relation $[\mathrm{Ind}_H^G \sigma : \pi] = [(\pi|_H) : \sigma]$, with $[\sigma] \in \widehat{H}$, $[\pi] \in \widehat{G}$, is called the Frobenius reciprocity theorem. In particular, all G-types in $\mathrm{Ind}_H^G \sigma$ are finite-dimensional; so we can define the distributional character of $\mathrm{Ind}_H^G \sigma$, as in (4.6.7), with σ replaced by $\mathrm{Ind}_H^G \sigma$. Writing:

$$\phi^{\mathrm{av}}(x) = \int_G \phi(yxy^{-1})\, dy,$$

we get:

$$\begin{aligned}
\operatorname{tr}\big(\operatorname{Ind}_H^G \sigma\big)(\phi) &= \operatorname{tr}\big(\operatorname{Ind}_H^G \sigma\big)(\phi^{\mathrm{av}}) = \operatorname{tr}\big(\operatorname{Ind}_H^G \sigma\big)\Big(\sum_{[\pi]\in\widehat{G}} \phi^{\mathrm{av}}(\chi_{\pi^\vee})\,\chi_\pi\Big)\\
&= \sum_{[\pi]\in\widehat{G}} \phi^{\mathrm{av}}(\chi_{\pi^\vee})\operatorname{tr}\big(\operatorname{Ind}_H^G \sigma\big)(\chi_\pi)\\
&= \sum_{[\pi]\in\widehat{G}} \phi^{\mathrm{av}}(\chi_{\pi^\vee})[\operatorname{Ind}_H^G \sigma : \pi^\vee] = \sum_{[\pi]\in\widehat{G}} \phi^{\mathrm{av}}(\chi_{\pi^\vee})[\pi^\vee|_H : \sigma]\\
&= \sum_{[\pi]\in\widehat{G}} \phi^{\mathrm{av}}(\chi_{\pi^\vee})\chi_{\pi^\vee|_H}(\chi_{\sigma^\vee}) = \sum_{[\pi]\in\widehat{G}} \phi^{\mathrm{av}}(\chi_{\pi^\vee})\chi_\sigma(\chi_\pi|_H)\\
&= \chi_\sigma\Big(\Big(\sum_{[\pi]\in\widehat{G}} \phi^{\mathrm{av}}(\chi_{\pi^\vee})\chi_\pi\Big)\Big|_H\Big) = \chi_\sigma(\phi^{\mathrm{av}}|_H).
\end{aligned}$$

Here we used in the first equality that distributional characters are conjugacy-invariant in the obvious sense. Formula (4.6.3) has been used in the second and the last identity, and (4.6.8) in the fourth and sixth one. The Frobenius reciprocity theorem entered in the fifth identity. We have obtained:

$$\operatorname{tr}\big(\operatorname{Ind}_H^G \sigma\big)(\phi) = \chi_\sigma(\phi^{\mathrm{av}}|_H).$$

The distributional character of the induced representation is therefore equal to the element $\chi_{\sigma,H}^G$ of $\mathrm{M}(G)^*$, defined by:

$$\chi_{\sigma,H}^G(\phi) := \chi_\sigma(\phi^{\mathrm{av}}|_H) = \int_H \Big(\int_G \phi(xhx^{-1})\,dx\Big)\chi_\sigma(h)\,dh, \tag{4.7.4}$$

for all $\phi \in \mathrm{M}(G)$. The distribution $\chi_{\sigma,H}^G$ extends to a measure on G (because the push-forward of a measure is a measure); it is defined entirely in terms of χ_σ, and is called the *character on* G*, induced by the character* χ_σ *on* H.

Conversely, evaluating (4.7.4) for $\phi = \chi_{\pi^\vee}$, we get $<\chi_\sigma, \chi_\pi|_H> = [(\pi|_H) : \sigma]$, so (4.7.4) is actually equivalent to the Frobenius reciprocity theorem $[\operatorname{Ind}_H^G \sigma : \pi] = [(\pi|_H) : \sigma]$.

(b) Note that, if $H = 1$, then σ can only be irreducible if $U = \mathbf{C}$; and the induced representation then is equal to the left regular one. Even in this seemingly trivial situation, the conclusion of the Frobenius reciprocity theorem is the nontrivial statement that the multiplicity of π in the left regular representation is equal to d_π.

(c) For any $L \in \mathrm{I}(\pi, \operatorname{Ind}_H^G \sigma)$, we have that $L(V)$ is a finite-dimensional, $\mathrm{L}(G)$-invariant subspace of $\mathrm{C}(G,U)^{\sigma\circ\mathrm{R}^*}$. Now the mapping: $f \mapsto (\mu \mapsto \mu\circ f)$ is an isomorphism: $\mathrm{C}(G,U) \to \mathbf{L}(U^*, \mathrm{C}(G))$; and $\mathrm{L}(x)(\mu\circ f) = \mu\circ\big(\mathrm{L}(x)(f)\big)$, for all $x \in G$, $\mu \in U^*$, $f \in \mathrm{C}(G,U)$. It follows that, for $[\pi] \in \widehat{G}$, the π-isotypical subspace of $\mathrm{C}(G,U)$, which is equal to the image of $\mathrm{d}_\pi\, L(\chi_{\pi^\vee})$, is mapped by this isomorphism onto $\mathbf{L}(U^*, \mathrm{M}_{\pi^\vee})$, cf. Theorem 4.4.5.(ii). The relation $\mu(f(xh)) = \mu(\sigma(h^{-1})(f(x)) = (\sigma^{\vee}(\mu))(f(x))$, for $h \in H$, shows that the isomorphism induces an isomorphism:

$$(4.7.5) \qquad \mathrm{C}(G,U)_{\pi}^{\sigma\circ\mathrm{R}^*} \xrightarrow{\sim} \mathrm{I}(\sigma^{\vee}, \mathrm{R}^* |_{M_{\pi^{\vee}}}), \quad \text{for } [\pi] \in \widehat{G}.$$

This can be used as an alternative approach to the Frobenius reciprocity theorem.

It also follows that the Frobenius reciprocity theorem remains true if $\mathrm{C}(G)$ is replaced by $\mathrm{M}(G)$, or by any complete, locally convex topological vector space that contains $\mathrm{M}(G)$ as a dense subspace, and to which the regular representation has a continuous extension. For instance, we could replace $\mathrm{C}(G)$ by $\mathrm{L}^2(G)$, or, if G is a compact Lie group, by $\mathrm{C}^{\infty}(G)$, or the space of distributions on G.

(d) The definition of induced representations makes sense for arbitrary groups, including noncompact Lie groups. Furthermore, it can be useful to restrict the induced representation to suitable invariant subspaces, leading to variations as "unitary induction" and "holomorphic induction", cf. Van den Ban [1997] and Vogan [1997-I]. Most explicitly known representations of Lie groups fall in this scheme.

(4.7.2) Corollary. *Let G be a compact group with a closed subgroup H. Let σ be a representation of H in the finite-dimensional vector space U. Then there exists a representation π of G in a finite-dimensional vector space V, which contains U as a $\pi(H)$-invariant subspace, such that $\sigma(x) = \pi(x)|_U$, for $x \in H$. The representation π can be chosen irreducible if σ is irreducible.*

Proof. Because σ is completely reducible (Corollary 4.2.2), and because the statement follows for a direct sum of representations if it is true for each of the summands, we may assume that σ is irreducible. Now the dimension of $\mathrm{Ind}_H^G \sigma$ is positive, and because the finite vectors form a dense subspace (Corollary 4.6.3.(i)), this space has a positive dimension too. It follows that $[\mathrm{Ind}_H^G \sigma : \pi] > 0$, for some $\pi \in \widehat{G}$; so by the Frobenius reciprocity theorem, $[(\pi|_H) : \sigma] > 0$, for some $\pi \in \widehat{G}$. □

4.8 Reality

Until now we have only worked with representations of G by means of complex linear transformations in a complex vector space. This was essential in Schur's lemma 4.1.3, and therefore also in all results based on that. However, in many situations one is confronted with representations $\sigma_{\mathbf{R}}$ of G by means of real linear transformations in a real linear vector space $U_{\mathbf{R}}$; it is the purpose of this section to discuss the theory for these.

For instance, the averaging principle 4.2.1 holds without change for real representations, and actually has already been used in this form in the proof of Bochner's linearization theorem 2.2.1. Replacing Hermitian inner products by real inner products, the analogue of Corollary 4.2.2 holds for real representations; in particular, real representations of a compact group are completely reducible as well.

In order to investigate the natural decompositions of $U_{\mathbf{R}}$, we will pass to the complexification $U = U_{\mathbf{R}} \otimes_{\mathbf{R}} \mathbf{C} = U_{\mathbf{R}} \oplus iU_{\mathbf{R}}$ of $U_{\mathbf{R}}$. For each $x \in G$, the transformation $\sigma_{\mathbf{R}}(x) \in \mathbf{GL}(U_{\mathbf{R}})$ extends to a unique complex linear transformation $\sigma(x)$ in U; so $\sigma_{\mathbf{R}}$ defines a complex representation σ of G in U. We have the complex conjugation $\mathfrak{C}\colon U \to U$, defined by $\mathfrak{C}(u+iv) = u-iv$, for $u, v \in U_{\mathbf{R}}$. This is a complex antilinear mapping from U onto itself, $\mathfrak{C}^2 = \mathrm{I}$, and $\mathfrak{C}$ commutes with $\sigma(x)$, for every $x \in G$. The real linear mapping $\mathrm{Re} := \frac{1}{2}(\mathrm{I} + \mathfrak{C})$ and $\mathrm{Im} := \frac{1}{2}(\mathrm{I} - \mathfrak{C})$ is a projection from U onto $U_{\mathbf{R}}$ and onto $iU_{\mathbf{R}}$ respectively; and both maps also commute with $\sigma(x)$, for all $x \in G$.

Before we proceed further, we need some facts about the "reality aspects" of the irreducible complex representations.

(4.8.1) Theorem. *Let π be an irreducible complex representation of G in V. Then the following conditions are equivalent:*

(i) χ_π *is real-valued on G.*
(ii) π *is equivalent to* $\pi^\vee$ *(as complex representations).*
(iii) There exists a $\pi(G)$-invariant, nonzero, complex bilinear form B on V, which is automatically nondegenerate and uniquely determined, up to a nonzero scalar factor. This form B is either symmetric, or antisymmetric. The first case is equivalent to: π is equal to the complexification of a real representation. The second case is equivalent to: V can be provided with the structure of an **H***-module, such that $\pi(x)(cv) = c\pi(x)(v)$, for all $x \in G$, $c \in \mathbf{H}$, $v \in V$. Here* **H** *is the system of quaternions, defined as in Section 1.2.b.*

Proof. (i) $\Leftrightarrow$ (ii). We have $\overline{\chi_\pi} = \chi_\pi \Leftrightarrow \chi_{\pi^\vee} = \chi_\pi \Leftrightarrow [\pi^\vee] = [\pi]$, cf. (4.3.4) and Corollary 4.3.5.(iii).

(ii) $\Leftrightarrow$ (iii). Because of Schur's lemma 4.1.1, $[\pi^\vee] = [\pi]$ if and only if there exists a nonzero complex linear mapping $B\colon V \to V^*$ that intertwines π with $\pi^\vee$; such a mapping is automatically bijective and unique up to a scalar factor. However, as we already observed in the proof of Corollary 4.1.3, such B can be identified with the $\pi(G)$-invariant complex bilinear forms on V. Note that the transposed B^* is also $\pi(G)$-invariant, so $B^* = cB$, for some $c \in \mathbf{C}$. But $B = (B^*)^* = (cB)^* = cB^* = ccB$; and this shows that $c^2 = 1$. The mapping B is symmetric, and antisymmetric, if and only if $c = 1$, and $c = -1$, respectively.

Now let A be the complex antilinear mapping: $V \to V^*$, related, via $(A(v))(u) = < u, v >$, for $u, v \in V$, with some $\pi(G)$-invariant Hermitian inner product in V. It follows that $\theta := A^{-1} \circ B$ is a bijective, complex antilinear mapping: $V \to V$, which intertwines π with itself. So $\theta \circ \theta$ is a bijective, complex linear mapping: $V \to V$, which intertwines π with itself. Because of Schur's lemma 4.1.1, $\theta \circ \theta = t\,\mathrm{I}$, for some $t \in \mathbf{C}$. On the other hand, the anti-linearity of θ yields that $(t\,\mathrm{I}) \circ \theta = \theta \circ \theta \circ \theta = \theta \circ (t\,\mathrm{I}) = (\bar{t}\,\mathrm{I}) \circ \theta$; so $t = \bar{t}$, or t is real. If we replace B by βB, with $\beta \in \mathbf{C}$, then, using the antilinearity of A^{-1}, we can replace $\theta \circ \theta$ by $(\overline{\beta}\beta)(\theta \circ \theta)$, so we can arrange that $t = \pm 1$.

If $\theta \circ \theta = \mathrm{I}$, then V is equal to the direct sum of the $\pi(G)$-invariant, $\mathbf{R}$-linear subspaces $V_{\pm} = \{ v \in V \mid \theta(v) = \pm v \}$; and the antilinearity of θ implies that $V_{-} = iV_{+}$. But this means that V can be viewed as the complexification of V_{+}, and π as the complexification of $x \mapsto \pi(x)|_{V_{+}}$.

If $\theta \circ \theta = -\,\mathrm{I}$, then, identifying θ, and $(i\,\mathrm{I}) \circ \theta$, with the multiplication by j, and k, respectively, we get an action of $\mathbf{H} = \mathbf{R} \cdot 1 \oplus \mathbf{R} \cdot i \oplus \mathbf{R} \cdot j \oplus \mathbf{R} \cdot k$ on V that is compatible with the quaternionic laws: $jj = -1$, $ji = -ij$ (that is, the antilinearity of θ), $ij = k$. The action also commutes with the $\pi(x)$, for $x \in G$, because all transformations involved do so.

Finally, the equality:

$$\begin{aligned} < \theta(v), \theta(v) > &= (A \circ \theta(v))(\theta(v)) = (B(v))(\theta(v)) = B^*(\theta(v))(v) \\ &= cB(\theta(v))(v) = c(A \circ \theta \circ \theta(v))(v) = ct < v, v > \end{aligned}$$

shows that c and t have the same sign. □

(4.8.2) Definition. *The irreducible complex representation π is said to be of complex type, if χ_π is not a real-valued function on G. Further π is said to be of real type, if it is the complexification of a real representation. Finally, π is said to be of quaternionic type, if V can be provided with the structure of an $\mathbf{H}$-module, for which the $\pi(x)$ are $\mathbf{H}$-linear mappings, for all $x \in G$.*

(4.8.3) Lemma. *An irreducible complex representation is reducible as a real representation, if and only if it is of real type.*

Proof. Let Y be a $\pi(G)$-invariant real linear subspace of V, not equal to 0 or V. Then $Y \cap i(Y)$ and $Y + i(Y)$ are $\pi(G)$-invariant complex linear subspaces, the first is not equal to V, and the second not equal to 0, so $Y \cap i(Y) = 0$, $Y + i(Y) = V$. But this means that π is equal to the complexification of the real representation $x \mapsto \pi(x)|_Y$. □

Let us resume our study of the general real representation σ. From $\overline{\chi_\pi} = \chi_{\pi^\vee}$, cf. (4.2.5), it follows, in the notation of Proposition 4.4.2, that

$$\mathfrak{C} \circ E_\pi = \mathfrak{C} \circ \big(\mathrm{d}_\pi\, \sigma(\chi_{\pi^\vee})\big) = \mathrm{d}_\pi\, \sigma(\overline{\chi_{\pi^\vee}}) \circ \mathfrak{C} = \mathrm{d}_{\pi^\vee}\, \sigma(\chi_\pi) \circ \mathfrak{C} = E_{\pi^\vee} \circ \mathfrak{C}.$$

In particular, $\mathfrak{C}(U_\pi) = U_{\pi^\vee}$, for all $[\pi] \in \widehat{G}$; and U^{fin} is invariant under $\mathfrak{C}$, so $\mathrm{Re}(U^{\mathrm{fin}}) = U_{\mathbf{R}} \cap U^{\mathrm{fin}}$. Because for each $u \in U_{\mathbf{R}}$, the real linear span of the $\sigma(x)(u)$, with $x \in G$, is finite-dimensional if and only if the complex linear span is finite-dimensional, we may identify $U_{\mathbf{R}} \cap U^{\text{fin}}$ with the space $U_{\mathbf{R}}^{\text{fin}}$ of $\sigma_{\mathbf{R}}(G)$-finite vectors in $U_{\mathbf{R}}$, in the sense of real representations. Because Re is a continuous real linear projection from U onto $U_{\mathbf{R}}$, the density of U^{fin} in U (Corollary 4.6.3.(i)) implies that $U_{\mathbf{R}}^{\text{fin}}$ **is dense in** $U_{\mathbf{R}}$. Also, $U_{\mathbf{R}}^{\text{fin}}$ **is equal to the direct sum of the** $\text{Re}(U_\pi) = \text{Re}(U_\pi + U_{\pi^\vee})$, for $[\pi] \in \widehat{G}$.

If π is of complex type, then $U_\pi \cap \mathfrak{C}(U_\pi) = U_\pi \cap U_{\pi^\vee} = 0$, and Re is a real linear isomorphism: $U_\pi \to \text{Re}(U_\pi)$, commuting with the action of $\sigma(G)$. Each splitting of U_π into irreducible subspaces $(V_j)_{j\in J}$ as in Proposition 4.4.4, leads to a corresponding splitting of $\text{Re}(U_\pi)$ as a direct sum of the nonzero, $\sigma_{\mathbf{R}}(G)$-invariant subspaces $\text{Re}(V_j)$, for $j \in J$. In particular, if $U_{\mathbf{R}}$ is an irreducible real representation and $U_\pi \neq 0$, then $U_{\mathbf{R}}$ is equivalent to π, considered as a real representation. Note that the latter is irreducible according to Lemma 4.8.3.

If π is of real type, then we may write π as the complexification of the real representation $\pi_{\mathbf{R}}$ in $V_{\mathbf{R}}$. Clearly, $\pi_{\mathbf{R}}$ is irreducible as a real representation. Note that $\pi^\vee = \pi$ implies that $\mathfrak{C}(U_\pi) = U_\pi$ in this case; so $U_\pi = U_{\pi,\mathbf{R}} \oplus i U_{\pi,\mathbf{R}}$, if we write $U_{\pi,\mathbf{R}} = \text{Re}(U_\pi) = U_\pi \cap U_{\mathbf{R}}$. Denote the complex conjugation in V also by $\mathfrak{C}$, then the mapping: $L \mapsto \mathfrak{C} \circ L \circ \mathfrak{C}$ maps $\text{I}(\pi, \sigma)$ to itself, is complex antilinear, and has square equal to the identity. That is, it defines a complex conjugation in $\text{I}(\pi, \sigma)$. For every real basis L_j in $\text{Re}(\text{I}(\pi, \sigma))$, the $\text{Re}(L_j(V))$ form a real direct sum decomposition of $U_{\pi,\mathbf{R}}$ into copies of $V_{\mathbf{R}}$.

If finally π is of quaternionic type, then again U_π is equal to the complexification of $U_{\pi,\mathbf{R}} = U_\pi \cap U_{\mathbf{R}}$. However, this time π is irreducible as a real representation, cf. Lemma 4.8.3. Now the mapping: $L \mapsto \mathfrak{C} \circ L \circ j$, where j denotes the multiplication with $j \in \mathbf{H}$, defines a complex antilinear mapping J from $\text{I}(\pi, \sigma)$ to itself, such that $\text{J} \circ \text{J} = -\text{I}$, making $\text{I}(\pi, \sigma)$ into an $\mathbf{H}$-module. Then, as in Proposition 4.4.4, one verifies that for any $\mathbf{H}$-basis $(L_j)_{j\in J}$ in $\text{I}(\pi, \sigma)$, the $\text{Re}(L_j(V))$, for $j \in J$, form a real direct sum decomposition of $U_{\pi,\mathbf{R}}$ into copies of π.

Specializing to the case that $\sigma_{\mathbf{R}}$ is irreducible, we get:

(4.8.4) Proposition. *Let $\sigma_{\mathbf{R}}$ be an irreducible real representation in $U_{\mathbf{R}}$ with complexification σ. Also write $\text{I}_{\mathbf{R}}(\sigma_{\mathbf{R}}, \sigma_{\mathbf{R}})$, for the algebra over $\mathbf{R}$ consisting of the real linear mappings: $U_{\mathbf{R}} \to U_{\mathbf{R}}$ that intertwine $\sigma_{\mathbf{R}}$ with itself. Then there are the following possibilities:*

(a) σ is an irreducible complex representation of real type, but reducible as a real representation. The character of $\sigma_{\mathbf{R}}$ is equal to χ_σ. Further $\text{I}_{\mathbf{R}}(\sigma_{\mathbf{R}}, \sigma_{\mathbf{R}})$ is isomorphic to $\mathbf{R}$.

(b) $\sigma = \pi \oplus \pi^\vee$, where π is an irreducible complex representation of complex type, equivalent to $\sigma_{\mathbf{R}}$ as a real representation. The character of $\sigma_{\mathbf{R}}$ is equal to $2\,\text{Re}\,\chi_\pi = \chi_\pi + \chi_{\pi^\vee}$. Further $\text{I}_{\mathbf{R}}(\sigma_{\mathbf{R}}, \sigma_{\mathbf{R}})$ is isomorphic to $\mathbf{C}$.

(c) $\sigma = \pi \oplus \pi$, where π is an irreducible complex representation of quaternionic type, equivalent to $\sigma_{\mathbf{R}}$ as a real representation. The character of $\sigma_{\mathbf{R}}$ is equal to $2\chi_\pi$. Further $\text{I}_{\mathbf{R}}(\sigma_{\mathbf{R}}, \sigma_{\mathbf{R}})$ is isomorphic to $\mathbf{H}$.

The equivalence classes of irreducible real representations are parametrized by $\widehat{G}$, if one identifies $[\pi]$ with $[\pi^\vee]$ in case π is of complex type.

Proof. What is left to prove is the characterization of the algebra $\mathrm{I}_{\mathbf{R}}(\sigma_{\mathbf{R}}, \sigma_{\mathbf{R}})$ in each case. Looking at the complex linear extensions of the elements of $\mathrm{I}_{\mathbf{R}}(\sigma_{\mathbf{R}}, \sigma_{\mathbf{R}})$, we obtain an injective, real linear mapping: $\mathrm{I}_{\mathbf{R}}(\sigma_{\mathbf{R}}, \sigma_{\mathbf{R}}) \to \mathrm{I}(\sigma, \sigma)$. Also, $\mathrm{I}_{\mathbf{R}}(\sigma_{\mathbf{R}}, \sigma_{\mathbf{R}}) \cap i\, \mathrm{I}_{\mathbf{R}}(\sigma_{\mathbf{R}}, \sigma_{\mathbf{R}}) = 0$, showing that $\dim_{\mathbf{R}} \mathrm{I}_{\mathbf{R}}(\sigma_{\mathbf{R}}, \sigma_{\mathbf{R}}) \leq \dim_{\mathbf{C}} \mathrm{I}(\sigma, \sigma)$.

In case (a), $\dim_{\mathbf{C}} \mathrm{I}(\sigma, \sigma) = 1$, because σ is irreducible. Hence we obtain $\dim_{\mathbf{R}} \mathrm{I}_{\mathbf{R}}(\sigma_{\mathbf{R}}, \sigma_{\mathbf{R}}) = 1$, or $\mathrm{I}_{\mathbf{R}}(\sigma_{\mathbf{R}}, \sigma_{\mathbf{R}}) = \mathbf{R}$.

In case (b), $\dim_{\mathbf{C}} \mathrm{I}(\sigma, \sigma) = 2$, because σ is the direct sum of 2 inequivalent irreducible representations and their isotypical subspaces remain invariant under each intertwining map. On the other hand we know that $U_{\mathbf{R}}$ admits a complex structure which commutes with $\sigma_{\mathbf{R}}$. Combining $\mathbf{C} \subset \mathrm{I}_{\mathbf{R}}(\sigma_{\mathbf{R}}, \sigma_{\mathbf{R}})$ with $\dim_{\mathbf{R}} \mathrm{I}_{\mathbf{R}}(\sigma_{\mathbf{R}}, \sigma_{\mathbf{R}}) \leq 2$, we get $\mathbf{C} = \mathrm{I}_{\mathbf{R}}(\sigma_{\mathbf{R}}, \sigma_{\mathbf{R}})$.

In case (c), we have $\dim_{\mathbf{C}} \mathrm{I}(\pi, \sigma) = [\sigma : \pi] = 2$, hence $\dim_{\mathbf{C}} \mathrm{I}(\sigma, \sigma) = 4$. On the other hand $\mathbf{H} \subset \mathrm{I}_{\mathbf{R}}(\sigma_{\mathbf{R}}, \sigma_{\mathbf{R}})$, so combined with $\dim_{\mathbf{R}} \mathrm{I}_{\mathbf{R}}(\sigma_{\mathbf{R}}, \sigma_{\mathbf{R}}) \leq 4$, we get $\mathbf{H} = \mathrm{I}_{\mathbf{R}}(\sigma_{\mathbf{R}}, \sigma_{\mathbf{R}})$. □

(4.8.5) Definition. *The irreducible real representation $\sigma_{\mathbf{R}}$ is said to be of* real *type, of complex type, or of quaternionic type, if $\mathrm{I}_{\mathbf{R}}(\sigma_{\mathbf{R}}, \sigma_{\mathbf{R}})$ is isomorphic to* $\mathbf{R}$, $\mathbf{C}$, *or* $\mathbf{H}$, *respectively.*

The point of Proposition 4.8.4 is of course that every irreducible representation is of one of these types, and that these types correspond to the type of (the irreducible pieces of) the complexification.

Remarks.

(a) Part (ii) of Schur's lemma 4.1.1 said that $\mathrm{I}_{\mathbf{R}}(\sigma_{\mathbf{R}}, \sigma_{\mathbf{R}})$ is a division algebra over $\mathbf{R}$. It is a classical result of Frobenius that **every** finite-dimensional division algebra $\mathcal{A}$ over $\mathbf{R}$ is isomorphic to either $\mathbf{R}$, $\mathbf{C}$, or $\mathbf{H}$.

We now can give the following proof. For each $a \in \mathcal{A}$, write $\mathrm{L}(a)\colon x \mapsto a\,x\colon \mathcal{A} \to \mathcal{A}$. Then $\mathrm{L}\colon a \mapsto \mathrm{L}(a)$ is an isomorphism from $\mathcal{A}$ onto a division subalgebra of $\mathbf{L}(\mathcal{A}, \mathcal{A})$.

Assume that $d := \dim \mathcal{A} > 1$. Let $a \in A$, $a \notin \mathbf{R} \cdot 1$. As in the beginning of Section 5.1, the mapping $\mathrm{ev}_{\mathrm{L}(a)}$ induces an isomorphism from $K := \mathbf{R}[X]/(m_{\mathrm{L}(a)}(X))$ onto a nonzero (Abelian) subalgebra of $\mathrm{L}(\mathcal{A})$, here $m_{\mathrm{L}(a)}(X)$ denotes the minimal polynomial of $\mathrm{L}(a)$. This implies that K is a field; or that $m_{\mathrm{L}(a)}(X)$ is irreducible. Because $a \notin \mathbf{R} \cdot 1$, we have $\deg m_{\mathrm{L}(a)}(X) > 1$, so $m_{\mathrm{L}(a)}(X) = X^2 + \beta X + \gamma$, with $\alpha, \beta \in \mathbf{R}$, and $\Delta := \beta^2 - 4\gamma < 0$.

In order to investigate the dependence of the coefficients β, γ on a, we apply Theorem 5.1.11; and this yields that $\det(X\,\mathrm{I} - \mathrm{L}(a)) = (m_{\mathrm{L}(a)}(X))^\mu$, with $\mu \in \mathbf{Z}_{>0}$, $d = 2\mu$. Using (5.1.14), we readily obtain:

(i) $$\gamma = \frac{1}{2\mu^2}(\operatorname{tr} \mathrm{L}(a))^2 - \frac{1}{2\mu}\operatorname{tr}(\mathrm{L}(a)^2);$$

(ii) $$\gamma^\mu = \det \mathrm{L}(a).$$

The right hand side of (i) defines a quadratic form γ on $\mathcal{A}$, and (ii) holds for all $a \in \mathcal{A}$, by continuity. Moreover, $\gamma(a) > 0$, whenever $a \neq 0$. It follows that $\gamma(a\,b) = \gamma(a)\,\gamma(b)$, for all $a, b \in \mathcal{A}$. In particular, the unit sphere

$$U := \{\, a \in \mathcal{A} \mid \gamma(a) = 1 \,\}$$

in $\mathcal{A}$ is a compact group with respect to the multiplication of $\mathcal{A}$. The representation $a \mapsto \mathrm{L}(a)$ of U in $\mathcal{A}$ is irreducible, because it acts transitively on the unit sphere U. Now the mappings $\mathrm{R}(b)\colon x \mapsto x\,b\colon \mathcal{A} \to \mathcal{A}$ commute with all $\mathrm{R}(a)$; and it follows from Proposition 4.8.4 that the algebra $\mathrm{R}(\mathcal{A})$ is isomorphic to $\mathbf{C}$ or $\mathbf{H}$.

If $\mathcal{A} \simeq \mathbf{H}$, then we end up with an anti-isomorphism: $\mathcal{A} \to \mathbf{H}$. This can be remedied by interchanging L and R in the proof. Or also by observing that the complex adjoint defines an anti-automorphism of $\mathbf{H}$, cf. (1.2.20). And this proves the theorem of Frobenius.

The standard circle action in the plane is an example of an irreducible real representation of complex type, whereas the action in $\mathbf{H}$ of $\mathbf{SU}(2)$, the unit sphere in $\mathbf{H}$, by means of quaternionic multiplications, is of quaternionic type. It is also easy to find finite subgroups of these groups, for which these representations remain irreducible, and are of the corresponding type. For example, in the second case, one can take the group $\{\, \pm 1, \pm i, \pm j, \pm k \,\}$ of 8 elements.

(b) We can apply the preceding theory of real representations to $U_{\mathbf{R}} = \mathrm{C}(G, \mathbf{R})$, the space of real-valued continuous functions on G, with $\sigma_{\mathbf{R}}$ equal to the right (or left) regular representation. The space of elements of finite type is equal to the space of matrix coefficients (of finite-dimensional complex representations), which happen to be real-valued on G; and this space is dense in $\mathrm{C}(G, \mathbf{R})$. However, in general the linear combinations of the characters of the irreducible real representations do not form a dense subspace of the space of all conjugacy-invariant elements in $\mathrm{C}(G, \mathbf{R})$. Orthogonal to that space are the imaginary parts of the characters of the irreducible complex representations of complex type. For instance, for the circle $G = \mathbf{R}/\mathbf{Z}$, all the characters $X \mapsto \mathrm{e}^{2\pi i n X}$, for $n \in \mathbf{Z}\backslash\{0\}$, are of complex type; and the functions $X \mapsto \sin(2\pi n X)$, for $n \in \mathbf{Z} \setminus \{0\}$, span the common orthogonal complement of the characters of the irreducible representations, which here consist of 1, the character of the trivial representation, and the $X \mapsto 2\cos(2\pi n X)$, for $n \in \mathbf{Z} \setminus \{0\}$.

As an intermezzo, we give a straightforward application of the theory of real representations to compact Lie algebras. See Theorem 3.6.2 for various equivalent characterizations of these.

If $\mathfrak{g}$ is a Lie algebra over $\mathbf{R}$, then an *ideal* in $\mathfrak{g}$ is defined as a linear subspace $\mathfrak{h}$ of $\mathfrak{g}$, such that $[\mathfrak{g}, \mathfrak{h}] := \{ [X, Y] \in \mathfrak{g} \mid X \in \mathfrak{g}, Y \in \mathfrak{h} \}$ is contained in $\mathfrak{h}$. If $\mathfrak{g} = \mathfrak{g}_1 \oplus \mathfrak{g}_2$ (as vector spaces), for two ideals $\mathfrak{g}_1$, $\mathfrak{g}_2$ in $\mathfrak{g}$, then $[\mathfrak{g}_1, \mathfrak{g}_2] \subset \mathfrak{g}_1 \cap \mathfrak{g}_2 = 0$ shows that the Lie algebra $\mathfrak{g}$ is isomorphic to $\mathfrak{g}_1 \times \mathfrak{g}_2$. The Lie algebra $\mathfrak{g}$ is called *simple* if this can only happen for $\mathfrak{g}_1 = \mathfrak{g}$ or $\mathfrak{g}_2 = \mathfrak{g}$.

The point is now that the linear subspace $\mathfrak{h}$ of $\mathfrak{g}$ is an ideal in $\mathfrak{g}$ if and only it is invariant under the adjoint group $\operatorname{Ad}\mathfrak{g}$ of $\mathfrak{g}$; and $\mathfrak{g}$ is simple if and only if it is irreducible for $\operatorname{Ad}\mathfrak{g}$. This leads to the following:

(4.8.6) Corollary. *Let $\mathfrak{g}$ be a Lie algebra, such that $\operatorname{Ad}\mathfrak{g}$ is compact. Then $\mathfrak{g}$ is isomorphic to a Cartesian product of simple, compact Lie algebras. If $\mathfrak{g}$ is simple, then $\mathfrak{g}_{\mathbf{C}}$ is simple, and of real type as irreducible complex representation space for $\operatorname{Ad}\mathfrak{g}$. If $\mathfrak{g}$ is simple and not equal to $\mathbf{R}$, then any $\operatorname{Ad}\mathfrak{g}$-invariant real bilinear form on $\mathfrak{g}$ is equal to a scalar multiple of the Killing form κ.*

Proof. The complete reducibility of $\mathfrak{g}$ under $\operatorname{Ad}\mathfrak{g}$ implies the first statement. Now assume that $\mathfrak{g}$ is simple, and not equal to $\mathbf{R}$. The irreducibility implies that $\mathfrak{z} = 0$, and the Killing form κ of $\mathfrak{g}$ is negative definite, cf. Theorem 3.6.2. But then the complex bilinear extension of κ to $\mathfrak{g}_{\mathbf{C}}$ is a nondegenerate, symmetric, complex bilinear form. Let $\mathfrak{g}_{\mathbf{C}} = \sum_j \mathfrak{g}^{(j)}$ be a splitting of $\mathfrak{g}_{\mathbf{C}}$ into irreducible, $\operatorname{Ad}\mathfrak{g}$-invariant, complex linear subspaces $\mathfrak{g}^{(j)}$. Then $[\mathfrak{g}, \mathfrak{g}^{(j)}] \subset \mathfrak{g}^{(j)}$ implies that $[\mathfrak{g}_{\mathbf{C}}, \mathfrak{g}^{(j)}] \subset \mathfrak{g}^{(j)}$; or the Lie algebra $\mathfrak{g}_{\mathbf{C}}$ is isomorphic to the Cartesian product of the $\mathfrak{g}^{(j)}$. In turn this implies that κ is equal to the direct sum of the $\kappa|_{\mathfrak{g}^{(j)}}$, each of which is nondegenerate. We conclude that the $\mathfrak{g}^{(j)}$ are of real type. We are therefore in case (a) in Proposition 4.8.4. So $\mathfrak{g}_{\mathbf{C}}$ is irreducible for $\operatorname{Ad}\mathfrak{g}$, and of real type. The complex bilinear extension to $\mathfrak{g}_{\mathbf{C}}$ of any real $\operatorname{Ad}\mathfrak{g}$-invariant bilinear form on $\mathfrak{g}$ must be equal to a scalar multiple of κ, because of Corollary 4.1.3.(i); and this proves the last statement. □

We conclude this section with an intriguing criterion for the types of the irreducible complex representations of G, which moreover is quite useful in explicit calculations.

(4.8.7) Proposition. *Let π be an irreducible complex representation of the compact group G. Then π is of real, complex, and quaternionic type, if and only if $\int_G \chi_\pi(x^2)\, dx$ is equal to 1, 0, and -1, respectively.*

Proof. Let $\mathcal{S}$, and $\mathcal{A}$ denote the space of symmetric, and antisymmetric, respectively, complex bilinear forms on V. In coordinates with respect to a given basis of V, the forms $(v, w) \mapsto v_j w_k + v_k w_j$, for $j \leq k$, and $(v, w) \mapsto v_j w_k - v_k w_j$, for $j < k$, form a basis of $\mathcal{S}$, and $\mathcal{A}$, respectively. Moreover, if $L \in \mathbf{L}(V, V)$, and the basis consists of eigenvectors for L, with eigenvalues c_j, then these forms are eigenvectors for the induced action of L on $\mathcal{S}$, and $\mathcal{A}$, respectively, with eigenvalues $c_j c_k$. It follows that:

$$\operatorname{tr} L|_{\mathcal{S}} - \operatorname{tr} L|_{\mathcal{A}} = \sum_{j\leq k} c_j c_k - \sum_{j<k} c_j c_k = \sum_j c_j^2 = \operatorname{tr} L^2.$$

Because the set of diagonalizable elements is dense in $\mathbf{L}(V, V)$, cf. Hirsch-Smale [1974], Theorem 7.3.2, the formula $\operatorname{tr} L|_{\mathcal{S}} - \operatorname{tr} L|_{\mathcal{A}} = \operatorname{tr} L^2$ is valid, for all $L \in \mathbf{L}(V, V)$, by continuity.

Now

$$\int_G \chi_\pi(x^2)\, dx = \int_G \operatorname{tr} \pi(x^2)\, dx = \int_G \operatorname{tr} \pi(x)^2\, dx$$
$$= \int_G \left(\operatorname{tr} \pi(x)|_{\mathcal{S}} - \operatorname{tr} \pi(x)|_{\mathcal{A}}\right) dx = \operatorname{tr} \int_G \pi(x)|_{\mathcal{S}}\, dx - \operatorname{tr} \int_G \pi(x)|_{\mathcal{A}}\, dx.$$

According to the averaging principle 4.2.1, $\int_G \pi(x)|_{\mathcal{S}}\, dx$, and $\int_G \pi(x)|_{\mathcal{A}}\, dx$, is equal to a linear projection from $\mathcal{S}$, and $\mathcal{A}$, onto the subpace of invariant elements for the action of $\pi(G)$ on $\mathcal{S}$, and $\mathcal{A}$, respectively, and the traces therefore are equal to the dimensions of these subspaces. According to Theorem 4.8.1, the dimensions of the spaces of invariant symmetric, and antisymmetric, complex bilinear forms on V are equal to 1, and 0, if π is of real type, equal to 0, and 0, if π is of complex type, and equal to 0, and 1, if π is of quaternionic type, respectively. □

4.9 Weyl's Character Formula

In this section, G is a connected, compact Lie group.

Fix a maximal torus T in G, with Lie algebra equal to the maximal Abelian subspace $\mathfrak{t}$ of $\mathfrak{g}$. Every element of G is conjugate to an element of T, cf. Theorem 3.7.1. Because characters χ_π of irreducible representations π of G are conjugacy-invariant functions on G, see the remark following (4.3.3), χ_π is determined by its restriction to T.

For the determination of $\chi_\pi|_T$, we start with the observation that $\pi|_T : t \mapsto \pi(t) : T \to \mathbf{GL}(V)$ is a finite-dimensional representation of T, which is a direct sum of finitely many one-dimensional representations, with characters $.^\mu : t \mapsto t^\mu$. Here μ is a linear form: $\mathfrak{t} \to i\mathbf{R}$, such that $\mu(\Lambda) \subset 2\pi i\mathbf{Z}$, where Λ denotes the lattice $\ker \exp|_{\mathfrak{t}}$ in $\mathfrak{t}$, see Corollary 4.2.2 and the example following Corollary 4.1.2. By a slight abuse of notation, we will use the letter μ both for such a linear form and for (the equivalence class of) the corresponding irreducible representation of T, also called a *weight of* T. The set of these μ will accordingly be denoted by $\widehat{T}$. So we have:

$$\chi_\pi(t) = \sum_{\mu \in \widehat{T}} m_\mu t^\mu, \quad \text{for } t \in T, \tag{4.9.1}$$

where the sum is only over finitely many of the $\mu \in \widehat{T}$, and

$$m_\mu := [\pi|_T : \mu] \tag{4.9.2}$$

is a positive integer for each of the occurring $\mu \in \widehat{T}$, the *multiplicity* of μ in π.

The next observation is that $\mathrm{N}(T)$, the normalizer of T in G, acts on T by conjugation; so the function $\chi_\pi|_T$ on T is invariant under the action on T of the Weyl group $W = \mathrm{N}(T)/T$. Now $s^*(.^\mu) = .^{s^*(\mu)}$, where on the right hand side we have used the action of $s \in W$ on $\mathfrak{t}^*$. The identity $\chi_\pi|_T = s^*(\chi_\pi|_T)$ therefore takes the form:

$$\sum_{\mu\in\widehat{T}} m_\mu \chi_\mu = \sum_{\mu\in\widehat{T}} m_\mu \chi_{s^*(\mu)} = \sum_{\mu\in\widehat{T}} m_{s\cdot\mu}\chi_\mu,$$

where we have written $s \cdot \mu = s^{*\,-1}(\mu)$ for the natural action of s on the weights of T. Because the χ_μ, for $\mu \in \widehat{T}$, are linearly independent (they form an L^2-orthonormal system of functions on T, as can be verified directly, or by reference to Theorem 4.3.4.(iv)), we conclude that:

$$m_{s\cdot\mu} = m_\mu, \quad \text{for all } s \in W, \mu \in \widehat{T}. \tag{4.9.3}$$

The third and last fact about χ_π that will be used is that the irreducibility of π implies:

$$\int_G \chi_\pi(x)\overline{\chi_\pi(x)}\, dx = 1, \tag{4.9.4}$$

cf. Theorem 4.3.4.(iv). Applying to the left hand side Weyl's integral formula (3.14.8) for conjugacy-invariant functions, we get:

$$\int_T \delta(t)\overline{\delta(t)}\, \chi_\pi(t)\overline{\chi_\pi(t)}\, dt = \#(W); \tag{4.9.5}$$

or $< \phi, \phi > = \#(W)$, if we write:

$$\phi := \delta(\chi_\pi|_T). \tag{4.9.6}$$

Here

$$\delta(t) = \delta_P(t) = \prod_{\alpha\in P}(1 - t^{-\alpha}), \tag{4.9.7}$$

for a choice P of positive roots, which we assume fixed from now on. Note that for each root α, the mapping $t \mapsto t^\alpha$ is a character of T, namely the one of the one-dimensional representation $t \mapsto (\mathrm{Ad}\, t)|_{\mathfrak{g}_\alpha}$. In other words, each root is a weight of T. It is also clear that the weights form an additive subgroup of $i\mathfrak{t}^*$. Combining this with (4.9.1), we get:

(4.9.8)
$$\phi(t) = \sum_{\mu \in \widehat{T}} c_\mu t^\mu;$$

here again the sum is finite, and $c_\mu \in \mathbf{Z}$, for all $\mu \in \widehat{T}$, although this time we expect also negative integers. Using that the characters of T form an orthonormal system, we see that (4.9.5) now takes the form:

(4.9.9)
$$\sum_{\mu \in \widehat{T}} {c_\mu}^2 = \#(W).$$

Because the ${c_\mu}^2$ are nonnegative integers, not more than $\#(W)$ of them can be nonzero.

In order to understand the consequences of the Weyl group invariance of $\chi_\pi|_T$ for the coefficients c_μ, we need to investigate how δ_P behaves under the action of $s \in W$.

If s is given by conjugation with $x \in \mathrm{N}(T)$, then, by definition,

$$(s \cdot \delta_P)(t) = \delta_P(s^{-1}(t)) = \delta_P(x^{-1}tx).$$

Now $\mathrm{Ad}(x^{-1}tx) = \mathrm{Ad}\, x^{-1} \circ \mathrm{Ad}\, t \circ \mathrm{Ad}\, x$ acts on $\mathfrak{g}_\alpha$ as multiplication by the same scalar as $\mathrm{Ad}\, t$ does on $(\mathrm{Ad}\, x)(\mathfrak{g}_\alpha) = \mathfrak{g}_{s \cdot \alpha}$, which is equal to $t^{s \cdot \alpha}$. So $s \cdot \delta_P = \delta_{s \cdot P}$. Here we have used the following general fact. If $\Phi \in \mathrm{Aut}\, \mathfrak{g}$, and $\Phi(\mathfrak{t}) = \mathfrak{t}$, then we have, for any $X \in \mathfrak{t}$, $Y \in \mathfrak{g}_\alpha$, that $[X, \Phi(Y)] = \Phi([\Phi^{-1}(X), Y]) = \Phi(\alpha(\Phi^{-1}(X))Y) = \alpha(\Phi^{-1}(X))\Phi(Y)$. Or, with the usual notation $(\Phi^{-1})^*(\alpha) = \Phi\alpha$,

$$\Phi(\mathfrak{g}_\alpha) = \mathfrak{g}_{\Phi\alpha}.$$

Because $s \cdot R = R$, and R is equal to the disjoint union of P and $-P$, cf. (3.5.10), we get that $s \cdot P$ is equal to the disjoint union of $P \cap s \cdot P$ and $-(P \setminus s \cdot P)$. It follows that:

$$\delta_{s \cdot P}(t) = \delta_P(t) \prod_{\alpha \in P \setminus s \cdot P} \frac{1 - t^\alpha}{1 - t^{-\alpha}}.$$

Now

(4.9.10)
$$1 - t^\alpha = -t^\alpha\,(1 - t^{-\alpha}), \quad \text{for } \alpha \in R,$$

so

$$\delta_{s \cdot P}(t) = \delta_P(t)\,(-1)^{\#(P \setminus s \cdot P)}\, t^{\Sigma(s)},$$

where we have written $\Sigma(s) = \sum_{\alpha \in P \setminus s \cdot P} \alpha$.

In Corollary 3.10.3, we have seen that s^{-1} can be written as a composition of $\#(P \setminus s \cdot P)$ many reflections in root hyperplanes. Because each of these reflections has determinant equal to -1, it follows that $(-1)^{\#(P \setminus s \cdot P)} = \det s^{-1} = \det s$.

A neat expression for $\Sigma(s)$ can be obtained by introducing:

(4.9.11)
$$\rho := \frac{1}{2}\sum_{\alpha\in P}\alpha;$$

the so-called *half the sum of the positive roots*. The idea is that

$$s\cdot\rho = \frac{1}{2}\sum_{\alpha\in P} s\cdot\alpha = \frac{1}{2}\sum_{\beta\in s\cdot P}\beta = \frac{1}{2}\Big(\sum_{\alpha\in P\cap s\cdot P}\alpha - \sum_{\alpha\in P\setminus s\cdot P}\alpha\Big) = \rho - \Sigma(s),$$

or $\Sigma(s) = \rho - s\cdot\rho$. Our desired transformation formula for δ_P now takes the form:

(4.9.12)
$$(s\cdot\delta_P)(t) = \det s\; t^{\rho - s\cdot\rho}\,\delta_P(t), \quad \text{for } s\in W.$$

Warning. The example of $\mathbf{SO}(3,\mathbf{R})$ (in Section 3.10) shows that the linear form ρ is not always a weight of T. However, the $\rho - s\cdot\rho$, for $s\in W$, always are weights of T, and we shall only use these if we consider functions on T.

The Weyl group invariance of $\chi_\pi|_T$ implies that (4.9.12) holds with δ replaced by ϕ, that is,

$$\sum_{\theta\in\widehat{T}} c_\theta t^{s^*(\theta)} = s^*(\phi) = (-1)^{l(s)}\,t^{\rho - s^*(\rho)}\phi = \sum_{\mu\in\widehat{T}} c_\mu t^{\mu+\rho-s^*(\rho)}.$$

Substituting $s^*(\theta) = \mu + \rho - s^*(\rho)$, or $\theta = s\cdot\mu + s\cdot\rho - \rho$, and using the linear independence of the $.^\mu$, for $\mu\in\widehat{T}$, we get:

(4.9.13)
$$c_{s\cdot\mu+s\cdot\rho-\rho} = \det s\; c_\mu, \quad \text{for } \mu\in\widehat{T}, s\in W.$$

In other words, the coefficients c_μ behave in an antisymmetric way under the *shifted action*

$$(s,\mu)\mapsto s\cdot\mu + (s\cdot\rho - \rho)$$

of the Weyl group W on the lattice $\widehat{T}$ of weights of T.

A weight μ of T is called a *weight* of π, if $m_\mu \neq 0$. Let us, temporarily, say that $\lambda\in\widehat{T}$ is a *highest weight* of π, if it is a weight of π , and if $\mu\in\widehat{T}$ is not a weight of π, whenever $\mu = \lambda + \Sigma(Q)$, with Q a nonvoid subset of P. Here we have written $\Sigma(Q) = \sum_{\alpha\in Q}\alpha$. Later we will give equivalent characterizations which agree more with the conventional definition of highest weights, see Proposition 4.9.5 and Section 4.11.

Because the set of weights of π is finite, and the convex cone in it^* generated by P is proper, cf. (3.5.13), there exists at least one highest weight λ. Otherwise we could continue adding nonzero sums of positive roots to weights of π, leading to infinitely many different weights of π.

Working out the product in (4.9.7), we get:

$$\delta_P(t) = \sum_{Q\subset P}(-1)^{\#(Q)}\,t^{-\Sigma(Q)}, \quad \text{or: } \phi(t) = \sum_{\mu,Q}(-1)^{\#(Q)}\,m_\mu\,t^{\mu-\Sigma(Q)},$$

or, comparing coefficients,

$$c_\theta = \sum_{\mu, Q} (-1)^{\#(Q)} \, m_{\theta + \Sigma(Q)}.$$

Again using that the convex cone generated by P is proper, the only possibility that $\Sigma(Q) = 0$ is that Q is void. This means that $c_\lambda = m_\lambda$, if λ is a highest weight.

Furthermore, if $s \in W$, and $s \cdot \lambda + s \cdot \rho - \rho = \lambda$, then the fact that $m_{s \cdot \lambda} = m_\lambda > 0$, cf. (4.9.3), implies that $\rho - s \cdot \rho = \Sigma(P \setminus s \cdot P) = 0$. That is, $s \cdot P = P$, or $s = 1$, because the Weyl group acts freely on the set of Weyl chambers, cf. Proposition 3.8.1. It follows that the shifted action of W is free on λ; or the orbit of λ has $\#(W)$ many elements. Applying (4.9.9) and the sentence following it, combined with (4.9.13), we find that ${c_\lambda}^2 = 1$, or $c_\lambda = m_\lambda = 1$, and $c_\theta = 0$, if θ is not in the orbit of λ for the shifted Weyl group action. We have proved:

(4.9.1) Theorem [Weyl's character formula]. *For each irreducible representation π of G, there is exactly one highest weight λ, which has multiplicity equal to 1. At $t \in T \cap G^{\mathrm{reg}}$, the character of π is given by the formula:*

(4.9.14)
$$\chi_\pi(t) = \frac{\sum_{s \in W} \det s \; t^{s \cdot \lambda + s \cdot \rho - \rho}}{\prod_{\alpha \in P} (1 - t^{-\alpha})}.$$

Remark. If π is the trivial representation, then $\chi_\pi = 1$, $\lambda = 0$, so (4.9.14) then yields:

(4.9.15)
$$\delta_P(t) := \prod_{\alpha \in P} (1 - t^{-\alpha}) = \sum_{s \in W} \det s \; t^{s \cdot \rho - \rho}, \quad \text{for } t \in T.$$

If we substitute $t = \exp X$, for $X \in \mathfrak{t}$, and multiply the numerator and the denominator in (4.9.14) both by $e^{\rho(X)}$, using the formula (4.9.15) for the denominator, then we get:

(4.9.16)
$$\chi_\pi(\exp X) = \frac{\Phi_{\lambda + \rho}(X)}{\Phi_\rho(X)}, \quad \text{for } X \in \exp^{-1}(G^{\mathrm{reg}}) \cap \mathfrak{t},$$

where we have written, for any $\theta \in \mathfrak{t}^*_{\mathbf{C}}$,

(4.9.16)
$$\Phi_\theta(X) = \sum_{s \in W} \det s \; e^{s \cdot \theta(X)}, \quad \text{for } X \in \mathfrak{t}.$$

The advantage of this form is that the functions Φ_θ are obviously W-anti-invariant on $\mathfrak{t}$, the slight disadvantage is that the numerator and denominator in (4.9.16) only are the pull back (under the exponential mapping) of single-valued functions on T if ρ is a weight of T.

(4.9.2) Theorem [Weyl's dimension formula]. *If the irreducible representation π of G has highest weight λ, then its dimension is given by:*

$$\text{(4.9.18)} \qquad \mathrm{d}_\pi = \chi_\pi(1) = \sum_{\mu \in \widehat{T}} m_\mu = \prod_{\alpha \in P} \frac{<\alpha, \lambda + \rho>}{<\alpha, \rho>} = \prod_{\alpha \in P} \frac{(\lambda+\rho)(\alpha^\vee)}{\rho(\alpha^\vee)},$$

where $<.,.>$ denotes any W-invariant inner product in $i\mathfrak{t}^*$. If $\mathfrak{z} = 0$, then $\{\,[\pi] \in \widehat{G} \mid \mathrm{d}_\pi \leq C\,\}$ is a finite subset of $\widehat{G}$, for each $C > 0$.

Proof. The inner product defines a real linear mapping $A\colon i\mathfrak{t}^* \to \mathfrak{t}$, such that $\mu(A(\theta)) = i <\mu, \theta>$, for all $\mu, \theta \in i\mathfrak{t}^*$. The W-invariance of the inner product implies that $s_\alpha \theta = \theta - 2\frac{<\theta,\alpha>}{<\alpha,\alpha>}\alpha$, or:

$$\text{(4.9.19)} \qquad \theta(\alpha^\vee) = 2\frac{<\theta, \alpha>}{<\alpha, \alpha>}, \quad \text{for } \theta \in i\mathfrak{t}^*, \alpha \in R.$$

Because $s_\alpha P = (P \setminus \{\alpha\}) \cup \{-\alpha\}$, see Lemma 5.5.10 below, $\rho - \alpha = s_\alpha \rho = \rho - \rho(\alpha^\vee)\alpha$, so:

$$\text{(4.9.20)} \qquad \rho(\alpha^\vee) = 1, \quad \text{for all simple roots } \alpha.$$

This in turn implies that:

$$\text{(4.9.21)} \qquad \rho(\alpha^\vee) > 0, \quad \text{for all } \alpha \in P,$$

because every positive root is a linear combination of simple roots with non-negative coefficients, cf. (3.8.6); the coefficients are actually integral and uniquely determined (cf. Theorem 5.5.9.(iv) below). In other words, ρ lies in the positive (dual) Weyl chamber.

For $t \in \mathbf{R}$, with $t \neq 0$ and t sufficiently small, $tA(\rho)$ now belongs to $\exp^{-1}(G^{\mathrm{reg}})$, so we may apply (4.9.16) with $X = tA(\rho)$; the idea is to evaluate the limit for $t \to 0$. Because $(s \cdot \mu)(tA(\rho)) = it <s \cdot \mu, \rho> = it <s^{-1}\rho, \mu> = (s^{-1}\rho)(tA(\mu))$, we get, using also $\det s = \det s^{-1}$, that $\Phi_\mu(tA(\rho)) = \Phi_\rho(tA(\mu))$. Now:

$$\text{(4.9.22)} \qquad \Phi_\rho(X) = \mathrm{e}^{\rho(X)}\,\delta_P(\exp X) = \prod_{\alpha \in P} \left(\mathrm{e}^{\alpha(X)/2} - \mathrm{e}^{-\alpha(X)/2}\right),$$

and the trick above shows that $\Phi_\mu(tA(\rho))$ gets a similar factorization. Therefore:

$$\chi_\pi(\exp tA(\rho)) = \frac{\Phi_{\lambda+\rho}(tA(\rho))}{\Phi_\rho(tA(\rho))} = \prod_{\alpha \in P} \frac{\sin \frac{1}{2}t <\alpha, \lambda+\rho>}{\sin \frac{1}{2}t <\alpha, \rho>}.$$

Taking the limit for $t \to 0$, we obtain the desired identities; for the last one, use (4.9.19).

For the last conclusion, we borrow from Proposition 4.9.4.(ii) below, that $\lambda(\alpha^\vee) \geq 0$, for all $\alpha \in P$. It follows that, for each $\beta \in P \setminus \{\alpha\}$, we have $(\lambda + \rho)(\beta^\vee)/\rho(\beta^\vee) \geq 1$, hence:

$$\mathrm{d}_\pi \geq \frac{(\lambda + \rho)(\alpha^\vee)}{\rho(\alpha^\vee)} = \lambda(\alpha^\vee) + 1, \quad \text{for each } \alpha \in S.$$

However, if $\mathfrak{z} = 0$, then R, hence S, spans $i\mathfrak{t}^*$ (the common kernel in $\mathfrak{t}$ of the roots is equal to $\mathfrak{z}$), and it follows that λ remains in a bounded subset of $i\mathfrak{t}^*$, if we keep $\mathrm{d}_\pi \leq C$. The discreteness of $\widehat{T}$ now yields that there remain only finitely many possibilities for the λ. Because $[\pi]$ is uniquely determined by χ_π, cf. Corollary 4.3.5.(iii), which in turn is uniquely determined by λ, the proof is complete. □

Remark. The last statement in 4.9.2 can be strengthened to a uniform estimate of the form:

$$c(|\lambda| + 1)^{\#(P) - \#(S) + 1} \leq m_\pi \leq C(|\lambda| + 1)^{\#(P)}. \tag{4.9.23}$$

Here c, C are positive constants, and $|\,.\,|$ denotes any norm in $i\mathfrak{t}^*$. The multiplicities grow as the left hand side if λ remains at a bounded distance of the intersection of $\#(S) - 1$ many coroot hyperplanes (the one-dimensional faces of the dual Weyl chamber, cf. the remarks following Theorem 3.8.3). On the other hand the multiplicities grow as the right hand side if the distance to each wall grows proportionally to $|\lambda|$.

(4.9.3) Lemma. *Suppose that there is only one positive root α for the group G. If π is an irreducible representation of G with highest weight λ, then $l := \lambda(\alpha^\vee) \in \mathbf{Z}_{\geq 0}$, and:*

$$\chi_\pi(t) = \sum_{k=0}^{l} t^{\lambda - k\alpha}.$$

Proof. We have $W = \{1, s\}$, with $s \cdot \mu = \mu - \mu(\alpha^\vee)\alpha$, for $\mu \in \widehat{T}$. Formula (4.9.14) now takes the form:

$$t^\lambda - t^{s \cdot \lambda - \alpha} = (1 - t^{-\alpha})\chi_\pi(t) = (1 - t^{-\alpha}) \sum_{\mu \in \widehat{T}} m_\mu \, t^\mu = \sum_{\mu \in \widehat{T}} (m_\mu - m_{\mu+\alpha}) \, t^\mu,$$

or $m_{\mu+\alpha} = m_\mu$, if $\mu \neq \lambda$, $\mu \neq s \cdot \lambda - \alpha$. Equivalently, $m_\mu = m_{\mu-\alpha}$, if $\mu \neq \lambda + \alpha$, $\mu \neq s \cdot \lambda$.

Let $m_\mu \neq 0$, and let N, and M, be the largest $k \in \mathbf{Z}_{\geq 0}$, such that $m_{\mu+k\alpha} \neq 0$, and $m_{\mu-k\alpha} \neq 0$, respectively. These numbers exist, because there are only finitely many weights of T with positive multiplicity. Then $\mu + N\alpha = s \cdot \lambda - \alpha$ cannot happen, because then the multiplicity of $\mu + (N+1)\alpha = s \cdot \lambda$ would be equal to 1. It follows that $\mu + N\alpha = \lambda$. Similarly $\mu - M\alpha = \lambda + \alpha$ cannot happen, because then $\mu - (M+1)\alpha = \lambda$ has positive multiplicity; so $\mu - M\alpha = s \cdot \lambda = \lambda - \lambda(\alpha^\vee)\alpha$. Combining, we get that

$$\lambda(\alpha^\vee) = N + M \in \mathbf{Z}_{\geq 0}.$$

Furthermore, $m_\theta = 0$, unless $\theta = s \cdot \lambda + k\alpha$, for some $k \in \mathbf{Z}$, $0 \le k \le \lambda(\alpha^\vee)$, in which case $m_\theta = 1$. □

Returning to our general connected, compact Lie group G, let us begin by observing that for any root α, we have $\exp 2\pi i\alpha^\vee = 1$, cf. Theorem 3.10.2.(iv). Hence, if $\mu \in \widehat{T}$, we have $1 = (\exp 2\pi i\alpha^\vee)^\mu = e^{\mu(2\pi i\alpha^\vee)} = e^{2\pi i\mu(\alpha^\vee)}$, or:

$$\mu(\alpha^\vee) \in \mathbf{Z}, \quad \text{for every } \mu \in \widehat{T}, \alpha \in R. \tag{4.9.24}$$

If $\mu \in \widehat{T}$, and $\alpha \in R$, then the *α-ladder from μ to $s_\alpha\mu$* is defined as the set:

$$\{\, \mu - k\alpha \mid k \in \mathbf{Z}, 0 \le k \le \mu(\alpha^\vee)\,\},$$

if $\mu(\alpha^\vee) \ge 0$; and as the same set with α replaced by $-\alpha$, if $\mu(\alpha^\vee) < 0$. From Lemma 4.9.3, we now get:

(4.9.4) Proposition.

(i) Let π be a finite-dimensional representation of G, and let μ be a weight of π. Then, for any root α, the α-ladder from μ to $s_\alpha\mu$ consists of weights of π.

(ii) If π is irreducible with highest weight λ, then $\lambda(\alpha^\vee) \ge 0$, for all positive roots α.

(iii) For a weight λ of π, the following conditions (a)–(c) are equivalent:
  - *(a) λ is the highest weight of π.*
  - *(b) If $\alpha \in P$, then $\lambda + \alpha$ is not a weight of π.*
  - *(c) For any weight μ of π, we have $\mu = \lambda - \sum_{\alpha\in P} n_\alpha\alpha$, for some $n_\alpha \in \mathbf{Z}_{\ge 0}$.*

Proof. (i) Let H be the Lie subgroup of G with Lie algebra equal to $\mathfrak{h} = \mathfrak{t} + (\mathfrak{g}_\alpha + \mathfrak{g}_{-\alpha}) \cap \mathfrak{g}$. As we have seen in the proof of Theorem 3.10.2.(i), H is compact, and T is a maximal torus of H. Also, α is the only positive root in the root system of H.

Decompose the representation space V of π into irreducible subspaces V_j for the representation $\pi|_H$, cf. Corollary 4.2.2. If μ is a weight of π, then μ is a weight of the irreducible representation $\sigma\colon x \mapsto \pi(x)|_{V_j}$, with $x \in H$, for some j. Let θ be the highest weight of σ. Applying Lemma 4.9.3 with π, and G, replaced by σ, and H, respectively, we get that the $\theta - k\alpha$, for $k \in \mathbf{Z}$, $0 \le k \le \theta(\alpha^\vee)$, are the weights of σ. So these are weights of π, and μ is one of them. In particular, if $\mu(\alpha^\vee) \ge 0$, then all $\mu - k\alpha$, for $k \in \mathbf{Z}$, $0 \le k \le \mu(\alpha^\vee)$, are weights of π.

(ii) Continuing the discussion above, we get that $\mu + \alpha$ is a weight of π, unless $\mu = \theta$. So if π is irreducible and μ is the highest weight of π, then $\mu = \theta$, which implies that $\mu(\alpha^\vee) \ge 0$.

(iii) (a) $\Rightarrow$ (c). From Weyl's character formula, (4.9.14), we read off that every weight of π is of the form:

$$\mu = \lambda + (s \cdot \lambda - \lambda) + (s \cdot \rho - \rho) - \sum_{\alpha \in P} k_\alpha \alpha, \quad \text{for some } s \in W, k_\alpha \in \mathbf{Z}_{\geq 0}.$$

Now write:

$$s = s_{\alpha_k} \cdots s_{\alpha_2} \cdot s_{\alpha_1},$$

as in the proof of Corollary 3.10.3. Iterating $s_\alpha(\lambda - \mu) = \lambda - \lambda(\alpha^\vee)\alpha - s_\alpha(\mu)$, we get:

$$s \cdot \lambda = \lambda - \lambda(\alpha_k^\vee)\alpha_k - \lambda(\alpha_{k-1}^\vee) s_{\alpha_k} \alpha_{k-1} - \ldots - \lambda(\alpha_1^\vee) s_{\alpha_k} \cdots s_{\alpha_2} \alpha_1.$$

Here the coefficients $\lambda(\alpha_j^\vee)$ all are nonnegative integers. In view of Proposition 5.5.11, we get that $\lambda - s \cdot \lambda$ is a linear combination of positive roots with nonnegative integers as coefficients. The same is true for $\rho - s \cdot \rho$, and this completes the proof of (c).

(c) $\Rightarrow$ (b) follows from the properness of the convex cone in $i\mathfrak{t}^*$, generated by the positive roots, cf. (3.5.13).

(b) $\Rightarrow$ (a). If $\lambda(\alpha^\vee) < 0$, then the α-ladder from λ to $s_\alpha \cdot \lambda$ contains $\lambda + \alpha$, so if (b) holds, then $\lambda(\alpha^\vee) \geq 0$, for all $\alpha \in P$. Now let $\mu = \lambda + \sum_{\alpha \in P} n_\alpha \alpha$ be a weight of π, with $n_\alpha \in \mathbf{Z}_{\geq 0}$, not all equal to 0. If $n_\alpha > 0$, $\mu(\alpha^\vee) > 0$, then $\mu - \alpha$ belongs to the α-ladder between μ and $s_\alpha \cdot \mu$, so it is a weight of μ, and we may decrease n_α by 1. We cannot end up with λ, because of (b). So we may keep decreasing some of the positive n_α, until we end up with the situation that $\mu(\alpha^\vee) \leq 0$, if $\alpha \in P$, $n_\alpha > 0$. From (4.9.19), we see that $\theta(\alpha^\vee)$ has the same sign as $< \theta, \alpha >$, for each $\theta \in i\mathfrak{t}^*$, $\alpha \in P$. Now

$$\begin{aligned} 0 << \sum_{\alpha \in P} n_\alpha \alpha, \sum_{\alpha \in P} n_\alpha \alpha > &= < \mu - \lambda, \sum_{\alpha \in P} n_\alpha \alpha > \\ &= \sum_{\alpha \in P} n_\alpha < \mu, \alpha > - \sum_{\alpha \in P} n_\alpha < \lambda, \alpha > \leq 0 \end{aligned}$$

leads to a contradiction. □

Remark. Because each irreducible summand of $\pi|_H$ yields its own α-ladder, and the α-ladders are either disjoint (and parallel), or stacked on top of each other with their middles at the same point in $i\mathfrak{t}^*$, it follows that the multiplicities along an α-ladder form a symmetric function under the reflection s_α, which moreover is monotonously nondecreasing towards the middle point.

In $\widehat{T}$, one introduces a *partial ordering* $\preceq$ by writing $\mu \preceq \theta$ if and only if $\mu = \theta - \sum_{\alpha \in P} n_\alpha \alpha$, for some $n_\alpha \in \mathbf{Z}_{\geq 0}$. The property that $\mu \preceq \theta$ and $\theta \preceq \mu$ can only happen if $\mu = \theta$; this follows from the fact that the convex cone generated by P is proper, cf. (3.5.13). The customary definition is to call a weight λ of T a *highest weight* of π, if it is a maximal element of the set of weights of π, with respect to the partial order $\preceq$; this is just condition (c) in Proposition 4.9.4.(iii).

A weight μ of T is called *dominant* if $\mu(\alpha^\vee) \geq 0$, for all $\alpha \in P$. That is, if μ belongs to the closure of the positive Weyl chamber $\{\, \mu \in it^* \mid \mu(\alpha^\vee) > 0$, for all $\alpha \in P \}$ in it^*, with respect to the dual root system of the $\alpha^\vee$, $\alpha \in R$, cf. the discussion following Corollary 3.10.3.

In the proof of Proposition 4.9.4.(iii), we have seen that if λ is dominant, then $s \cdot \lambda \preceq \lambda$, for all $s \in W$, whereas conversely the condition that $s_\alpha \cdot \lambda \preceq \lambda$, for all $\alpha \in S$, already implies that λ is dominant. This is the origin of the name "dominant".

Remark. If $\mathfrak{z} = 0$, then $\mu \succeq 0$, for all dominant μ. Indeed, $\mu = \sum_{\alpha \in S} c_\alpha \alpha$, for some $c_\alpha \in \mathbf{R}$. Now apply Lemma 3.10.4.(iii).

One should be warned however, that in general the condition that μ is dominant is not equivalent to the condition that $\mu \succeq 0$. For instance, for $G = \mathbf{SU}(3)$, we have $\alpha(\beta^\vee) = -1$, for the two simple roots α, β, so none of them belongs to the dominant chamber.

The following theorem contains a converse to Proposition 4.9.4.(ii).

(4.9.5) Theorem. *The mapping that assigns to each irreducible representation of G its highest weight, induces a bijection from $\widehat{G}$ onto the set of dominant weights of T.*

Proof. The injectivity of the induced mapping means that if π and π' are irreducible representations of G with the same highest weight λ, then π' is equivalent to π. This follows because (4.9.14) then implies that $\chi_\pi = \chi_{\pi'}$ on $T \cap G^{\text{reg}}$. The conjugacy-invariance of both characters now yields that $\chi_\pi = \chi_{\pi'}$ on G^{reg}, cf. Theorem 3.7.2. But G^{reg} is dense in G and characters of finite-dimensional representations of G are continuous, so $\chi_\pi = \chi_{\pi'}$, and therefore $\pi \sim \pi'$, cf. Corollary 4.3.5.(iii).

We now prove the surjectivity. Let $\lambda \in \widehat{T}$ be dominant. Define the function f_λ on $T \cap G^{\text{reg}}$ by the right hand side of (4.9.14). This function is real-analytic on its domain of definition. For the numerator ϕ we get:

$$s_\alpha^* \phi = \sum_{s \in W} \det s \; t^{s_\alpha \cdot s \cdot \lambda + s_\alpha \cdot s \cdot \rho - \rho + \alpha} = -t^\alpha \phi,$$

for any $\alpha \in P$. Because this transformation rule holds for both the numerator and the denominator of f_λ, it follows that $s_\alpha^* f_\lambda = f_\lambda$. Because this is valid for all $\alpha \in P$, the conclusion is that f_λ is a Weyl group invariant function on T. In view of the Weyl covering: $G/T \times (T \cap G^{\text{reg}}) \to G^{\text{reg}}$, cf. Theorem 3.7.2, the function f_λ has a unique extension to a conjugacy-invariant function f_λ on G^{reg}, which is real-analytic there.

It follows from (4.9.23) that ρ belongs to the positive Weyl chamber in it^*, so $\lambda + \rho$ belongs to the positive chamber as well. If now $s \in W$ and $s \cdot \lambda + s \cdot \rho - \rho = \lambda$, then $s \cdot (\lambda + \rho) = \lambda + \rho$, and therefore $s = 1$, because

W acts freely on the set of Weyl chambers, cf. Proposition 3.8.1. That is, the shifted Weyl group action is free on λ.

This means that the $s \cdot \lambda + s \cdot \rho - \rho$, for $s \in W$, are all distinct. Using the Weyl integration formula, cf. Corollary 3.14.2, we get:

$$\begin{aligned} < f_\lambda, f_\lambda > &= \#(W)^{-1} < \delta_P f_\lambda|_T, \delta_P f_\lambda|_T > \\ &= \#(W)^{-1} < \sum_{s \in W} (\det s)\, t^{s\cdot\lambda + s\cdot\rho - \rho}, \sum_{s \in W} (\det s)\, t^{s\cdot\lambda + s\cdot\rho-\rho} > = 1, \end{aligned}$$

using the orthonormality of the characters on T. In particular, f_λ defines an element of $\mathrm{L}^2(G)$, nonzero and conjugacy-invariant.

The same argument as above shows that if λ, λ' are dominant weights, and $s \in W$, then the equality $s \cdot \lambda + s \cdot \rho - \rho = \lambda'$ can only occur for $s = 1$; and then also $\lambda = \lambda'$. That is, if λ, λ' are distinct dominant weights, then their shifted Weyl group orbits are disjoint, and we get $< f_\lambda, f_{\lambda'} > = 0$. In particular, if the dominant weight λ is not equal to the highest weight of an irreducible representation of G, then $< f_\lambda, \chi_\pi > = 0$, for all $[\pi] \in \widehat{G}$. Applying (4.6.3), a direct consequence of the Peter-Weyl theorem, we would obtain that $f_\lambda = 0$, a contradiction. □

With a slight abuse of notation, we shall write $[\pi] = [\pi(\lambda)]$, and $\chi_\pi = \chi_\lambda$, if λ is the highest weight of π. That is, χ_λ is the conjugacy-invariant function on G, whose restriction to T is given by the right hand side in (4.9.14), the function, which was called f_λ in the proof of Theorem 4.9.5.

Remark. Using an $\mathrm{Ad}\, G = \mathrm{Ad}\, \mathfrak{g}$-invariant inner product on $\mathfrak{g}$, we get an identification of $\mathfrak{g}$ with $i\mathfrak{g}^* \subset \mathfrak{g}_C^*$, which intertwines the adjoint action with the *coadjoint* (that is, the contragredient of the adjoint) action on $\mathfrak{g}_C^*$. Under this identification, $\mathfrak{t}$ is mapped to a linear subspace of $i\mathfrak{g}^*$, which can be identified with $i\mathfrak{t}^*$. According to Theorem 3.8.3.(iii), the adjoint orbits have a unique intersection point with the closure of the Weyl chamber; this implies that each coadjoint orbit in $i\mathfrak{g}^*$ has a unique intersection point λ with the dominant chamber in $i\mathfrak{t}^*$. In this way, one may think of the equivalence classes of irreducible representations of G as being parametrized by certain coadjoint orbits, a picture that is less dependent on choices of maximal Abelian subspaces and Weyl chambers. In the Borel-Weil theorem 4.12.5,7, representations are constructed in each equivalence class, in terms of geometric data related to the coadjoint orbits (which may be identified with the spaces G/G_λ of 4.12.7). It is a *philosophy of Kirillov* that, in very great generality, irreducible representations of Lie groups are parametrized by, or even constructed geometrically from, coadjoint orbits. For more comments in this direction, see Vogan [1997-II].

4.10 Weight Exercises

```
•   ·   ○       •   ···   ○   •   ○   •   ○   ···   •       ○       •
-λ  ···  -λ+α  ···      -α  -ρ  0   ρ   α  ···   λ-α  ···     λ
```
quaternionic type

```
•       ○       •   ···   •   ○   •   ○   •   ···   •       ○       •
-λ  ···  -λ+α  ···      -α  -ρ  0   ρ   α  ···   λ-α  ···     λ
```
real type

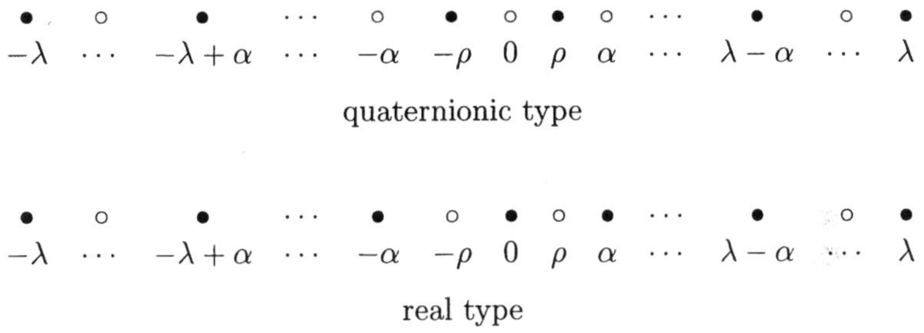

Fig. 4.10.1. Multiplicities for $\mathbf{SU}(2)$

(A) Exercise. Let us start with the simple, but instructive and important example of $G = \mathbf{SU}(2)$. There is only one positive root, α; and $\Lambda = \mathbf{Z} \cdot 2\pi i \alpha^\vee$. Recalling that $\alpha(\alpha^\vee) = 2$, we find:

$$(4.10.1) \qquad \widehat{T} = \mathbf{Z} \cdot \tfrac{1}{2}\alpha = \{\, \mu \in i\mathfrak{t}^* \mid \mu(\alpha^\vee) \in \mathbf{Z}\,\}.$$

On the other hand, if $G = \mathbf{SO}(3, \mathbf{R}) = \operatorname{Ad} \mathbf{SU}(2)$, then $\Lambda = \mathbf{Z} \cdot \pi i \alpha^\vee$, so:

$$(4.10.2) \qquad \widehat{T} = \mathbf{Z} \cdot \alpha = \{\, \mu \in i\mathfrak{t}^* \mid \mu(\alpha^\vee) \in 2\mathbf{Z}\,\}.$$

In particular, $\rho = \tfrac{1}{2}\alpha$ is not a weight, if $G = \mathbf{SO}(3, \mathbf{R})$.

The Weyl group action on $\widehat{T}$ consists of the identity and the reflection $\mu \mapsto -\mu = \mu - \mu(\alpha^\vee)\alpha$. So (4.9.14) yields:

$$(4.10.3) \qquad \chi_\pi(t) = \frac{t^\lambda - t^{-\lambda-\alpha}}{1 - t^{-\alpha}} = \sum_{k=0}^{\lambda(\alpha^\vee)} t^{\lambda - k\alpha}.$$

In other words, the weights which occur in an irreducible representation of $\mathbf{SU}(2)$ or $\mathbf{SO}(3,\mathbf{R})$, form a sequence $\lambda, \lambda-\alpha, \lambda-2\alpha, \ldots, \lambda-\lambda(\alpha^\vee)\alpha = -\lambda$, while each has multiplicity equal to 1. Also, the dimension of π is equal to:

$$(4.10.4) \qquad d_\pi = d = \lambda(\alpha^\vee) + 1,$$

which therefore always is odd if $G = \mathbf{SO}(3,\mathbf{R})$. In the form (4.9.16), the character formula looks like:

$$\chi_\pi(\exp X) = \frac{\sin d\pi x}{\sin \pi x}, \tag{4.10.5}$$

if we write $X = x\pi i\alpha^\vee$, for $x \in \mathbf{R} \setminus \mathbf{Z}$. See Figures 4.10.1,2. (For the determination of the types in Fig.4.10.1, we have used Proposition 4.8.7. In view of Weyl's integration formula, the integral over G of $\chi_\pi(x^2)$ is equal to the constant term in:

$$\tfrac{1}{2}(1-t^\alpha)(1-t^{-\alpha}) \sum_{k=0}^{\lambda(\alpha^\vee)} t^{2(\lambda-k\alpha)}.)$$

(B) Exercise. In general there is a simple recursive procedure to find the multiplicities $m_{\lambda,\mu} := m_\mu = [\pi|_T : \mu]$, if π is an irreducible representation of G with highest weight equal to λ.

Let $\alpha_1, \dots, \alpha_p$ be an enumeration of P. Define:

$$f_k(t) = \prod_{j=1}^{k}(1-t^{-\alpha_j}) \sum_{\mu\in\widehat{T}} m_\mu t^\mu. \tag{4.10.6}$$

Then $f_k(t) = \sum_{\mu\in\widehat{T}} c_{k,\mu} t^\mu$, where $c_k : \mu \mapsto c_{k,\mu}$, the Fourier series of f_k, is a function: $\widehat{T} \to \mathbf{Z}$ with finite support. Furthermore,

$$c_{0,\mu} = m_\mu, \quad \text{for } \mu \in \widehat{T}, \tag{4.10.7}$$

whereas at the other extreme we have:

$$c_{p,\mu} = \begin{cases} \det s, & \text{if } \mu = s\cdot\lambda + s\cdot\rho - \rho, \text{ with } s \in W; \\ 0, & \text{else.} \end{cases} \tag{4.10.8}$$

The formula $f_k = (1-t^{-\alpha_k})f_{k-1}$ leads to the recursive relation:

$$c_{k,\mu} = c_{k-1,\mu} - c_{k-1,\mu+\alpha_k}; \tag{4.10.9}$$

in the space of finitely supported functions on $\widehat{T}$, this relation has a unique solution c_{k-1}, for given c_k. Compare with the proof of Lemma 4.9.3.

For "big" Weyl groups W, this procedure is not very efficient. An improvement is given by Demazure's character formula, cf. Demazure [1974]. There one obtains the character by applying difference operators like the ones above $\#(P)$ many times, but one starts with t^λ only, instead of the Weyl numerator $\sum_{s\in W}(\det s)\, t^{s\cdot\lambda+s\cdot\rho-\rho}$. However, its general proof needs extensive calculations with rank 2 root systems; this is the reason why we do not treat this here.

However, for $G = \mathbf{SU}(3)$, (4.9.7) leads quickly to a complete description of the multiplicities. From Section 3.12 we see that there are two simple roots $\alpha := \alpha_{1,2}$ and $\beta := \alpha_{2,3}$; and there is only one other positive root, namely

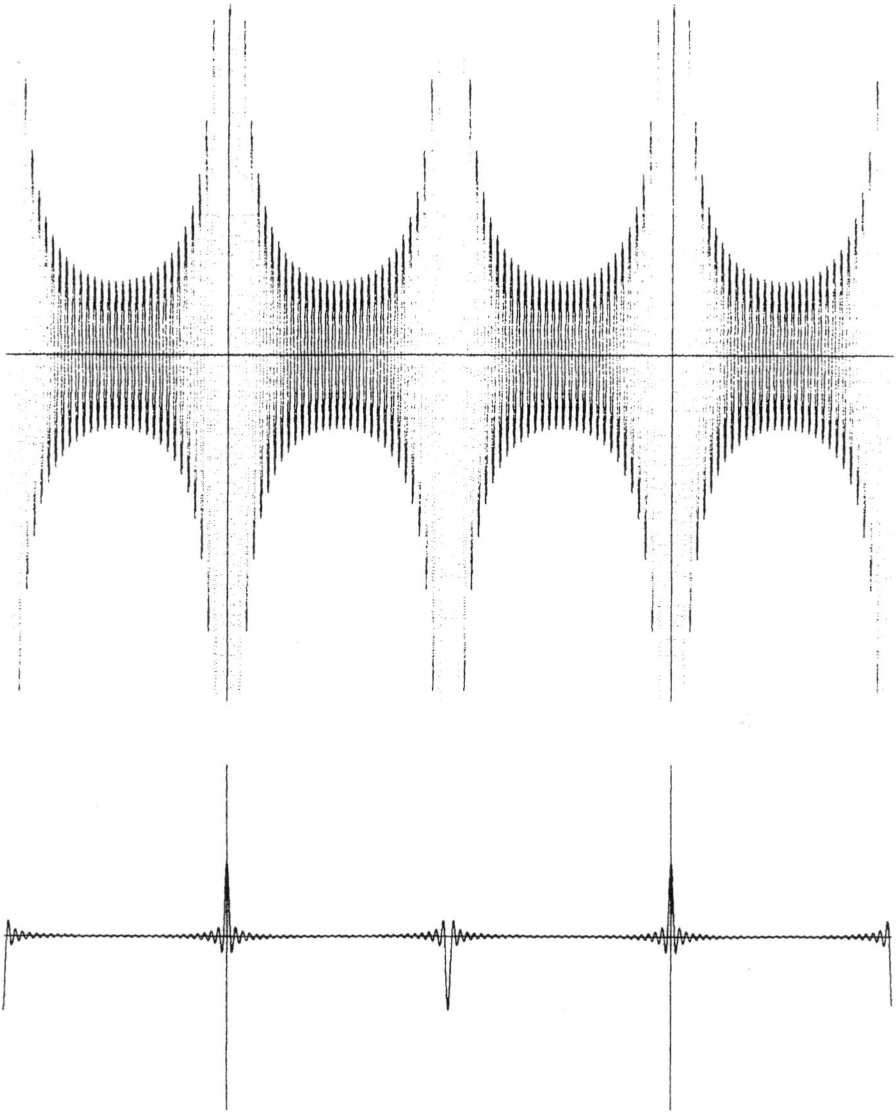

Fig. 4.10.2. Graph of $\chi_\pi \circ \exp|_{\mathfrak{t}}$

$\alpha_{1,3} = \alpha + \beta$. It follows that $\rho = \frac{1}{2}(\alpha + \beta + (\alpha + \beta)) = \alpha + \beta$. Finally, $\alpha(\beta^\vee) = -1$. Using a W-invariant inner product on $i\mathfrak{t}^*$, we therefore have:

$$< \alpha, \beta > = \frac{1}{2} < \beta, \beta > \alpha(\beta^\vee) = -\frac{1}{2} < \beta, \beta >,$$

cf. (4.9.19). Because the Weyl group permutes all the roots, they all have the same length; so $< \alpha, \beta > = -\frac{1}{2}|\alpha||\beta|$, that is, α and β make an angle of $\frac{2}{3}\pi$. The conclusion is that the 6 roots form a regular hexagon around the origin.

Moving from one Weyl chamber to the next adjacent one, starting with the positive one, we get the Weyl group elements:

$$1, \quad s_\alpha, \quad s_\alpha s_\beta, \quad s_\beta s_\alpha s_\beta = s_\alpha s_\beta s_\alpha, \quad s_\beta s_\alpha, \quad s_\beta,$$

written as in Corollary 3.10.3. In the computations it will be convenient, to keep in mind that:

$$s_\alpha \alpha = -\alpha, \quad s_\beta \beta = -\beta, \quad s_\alpha \beta = \alpha + \beta = s_\beta \alpha.$$

Writing:

$$\lambda_\alpha = \tfrac{2}{3}\alpha + \tfrac{1}{3}\beta, \quad \lambda_\beta = \tfrac{1}{3}\alpha + \tfrac{2}{3}\beta,$$

we have:

$$\lambda_\alpha(\alpha^\vee) = 1, \quad \lambda_\alpha(\beta^\vee) = 0, \quad \lambda_\beta(\alpha^\vee) = 0, \quad \lambda_\beta(\beta^\vee) = 1;$$

so the weights of $\mathfrak{g}_{\mathbf{C}} = \mathfrak{sl}(3, \mathbf{C})$ are precisely the:

$$\lambda = n_\alpha \lambda_\alpha + n_\beta \lambda_\beta,$$

with $n_\alpha, n_\beta \in \mathbf{Z}_{\geq 0}$. Figures 4.10.3-6 below contain plots of the image points in $\mathbf{C}$ of the character of $\mathbf{SU}(3)$ with highest weight $n_\alpha \lambda_\alpha + n_\beta \lambda_\beta$, for various values of n_α, n_β, evaluated at the points of a Weyl group invariant finite subgroup of T.

It follows that:

$$\begin{aligned}
s_\alpha \lambda &= \lambda - \lambda(\alpha^\vee)\alpha = \lambda - n_\alpha \alpha, \\
s_\beta \lambda &= \lambda - n_\beta \beta \quad \text{(by symmetry under interchanging } \alpha \text{ and } \beta), \\
s_\alpha s_\beta \lambda &= (\lambda - n_\alpha \alpha) - n_\beta(\alpha + \beta) = \lambda - (n_\alpha + n_\beta)\alpha - n_\beta \beta, \\
s_\beta s_\alpha \lambda &= \lambda - n_\alpha \alpha - (n_\alpha + n_\beta)\beta, \\
s_\alpha s_\beta s_\alpha \lambda &= (\lambda - n_\alpha \alpha) + n_\alpha \alpha - (n_\alpha + n_\beta)(\alpha + \beta) = \lambda - (n_\alpha + n_\beta)(\alpha + \beta).
\end{aligned}$$

Because $\rho = \alpha + \beta = \lambda_\alpha + \lambda_\beta$, we get for the $s \cdot \rho$ the same formulae, with $\lambda, n_\alpha, n_\beta$ replaced by $\rho, 1, 1$, respectively. So (4.10.8) gives that:

$$c_{3,\mu} = \begin{cases} 1, & \text{for } \mu = \begin{cases} \lambda, \ \lambda - (n_\alpha + n_\beta + 2)\alpha - (n_\beta + 1)\beta, \\ \lambda - (n_\alpha + 1)\alpha - (n_\alpha + n_\beta + 2)\beta; \end{cases} \\ -1, & \text{for } \mu = \begin{cases} \lambda - (n_\alpha + 1)\alpha, \ \lambda - (n_\beta + 1)\beta, \\ \lambda - (n_\alpha + n_\beta + 2)(\alpha + \beta); \end{cases} \\ 0, & \text{for all other } \mu. \end{cases}$$

Applying (4.10.9) for $k = 3$, $\alpha_3 = \alpha$, we get:

$$c_{2,\mu} = \begin{cases} 1, & \text{for } \mu = \lambda - j\alpha, \quad 0 \leq j \leq n_\alpha; \\ -1, & \text{for } \mu = \lambda - (n_\beta + 1)\beta - j\alpha, \quad 0 \leq j \leq n_\alpha + n_\beta + 1; \\ 1, & \text{for } \mu = \lambda - (n_\alpha + 1 + j)\alpha - (n_\alpha + n_\beta + 2)\beta, \quad 0 \leq j \leq n_\beta; \\ 0, & \text{for all other } \mu. \end{cases}$$

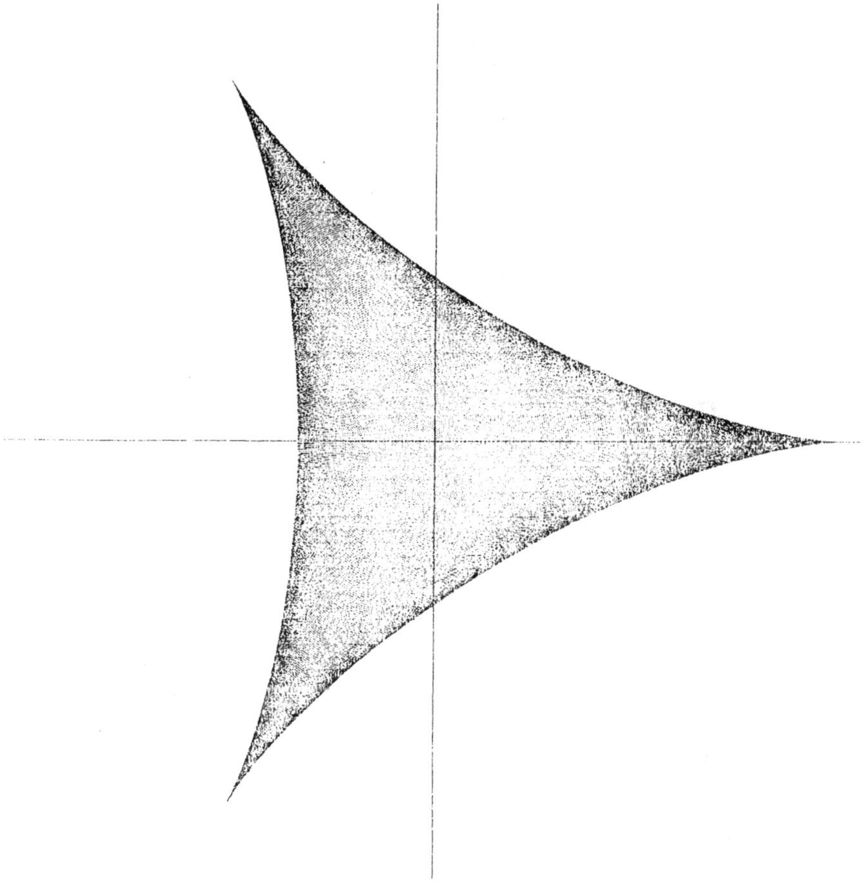

Fig. 4.10.3. $n_\alpha = 1, n_\beta = 0$

Next applying (4.10.9) for $k = 2$, $\alpha_2 = \beta$, we get:

$$c_{1,\mu} = \begin{cases} 1, & \text{for } \mu = \lambda - j\alpha - k\beta, \quad 0 \le j \le n_\alpha, \quad 0 \le k \le n_\beta; \\ -1, & \text{for } \mu = \begin{cases} \lambda - (n_\alpha + 1 + j)\alpha - (n_\beta + 1 + k)\beta, \\ 0, \le j \le n_\beta, \quad 0 \le k \le n_\alpha; \end{cases} \\ 0, & \text{for all other } \mu. \end{cases}$$

For the last step, we apply (4.10.9) for $k = 1$, $\alpha_1 = \alpha + \beta$. We make the assumption that $n_\alpha \ge n_\beta$; the answer in the other case is obtained by interchanging the role of α and β. We get:

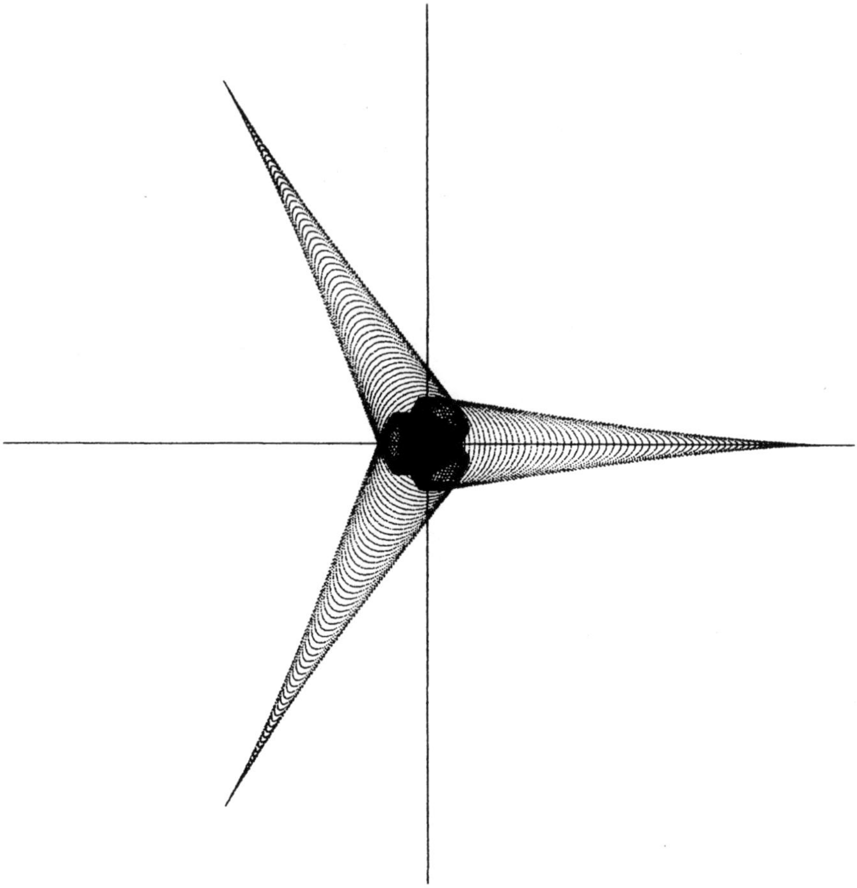

Fig. 4.10.4. $n_\alpha = 2, n_\beta = 1$

$$m_\mu = \begin{cases} l+1, & \text{for } \mu = \lambda - k\beta - l(\alpha+\beta), \quad 0 \leq k \leq n_\beta - l, \quad 0 \leq l; \\ l+1, & \text{for } \mu = \begin{cases} \lambda - j\alpha - l(\alpha+\beta), \\ 0 \leq j \leq n_\alpha - l, \quad 0 \leq l \leq n_\beta; \end{cases} \\ n_\beta + 1, & \text{for } \mu = \lambda - j\alpha - k\beta, \quad n_\beta \leq k \leq j \leq n_\alpha; \\ 0, & \text{for all other dominant } \mu. \end{cases}$$

Here it has been used that the dominant $\mu = \lambda - j\alpha - k\beta$ are characterized by the inequalities $n_\alpha - 2j + k \geq 0$, $n_\beta + j - 2k \geq 0$; combined with $j \geq 0$, $k \geq 0$, this is contained in the union of the domain $0 \leq j \leq n_\alpha$, $0 \leq k \leq n_\beta$, and the domain $n_\beta \leq k \leq j \leq n_\alpha$.

Because of the Weyl group invariance of the multiplicities, the multiplicities are equal to $l+1$ on the root ladders which successively connect the points on the Weyl group orbit of $\lambda - l(\alpha+\beta)$, if $0 \leq l \leq n_\beta$. This is a hexagon, which is not equilateral, unless $n_\alpha = n_\beta$, and degenerates into a

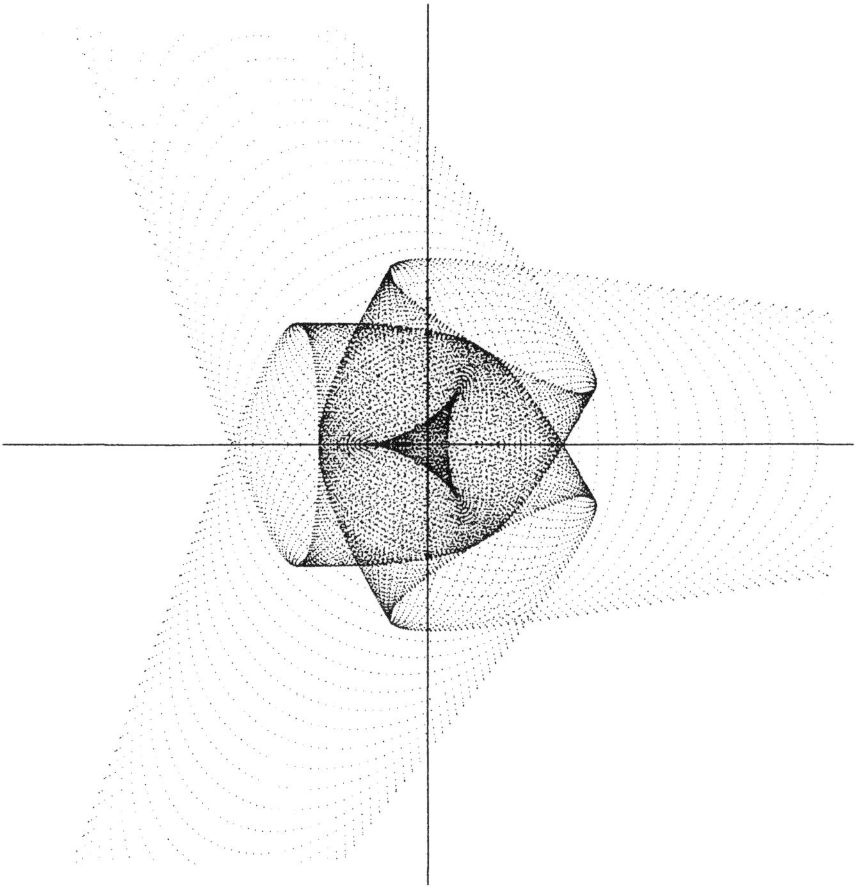

Fig. 4.10.5. $n_\alpha = 2, n_\beta = 1$, magnified 4 times

triangle (a point if $n_\alpha = n_\beta$), if $l = n_\beta$. The multiplicities are constant equal to $n_\beta + 1$, for the $\mu = \lambda - j\alpha - k\beta$, with $j, k \in \mathbf{Z}_{\geq 0}$, which belong to the convex hull of the triangular Weyl group orbit of $\lambda - n_\beta(\alpha + \beta)$. See Fig.4.10.3.

It is also fun to verify the formula:

$$\mathrm{d}_\pi = \tfrac{1}{2}(n_\alpha + 1)(n_\beta + 1)(n_\alpha + n_\beta + 2),$$

for the dimension of π (that is, the sum of the multiplicities), both from the calculation above, and from the Weyl dimension formula (4.9.18).

Remark. Already in the case above, the cancellations that do occur when $\chi_\pi(t) = \sum_{\mu \in \widehat{T}} m_\mu \, t^\mu$ is multiplied by $\delta_P(t)$ to yield $\sum_{s \in W} \det s \; t^{s \cdot \lambda + s \cdot \rho - \rho}$, is quite impressive; this phenomenon increases rapidly with more complicated Weyl groups.

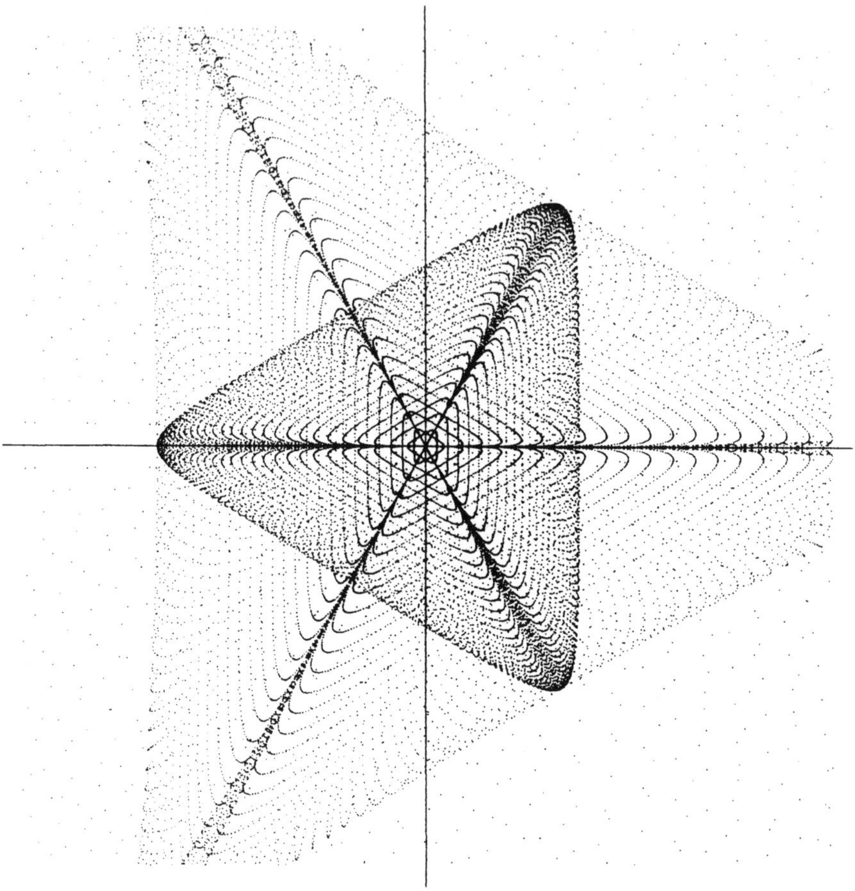

Fig. 4.10.6. $n_\alpha = 3, n_\beta = 1$, magnified 16 times

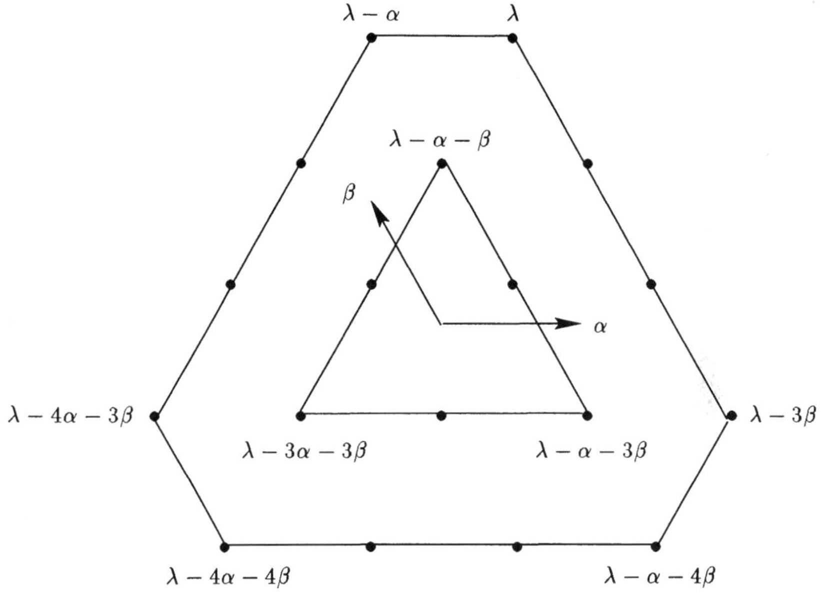

Fig. 4.10.7. Multiplicities for **SU**(3). The highest weight λ equals $\lambda_\alpha + 3\lambda_\beta = \frac{1}{3}(5\alpha + 7\beta)$, the multiplicities are 1, and 2, on the outer, and inner ring, respectively.

(C) Exercise. We now want to give general statements concerning some of the phenomena that we have observed in the examples in (A) and (B). In Section 3.11 we have met several lattices in $\mathfrak{t}$, the Lie algebra of T, viz.:

(i) the coroot lattice $\Lambda_{R^\vee}$, the set of sums $\sum_{\alpha \in R} 2\pi i n_\alpha \alpha^\vee$, for $n_\alpha \in \mathbf{Z}$;
(ii) the "T-lattice" $\Lambda = \ker \exp|_{\mathfrak{t}}$;
(iii) the central lattice $\Lambda_Z = \{ X \in \mathfrak{t} \mid \alpha(X) \in 2\pi i \mathbf{Z} \text{ for all } \alpha \in R \}$.

(Here the word "lattice" is not quite appropriate if $\mathfrak{g}$ has a nonzero center, but we ignore this for the moment.)

For the purposes of representation theory, it is natural to define the following lattices in $i\mathfrak{t}^*$:

(i) the *root lattice* Λ_R, the set of linear combinations of roots with integral coefficients, that is,

$$\{ \mu \in i\mathfrak{t}^* \mid \mu(\Lambda_Z) \subset 2\pi i \mathbf{Z} \};$$

(ii) the *weight lattice* $\widehat{T}$ of T, given by:

$$\widehat{T} = \{ \mu \in i\mathfrak{t}^* \mid \mu(\Lambda) \subset 2\pi i \mathbf{Z} \};$$

(iii) the *weight lattice* Λ_{weight} of $\mathfrak{g}_{\mathbf{C}}$ defined by:

$$\{\mu \in \mathfrak{t}_{\mathbf{C}}^* \mid \mu(\alpha^\vee) \in \mathbf{Z}, \text{ for all } \alpha \in R\} = \{\mu \in \mathfrak{t}_{\mathbf{C}}^* \mid \mu(\Lambda_{R^\vee}) \subset 2\pi i\mathbf{Z}\}.$$

The inclusions $\Lambda_{R^\vee} \subset \Lambda \subset \Lambda_Z$, cf. (3.11.5), imply:

$$\Lambda_R \subset \widehat{T} \subset \Lambda_{\text{weight}}. \tag{4.10.10}$$

The first inclusion also follows from the observation that each root is a weight of T, for the adjoint representation in $\mathfrak{g}_{\mathbf{C}}$, whereas the second one also follows from (4.9.24).

Furthermore, using Proposition 3.11.1.(i), we get that G is simply connected $\Leftrightarrow \Lambda = \Lambda_{R^\vee} \Leftrightarrow \widehat{T} = \Lambda_{\text{weight}}$.

This also gives the following explanation why Λ_{weight} is called the weight lattice of $\mathfrak{g}_{\mathbf{C}}$. If π is a representation of G in a finite-dimensional vector space V, then $\pi' := \mathrm{T}_1 \pi$ is a homomorphism of Lie algebras: $\mathfrak{g} \to \mathbf{L}(V, V)$, which extends in a unique way to a homomorphism of complex Lie algebras: $\mathfrak{g}_{\mathbf{C}} \to \mathbf{L}(V, V)$. Such a homomorphism is called a *representation of* $\mathfrak{g}_{\mathbf{C}}$ *in* V. Irreducibility of Lie algebra representations is defined in the same way as irreducibility of group representations. Because a linear subspace U of V is $\pi(G)$-invariant if and only if it is $\pi'(\mathfrak{g}_{\mathbf{C}})$-invariant, we get that π is irreducible if and only if π' is irreducible. Applying the passage from π to π' to the characters of T, we get the identification of the set of characters of T with the set of weights of T, as discussed in the beginning of Section 4.9. So weights of π can be identified with weights of π', where $\mu \in \mathfrak{t}_{\mathbf{C}}^*$ is called a weight of π' if there is a $v \in V \setminus \{0\}$ such that $\pi'(X)(v) = \mu(X)v$, for all $X \in \mathfrak{t}_{\mathbf{C}}$. (The roots are just the weights of ad $=$ Ad$'$.) If now G is simply connected, then the mapping $\pi \mapsto \pi'$ is bijective from the set of finite-dimensional representations of G onto the set of finite-dimensional representations of $\mathfrak{g}$, cf. Corollary 1.10.5; and the latter are identified with their complex linear extensions: $\mathfrak{g}_{\mathbf{C}} \to \mathbf{L}(V, V)$. So in this case, Λ_{weight} is equal to the set of weights of the finite-dimensional representations of $\mathfrak{g}_{\mathbf{C}}$.

In general, the equation $\pi_1(G) = \Lambda/\Lambda_{R^\vee}$ yields that $\Lambda_{\text{weight}}/\widehat{T}$ is canonically isomorphic to the dual $\mathrm{Hom}(\pi_1(G), 2\pi i\mathbf{Z})$ of $\pi_1(G)$.

At the other extreme, we have $G \xrightarrow{\sim} \mathrm{Ad}\, G \Leftrightarrow \Lambda = \Lambda_Z \Leftrightarrow \widehat{T} = \Lambda_R$. Here the more general statement is that $\widehat{T}/\Lambda_R$ is canonically dual to $\ker \mathrm{Ad} = \mathrm{Z}(G)$, the center of G. In Proposition 4.9.4.(iii), we have seen that the weights of π are contained in $\lambda + \Lambda_R$; so if G is not isomorphic to $\mathrm{Ad}\, G$, then we find, in between the weights of π, a lot of "holes": weights of T with multiplicity equal to 0. See Fig.4.10.1 and 7. Finally, note that, if $\mathfrak{z} = 0$, then

$$\begin{aligned} \#(\Lambda_{\text{weight}}/\Lambda_R) &= \#(\Lambda_Z/\Lambda_{R^\vee}) = \det(\alpha(\beta^\vee))_{\alpha,\beta \in S} \\ &= \text{the determinant of the Cartan matrix } C, \end{aligned}$$

cf. the text after Corollary 5.5.13.

(D) Exercise. The result in Proposition 4.10.1 below is suggested by Figure 4.10.7, see also Fig.4.10.8. The *convex hull* of a subset A of a vector space is defined as the set of finite linear combinations $\sum_{a\in A} c_a a$, with $c_a \in \mathbf{R}_{\geq 0}$, $\sum_{a\in A} c_a = 1$; we shall denote the convex hull of A by $\operatorname{conv} A$. The Weyl group orbit in it^* of $\mu \in \widehat{T}$ will be denoted by $W \cdot \mu = \{ s \cdot \mu \mid s \in W \}$. Finally, recall the partial ordering $\preceq$, introduced after Proposition 4.9.4.

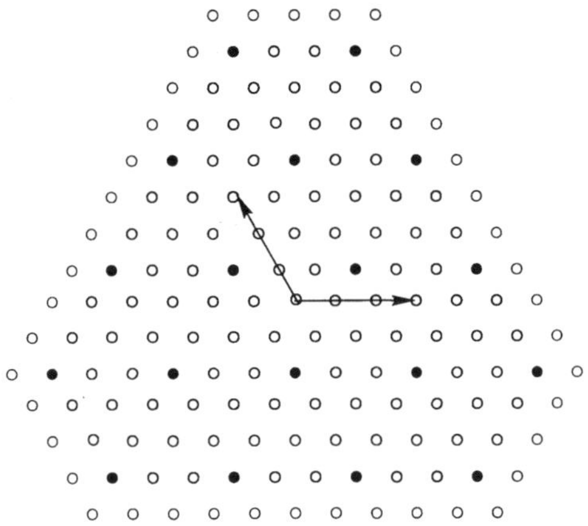

Fig. 4.10.8. G and λ are as in Fig.4.10.7. • denotes a weight of the representation of highest weight λ and ∘ an element in $\widehat{T} = \Lambda_{\text{weight}}$.

(4.10.1) Proposition. *Let π be an irreducible representation of G with highest weight λ. Then the set of weights $\Omega(\pi)$ of π is equal to each of the following sets A–C:*

$$A := \bigcap_{s\in W} s\cdot\{\mu\in\widehat{T} \mid \mu\preceq\lambda\}, \quad B := (\lambda+\Lambda_R)\cap\operatorname{conv}(W\cdot\lambda),$$

$$C := \bigcup_{s\in W} s\cdot\{\mu\in\widehat{T} \mid \mu \textit{ is dominant and } \mu\preceq\lambda\}.$$

Proof. $\Omega(\pi) \subset A$ follows from Proposition 4.9.4.(iii),(c) and the Weyl group invariance of $\Omega(\pi)$.

$A \subset B$. In the proof of Proposition 4.9.4.(iii), it has been observed that $s\cdot\lambda \preceq \lambda$, implying that $s\cdot\lambda \in \lambda+\Lambda_R$, for any $s \in W$. It follows that A is W-invariant; so it suffices to prove that $\mu \in A$ if $s\cdot\mu \preceq \lambda$, for all $s \in W$. Again, note that $\mu \preceq \lambda$ implies that $\mu \in \lambda+\Lambda_R$. Now assume that μ does not belong to $\operatorname{conv}(W\cdot\lambda)$. Then there exists a real linear form X on it^* which separates μ from $W\cdot\lambda$, that is, such that $<\mu, X> > <s\cdot\lambda, X> = <\lambda, s^{-1}X>$, for all $s \in W$. By a slight perturbation we can arrange that X does not lie on any root hyperplane $\ker\alpha$; let $s \in W$ be such that $< \alpha, Y >> 0$, for all $\alpha \in P$, here we write $Y = s^{-1}X$. Now we get the following contradiction:

$$< \lambda, Y > < < \mu, sY > = < s^{-1} \cdot \mu, Y > \leq < \lambda, Y > .$$

$B \subset C$. Every W-orbit meets the closure of the positive Weyl chamber (use the transitivity of W on the set of Weyl chambers, combined with a limit argument). Because both B and C are W-invariant, it suffices therefore to prove that $\mu \preceq \lambda$, if $\mu \in \lambda + \Lambda_R$ and $\mu = \sum_{s \in W} c_s \, s \cdot \lambda$, for some $c_s \in \mathbf{R}_{\geq 0}$ such that $\sum_{s \in W} c_s = 1$, and moreover μ is dominant.

Before proceeding with the proof, we recall from Remark (C) after Proposition 3.11.1, that every root, and therefore also every element of Λ_R, is an integral linear combination of simple roots. Moreover, because the simple roots are linearly independent, the coefficients are uniquely determined.

Now use that $s \cdot \lambda \preceq \lambda$ implies that $s \cdot \lambda = \lambda - \sum_{\alpha \in S} n_{s,\alpha}\alpha$, for certain $n_{s,\alpha} \in \mathbf{Z}_{\geq 0}$. Substituting this, we get $\mu = \lambda - \sum_{\alpha \in S} k_\alpha \alpha$, with $k_\alpha := \sum_{s \in W} c_s n_{s,\alpha} \geq 0$. The previous paragraph now yields that $k_\alpha \in \mathbf{Z}$, for all $\alpha \in S$; or $\mu \preceq \lambda$.

$C \subset \Omega(\pi)$. In view of the W-invariance of $\Omega(\pi)$, it suffices to prove that $\mu \in \Omega(\pi)$, if μ is dominant and $\mu \preceq \lambda$. Write $\mu = \lambda - \sum_{\alpha \in S} n_\alpha \alpha$, with $n_\alpha \in \mathbf{Z}_{\geq 0}$. Suppose that $\mu + \theta \in \Omega(\pi)$, for $\theta = \sum_{\alpha \in S} k_\alpha \alpha$, for $k_\alpha \in \mathbf{Z}$, $0 \leq k_\alpha \leq n_\alpha$; we may assume that $\sum_{\alpha \in S} k_\alpha > 0$. (Note that we do not require here that $\mu + \theta$ is dominant.) Take a Weyl group invariant inner product $< .,. >$ on $i\mathfrak{t}^*$. We then have that $0 < < \theta, \theta > = \sum_{\alpha \in S} k_\alpha < \theta, \alpha >$; so there is at least one $\alpha \in S$ such that $k_\alpha < \theta, \alpha >> 0$, and thus $k_\alpha > 0$ and $< \theta, \alpha >> 0$. In view of (4.9.19), the latter implies that $\theta(\alpha^\vee) > 0$; and because μ is dominant, this in turn yields that $(\mu + \theta)(\alpha^\vee) > 0$. Now Proposition 4.9.4.(i) gives that $\mu + \theta - \alpha \in \Omega(\pi)$. Noting that $\theta - \alpha = \sum_{\beta \neq \alpha} k_\beta \beta + (k_\alpha - 1)\alpha$, with $k_\alpha - 1 \in \mathbf{Z}_{\geq 0}$, we get $\mu \in \Omega(\pi)$ by means of an induction on decreasing $\sum_{\alpha \in S} k_\alpha$. □

As a consequence, if π, and π', is an irreducible representation of G with highest weight equal to λ, and λ', respectively, then $\lambda \preceq \lambda'$ if and only if $\Omega(\pi) \subset \Omega(\pi')$. So the partial ordering between the highest weights induces a quite natural partial ordering in $\widehat{G}$, whose definition does not involve a choice of positive roots. Note that $\widehat{G}$ can be identified with $\{ \mu \in \widehat{T} \mid \mu \text{ is dominant} \}$, cf. Theorem 4.9.5.

(E) Exercise. In Theorem 4.4.5.(iv), it has been observed that the space $\mathrm{M}(G)$ of matrix coefficients is a commutative algebra, when provided with the "ordinary" product of pointwise multiplication of functions. In Corollary 14.6.9 it is shown furthermore that G has a unique structure of a real linear algebraic group, for which $\mathrm{M}(G)$ becomes the algebra of complex-valued polynomial functions on G. The points x of G are identified, via the evaluation homomorphisms: $f \mapsto f(x)$, with the homomorphisms: $\mathrm{M}(G) \to \mathbf{C}$ that take real values on the $\mathbf{R}$-subalgebra $\mathrm{M}(G)_{\mathbf{R}}$ of real-valued elements of $\mathrm{M}(G)$.

The space $\mathrm{M}(G)^{\mathrm{Ad}\,G}$ of conjugacy-invariant elements in $\mathrm{M}(G)$ is equal to the linear span of the characters χ_π, for $\pi \in \widehat{G}$, because the projections $E_\pi \colon \mathrm{M}(G) \to \mathrm{M}_\pi$ onto the π-isotypical subspaces M_π map $f \in \mathrm{M}(G)^{\mathrm{Ad}\,G}$ to $< f, \chi_\pi > \chi_\pi$, see (4.6.3). In Section 14.7 it is shown that the orbit space $(\mathbf{Ad}\,G)\backslash G$ of the conjugacy classes in G can be provided with the structure of a real affine algebraic variety, for which $\mathrm{M}(G)^{\mathrm{Ad}\,G}$ becomes the algebra of complex-valued polynomial functions on $(\mathbf{Ad}\,G)\backslash G$.

In a similar fashion, T is a real affine algebraic subgroup of G, with algebra of complex-valued polynomials equal to $\mathrm{M}(T)$ = the space of finite linear combinations of characters of T = the space of "finite Fourier series" on T. The quotient space $W\backslash T$ of T under the action of the Weyl group, is a real affine algebraic variety with algebra of complex-valued polynomials equal to $\mathrm{M}(T)^W$. The injection: $T \to G$ induces a mapping $W\backslash T \to (\mathbf{Ad}\,G)\backslash G$; Lemma 3.9.2.(iii) implies that this mapping is a homeomorphism. Dual to the injection: $T \to G$ is the restriction mapping: $f \mapsto f|_T \colon \mathrm{M}(G) \to \mathrm{M}(T)$; it induces the restriction mapping: $\mathrm{M}(G)^{\mathrm{Ad}\,G} \to \mathrm{M}(T)^W$. We now have the following strengthening of the statement that $W\backslash T$ is homeomorphic to $(\mathbf{Ad}\,G)\backslash G$, with an independent, representation theoretic, proof.

(4.10.2) Proposition. *The restriction mapping: $f \mapsto f|_T$ induces an isomorphism of algebras:*

$$\mathrm{M}(G)^{\mathrm{Ad}\,G} \to \mathrm{M}(T)^W,$$

preserving the real structures. That is, the injection: $T \to G$ induces an isomorphism of real affine algebraic varieties: $W\backslash T \to (\mathbf{Ad}\,G)\backslash G$.

Proof. That the restriction mapping is injective: $\mathrm{M}(G)^{\mathrm{Ad}\,G} \to \mathrm{M}(T)^W$ follows from the fact that each conjugacy class in G meets T, cf. Theorem 3.7.1.(iv). Its image $\mathcal{I}$ consists of the finite linear combinations of the χ_λ, where λ runs over:

$$\widehat{T}_{\mathrm{dom}} := \{\, \lambda \in \widehat{T} \mid \lambda \text{ is dominant} \,\}.$$

On the other hand, as an application of the averaging principle 4.2.1, we get that $\mathrm{M}(T)^W$ consists of the finite linear combinations of the $a_\mu := \#(W)^{-1} \sum_{s \in W} t^{s \cdot \mu}$, with $\mu \in \widehat{T}_{\mathrm{dom}}$. Note that each Weyl group orbit in it^* intersects the closure of the Weyl chamber in exactly one point, cf. Lemma 3.8.2.

Write $\mathcal{I}_{\preceq \mu}$ for the linear subspace of $\mathcal{I}$ spanned by the χ_λ, with $\lambda \in \widehat{T}_{\mathrm{dom}}$ and $\lambda \preceq \mu$. Similarly, let $\mathrm{M}(T)^W_{\preceq \mu}$ be the linear subspace of $\mathrm{M}(T)^W$ spanned by the a_θ, with $\theta \in \widehat{T}_{\mathrm{dom}}$ and $\theta \preceq \mu$. Then Proposition 4.9.4.(iii),(c) shows that $\mathcal{I}_{\preceq \mu} \subset \mathrm{M}(T)^W_{\preceq \mu}$. On the other hand, both:

$$\mathcal{I}_{\preceq \mu} \Big/ \sum_{\theta \prec \mu} \mathcal{I}_{\preceq \theta} \quad \text{and} \quad \mathrm{M}(T)^W_{\preceq \mu} \Big/ \sum_{\theta \prec \mu} \mathrm{M}(T)^W_{\preceq \theta}$$

are one-dimensional (represented by χ_μ, and by a_μ), respectively; so by induction on the height of μ we get that $\mathcal{I}_{\preceq\mu} = \mathrm{M}(T)^W_{\preceq\mu}$, for all $\mu \in \widehat{T}_{\text{dom}}$. Passing to the sum over all $\mu \in \widehat{T}_{\text{dom}}$, we get $\mathcal{I} = \mathrm{M}(T)^W$. □

According to Sections 14.6 and 7, the algebras $\mathrm{M}(G)$, $\mathrm{M}(G)^{\mathrm{Ad}\,G}$, $\mathrm{M}(T)$, and $\mathrm{M}(T)^W$, can be viewed as the algebras of the complex polynomial functions on the complex affine algebraic varieties $G_{\mathbf{C}}$, $(\mathbf{Ad}\,G_{\mathbf{C}})\backslash G_{\mathbf{C}}$, $T_{\mathbf{C}}$, and $W\backslash T_{\mathbf{C}}$, respectively. Here $G_{\mathbf{C}}$, and $T_{\mathbf{C}}$, are actually complex affine algebraic groups, the complexification of G, and T, respectively. Furthermore, $(\mathbf{Ad}\,G_{\mathbf{C}})\backslash G_{\mathbf{C}}$ should be interpreted as the space of the closed conjugacy classes in $G_{\mathbf{C}}$, rather than the space of all conjugacy classes. With these interpretations however, Proposition 4.10.2 also implies that $(\mathbf{Ad}\,G_{\mathbf{C}})\backslash G_{\mathbf{C}}$ and $W\backslash T_{\mathbf{C}}$ are isomorphic as complex affine algebraic varieties.

(F) Exercise. Our next purpose is to show that the structure of the algebra $\mathrm{M}(T)^W$ is particularly simple, if G **is simply connected.**

In this case we have:

$$\widehat{T} = \Lambda_{\text{weight}} = \{\, \mu \in i\mathfrak{t}^* \mid \mu(\alpha^\vee) \in \mathbf{Z}, \text{ for all } \alpha \in R \},$$

cf. the discussion following (4.10.1). Using the analogue for the coroot system of Lemma 3.10.4.(iii) and an assertion following (3.5.10), we get that the $\alpha^\vee$, for $\alpha \in S$, are linearly independent, and that $\beta^\vee$ is an integral linear combination of the $\alpha^\vee$, with $\alpha \in S$, for every positive root β; hence for every root β, because $R = P \cup -P$, cf. (3.5.10).

Now define the *fundamental weights* λ_α, for $\alpha \in S$, by:

$$\lambda_\alpha(\beta^\vee) = \begin{cases} 0, & \text{if } \beta \in S, \beta \neq \alpha; \\ 1, & \text{if } \beta = \alpha. \end{cases} \tag{4.10.11}$$

The previous paragraph yields that $\lambda_\alpha \in \Lambda_{\text{weight}}$, for each $\alpha \in S$; and actually the mapping $(n_\alpha)_{\alpha\in S} \mapsto \sum_{\alpha \in S} n_\alpha \lambda_\alpha$ defines an isomorphism of additive semi-groups: $(\mathbf{Z}_{\geq 0})^S \to \widehat{T}_{\text{dom}}$. The irreducible representations with highest weight equal to the λ_α, with $\alpha \in S$, uniquely determined up to equivalence, are called the *fundamental representations* of G. For $G = \mathbf{SU}(n)$, these will be described after the next proposition; the study of the fundamental representations for the general simple Lie algebras will be postponed until we have obtained a more detailed understanding of root systems.

(4.10.3) Proposition. *If G is simply connected, then $\mathrm{M}(G)^{\mathrm{Ad}\,G} \xrightarrow{\sim} \mathrm{M}(T)^W$ is a free polynomial algebra, with generators equal to the χ_{λ_α}, for $\alpha \in S$. It follows that $W\backslash T_{\mathbf{C}}$ is isomorphic, as a complex affine algebraic variety, to $\mathbf{C}^l$, where $l = \operatorname{rank} G$.*

Proof. It will be convenient to use the partial ordering $\preceq'$ in $\widehat{T}_{\text{dom}}$, defined by:

$$\theta \preceq' \mu \Leftrightarrow \theta = \mu - \sum_{\alpha \in S} c_\alpha \alpha, \quad \text{with } c_\alpha \in \mathbf{R}_{\geq 0}.$$

Note that $\theta \preceq \mu$ means that $\theta \preceq' \mu$ and $\theta \in \mu + \Lambda_R$.

Let $\mu = \sum_{\alpha \in S} n_\alpha \lambda_\alpha$, with $n_\alpha \in \mathbf{Z}_{\geq 0}$. Then the restriction to T of $g_\mu := \prod_{\alpha \in S} (\chi_{\lambda_\alpha})^{n_\alpha}$ is equal to a linear combination of a_θ, for $\theta \preceq \mu$, with a nonzero coefficient for $\theta = \mu$. This follows from:

$$\chi_{\lambda_\alpha}|_T = \sum_{\theta \preceq \lambda_\alpha} m_{\alpha,\theta}\, t^\theta, \quad \text{and } m_{\alpha,\lambda_\alpha} = 1.$$

It follows, by induction over the height of μ, that $(c_\theta)_{\theta \preceq' \mu} \mapsto \sum_{\theta \preceq \mu} c_\theta g_\theta|_T$ is a linear isomorphism onto the linear span $\mathrm{M}(T)^W_{\preceq' \mu}$ of the a_θ with $\theta \preceq' \mu$.

Taking the union over all dominant weights μ, we get a linear isomorphism from the space of (c_θ), with $c_\theta \neq 0$ only for finitely many dominant θ, onto the linear span of the a_θ, with θ dominant, which is equal to $\mathrm{M}(T)^W$. This proves that the algebra A generated by the χ_{λ_α} is free, and that restriction to T yields an isomorphism: $A \to \mathrm{M}(T)^W$. From Proposition 4.10.2 we now get also that $A = \mathrm{M}(G)^{\mathrm{Ad}\, G}$. □

Remarks.

(i) The case that G is simply connected is also the case that the real orbit space $(\mathbf{Ad}\, G)\backslash G \cong W\backslash T$ is homeomorphic to the closure of the alcove $\mathfrak{b}$, a compact, convex polytope, cf. Corollary 3.9.6.(iv).

(ii) The proof of Proposition 4.10.3 shows that, if we drop the assumption that G is simply connected, then it is still true that $\mathrm{M}(G)^{\mathrm{Ad}\, G}$ is generated by the χ_μ, for $\mu \in F$, if F is a subset of dominant weights that generates the set of dominant weights as a semigroup. One can always find a finite set F with these properties, cf; here we don't enter into the combinatorics of finding minimal such F.

Also, select, for each $\mu \in F$, a basis $f_{\mu,j}$, with $j \in \{1, \dots, \chi_\mu(1)^2\}$, in the space of matrix coefficients of the representation $\pi(\mu)$ with character χ_μ. Then the $f_{\mu,j}$ together generate the algebra $\mathrm{M}(G)$. This follows from the observation that, for each $[\pi] \in \widehat{G}$, the space M_π is spanned by the $\mathrm{R}(x)(\chi_\pi)$, for $x \in G$, cf. Theorem 4.4.5.(i). Writing χ_π as a linear combination of $\prod_{\mu \in F} (\chi_\mu)^{n_\mu}$, with $n_\mu \in \mathbf{Z}_{\geq 0}$, and observing that $\mathrm{R}(x)\big(\prod_{\mu \in F} (\chi_\mu)^{n_\mu}\big) = \prod_{\mu \in F} \big(\mathrm{R}(x)(\chi_\mu)\big)^{n_\mu}$ belongs to the algebra generated by the $f_{\mu,j}$, we get the assertion. Conversely, averaging over the action of conjugation shows that if the $M_{\pi(\mu)}$, for $\mu \in F$, generate $\mathrm{M}(G)$, then the χ_μ, with $\mu \in F$, have to generate $\mathrm{M}(G)^{\mathrm{Ad}\, G}$.

(iii) If $G = \mathbf{R}/\mathbf{Z}$ is the circle, then the unique minimal set of generators of $\mathrm{M}(G) = \mathrm{M}(G)^{\mathrm{Ad}\, G}$ is given by $z\colon x \mapsto \mathrm{e}^{2\pi i x}$ and $w\colon x \mapsto \mathrm{e}^{-2\pi i x}$, with the relation $zw = 1$. This makes $G_{\mathbf{C}}$ isomorphic to the complex algebraic subvariety $\{ (z,w) \in \mathbf{C}^2 \mid zw = 1 \}$ of $\mathbf{C}^2$. By taking the multiplication $(z,w)(z',w') = (zz', ww')$, we see that this is a polynomial operation, making $G_{\mathbf{C}}$ into a complex affine algebraic group, obviously isomorphic to the multiplicative group $\mathbf{C}^\times$ of the nonzero complex numbers. We find back G in $\mathbf{C}^\times$ as the subgroup $\{ z \in \mathbf{C}^\times \mid |z| = 1 \}$. At this moment we would like to emphasize that $\mathrm{M}(G)^{\mathrm{Ad}\, G}$ is **not** a free polynomial algebra in $1(= \operatorname{rank} G)$ generator.

Example. For $G = \mathbf{SU}(n)$, the standard choice of $\mathfrak{t}$ is the set of diagonal matrices X with $i\theta_j$, for $j = 1, \dots, n$, on the diagonal. Here $\theta_j \in \mathbf{R}$ and $\sum_j \theta_j = 0$. The roots are the linear forms $\alpha_{p,q}$ which assign to X the number $i\theta_p - i\theta_q$, for $p \neq q$. The simple roots are the $\alpha_{p,p+1}$, and the corresponding simple coroots are the elements of $i\mathfrak{t}$ which, on the diagonal, have 1 at the p-th place, -1 at the $(p+1)$-th place, and zeros everywhere else. See Section 3.12.

Now consider the induced action of $\mathbf{SU}(n)$ on the k-th exterior power $\bigwedge^k(\mathbf{C}^n)$ of $\mathbf{C}^n$, for $1 \leq k \leq n-1$. Then X acts on the base vector $e_{i_1} \wedge \dots \wedge e_{i_k}$, with $i_1 < \dots < i_k$, as:

$$\left.\frac{d}{dt}\right|_{t=0} (\exp tX)(e_{i_1}) \wedge \dots \wedge (\exp tX)(e_{i_k})$$
$$= \sum_{j=1}^{k} e_{i_1} \wedge \dots \wedge X e_{i_j} \wedge \dots \wedge e_{i_k} = \left(\sum_{j=1}^{k} i\theta_{i_j} \right) e_{i_1} \wedge \dots \wedge e_{i_k}.$$

That is, this representation has weights equal to $\sum_{j=1}^{k} i\theta_{i_j}$. This weight is dominant, if and only if p occurs in the sequence i_j, whenever $p+1$ occurs. There is only one such sequence, namely $i_j = j$, for $1 \leq j \leq k$. But this means that the representation is irreducible. Moreover, this weight is equal to 1 on $\alpha_{k,k+1}^\vee$, and equal to 0 on the other simple coroots; so these are precisely the fundamental representations of $\mathbf{SU}(n)$.

The Weyl group acts, via the permutations of the indices, transitively on the weights, so all weights have multiplicity equal to 1. The character, evaluated at the diagonal matrix $t \in \mathbf{SU}(n)$ with entries $z_1, \dots, z_n$, is therefore equal to:

$$\epsilon_k(z_1, \dots, z_n) = \sum_{i_1 < \dots < i_k} \prod_{j=1}^{k} z_{i_j};$$

a formula which follows also directly by computing the trace of the action of t on $\bigwedge^k(\mathbf{C}^n)$. Proposition 4.10.3 says in this context that, on the algebraic variety $V = \{ z \in \mathbf{C}^n \mid \prod_{j=1}^{n} z_j = 1 \}$, the elementary symmetric functions ϵ_k, for $1 \leq k \leq n-1$, are algebraically independent, and generate the algebra of all symmetric polynomials on V as a free polynomial algebra in $n-1$ variables. But this means that every symmetric polynomial on $\mathbf{C}^n$ can be written as a polynomial in $(\prod_{j=1}^{n} z_j - 1)$, with coefficients which are unique polynomials in the ϵ_k, with $1 \leq k \leq n-1$; so Proposition 4.10.3 is equivalent to the classical theorem that the symmetric polynomials on $\mathbf{C}^n$ form a free polynomial algebra, with the ϵ_k, with $1 \le k \le n$, as generators. See Van der Waerden [1966], §33.

The action on the exterior powers is a polynomial mapping; it follows that every finite-dimensional continuous representation of $\mathbf{SU}(n)$ is given by a polynomial mapping. The complex-analytic representations of the complexification $\mathbf{SL}(n, \mathbf{C})$ are given by the corresponding complex polynomial mappings, in this way the representation theory of $\mathbf{SU}(n)$ coincides with the classification of the complex polynomial representations of $\mathbf{SL}(n, \mathbf{C})$ of Schur [1901]. In fact all finite-dimensional representations of all compact Lie groups are polynomial, cf. Corollary 14.6.2.

(G) Exercise. Let π, and π', be irreducible representations of G, with highest weights equal to λ, and λ', respectively. Then $\lambda + \lambda' \in \widehat{T}$, and $\lambda + \lambda'$ is dominant. So, according to Theorem 4.9.5, it is the highest weight of some irreducible representation of G. We shall now describe one which is naturally defined in terms of π and π'.

Recall that the product $\chi_\pi \chi_{\pi'}$ of the characters of π and π' is equal to the character of the tensor product representation $\pi \otimes \pi' := (\pi^\vee)^\vee \otimes \pi'$ of π and π', cf. (4.2.7) and (4.3.12). Writing χ_π, and $\chi_{\pi'}$, respectively, as in (4.9.1), we get that:

(4.10.12) $$m_{\pi\otimes\pi',\theta} = \sum_{\mu+\mu'=\theta} m_{\pi,\mu} m_{\pi',\mu'}.$$

Using Proposition 4.9.4.(iii),(c), we get that:

(4.10.13) $$m_{\pi\otimes\pi',\theta} \neq 0 \Rightarrow \theta \preceq \lambda + \lambda' \text{ and } m_{\pi\otimes\pi',\lambda+\lambda'} = 1.$$

In turn, this implies that if $[\tau] \in \widehat{G}$ has highest weight equal to θ, then:

(4.10.14) $$[\pi \otimes \pi' : \tau] \neq 0 \Rightarrow \theta \preceq \lambda + \lambda'.$$

If $\theta \neq \lambda + \lambda'$, then $\mu \preceq \theta \prec \lambda + \lambda'$, for all weights μ of τ; so there must be a $[\tau] \in \widehat{G}$ with $[\pi \otimes \pi' : \tau] \neq 0$ and with highest weight equal to $\lambda + \lambda'$. Because (4.10.12) shows that $[\pi \otimes \pi' : \tau] \le 1$, we have arrived at:

(4.10.15) $$\theta = \lambda + \lambda' \Rightarrow [\pi \otimes \pi' : \tau] = 1.$$

(4.10.4) Definition. *Let π, and π', be irreducible representations of the connected, compact Lie group G, with highest weights equal to λ, and λ', respectively. Then the Cartan product $\pi \otimes^{\mathrm{c}} \pi'$ of π and π' is the τ-isotypical component in $\pi \otimes \pi'$, where $[\tau] \in \widehat{G}$ has highest weight equal to $\lambda + \lambda'$.*

For the definition and properties of isotypical components, see the text up to Proposition 4.4.2.

(4.10.14) shows that $\pi \otimes^{\mathrm{c}} \pi'$ itself is irreducible and has highest weight equal to $\lambda + \lambda'$. Moreover, $[\pi \otimes \pi' : \pi \otimes^{\mathrm{c}} \pi'] = 1$. That is, the definition chooses a natural representative in the equivalence class $[\tau] \in \widehat{G}$.

(H) Exercise. The next exercise in this section is the question, how to read off the type of the irreducible representation π from its highest weight λ. Here the type of π is real, complex, or quaternionic, cf. Definition 4.8.2.

In the Lemma 4.10.5 below, we use that the Weyl group acts freely and transitively on the set of Weyl chambers, cf. Proposition 3.8.1; so there exists a unique element $s_P \in W$ which maps P to $-P$. In the terminology of the remark after Corollary 3.10.3, s_P is the unique element $s \in W$ such that $l_P(s) = \#(P) \geq l_P(s')$, for all $s' \in W$. Note that $s_P{}^2(P) = P$, hence $s_P{}^2 = \mathrm{I}$. In general, s_P is neither a reflection in a root hyperplane, nor equal to $-\mathrm{I}$; cf. the example of $\mathbf{SU}(n)$, after Proposition 4.10.8.

(4.10.5) Lemma. *The element $-s_P$ maps $\widehat{T}_{\mathrm{dom}}$ into itself, and permutes the fundamental weights. If π is an irreducible representation of G with highest weight equal to $\lambda = \sum_{\alpha \in S} n_\alpha \lambda_\alpha$, then $\pi^\vee$ has highest weight equal to $-s_P(\lambda) = \sum_{\alpha \in S} n_\alpha \cdot -s_P(\lambda_\alpha)$; it is of complex type, unless $\lambda = -s_P(\lambda)$, or $n_\alpha = n_\beta$, whenever $\lambda_\beta = -s_P(\lambda_\alpha)$.*

Proof. Definition 4.8.2, combined with Theorem 4.8.1, shows that π is not of complex type if and only if χ_π is real-valued, or π is equivalent to $\pi^\vee$. Now $\chi_{\pi^\vee}(t) = \chi_\pi(t^{-1})$, cf. (4.3.4); so μ is a weight of $\pi^\vee$ if and only if $-\mu$ is a weight of π. We obtain, using the Weyl group invariance of the set $\Omega(\pi)$ of weights of π, that $-s_P(\Omega(\pi)) = s_P(-\Omega(\pi)) = \Omega(\pi)$; and using that $-s_P(P) = s_P(-P) = P$, we get that $-s_P(\lambda)$ is a highest weight of π, or $-s_P(\lambda) = \lambda$, if and only if $[\pi] = [\pi^\vee]$.

The argument above also yield that $-s_P$ permutes the set of simple coroots, so its action on $i\mathfrak{t}^*$ permutes the fundamental weights as well. □

Remark. The transformation $-s_P$ is called the *opposition involution.* It follows from Corollary 5.6.5 that it is trivial, if and only if no component of the Dynkin diagram is of type A_l with $l \geq 2$, D_l with $l \geq 4$ and l odd, or E_6.

(4.10.6) Lemma. *For any irreducible representation π of G, $\pi^\vee \otimes^{\mathrm{c}} \pi$ is of real type.*

Proof. If π is a representation of G in the vector space V, then $\theta := \pi^\vee \otimes \pi$ is the representation of G in $\mathbf{L}(V,V)$, defined by: $(\theta(x))(L) = \pi(x) \circ L \circ \pi(x)^{-1}$, for $x \in G$, $L \in \mathbf{L}(V,V)$, cf. (4.2.7). The trace form $\tau\colon (L,M) \mapsto \mathrm{tr}(L \circ M)$, cf. (4.3.1), is a nondegenerate, symmetric bilinear form on $\mathbf{L}(V,V)$, invariant under θ.

We claim that the isotypical subspaces $\mathbf{L}(V,V)_\sigma$, and $\mathbf{L}(V,V)_\mu$, respectively, of $\mathbf{L}(V,V)$, for $[\sigma], [\mu] \in \widehat{G}$, are mutually orthogonal with respect to τ, unless $[\mu] = [\sigma^\vee]$. Indeed, a variation of the proof of Lemma 4.4.3 yields that $\tau(\theta(f)(L), M) = \tau(L, \theta(f^\vee)(M))$; hence, in view of (4.4.6),

$$\tau(\theta(f)(L), \theta(g)(M)) = \tau(L, \theta(f^\vee) \circ \theta(g)(M)) = \tau(L, \theta(f^\vee * g)(M)),$$

if $f, g \in \mathrm{C}(G)$, $L, M \in \mathbf{L}(V, V)$. Applying this to the linear projections $E_\sigma = \theta(d_\sigma \chi_\sigma)$, and $E_\mu = \theta(d_\mu \chi_\mu)$, onto $\mathbf{L}(V,V)_\sigma$, and $\mathbf{L}(V,V)_\mu$, respectively, we can establish the assertion.

The Cartan product $\pi^\vee \otimes^{\mathrm{c}} \pi$ is defined as the σ-isotypical component of $\mathbf{L}(V,V)$, where σ has highest weight equal to $-s_P(\lambda) + \lambda$, cf. Lemma 4.10.5. Because this is fixed under $-s_P$, another application of Lemma 4.10.5 shows that $[\sigma] = [\sigma^\vee]$. It follows that $\mathbf{L}(V,V)_\sigma$ is τ-orthogonal with respect to all other isotypical subspaces of $\mathbf{L}(V,V)$, which then implies that the restriction of τ to $\mathbf{L}(V,V)_\sigma$ is nondegenerate. Theorem 4.8.1 now yields that σ is of real type. □

If π is an irreducible representation of G, then the *type index* of π is defined to be equal to 1, 0, and -1, if π is of real, complex, and quaternionic type, respectively.

(4.10.7) Lemma. *Let π, and σ, be irreducible representations of G, with nonzero type index ϵ, and δ, respectively. Then $\pi \otimes^{\mathrm{c}} \sigma$ has type index equal to $\epsilon\delta$.*

Proof. Note that $\epsilon \neq 0$ implies that $[\pi] = [\pi^\vee]$. Let V, and U, be the representation space of π, and σ, respectively, so that $\mathbf{L}(V,U)$ is the representation space of $\pi^\vee \otimes \sigma \cong \pi \otimes \sigma$. Let B, and A, be the linear isomorphism: $V \to V^*$, and $U \to U^*$, representing the invariant nondegenerate bilinear form $(v_1, v_2) \mapsto B(v_2)(v_1)$, and $(u_1, u_2) \mapsto A(u_2)(u_1)$, on V, and U, respectively. That is, B, and A, intertwines π with $\pi^\vee$, and σ with $\sigma^\vee$, respectively; and $B^* = \epsilon B$, $A^* = \delta A$. Now the form $\theta(L, M) = \mathrm{tr}(B^{-1} \circ M^* \circ A \circ L)$, for $L, M \in \mathbf{L}(V,U)$, is clearly $\pi^\vee \otimes \sigma$-invariant, and nondegenerate. Moreover, the highest weights λ, and κ, of π, and σ, respectively, are fixed under $-s_P$, so the same is true for $\lambda + \kappa$. Applying the same argument as in the proof of Lemma 4.10.6, we get that the restriction of θ to $\pi^\vee \otimes^{\mathrm{c}} \sigma$ is nondegenerate. The proof is completed by the observation that, for all $L, M \in \mathbf{L}(V,U)$:

$$\begin{aligned}\theta(M, L) &= \mathrm{tr}(B^{-1} \circ L^* \circ A \circ M) = \mathrm{tr}(L^* \circ A \circ M \circ B^{-1}) \\ &= \mathrm{tr}(L^* \circ A \circ M \circ B^{-1})^* = \mathrm{tr}(B^{-1*} \circ M^* \circ A^* \circ L) \\ &= \epsilon\delta\, \mathrm{tr}(B^{-1} \circ M^* \circ A \circ L) = \epsilon\delta\, \theta(L, M).\end{aligned}$$

□

(4.10.8) Proposition. *Let π be an irreducible representation of G with highest weight equal to $\lambda = \sum_{\alpha \in S} n_\alpha \lambda_\alpha$.*

Then π is of quaternionic type, if and only if: (i) $n_\alpha = n_\beta$, whenever $[\pi(\lambda_\alpha)] = [\pi(\lambda_\beta)^\vee]$; and (ii) $\sum_{\alpha \in Q} n_\alpha$ is odd; here Q denotes the set of $\alpha \in S$ such that $[\pi(\lambda_\alpha)]$ is of quaternionic type.

Further, π is of real type if (i) holds, but not (ii); and π is of complex type if (i) does not hold.

Proof. Lemma 4.10.5 gave that π is of complex type if and only if (i) does not hold. Assuming that (i) holds, we let C be the set of $\alpha \in S$ such that $[\pi(\lambda_\alpha)]$ is of complex type. Then $-s_P$ induces an involution without fixed points in C; we can therefore write C as the disjoint union of C^+ and C^-, with $C^- = -s_P(C^+)$. Let π_0, and π_1, be an irreducible representation of G with highest weight equal to $\sum_{\alpha\in C^+} n_\alpha\lambda_\alpha$, and $\sum \alpha \in C n_\alpha\lambda_\alpha$, respectively. Then $[\pi_1] = [\pi_0^\vee \otimes^c \pi_0]$, which is of real type, according to Lemma 4.10.6. Repeated application of Lemma 4.10.7, starting with π_1, and taking successive Cartan products with the fundamental representations with highest weights λ_α, and this n_α times for each $\alpha \in S \setminus C$, leads to the desired result. □

Example. For $G = \mathbf{SU}(n)$, continuing the discussion following Proposition 4.10.3, let us denote the weight of the representation $\bigwedge^k$ of G in $\bigwedge^k(\mathbf{C}^n)$ by λ_k, for $1 \le k \le n-1$. The pairing:

$$(a, b) \mapsto a \wedge b\colon \bigwedge^{k}(\mathbf{C}^n) \times \bigwedge^{n-k}(\mathbf{C}^n) \to \bigwedge^{n}(\mathbf{C}^n),$$

combined with the triviality of the action of $\mathbf{SU}(n)$ on the right hand side (because $\det x = 1$, for all $x \in \mathbf{SU}(n)$), makes that $[\bigwedge^k] = [(\bigwedge^{n-k})^\vee]$, or $\lambda_k = -s_P(\lambda_{n-k})$. So there exist only fundamental representations of noncomplex type, if n is even, say $n = 2p$; and then $\bigwedge^k$ is of noncomplex type if and only if $k = p$. In this case the pairing is symmetric, and anti-symmetric, if and only if p is even, and odd, respectively.

The conclusion is that the irreducible representation of $\mathbf{SU}(n)$ with highest weight equal to:

$$\lambda = \sum_{k=1}^{n-1} n_k\lambda_k$$

is of complex type, if $n_k \ne n_{n-k}$, for some $1 \le k \le n/2$. It is of quaternionic type, if $n = 2 \pmod 4$, $n_k = n_{n-k}$, for $1 \le k < p$, and n_p is odd, if we write $p = n/2$. In all other cases it is of real type.

(I) Exercise. Another approach is to use the criterion of Proposition 4.8.7. For this, we have to compute:

$$\begin{aligned}\int_G \chi_\lambda(x^2)\,dx &= \#(W)^{-1}\int_T \delta(t)\overline{\delta(t)}\,\chi_\lambda(t^2)dt\\ &= \#(W)^{-1}\int_T \Big(\sum_{s\in W}\det s\; t^{s\cdot\rho-\rho}\Big)\Big(\sum_{s\in W}\det s\; t^{\rho-s\cdot\rho}\Big)\Big(\sum_\mu m_\mu\, t^{2\mu}\Big)\,dt\\ &= \#(W)^{-1}\int_T \sum_{s,s',\mu}(\det s'\det s)\; m_\mu\, t^{s'\cdot\rho-s\cdot\rho+2\mu}\,dt\\ &= \#(W)^{-1}\sum_{\{\,s,s',\mu\mid s'\cdot\rho-s\cdot\rho+2\mu=0\,\}}(\det s')(\det s)m_\mu\\ &= \sum_{w\in W}(\det w)\; m_{\frac12(\rho-w\cdot\rho)}.\end{aligned}$$

Here we have used that the function $x\mapsto\chi_\lambda(x^2)$ on G is conjugacy-invariant, so that we could apply Weyl's integral formula (3.14.8) for conjugacy-invariant functions. We have also substituted the formula (4.9.15) for $\delta(t)$; and we have put $s'=s\cdot w$, thereby using that $(\det s)^2=1$ and that $m_{s\cdot\theta}=m_\theta$, for all $s\in W$. This leads to the following generalization of the criterion which we used in Fig.4.10.1.

(4.10.9) Lemma. *The irreducible representation π with highest weight equal to λ is of real, complex, and quaternionic type, if $\sum_{w\in W}\det w\; m_{\frac12(\rho-w\cdot\rho)}$ is equal to 1, 0, and -1, respectively. Here $m_\theta=0$, if θ is not a weight.*

In general, this criterion does not look efficient, because in any computational scheme the multiplicities of weights μ close to 0, such as the $\frac12(\rho-w\cdot\rho)$, for $w\in W$, come late if λ is large. However, in special cases, Lemma 4.10.9 can be useful; for instance, it seems reasonable that it can help in the determination of the types of the fundamental representations of arbitrary simple Lie algebras.

As an exercise, let us once more determine the types for $G=\mathbf{SU}(3)$, starting from the discussion in the beginning of Exercise B.

The $\frac12(\rho-w\cdot\rho)$, for $w\in W$, are equal to:

$$0,\quad \tfrac12\alpha=\lambda_\alpha-\tfrac12\lambda_\beta,\quad \alpha+\frac12\beta=\frac23\lambda_\alpha,\quad \alpha+\beta=\lambda_\alpha+\lambda_\beta,$$
$$\tfrac12\alpha+\beta=\tfrac23\lambda_\beta,\quad \tfrac12\beta=\lambda_\beta-\tfrac12\lambda_\alpha.$$

Of these, only 0, and $\alpha+\beta$, are weights of $\mathfrak{g}$; these correspond to $\det w=1$, and $\det w=-1$, respectively. So the irreducible representation π of $\mathbf{SU}(3)$ with highest weight equal to λ is of real, complex, and quaternionic type, if $m_0-m\alpha+\beta$ is equal to 1, 0, and -1, respectively. From the description of the multiplicities preceding Fig.4.10.7, we see now that π is of complex type, unless $\lambda=k(\alpha+\beta)$, for some $k\in\mathbf{Z}_{\geq 0}$; and then π is of real type. Of course, one should also compare the answer with the result of the more general discussion following Proposition 4.10.8.

(J) Exercise. *Kostant's partition function* is defined as the function that assigns to each μ the number $P(\mu)$ of distinct functions:

$$\alpha \mapsto n_\alpha : P \to \mathbf{Z}_{\geq 0}, \qquad \text{such that } \mu = \sum_{\alpha \in P} n_\alpha \alpha,$$

cf. the definition of the ordering in $\widehat{T}$ preceding Theorem 4.9.5. One has the formal identity:

$$\delta_P(t)^{-1} = \sum_{\mu \succeq 0} P(\mu) t^{-\mu}.$$

Using Weyl's character formula (4.9.14), this in turn implies that the multiplicity of μ in the irreducible representation of G with highest weight λ is equal to:

$$m_\mu = \sum_{s \in W} \det s \; P(s \cdot \lambda + s \cdot \rho - \rho - \mu). \tag{4.10.16}$$

(K) Exercise. Let π_λ, π_μ, and π_θ, be irreducible representations of G with highest weight equal to λ, μ, and θ. Then:

$$\chi_\lambda(t)\chi_\mu(t) = \sum_\theta [\pi_\lambda \otimes \pi_\mu : \pi_\theta] \chi_\theta(t),$$

or, multiplying both sides with $\delta_P(t)$, we get:

$$\Big(\sum_{\delta \in \widehat{T}} m_{\lambda,\delta}\, t^\delta\Big)\Big(\sum_{s \in W} \det s \; t^{s \cdot \mu + s \cdot \rho - \rho}\Big)$$
$$= \sum_\theta [\pi_\lambda \otimes \pi_\mu : \pi_\theta] \Big(\sum_{s \in W} \det s \; t^{s \cdot \theta + s \cdot \rho - \rho}\Big).$$

Comparing coefficients of dominant powers, we obtain:

$$[\pi_\lambda \otimes \pi_\mu : \pi_\theta] = \sum_{s \in W} \det s \; m_{\lambda, \theta - s \cdot \mu - s \cdot \rho + \rho}. \tag{4.10.17}$$

Inserting (4.10.16), we get the more symmetric expression:

$$[\pi_\lambda \otimes \pi_\mu : \pi_\theta] = \sum_{s, w \in W} \det s \, \det w \; P(w \cdot \lambda + s \cdot \mu - \theta + w \cdot \rho + s \cdot \rho - 2\rho). \tag{4.10.18}$$

For $G = \mathbf{SU}(2)$, the result is the classical *formula of Clebsch-Gordan*; this name has become customary for any computation of the multiplicities of an irreducible representation of a compact connected Lie group in a tensor product of such ones.

4.11 Highest Weight Vectors

If π is a representation of a Lie group G in a finite-dimensional complex vector space V, then, as already has been observed in Introduction 4.0, the Corollary 1.10.9 implies that $\pi\colon G \to \mathbf{GL}(V)$ is real-analytic. That is, **all matrix coefficients of finite-dimensional representations of G are real-analytic functions on G.** This suggests to look for differential equations satisfied by the matrix coefficients, an idea that will turn out to be very fruitful.

Because π is a homomorphism of Lie groups, $\mathrm{T}_1\,\pi$ is a homomorphism of Lie algebras: $\mathfrak{g} \to \mathbf{L}(V, V)$. Here the Lie algebra structure of $\mathbf{L}(V, V)$ is given by the commutator $[A, B] = A \circ B - B \circ A$, cf. (1.1.18). With a slight abuse of notation, this homomorphism of Lie algebras will also be denoted by π.

Since V is a complex vector space, $\mathbf{L}(V, V)$ is a complex Lie algebra, and π extends in a unique way to a homomorphism of complex Lie algebras: $\mathfrak{g}_{\mathbf{C}} \to \mathbf{L}(V, V)$, which will again be denoted by π. We have to be careful with the notation, if $\mathfrak{g}$ already has a structure of complex Lie algebra: then $\mathfrak{g}_{\mathbf{C}} = \mathfrak{g} \oplus i\mathfrak{g}$ possibly gets two complex structures. However, such collisions of notation will not happen in this section, because we shall concentrate on the case that G is compact; the fact that the roots only take purely imaginary values, cf. Lemma 3.5.1, then shows that $\mathfrak{g}$ can only be a complex Lie algebra if it is Abelian.

The study of homomorphisms of complex Lie algebras $\pi\colon \mathfrak{g}_{\mathbf{C}} \to \mathbf{L}(V, V)$, also called *representations of complex Lie algebras*, is a purely algebraic matter. Frequently we shall refer to the results for compact Lie groups that previously we have obtained with partially analytic methods, nevertheless the reader will have no problems in identifying which of the arguments given below are purely algebraic; moreover some light is shed on the representation theory for compact Lie groups from a new direction.

Assume from now on that G is a connected, compact Lie group, with Lie algebra equal to $\mathfrak{g}$. As in Section 3.5, we start by selecting a maximal Abelian subspace $\mathfrak{t}$ of $\mathfrak{g}$. Because $[\pi(X), \pi(Y)] = \pi([X, Y]) = 0$, whenever $X, Y \in \mathfrak{t}$, and the $\mathrm{e}^{t\pi(X)} = \pi(\exp tX)$, for $t \in \mathbf{R}$, remain in a compact subset of $\mathbf{L}(V, V)$, for every $X \in \mathfrak{g}$, the argument leading to the root space decomposition (3.5.7) now gives a direct sum decomposition:

(4.11.1)
$$V = \bigoplus_{\mu} V_{\mu}.$$

Here the sum is over finitely many distinct real linear forms $\mu\colon \mathfrak{t} \to i\mathbf{R}$, and the corresponding *weight spaces* V_{μ} are defined by:

(4.11.2)
$$V_{\mu} := \{\, v \in V \mid \pi(X)(v) = \mu(X)v, \text{ for all } X \in \mathfrak{t} \,\}.$$

The $\mu \in i\mathfrak{t}^*$, for which $V_\mu \neq 0$, are called the *weights* of π. Extending μ to a complex linear form on $\mathfrak{h} := \mathfrak{t} \oplus i\mathfrak{t}$, which will also be denoted by μ, it follows that:

$$\pi(X)(v) = \mu(X)v, \quad \text{for } X \in \mathfrak{h}, v \in V_\mu. \tag{4.11.3}$$

If π is the adjoint representation ad of $\mathfrak{g}_{\mathbf{C}}$ in $\mathfrak{g}_{\mathbf{C}}$, then (4.11.1) is just the root space decomposition (3.5.7); the roots are the nonzero weights of ad, whereas $\mathfrak{h}$ is equal to the weight space for $\mu = 0$. Also, writing $t = \exp X$, for $X \in \mathfrak{t}$, we get that $\pi(t)$ acts on V_μ as multiplication by $e^{\mu(X)} = t^\mu$; so we recognize the weights of π as being the same as the weights of π, introduced in (4.9.1).

(4.11.1) Lemma. *For each weight μ of π, and root α of $\mathfrak{g}_{\mathbf{C}}$, we have:*

$$\pi(X)(V_\mu) \subset V_{\mu+\alpha}, \quad \textit{for } X \in \mathfrak{g}_\alpha. \tag{4.11.4}$$

Proof. If $v \in V_\mu$, $H \in \mathfrak{h}$, then:

$$\begin{aligned}\pi(H) \circ \pi(X)(v) &= \pi(X) \circ \pi(H)(v) + [\pi(H), \pi(X)](v) \\ &= \pi(X)(\mu(H)v) + \pi([H, X])(v) = \mu(H)\pi(X)(v) + \pi(\alpha(H)X)(v) \\ &= (\mu(H) + \alpha(H))\pi(X)(v).\end{aligned}$$

□

Remark. In view of (4.11.4), $\pi(X)$ is called a *raising operator*, and a *lowering operator*, if $X \in \mathfrak{g}_\alpha$, for $\alpha \in P$, and $\alpha \in -P$, respectively.

If π equals the adjoint representation, then (4.11.4) takes the form:

$$[\mathfrak{g}_\alpha, \mathfrak{g}_\beta] \subset \mathfrak{g}_{\alpha+\beta}, \quad \text{for all } \alpha, \beta \in R, \tag{4.11.5}$$

a fact which we have mentioned before in (3.10.2), but not used at all, yet.

From now on, P denotes a fixed choice of positive roots. Recall the partial order $\preceq$ in the set of weights, which was defined after Proposition 4.9.4, in terms of the choice P.

(4.11.2) Lemma. *Let λ be a weight of π, and v a nonzero element of V_λ, such that $\pi(X)(v) = 0$, whenever $X \in \mathfrak{g}_\alpha$, with $\alpha \in P$. Let U be the linear subspace of V, spanned by v, and the vectors $\pi(X_k) \circ \ldots \circ \pi(X_1)(v)$, where $k \in \mathbf{Z}_{>0}$, and $X_j \in \mathfrak{g}_{-\alpha_j}$, with $\alpha_j \in P$, for all $1 \leq j \leq k$. Then:*
(i) $U = \mathbf{C} \cdot v \oplus_{\mu \prec \lambda} U \cap V_\mu$;
(ii) U is $\pi(\mathfrak{g}_{\mathbf{C}})$-invariant.

Proof. (i) From (4.11.4), we get $\pi(X_k) \circ \ldots \circ \pi(X_1)(v) \in V_{\mu-\alpha_1-\ldots-\alpha_k}$, by induction over k.

(ii) a) That U is $\pi(\mathfrak{g}_{-\alpha})$-invariant for every $\alpha \in P$, is obvious from the definition.

b) If $X \in \mathfrak{h}$, then, by induction over k, we get:

$$\pi(X) \circ \pi(X_k) \circ \ldots \circ \pi(X_1)(v) = \pi(X_k) \circ \ldots \circ \pi(X_1) \circ \pi(X)(v)$$
$$+ \sum_{j=1}^{k} \pi(X_k) \circ \ldots \circ \pi(X_{j+1}) \circ [\,\pi(X), \pi(X_j)\,] \circ \pi(X_{j-1}) \circ \ldots \circ \pi(X_1)(v).$$

Because $\pi(X)(v) = \mu(X)v$, and

$$[\,\pi(X), \pi(X_j)\,] = \pi([\,X, X_j\,]) = \pi(-\alpha_j(X)X_j) = -\alpha_j(X)\pi(X_j),$$

all terms belong to U.

c) If $X \in \mathfrak{g}_\alpha$, with $\alpha \in P$, then $\pi(X)(v) = 0$ causes the first term in the sum above to be equal to zero. Furthermore, $[\,X, X_j\,] \in \mathfrak{g}_{\alpha-\alpha_j}$, cf. (4.11.5), shows that the j-th term belongs to U, if $\alpha = \alpha_j$, or $\alpha_j - \alpha \in P$. On the other hand, if $\alpha - \alpha_j \in P$, then replacing k by $j-1$, α by $\alpha - \alpha_j$ and X by $[\,X, X_j\,]$, we get by induction on k that

$$[\,\pi(X), \pi(X_j)\,] \circ \pi(X_{j-1}) \circ \ldots \circ \pi(X_1)(v)$$

belongs to U, and therefore also the j-th term in the sum, in view of a). □

In order to simplify notation, it will be convenient to introduce the complex linear subspaces:

(4.11.6)
$$\mathfrak{n} := \sum_{\alpha \in P} \mathfrak{g}_\alpha; \quad \overline{\mathfrak{n}} := \sum_{\alpha \in P} \mathfrak{g}_{-\alpha}$$

of $\mathfrak{g}_{\mathbf{C}}$. In view of (4.11.5), these are Lie subalgebras of $\mathfrak{g}_{\mathbf{C}}$. Note that $\overline{\mathfrak{n}}$ is equal to the complex conjugate of $\mathfrak{n}$. The choice of the letter $\mathfrak{n}$ is explained by the following property of nilpotency in any finite-dimensional representation of G:

(4.11.3) Lemma. *Let π be a representation of G in a finite-dimensional vector space V. Then there exists a basis of V on which $\pi(X)$ is a strictly lower triangular, diagonal, and strictly upper triangular matrix, if $X \in \mathfrak{n}$, $X \in \mathfrak{h}$, and $X \in \overline{\mathfrak{n}}$, respectively.*

Proof. Let $U \in \mathfrak{h}$ be chosen such that $\alpha(U) > 0$, for all $\alpha \in P$. For instance, one may choose $U = \rho^\vee := \frac{1}{2}\sum_{\alpha \in P} \alpha^\vee$, and apply the dual version of (4.9.21). Take a basis e_j of V, consisting of successive bases of the weight spaces V_μ, ordered such that $j < k$, if $e_j \in V_\mu$, $e_k \in V_\theta$, $\mu(U) < \theta(U)$. If $e_j \in V_\mu$, and $X \in \mathfrak{n}$, or in $\overline{\mathfrak{n}}$, then (4.11.4) shows that $\pi(X)(e_j)$ is a linear combination of basis elements $e_k \in V_{\mu+\alpha}$, with $\alpha \in P$, or with $\alpha \in -P$, respectively; hence $(\mu+\alpha)(H) > \mu(H)$, and $(\mu+\alpha)(H) < \mu(H)$, and this implies that $k > j$, and

$k < j$, respectively. On the other hand, if $X \in \mathfrak{h}$, then $\pi(X)(e_j) = \mu(X)e_j$. □

On the other hand the notation is not so informative, because it does not contain the choice of positive roots P. Lemma 4.11.2.(ii) can now be phrased as: "If v is a weight vector (that is, an element of a weight space) that is annihilated by $\pi(\mathfrak{n})$, then the smallest $\pi(\overline{\mathfrak{n}})$-invariant linear subspace of V which contains v is actually $\pi(\mathfrak{g}_{\mathbf{C}})$-invariant".

(4.11.4) Theorem. *Let π be an irreducible representation of G in the finite-dimensional vector space V. Then:*

$$V^{\mathfrak{n}} := \{ v \in V \mid \pi(X)(v) = 0, \text{ if } X \in \mathfrak{n} \}$$

is a one-dimensional linear subspace of V, equal to V_λ, for some weight λ of π; and V is equal to the space U described in Lemma 4.11.2. The weight λ is uniquely determined by each of the following conditions (i), (ii):

(i) $\lambda + \alpha$ is not a weight of π, for every $\alpha \in P$.
(ii) $\mu \preceq \lambda$, for each weight μ of π.

Proof. Because there are only finitely many weights of π, and the convex cone generated by P is proper, cf. (3.5.13), there is at least one weight λ satisfying (i). Lemma 4.11.1 yields that $V_\lambda \subset V^{\mathfrak{n}}$, so $V^{\mathfrak{n}} \neq 0$.

Now let $v \in V^{\mathfrak{n}}$, $v \neq 0$. Because of (4.11.1), we can write $v = \sum_{\mu \in \Omega(\pi)} v_\mu$, with $v_\mu \in V_\mu$, for all $\mu \in \Omega(\pi)$, the set of weights of π. For any $X \in \mathfrak{g}_\alpha$, with $\alpha \in P$, this implies that $0 = \pi(X)(v) = \sum_{\mu \in \Omega(\pi)} \pi(X)(v_\mu)$. Now $\pi(X)(v_\mu) \in V_{\mu+\alpha}$, cf. (4.11.4); and because the $\mu + \alpha$ are different for different μ, the directness of the sum decomposition (4.11.1) implies that $\pi(X)(v_\mu) = 0$, for every $\mu \in \Omega(\pi)$, $X \in \mathfrak{g}_\alpha$, $\alpha \in P$. That is, $v_\mu \in V^{\mathfrak{n}}$, for each $\mu \in \Omega(\pi)$.

Let $\lambda \in \Omega(\pi)$ be such that $v_\lambda \neq 0$. There is at least one such λ, because $v \neq 0$. Because G is connected, the representation π of $\mathfrak{g}_{\mathbf{C}}$ is also irreducible; so the space U which we obtain in Lemma 4.11.2 (with v replaced by v_λ), is equal to V. That is, $V_\lambda = \mathbf{C} \cdot v_\lambda$, and $\mu \prec \lambda$, for each other weight μ of π. Or, λ satisfies (ii). Because there can only be one weight λ satisfying (ii), we get that $v = v_\lambda$ for that one; and because the reasoning remains valid for any $v \in V^{\mathfrak{n}} \setminus \{0\}$, it follows that $V^{\mathfrak{n}} = V_\lambda$.

Finally, if λ satisfies the weaker condition (i), then another look at (4.11.4) shows that $V_\lambda \subset V^{\mathfrak{n}}$, so if λ is also a weight of π, then $V_\lambda = V^{\mathfrak{n}}$ and λ satisfies (ii). □

In terms of the definition following Proposition 4.9.4, the weight λ which appears in Theorem 4.11.4, is equal to the highest weight of π. The nonzero $v \in V_\lambda$ are therefore said to be the *highest weight vectors* of the representation π. Theorem 4.11.4 will be referred to as **Cartan's highest weight theorem**.

Because the proof above is independent of the proof of Weyl's character formula in Section 4.9, it leads to a new proof of the equivalence between b) and c) in Proposition 4.9.4.(iii), and the one-dimensionality of the corresponding weight space V_λ. Also, applying Cartan's highest weight theorem in the situation that there is only one positive root, and using Lemma 4.11.6 below, we recover that $\lambda(\alpha^\vee) \in \mathbf{Z}_{\geq 0}$, and that the weights of π form a ladder $\lambda - k\alpha$, with $k \in \mathbf{Z}$, $0 \leq k \leq \lambda(\alpha^\vee)$. That is, we obtain a new proof of Lemma 4.9.3 and Proposition 4.9.4.(i).(ii). In turn, Proposition 4.9.4.(i) implies that $\Omega(\pi)$, the set of weights of π, is invariant under the Weyl group W, which here is defined as the group generated by the reflections s_α, for $\alpha \in R$.

Now let σ be any representation of G in the complete, locally convex topological vector space U, not necessarily irreducible or even finite-dimensional. Recall the definition of the space U^{fin} of $\sigma(G)$-finite vectors in Section 4.4. Each $u \in U^{\mathrm{fin}}$ belongs to a finite-dimensional $\sigma(G)$-invariant linear subspace U_0, on which $\sigma(X)$ is defined, for any $X \in \mathfrak{g}_{\mathbf{C}}$. The *highest weight space of finite type* of σ is defined as:

$$U^{\mathrm{fin},\mathfrak{n}} = \{\, u \in U^{\mathrm{fin}} \mid \sigma(X)(u) = 0, \text{ for all } X \in \mathfrak{n} \,\}. \tag{4.11.7}$$

From Theorem 4.11.4, it now follows that for any direct sum decomposition of U^{fin} into irreducible subspaces U_j, cf. Corollary 4.2.2 and Proposition 4.4.4, the intersection $U_j \cap U^{\mathrm{fin},\mathfrak{n}}$ is one-dimensional. We get:

(4.11.5) Corollary. *For an arbitrary representation σ of G, the highest weight space of finite type is finite-dimensional if and only if σ itself is finite-dimensional. In this case, the dimension of the highest weight space is equal to $\sum_{[\pi]\in\widehat{G}}[\sigma : \pi]$; and for each $[\pi] \in \widehat{G}$ with highest weight equal to λ, we have: $[\sigma : \pi] = [(\sigma|_T)|_{U^{\mathfrak{n}}} : \lambda]$.*

In terms of Remark (A) after Proposition 4.4.4, the highest weight space of finite type is equal to the direct sum over the $[\pi] \in \widehat{G}$ of the spaces $Z_\pi = \{\, A(v_\pi) \mid A \in \mathrm{I}(\pi, \sigma) \,\}$, where the v_π are highest weight vectors for the irreducible representations π. The remarkable fact about the whole matter is the uniformity in the choice for all, usually infinitely many, $[\pi] \in \widehat{G}$.

(4.11.6) Proposition. *Let π, and π', be an irreducible representation of $\mathfrak{g}_{\mathbf{C}}$ in V, and V', with highest weight equal to λ, and λ', respectively. Then π' is equivalent to π if and only if $\lambda' = \lambda$.*

Proof. The graphs V'' of (π, π')-intertwining operators: $V \to V'$ are the $(\pi \times \pi')(\mathfrak{g}_{\mathbf{C}})$-invariant subspaces of $V \times V'$ for which the projection: $V'' \to V$ is bijective. Now let $v \in V^{\mathfrak{n}} \setminus \{0\}$, $(v, v') \in V''$. If $X \in \mathfrak{n}$, $H \in \mathfrak{h}$, then $\pi(X)(v) = 0$, $\pi(H)(v) = \lambda(H)v$; hence $\pi'(X)(v') = 0$, $\pi'(H)(v') = \lambda(H)v'$. So $v' \in V'^{\mathfrak{n}} \cap V'_\lambda$; and we get $\lambda' = \lambda$, if $v' \neq 0$. If conversely $\lambda' = \lambda$, choose $v' \in V'^{\mathfrak{n}} = V'_\lambda$, $v' \neq 0$. Then $(v, v') \in (V \times V')^{\mathfrak{n}} \cap (V \times V')_\lambda$; let V'' be the smallest $\pi(\mathfrak{g}_{\mathbf{C}})$-invariant subspace of $V \times V'$ that contains (v, v'). It follows from Lemma 4.11.2 that $V''^{\mathfrak{n}}$ is one-dimensional; so by Corollary 4.11.5, V'' is $\pi \times \pi'$-irreducible. The projection $V'' \to V$, and $V'' \to V'$, intertwines $(\pi \times \pi')|_{V''}$ with π, and π', respectively, and is nonzero; so by Schur's lemma 4.1.1.(ii), it is an equivalence. The conclusion is that $\pi \sim \pi'$. □

4.12 The Borel-Weil Theorem

Looking at the highest weight vectors of the right regular representation of G in $\mathrm{C}(G)$, we shall arrive at a canonical way of singling out irreducible representations of G, one for each equivalence class, in the space of matrix coefficients. We then shall look at these representations in several different ways. Although the transition from one to another is quite straightforward, we have taken our time for extensive discussions of these models.

(4.12.1) Proposition.

(i) *The space* $\mathrm{M}(G)^{\mathrm{R}^*\mathfrak{n}}$ *of the matrix coefficients that are annihilated by the first-order linear partial differential operators* $\mathrm{R}^*(X)$*, with* $X \in \mathfrak{n}$*, is* $\mathrm{L}(G)$*-invariant.*

(ii) *Write:*

$$\mathrm{L}^{\mathrm{R}^*\mathfrak{n}} : x \mapsto \mathrm{L}(x)|_{\mathrm{M}(G)^{\mathrm{R}^*\mathfrak{n}}}.$$

Then each $[\pi] \in \widehat{G}$ *occurs in* $\mathrm{L}^{\mathrm{R}^*\mathfrak{n}}$ *with multiplicity equal to* 1.

(iii) *For each* $[\pi] \in \widehat{G}$*, the space* $\mathrm{M}_\pi \cap \mathrm{M}(G)^{\mathrm{R}^*\mathfrak{n}}$ *is equal to the* $\pi^\vee$*-isotypical subspace of* $\mathrm{L}^{\mathrm{R}^*\mathfrak{n}}$*, and is also equal to:*

$$\mathrm{M}(G)^{\mathrm{R}^*\mathfrak{n}}_\lambda := \{\, f \in \mathrm{M}(G)^{\mathrm{R}^*\mathfrak{n}} \mid \mathrm{R}^*(H)(f) = \lambda(H)f, \text{ for all } H \in \mathfrak{h} \,\}, \tag{4.12.1}$$

if λ *denotes the highest weight of* π.

(iv) *The representation of* G*:*

$$\mathrm{L}^{\mathrm{R}^*\mathfrak{n}}_\lambda : x \mapsto \mathrm{L}(x)|_{\mathrm{M}(G)^{\mathrm{R}^*\mathfrak{n}}_\lambda}$$

is irreducible, and has highest weight equal to $-s_P \cdot \lambda$.

Proof. (i) This follows from the fact that $\mathrm{L}(x)$ and $\mathrm{R}^*(y)$ commute for all $x, y \in G$, cf. (4.3.8,9). In turn this implies that $\mathrm{L}(x)$ and $\mathrm{R}^*(Y)$ commute on $\mathrm{M}(G)$ for all $x \in G$, $Y \in \mathfrak{g}$; so by complex-linear extension, $\mathrm{L}(x)$ and $\mathrm{R}^*(Y)$ commute for all $x \in G$, $Y \in \mathfrak{g}_{\mathbf{C}}$.

(ii) Corollary 4.11.5 shows that:

$$\dim \mathrm{M}_\pi = [\mathrm{R}^* : \pi]\, \mathrm{d}_\pi = \dim({\mathrm{M}_\pi}^{\mathrm{R}^*\mathfrak{n}})\, \mathrm{d}_\pi\,.$$

On the other hand,

$$\dim({\mathrm{M}_\pi}^{\mathrm{R}^*\mathfrak{n}}) = m\,\mathrm{d}_{\pi^\vee} = m\,\mathrm{d}_\pi,$$

with $m = [\mathrm{L}\,|_{{\mathrm{M}_\pi}^{\mathrm{R}^*\mathfrak{n}}} : \pi^\vee]$. Now $0 < \dim \mathrm{M}_\pi \leq {\mathrm{d}_\pi}^2$, as proved for arbitrary compact groups in Lemma 4.3.1. It follows that $m = 1$; and this also verifies that $\dim \mathrm{M}_\pi = {\mathrm{d}_\pi}^2$, cf. Corollary 4.3.6.

(iii) M_π, the space of matrix coefficients of π, is equal to the $\pi^\vee$-isotypical subspace for L, cf. Lemma 4.3.1; and this proves the first part of the sentence. This space is also equal to the π-isotypical subspace of $\mathrm{M}(G)$ for R^*; so Theorem 4.11.4 implies that ${\mathrm{M}_\pi}^{\mathrm{R}^*\mathfrak{n}} \subset \mathrm{M}(G)_\lambda^{\mathrm{R}^*\mathfrak{n}}$. On the other hand, Proposition 4.11.6 shows that $\mathrm{M}_{\pi'} \cap \mathrm{M}(G)_\lambda^{\mathrm{R}^*\mathfrak{n}} = 0$, if $[\pi'] \in \widehat{G}$, and $[\pi'] \neq [\pi]$. The space $\mathrm{M}(G)$ is equal to the direct sum of the R^*-invariant subspaces $M_{\pi'}$; so $\mathrm{M}(G)_\lambda^{\mathrm{R}^*\mathfrak{n}}$ is equal to the direct sum of the $\mathrm{M}_{\pi'} \cap \mathrm{M}(G)_\lambda^{\mathrm{R}^*\mathfrak{n}}$, for $[\pi'] \in \widehat{G}$.

(iv) The representation $\pi^\vee$ of $\mathfrak{g}_{\mathbf{C}}$ which is contragredient to π, is defined by $\pi^\vee(X) = -\pi(X)^*$, for $X \in \mathfrak{g}_{\mathbf{C}}$. It follows that $\Omega(\pi^\vee) = -\Omega(\pi)$. Also, $\Omega(\pi)$ is invariant under the Weyl group W, as observed after Cartan's highest weight theorem 4.11.4. Recalling the element $s_P \in W$ which sends P to $-P$, cf. the discussion preceding Lemma 4.10.5, we see that $-s_P$ maps $\Omega(\pi)$ onto $\Omega(\pi^\vee)$, and leaves the ordering invariant. It follows that $-s_P(\lambda)$ is equal to the highest weight of $\pi^\vee \sim \mathrm{L}_\lambda^{\mathrm{R}^*\mathfrak{n}}$. □

Our next purpose is to show that the space (4.12.1) remains the same if, in the right hand side, the space $\mathrm{M}(G)$ of matrix coefficients is replaced by any reasonable space of functions, or generalized functions, on G. Suppose that $\mathcal{F}$ is a dense linear subspace of $\mathrm{C}(G)$, provided with a topology stronger than the topology of uniform convergence, and for which $\mathcal{F}$ is a complete, locally convex topological vector space. Also suppose that $\mathcal{F}$ is $\mathrm{L}(G)$-, and $\mathrm{R}^*(G)$-invariant, and that the action $(x, f) \mapsto \mathrm{L}(x)(f)$, and $(y, f) \mapsto \mathrm{R}^*(y)(f)$, respectively, is continuous: $G \times \mathcal{F} \to \mathcal{F}$.

Because $\mathcal{F}$ is an $\mathrm{L}(G)$-, and $\mathrm{R}^*(G)$-invariant subspace of $\mathrm{C}(G)$, we get that the space $\mathcal{F}^{\mathrm{fin}}$ of $\mathrm{L}(G)$-, and $\mathrm{R}^*(G)$-finite vectors, in $\mathcal{F}$ is contained in $\mathrm{M}(G)$, the space of $\mathrm{L}(G)$-, and $\mathrm{R}^*(G)$-finite vectors, respectively, in $\mathrm{C}(G)$, cf. Theorem 4.4.5.(iii). On the other hand, it is a consequence of the Peter-Weyl theorem that $\mathcal{F}^{\mathrm{fin}}$ is dense in $\mathcal{F}$, cf. Corollary 4.6.3.(i); so $\mathcal{F}^{\mathrm{fin}}$ is dense in $\mathrm{C}(G)$ as well. Applying the continuous linear projection $E_\pi \colon \mathrm{C}(G) \to \mathrm{M}_\pi$, it follows that $\mathcal{F}^{\mathrm{fin}} \cap \mathrm{M}_\pi = E_\pi(\mathcal{F}^{\mathrm{fin}})$ is dense in M_π, and thus equal to M_π, since M_π is finite-dimensional. Because this holds for every $[\pi] \in \widehat{G}$, the conclusion is:

$$\mathcal{F}^{\mathrm{fin}} = \mathrm{M}(G). \tag{4.12.2}$$

We now view $\mathcal{F}$ as a space of "test functions", so that its topological dual $\mathcal{F}'$, the space of continuous linear forms on $\mathcal{F}$, is the corresponding space of "distributions". The continuous dense embeddings $\mathrm{M}(G) \subset \mathcal{F} \subset \mathrm{C}(G)$, lead, by restriction of the continuous linear forms, to continuous embeddings:

$\mathrm{Meas}(G) \hookrightarrow \mathcal{F}' \hookrightarrow \mathrm{M}(G)^*$. With the usual continuous embedding: $f \mapsto (g \mapsto \int_G f(x)g(x)\,dx)\colon \mathrm{C}(G) \hookrightarrow \mathrm{Meas}(G)$, this leads to the following sequence of continuous embeddings (with dense images):

$$\mathrm{M}(G) \subset \mathcal{F} \subset \mathrm{C}(G) \subset \mathrm{Meas}(G) \subset \mathcal{F}' \subset \mathrm{M}(G)^*, \tag{4.12.3}$$

exhibiting $\mathrm{M}(G)^*$ as the "largest reasonable space of distributions on G".

All these remarks hold for arbitrary compact groups G, being just a continuation of Remark (C) after Corollary 4.6.2. For compact Lie groups G, we think of course of examples like $\mathcal{F} = \mathrm{C}^\infty(G)$, and $\mathcal{F} = \mathrm{C}^\omega(G)$, for which $\mathcal{F}' = \mathcal{D}'(G)$, the space of *Schwartz distributions* on G, and $\mathcal{F}' = \mathcal{H}(G)$, the space of *hyperfunctions* on G, respectively.

For any $u \in \mathrm{M}(G)^*$, and $X \in \mathfrak{g}_{\mathbf{C}}$, the distribution $\mathrm{R}^*(X)u \in \mathrm{M}(G)^*$ is defined by transposition:

$$(\mathrm{R}^*(X)u)(f) := u({}^t\mathrm{R}^*(X)f) = u(-\mathrm{R}^*(X)f), \quad \text{for } f \in \mathrm{M}(G); \tag{4.12.4}$$

the definition is chosen in such a way that it agrees with the usual one if $u \in \mathrm{M}(G)$, included in $\mathrm{M}(G)^*$ as in (4.12.3). In this fashion, for any space $\mathcal{F}$ as above, $u \in \mathcal{F}$, and $\in \mathcal{F}'$, and $X \in \mathfrak{g}_{\mathbf{C}}$, the element $\mathrm{R}^*(X)u$ at least is defined as an element of $\mathrm{M}(G)^*$; and any reasonable definition of "u is a solution in $\mathcal{F}$, and in $\mathcal{F}'$, respectively, of the linear partial differential equation $\mathrm{R}^*(X)u = cu$" will imply that the identity $\mathrm{R}^*(X)u = cu$ holds in $\mathrm{M}(G)^*$. For the examples $\mathcal{F} = \mathrm{C}^\infty(G)$, and $\mathrm{C}^\omega(G)$, respectively, this is obvious.

Because M_π is $\mathrm{R}^*(G)$-invariant, for every $[\pi] \in \widehat{G}$, we have $\mathrm{R}^*(X)u = cu$ if and only if each component $u|_{\mathrm{M}_\pi} \in (\mathrm{M}_\pi)^* \cong \mathrm{M}_{\pi^\vee}$, for $[\pi] \in \widehat{G}$, satisfies this differential equation, as an element of $\mathrm{M}(G)$. It follows that **if the space $\mathcal{S}$ of solutions in $\mathrm{M}(G)$ of a system of left invariant differential equations is finite-dimensional, then the space of solutions in $\mathrm{M}(G)^*$ of the same equations is equal to $\mathcal{S}$.**

Applying this to the system of differential equations that define the space:

$$\Gamma := \mathrm{M}(G)_\lambda^{\mathrm{R}^*\mathfrak{n}}$$

in (4.12.1), we see that it is also equal to the space of solutions f in $\mathrm{M}(G)^*$, hence in any "reasonable" space $\mathcal{F}$, and $\mathcal{F}'$, of functions, and distributions, respectively, on G, of the system of linear partial differential equations:

$$\mathrm{R}^*(X)(f) = 0, \quad \text{for all } X \in \mathfrak{n}; \tag{4.12.5}$$

and

$$\mathrm{R}^*(H)(f) = \lambda(H)f, \quad \text{for all } H \in \mathfrak{h}. \tag{4.12.6}$$

In this way, Proposition 4.12.1 provides an "explicit realization" of each irreducible representation of G, as the space of solutions f on G of the system (4.12.5,6), on which space the group G then acts via left translations (that is, via the left regular representation). Here λ is a complex linear form on $\mathfrak{h}$; the system (4.12.5,6) has nonzero solutions in $\mathcal{F}$, and $\mathcal{F}'$, respectively, if and only if λ is equal to the highest weight of some irreducible representation π of G. Finally, if this is the case, then the representation above is equivalent to $\pi^\vee$, and has highest weight equal to $-s_P \cdot \lambda$.

An infinitesimal formulation, with a strong algebraic flavor, can be given in terms of the *universal enveloping algebra* $\mathrm{U}(\mathfrak{g}_\mathbf{C})$ of $\mathfrak{g}_\mathbf{C}$. This is the "smallest associative algebra over $\mathbf{C}$, with unit, which contains $\mathfrak{g}_\mathbf{C}$ in such a way that $[X,Y] = XY - YX$ in $\mathrm{U}(\mathfrak{g}_\mathbf{C})$, for all $X, Y \in \mathfrak{g}_\mathbf{C}$". See Helgason [1962], Section II.1.2 and Proposition II.1.9 for the definition, and for the observation that the mapping $X \mapsto \mathrm{R}^*(X)$ extends in a unique way to an isomorphism R^* from $\mathrm{U}(\mathfrak{g}_\mathbf{C})$ onto the algebra of all linear partial differential operators on G which commute with the $\mathrm{L}(x)$, for $x \in G$.

The proof of the following proposition is largely independent of Proposition 4.12.1, but the reader will have no trouble in recognizing the space Γ as being the same as $\mathrm{M}(G)_\lambda^{\mathrm{R}^*\mathfrak{n}}$.

(4.12.2) Proposition. *Let $\lambda \in \mathfrak{h}^*$, and let Γ be a finite-dimensional, $\mathrm{L}(G)$-invariant vector space of continuous solutions of (4.12.5,6). Introduce the following annihilators in $\mathrm{U}(\mathfrak{g}_\mathbf{C})$:*

$$f^0 := \{ u \in \mathrm{U}(\mathfrak{g}_\mathbf{C}) \mid (\mathrm{R}^* u) f = 0 \}, \quad \Gamma^0 := \bigcap_{f \in \Gamma} f^0,$$

$$\Gamma_x^0 := \{ u \in \mathrm{U}(\mathfrak{g}_\mathbf{C}) \mid \big(\mathrm{R}^* u\big) f(x) = 0, \text{ for all } f \in \Gamma \}.$$

Then we have the following:

(i) $f^0 = \Gamma^0 = \Gamma_x^0$, for every $f \in \Gamma \setminus \{0\}$ and every $x \in G$.
(ii) Γ^0 is a left ideal in $\mathrm{U}(\mathfrak{g}_\mathbf{C})$, containing the sets:

$$\mathfrak{n}; \quad \{ H - \lambda(H)1 \mid H \in \mathfrak{h} \}; \quad \{ Y^{\lambda(\alpha^\vee)+1} \mid Y \in \mathfrak{g}_{-\alpha}, \alpha \in P \}.$$

(iii) Write

$$V = \mathrm{U}(\mathfrak{g}_\mathbf{C})/\Gamma^0,$$

then the left multiplication $(X, u) \mapsto Xu \colon \mathfrak{g}_\mathbf{C} \times \mathrm{U}(\mathfrak{g}_\mathbf{C}) \to \mathrm{U}(\mathfrak{g}_\mathbf{C})$ induces a representation π of $\mathfrak{g}_\mathbf{C}$ in V. This representation is finite-dimensional and irreducible, with highest weight equal to λ. Finally, the mapping $u \mapsto \big(f \mapsto (\mathrm{R}^* u) f(1)\big)$ defines an equivalence of representations: $V \to \Gamma^*$; so Γ is irreducible as well.

Proof. Note that the representation $x \mapsto \mathrm{L}(x)|_\Gamma \colon G \to \mathbf{L}(\Gamma, \Gamma)$ is real-analytic, so Γ actually consists of real-analytic functions. Let $f \in \Gamma$, and consider:

$$\mathrm{T}_1 f \colon u \mapsto (\mathrm{R}^* u) f(1) \in \mathrm{U}(\mathfrak{g}_\mathbf{C})^*,$$

the "Taylor expansion of f at 1". If $\mathrm{T}_1 f = 0$, then, using that f is real-analytic and G is connected, we get $f = 0$. That is, the mapping $f \mapsto \mathrm{T}_1 f \colon \Gamma \to \mathrm{U}(\mathfrak{g}_{\mathbf{C}})^*$ is injective, so its transposed linear mapping $u \mapsto (f \mapsto (\mathrm{R}^* u) f(1)) \colon \mathrm{U}(\mathfrak{g}_{\mathbf{C}}) \to \Gamma^*$ is surjective; and hence induces a bijective linear mapping:

$$\delta \colon \mathrm{U}(\mathfrak{g}_{\mathbf{C}})/\Gamma_1^0 \to \Gamma^*.$$

If $f \in \Gamma$ and $X \in \mathfrak{g}$, then:

$$\begin{aligned}\big(\mathrm{R}^*(Xu)\big) f(1) &= \big(\mathrm{R}^* X \circ \mathrm{R}^* u\big) f(1) = -\big(\mathrm{L}(X) \circ \mathrm{R}^* u\big) f(1) \\ &= -\big(\mathrm{R}^* u \circ \mathrm{L}(X)\big) f(1).\end{aligned}$$

This shows simultaneously that Γ_1^0 is a left ideal in $\mathrm{U}(\mathfrak{g}_{\mathbf{C}})$ and that δ intertwines the left action of $\mathfrak{g}$, hence of $\mathfrak{g}_{\mathbf{C}}$, on $\mathrm{U}(\mathfrak{g}_{\mathbf{C}})/\Gamma_1^0$ with the contragredient action on Γ^* induced by the left representation of $\mathfrak{g}_{\mathbf{C}}$ in Γ.

Using the left invariance of Γ, and the fact that the $\mathrm{R}^* u$, for $u \in \mathrm{U}(\mathfrak{g}_{\mathbf{C}})$, commute with the $\mathrm{L}(x)$, for $x \in G$, we obtain that $\Gamma_1^0 = \Gamma_x^0$, for all $x \in G$. This in turn implies that these annihilators are equal to Γ^0. Finally, for the same reason, $f^0 = (\mathrm{L}(x) f)^0$, for all $x \in G$; and because Γ is spanned by the $\mathrm{L}(x) f$, for $x \in G$, the conclusion is that $f^0 = \Gamma^0$, if $f \in \Gamma$, $f \neq 0$.

The vector $v = 1 + \Gamma^0$ generates the finite-dimensional, nonzero $\mathfrak{g}_{\mathbf{C}}$-representation space $V = \mathrm{U}(\mathfrak{g}_{\mathbf{C}})/\Gamma^0$. The fact that Γ consists of solutions of (4.12.5,6) just means that $\mathfrak{n} \subset \Gamma^0$, and $H - \lambda(H)1 \in \Gamma^0$, for all $H \in \mathfrak{h}$. But this implies that v is a highest weight vector; so Lemma 4.11.2 implies that V, which is generated by v, is irreducible. Furthermore, $\pi(H)(v) = \lambda(H)v$, for all $H \in \mathfrak{h}$; hence the highest weight of π is equal to λ.

The observation made after Cartan's highest weight theorem 4.11.4, finally yields that $Y^{\lambda(\alpha^\vee)+1} v = 0$, or $Y^{\lambda(\alpha^\vee)+1} \in \Gamma^0$, if $Y \in \mathfrak{g}_{-\alpha}$, $\alpha \in P$. □

(A) *Remark.* If we view $\mathrm{U}(\mathfrak{g}_{\mathbf{C}})^*$ as the space of formal power series at 1 in G, as in the proof of Proposition 4.12.2, then the last statement in Proposition 4.12.2.(iii) implies that every formal power series in $\mathrm{U}(\mathfrak{g}_{\mathbf{C}})^*$ which is annihilated by all $u \in \Gamma^0$, actually is equal to the Taylor expansion $\mathrm{T}_1 f$ at 1 of some globally defined solution $f \in \mathrm{C}^\infty(G)$ of (4.12.5,6). Quite miraculous, although a matter of fact if one looks at the proof.

If π denotes the representation of G in V, whose infinitesimal representation is described in Proposition 4.12.2.(iii), then the solution f is given in terms of the formal power series s by means of the formula:

$$f(x) = s(\pi(x)1) = (\pi(x)^* s)(1) = \text{ the constant term in } \pi(x)^* s, \quad \text{for } x \in G.$$

Let $\mathrm{L}(\lambda)$ be the left ideal in $\mathrm{U}(\mathfrak{g}_{\mathbf{C}})$ generated by $\mathfrak{n}$ and the $H - \lambda(H)1$, for $H \in \mathfrak{h}$. The proof of Proposition 4.12.2 actually shows that if M is a left ideal in $\mathrm{U}(\mathfrak{g}_{\mathbf{C}})$ which contains $\mathrm{L}(\lambda)$ and has finite codimension in $\mathrm{U}(\mathfrak{g}_{\mathbf{C}})$, then $\mathrm{U}(\mathfrak{g}_{\mathbf{C}})/M$ is an irreducible $\mathfrak{g}_{\mathbf{C}}$-module, with highest weight vector $1 + M$ and highest weight equal to λ. Now define M as the left ideal in $\mathrm{U}(\mathfrak{g}_{\mathbf{C}})$, generated by $\mathrm{L}(\lambda)$ and the $Y^{\lambda(\alpha^\vee)+1}$, for $Y \in \mathfrak{g}_{-\alpha}$, $\alpha \in P$. Using principal symbols, cf. Hörmander [1983] p.151, we easily verify that M has finite codimension in $\mathrm{U}(\mathfrak{g}_{\mathbf{C}})$. The inclusion $M \subset \Gamma^0$ induces a surjective intertwining map: $\mathrm{U}(\mathfrak{g}_{\mathbf{C}})/M \to \mathrm{U}(\mathfrak{g}_{\mathbf{C}})/\Gamma^0$ between irreducible $\mathfrak{g}_{\mathbf{C}}$-modules; so by Schur's lemma it is an isomorphism, or $\Gamma^0 = M$.

(B) *Remark.* A purely *algebraic construction of irreducible representations* of $\mathfrak{g}_{\mathbf{C}}$ with highest weight equal to λ has been given by Harish-Chandra [1951], for an arbitrary complex reductive Lie algebra $\mathfrak{g}_{\mathbf{C}}$. The assertions are the following.

(i) Let $\mathrm{L}(\lambda)$ be the left ideal in $\mathrm{U}(\mathfrak{g}_{\mathbf{C}})$, generated by $\mathfrak{n}$ and the elements $H - \lambda(H)1$, for $H \in \mathfrak{h}$, as above. Then there is a left ideal $\mathrm{M}(\lambda)$ in $\mathrm{U}(\mathfrak{g}_{\mathbf{C}})$, not equal to $\mathrm{U}(\mathfrak{g}_{\mathbf{C}})$, such that $I \subset \mathrm{M}(\lambda)$, for every other left ideal I in $\mathrm{U}(\mathfrak{g}_{\mathbf{C}})$ such that $\mathrm{L}(\lambda) \subset I \neq \mathrm{U}(\mathfrak{g}_{\mathbf{C}})$. Clearly $\mathrm{M}(\lambda)$ is uniquely determined by these properties.

(ii) Write $V = \mathrm{U}(\mathfrak{g}_{\mathbf{C}})/\mathrm{M}(\lambda)$. The left multiplication in $\mathrm{U}(\mathfrak{g}_{\mathbf{C}})$ induces a representation π of $\mathfrak{g}_{\mathbf{C}}$ in V, which is irreducible, and has $v := 1 + \mathrm{M}(\lambda)$ as a nonzero highest weight vector, with highest weight equal to λ.

(iii) V is finite-dimensional, if and only if $\lambda(\alpha^\vee) \in \mathbf{Z}_{\geq 0}$, for all $\alpha \in P$. That is, λ is a "dominant weight of $\mathfrak{g}_{\mathbf{C}}$", in the terminology of Section 4.10.(c) and the definition preceding Theorem 4.9.5. If this is the case, then $\mathrm{M}(\lambda)$ is equal to the left ideal in $\mathrm{U}(\mathfrak{g}_{\mathbf{C}})$ generated by $\mathfrak{n}$, the $H - \lambda(H)1$, for $H \in \mathfrak{h}$, and the $Y^{\lambda(\alpha^\vee)+1}$, for $Y \in \mathfrak{g}_{-\alpha}$, $\alpha \in S$.

Now the left ideal Γ^0 of Proposition 4.12.2 is contained in $\mathrm{M}(\lambda)$ (if $\Gamma \neq 0$); and we get a surjective intertwining map: $\mathrm{U}(\mathfrak{g}_{\mathbf{C}})/\Gamma^0 \to V$ between irreducible $\mathfrak{g}_{\mathbf{C}}$-modules. So again using Schur's lemma, we conclude that $\Gamma^0 = \mathrm{M}(\lambda)$. That is, **Harish-Chandra's construction is canonically dual to the solution space of (4.12.5,6)**. Or, in other words, Γ^0 is already generated by $\mathrm{L}(\lambda)$ and the $Y^{\lambda(\alpha^\vee)+1}$, for $Y \in \mathfrak{g}_{-\alpha}$, $\alpha \in S$, which is somewhat less than the generators mentioned at the end of Remark (A).

(C) *Remark.* We return to the situation that $\mathfrak{g}_{\mathbf{C}}$ is the complexification of the Lie algebra $\mathfrak{g}$ of a connected, compact Lie group G. Then π is equal to $\mathrm{T}_1 \widetilde{\pi}$, for a unique representation $\widetilde{\pi}$ of the universal covering $\widetilde{G}$ of G. As we have seen in the remark after Corollary 3.9.5, $\widetilde{G} = \widetilde{G'} \times \mathfrak{z}$, where G', the derived group of G, has a compact universal covering $\widetilde{G'}$. It follows that the center of $\widetilde{G'}$ is contained in the maximal torus of $\widetilde{G'}$ with Lie algebra equal to $\mathfrak{t} \cap [\mathfrak{g}, \mathfrak{g}]$, and that the center of $\widetilde{G}$, which contains the kernel of the projection $p\colon \widetilde{G} \to G$, is contained in $\exp \mathfrak{t} \subset \widetilde{G}$. Now we have: $\widetilde{\pi} = \pi \circ p$, for a (unique) representation π of $G \Leftrightarrow \widetilde{\pi} = 1$ on $\ker p \Leftrightarrow$ all weights of π take values in $2\pi i\mathbf{Z}$ on $\ker \exp(\mathfrak{t} \to G) \Leftrightarrow \lambda \in \widehat{T}$. (For the last equivalence, use that $\Omega(\pi) \subset \lambda + \Lambda_R$, cf. Theorem 4.11.4.(ii).)

In other words, Harish-Chandra's construction provides an alternative, completely algebraic proof of Theorem 4.9.5, or: **the solution space $\Gamma := \mathrm{M}(G)_\lambda^{\mathrm{R}^*\mathfrak{n}}$ of (4.12.5,6) is nonzero, if and only if λ is a dominant weight of T.**

(D) *Remark.* For any $u \in \mathrm{U}(\mathfrak{g}_{\mathbf{C}})$, the (left invariant) linear partial differential operator $\mathrm{R}^*\, u$ leaves invariant each finite-dimensional vector space M_π; actually the same is valid for each irreducible right invariant subspace V of M_π. So, if f_j denotes a basis of V, one gets a system of linear partial differential equations of the form:

$$(\mathrm{R}^*\, u) f_j = \sum_k A_{kj}(u) f_k,$$

where $A_{kj}(u)$ is a scalar matrix, depending on u. If u commutes with all $X \in \mathfrak{g}_{\mathbf{C}}$, then $\mathrm{R}^*\, u$ commutes both with the left and the right regular representation. Because M_π is an irreducible subspace for the regular representation of $G \times G$, cf. Corollary 4.3.6, it follows from Schur's lemma 4.1.1.(iii) that there is a constant $c = c(\pi, u)$ such that $(\mathrm{R}^*\, u) f = cf$, for all $f \in \mathrm{M}_\pi$. The algebra of $u \in \mathrm{U}(\mathfrak{g}_{\mathbf{C}})$ that commute with all $X \in \mathfrak{g}_{\mathbf{C}}$, is called the *center* $\mathbf{Z}(\mathfrak{g}_{\mathbf{C}})$ *of the universal enveloping algebra* $\mathrm{U}(\mathfrak{g}_{\mathbf{C}})$; and clearly $u \mapsto c(\pi, u)$ is an algebra homomorphism: $\mathbf{Z}(\mathfrak{g}_{\mathbf{C}}) \to \mathbf{C}$, which depends only on the highest weight λ of $[\pi] \in \widehat{G}$. The differential operators in $\mathbf{Z}(\mathfrak{g}_{\mathbf{C}})$ form an important tool, with a strong algebraic flavor, in the analysis on semisimple Lie groups. (See for results of Beilinson-Bernstein and Brylinski-Kashiwara the survey of Schmid [1985])

Our next purpose is to give a differential geometric interpretation for the space of solutions of (4.12.5,6); this is to be expected, because the differential operators are the actions of certain left invariant vector fields.

The equations (4.12.6) are equivalent to the same equations for $H \in \mathfrak{t}$. Or:

$$\frac{d}{dt}\, \mathrm{e}^{-t\lambda(H)}\, \mathrm{R}^*(\exp tH) f = 0,$$

or: $t \mapsto \mathrm{e}^{-t\lambda(H)}\, \mathrm{R}^*(\exp tH) f$ is constant, equal to its value f at $t = 0$; or: $\mathrm{R}^*(\exp H) f = \mathrm{e}^{\lambda(H)}\, f$, for all $H \in \mathfrak{t}$. In other words, (4.12.6) is equivalent to:

$$\mathrm{R}^*(t) f = t^\lambda f, \qquad \text{for all } t \in T. \tag{4.12.7}$$

In terms of (4.7.2) and the comments thereafter this means that f **is a (differentiable) section of the homogeneous line bundle** $\mathrm{L}_{-\lambda}$ **over** G/T**, defined by the weight** $-\lambda$ **of** T, that is, by the representation $(t, c) \mapsto t^{-\lambda} c$ of T in $\mathbf{C}$. In still other terms, this means that $\mathrm{L}_\lambda^{\mathrm{R}^*\mathfrak{n}}$ **is a subrepresentation of the induced representation** $\mathrm{Ind}_T^G(-\lambda)$; to be precise, the $\mathrm{L}(G)$-invariant subspace of elements of finite type that are highest weight vectors for the right regular representation.

Given (4.12.6), it will turn out that the equations (4.12.5) mean that the section f of the line bundle $\mathrm{L}_{-\lambda}$ is **complex-analytic** (that is, **holomorphic**) with respect to a natural complex-analytic structure on G/T and $\mathrm{L}_{-\lambda}$.

In order to introduce the complex structure on G/T, we start by embedding G as a closed subgroup in $\mathbf{GL}(V)$, for some finite-dimensional complex vector space V, so that $\mathfrak{g}$ can be viewed as a Lie subalgebra of $\mathbf{L}(V,V)$. Let $\mathfrak{g}_{\mathbf{C}} = \mathfrak{g} + i\mathfrak{g}$ be the complexification of $\mathfrak{g}$ in $\mathbf{L}(V,V)$, then $\mathfrak{g}_{\mathbf{C}}$ is a complex Lie subalgebra of $\mathbf{L}(V,V)$. Define $G_{\mathbf{C}}$ to be the connected Lie subgroup of $\mathbf{GL}(V)$ with Lie algebra equal to $\mathfrak{g}_{\mathbf{C}}$. Then $G_{\mathbf{C}}$ is a complex Lie subgroup of $\mathbf{GL}(V)$, cf. 1.9.4, containing G as a real Lie subgroup. In Section 14.6, we shall see that the algebra $\mathrm{M}(G)$ is finitely generated, and equal to the algebra of polynomial functions of a complex affine algebraic group $G_{\mathbf{C},\mathrm{alg}}$. The group $G_{\mathbf{C}}$ is isomorphic, as a complex Lie group, to $G_{\mathbf{C},\mathrm{alg}}$; this fact removes the flaw of the arbitrary choice of the injective representation: $G \to \mathbf{GL}(V)$.

Next consider the complex linear subspace:

$$\mathfrak{b} := \mathfrak{h} + \mathfrak{n} \tag{4.12.8}$$

of $\mathfrak{g}_{\mathbf{C}}$. Because $\mathfrak{h} = \mathfrak{t} + i\mathfrak{t}$, formula (3.5.4) shows that $[\mathfrak{h}, \mathfrak{n}] = \mathfrak{n}$ (note that roots are nonzero!), and after (4.11.6) we observed that $[\mathfrak{n}, \mathfrak{n}] \subset \mathfrak{n}$. It follows that $\mathfrak{b}$ is a complex Lie subalgebra of $\mathfrak{g}_{\mathbf{C}}$, in which $\mathfrak{n}$ actually is an ideal. ($\mathfrak{b}$ is a *Borel subalgebra*, that is, a *maximal solvable subalgebra* of $\mathfrak{g}_{\mathbf{C}}$, cf. Section 5.2.) Furthermore, we introduce:

$$B := \{ x \in G_{\mathbf{C}} \mid \mathrm{Ad}\, x(\mathfrak{b}) = \mathfrak{b} \}, \tag{4.12.9}$$

the normalizer of $\mathfrak{b}$ in $G_{\mathbf{C}}$. Clearly, B is a closed, hence Lie subgroup of $G_{\mathbf{C}}$. Because its Lie algebra,

$$\mathrm{T}_1 B = \{ X \in \mathfrak{g}_{\mathbf{C}} \mid [X, \mathfrak{b}] \subset \mathfrak{b} \}, \tag{4.12.10}$$

the normalizer of $\mathfrak{b}$ in $\mathfrak{g}_{\mathbf{C}}$, is a complex Lie subalgebra of $\mathfrak{g}_{\mathbf{C}}$, it follows that B is a closed, complex Lie subgroup of $G_{\mathbf{C}}$. Also note that $\mathfrak{t} \subset \mathfrak{b} \subset \mathrm{T}_1 B$, so $T \subset B$.

As a variation on the same theme, we will also make use of:

$$\mathcal{B} := \{ \Phi \in \mathrm{Ad}\, \mathfrak{g}_{\mathbf{C}} \mid \Phi(\mathfrak{b}) = \mathfrak{b} \}, \tag{4.12.11}$$

the normalizer of $\mathfrak{b}$ in $\mathrm{Ad}\, \mathfrak{g}_{\mathbf{C}}$, the adjoint group of $\mathfrak{g}_{\mathbf{C}}$. Note that $\mathrm{Ad}\, G_{\mathbf{C}} = \mathrm{Ad}\, \mathfrak{g}_{\mathbf{C}}$, $\mathrm{Ad}\, B = \mathcal{B}$, and $\mathrm{ad}\, \mathfrak{b} \subset \mathrm{T}_1 \mathcal{B}$.

(4.12.3) Lemma. *The inclusions $G \subset G_{\mathbf{C}}$ and $T \subset B$ induce a real-analytic diffeomorphism from G/T onto $G_{\mathbf{C}}/B$. Similarly, the mappings $\mathrm{Ad}\colon G_{\mathbf{C}} \to \mathrm{Ad}\, \mathfrak{g}_{\mathbf{C}}$ and $\mathrm{Ad}\colon B \to \mathcal{B}$ induce a complex-analytic diffeomorphism from $G_{\mathbf{C}}/B$ onto $\mathrm{Ad}\, \mathfrak{g}_{\mathbf{C}}/\mathcal{B}$.*

The Lie group B, and $\mathcal{B}$, is connected, and has Lie algebra equal to $\mathfrak{b}$, and $\mathrm{ad}\, \mathfrak{b}$, respectively.

Proof. Consider the action $(x, yB^{\circ}) \mapsto xyB^{\circ}$, and $(x, \Phi\mathcal{B}^{\circ}) \mapsto \mathrm{Ad}\, x \circ \Phi\mathcal{B}^{\circ}$, of G on the space $G_{\mathbf{C}}/B^{\circ}$, and $\mathrm{Ad}\, \mathfrak{g}_{\mathbf{C}}/\mathcal{B}^{\circ}$, respectively. If $Z \in \overline{\mathfrak{n}}$, then $Z = (Z + \overline{Z}) - \overline{Z} \in \mathfrak{g} + \mathfrak{n}$, so:

$$\mathfrak{g}_{\mathbf{C}} = \bar{\mathfrak{n}} + \mathfrak{n} + \mathfrak{h} \subset \mathfrak{g} + \mathfrak{n} + \mathfrak{h} = \mathfrak{g} + \mathfrak{b}.$$

Because $\mathfrak{b} \subset \mathrm{T}_1 B$, and $\mathrm{ad}\,\mathfrak{b} \subset \mathrm{T}_1 \mathcal{B}$, it follows that the tangent mapping at 1 of the mapping $x \mapsto xB^\circ$, and $x \mapsto \mathrm{Ad}\, x\mathcal{B}^\circ$, from G to $G_{\mathbf{C}}/B^\circ$, and to $\mathrm{Ad}\,\mathfrak{g}_{\mathbf{C}}/\mathcal{B}^\circ$, respectively, is surjective. But this means that the orbit of the G-action through $1B^\circ$, and $1\mathcal{B}^\circ$, respectively, is open. On the other hand the orbit is compact, hence closed, as the image of the compact group G under the mapping mentioned above. Because the space $G_{\mathbf{C}}/B^\circ$, and $\mathrm{Ad}\,\mathfrak{g}_{\mathbf{C}}/\mathcal{B}^\circ$, is connected, the conclusion is that G/T is mapped surjectively to $G_{\mathbf{C}}/B^\circ$, and $\mathrm{Ad}\,\mathfrak{g}_{\mathbf{C}}/\mathcal{B}^\circ$, respectively.

Now let $x \in G$, and let $\mathrm{Ad}\, x(\mathfrak{b}) = \mathfrak{b}$. This implies that $\mathrm{Ad}\, x(\mathfrak{t}) \subset \mathfrak{b} \cap \mathfrak{g} = \mathfrak{t}$, the latter because $\bar{\mathfrak{b}} = \bar{\mathfrak{h}} \oplus \bar{\mathfrak{n}} = \mathfrak{h} \oplus \bar{\mathfrak{n}}$ has intersection with $\mathfrak{b} = \mathfrak{h} \oplus \mathfrak{n}$ equal to $\mathfrak{h}$, and $\mathrm{Re}\,\mathfrak{h} = \mathfrak{t}$. In other words, $x \in \mathrm{N}(\mathfrak{t}) = \mathrm{N}(T)$; or $s = \mathrm{Ad}\, x|_{\mathfrak{t}}$ belongs to the Weyl group W. Now:

$$\mathfrak{h} + \sum_{\alpha \in P} \mathfrak{g}_\alpha = (\mathrm{Ad}\, x)(\mathfrak{h} + \sum_{\alpha \in P} \mathfrak{g}_\alpha) = \mathfrak{h} + \sum_{\alpha \in P} (\mathrm{Ad}\, x)(\mathfrak{g}_\alpha) = \mathfrak{h} + \sum_{\alpha \in P} \mathfrak{g}_{s.\alpha},$$

since an automorphism of $\mathfrak{g}$ that normalizes $\mathfrak{t}$, permutes the root spaces; in turn this implies that $P = s \cdot P$, or $s = 1$, cf. Proposition 3.8.1. That is, $x \in T$; and we have proved that the mapping: $G/T \to \mathrm{Ad}\,\mathfrak{g}_{\mathbf{C}}/\mathcal{B}$ is injective. Because it is equal to the composition of $G/T \to G_{\mathbf{C}}/B^\circ$, $G_{\mathbf{C}}/B^\circ \to G_{\mathbf{C}}/B$ and $G_{\mathbf{C}}/B \to \mathrm{Ad}\,\mathfrak{g}_{\mathbf{C}}/\mathcal{B}$, all of which are surjective, the conclusion is that all mappings are bijective, implying also that $B^\circ = B$; or B is connected. Because $G/T \to G_{\mathbf{C}}/B$, and $G_{\mathbf{C}}/B \to \mathrm{Ad}\,\mathfrak{g}_{\mathbf{C}}/\mathcal{B}$, is a real-, and complex-analytic fibration, it is a real-, and a complex-analytic diffeomorphism, respectively.

Finally, the injectivity of the tangent mapping at $1T$ of the mapping $G/T \to G_{\mathbf{C}}/B$ means that the mapping $\mathfrak{g}/\mathfrak{t} \to \mathfrak{g}_{\mathbf{C}}/\mathrm{T}_1 B$ is injective. However, the latter is equal to the composition of the surjective mapping $\mathfrak{g}/\mathfrak{t} \to \mathfrak{g}_{\mathbf{C}}/\mathfrak{b}$ and $\mathfrak{g}_{\mathbf{C}}/\mathfrak{b} \to \mathfrak{g}_{\mathbf{C}}/\mathrm{T}_1 B$. This implies that $\mathfrak{b} = \mathrm{T}_1 B$; and therefore also $\mathrm{T}_1 \mathcal{B} = \mathrm{T}_1(\mathrm{Ad}\, B) = \mathrm{ad}\,\mathrm{T}_1 B = \mathrm{ad}\,\mathfrak{b}$. □

Clearly, $G_{\mathbf{C}}/B$ is a complex-analytic manifold, and we will use the complex-analytic structure on G/T that comes from $G_{\mathbf{C}}/B$ via the diffeomorphism $G/T \to G_{\mathbf{C}}/B$. On it, the action of $G_{\mathbf{C}}$ is a complex-analytic one. In particular, G acts on G/T by means of complex-analytic diffeomorphisms with respect to the complex-analytic structure on G/T. However, it is worthwhile to note that the full automorphism group of the compact, complex analytic manifold G/T actually is much larger, containing at least also the action of the complex Lie group $G_{\mathbf{C}}$.

Even stronger, $\mathrm{Ad}\,\mathfrak{g}_{\mathbf{C}}$, being the identity component of $\{\Phi \in \mathrm{Aut}\,\mathfrak{g}_{\mathbf{C}} \mid \Phi|_{\mathfrak{z}} = \text{identity on } \mathfrak{z}\}$, is a complex algebraic group, with $\mathcal{B}$ as an algebraic subgroup. In other words, the complex-analytic manifold G/T is isomorphic to an algebraically defined one. Moreover, G, and $G_{\mathbf{C}}$, respectively, acts via $\mathrm{Ad}\,\mathfrak{g}_{\mathbf{C}}$, which is an algebraic action.

Another remarkable conclusion of Lemma 4.12.3 is that both $G_{\mathbf{C}}/B$ and $\mathrm{Ad}\,\mathfrak{g}_{\mathbf{C}}/\mathcal{B}$ are compact, which a priori is not so obvious, because neither $G_{\mathbf{C}}$, nor $\mathrm{Ad}\,\mathfrak{g}_{\mathbf{C}}$ is compact if $\mathfrak{g}$ is not Abelian. We shall see below that G/T is actually isomorphic (as a complex-analytic manifold) to a complex projective algebraic variety, and that the action of $G_{\mathbf{C}}$ on it is projective linear.

In order to recognize the complex line bundle $\mathrm{L}_{-\lambda}$ over $G/T \xrightarrow{\sim} G_{\mathbf{C}}/B$ as a complex-analytic line bundle over $G_{\mathbf{C}}/B$, we need some more insight in the structure of the group B.

(4.12.4) Lemma. *Let A, and N, be the connected Lie subgroup of $G_{\mathbf{C}}$ with Lie algebra equal to $\mathfrak{a} := i\mathfrak{t}$, and $\mathfrak{n}$, respectively. Then $N = \exp\mathfrak{n}$, and the mapping $(t, X, Y) \mapsto t\exp X \exp Y$ is a real-analytic diffeomorphism from $T \times \mathfrak{a} \times \mathfrak{n}$ onto B.*

Proof. If $n = \dim V$, then

$$\mathcal{N} := \{\, A \in \mathbf{L}(V,V) \mid A^n = 0 \,\}$$

is the set of nilpotent elements in $\mathbf{L}(V,V)$; it is a complex algebraic variety in $\mathbf{L}(V,V)$. It contains the space $\mathcal{L}$ of strictly lower triangular matrices, and $\mathfrak{n} \subset \mathcal{L} \subset \mathcal{N}$, cf. Lemma 4.11.3. The mapping:

$$A \mapsto \mathrm{e}^A - \mathrm{I} = \sum_{k=1}^{n-1} \frac{1}{k!} A^k$$

is a polynomial mapping: $\mathcal{N} \to \mathcal{N}$, with a two-sided polynomial inverse $B \mapsto \log(\mathrm{I} + B) = \sum_{k=1}^{n-1} \frac{(-1)^{k-1}}{k} B^k$. That is, the product in logarithmic coordinates, cf. Section 1.6, for the Lie subalgebra $\mathcal{L}$ of $\mathbf{L}(V,V)$, is a polynomial mapping: $\mathcal{L} \times \mathcal{L} \to \mathcal{L}$. Further, for the Lie subalgebra $\mathfrak{n}$ of $\mathcal{L}$ we get that $N = \exp\mathfrak{n}$ is a closed Lie subgroup of the subgroup $\mathrm{I} + \mathcal{L}$ of $\mathbf{GL}(V)$, the exponential mapping: $\mathfrak{n} \to N$ being a polynomial diffeomorphism with a polynomial inverse.

$H := \exp\mathfrak{h}$ is a subgroup of B consisting of diagonal matrices, and because $[\,\mathfrak{h}, \mathfrak{n}\,] = \mathfrak{n}$, as observed after (4.12.8), we get that $H \cdot N$ is a subgroup of B, with N as a normal subgroup. Because $H \cdot N$ is generated by $\exp\mathfrak{b}$, and B is connected, cf. Lemma 4.12.3, the conclusion is that $H \cdot N = B$, cf. Theorem 1.9.1. Now the inverse of the mapping: $(t, X, Y) \mapsto b = t\exp X \exp Y$ is obtained as follows: b has a unique splitting $b = d(\mathrm{I} + L)$, where d is diagonal and L is strictly lower triangular. The element d has a unique splitting $d = t\exp X$, with t, X diagonal, and the diagonal entries of t on the unit circle in $\mathbf{C}$, and the entries of X real. Finally, $\mathrm{I} + L = \exp Y$, for a unique strictly lower diagonal Y. Clearly $t \in T$, $X \in \mathfrak{a}$, $Y \in \mathfrak{n}$, and these depend analytically on b. □

(4.12.5) [Borel-Weil theorem].

(i) *For each $\mu \in \widehat{T}$, there is a unique homomorphism of complex Lie groups: $B \to \mathbf{C}^\times$, extending the character: $t \mapsto t^\mu$ of T; this homomorphism will also be denoted by μ, and viewed as a representation of B in $\mathbf{C}$.*

(ii) *The inclusions $G \subset G_{\mathbf{C}}$, and $T \subset B$ induce a real-analytic isomorphism of homogeneous line bundles:*

$$\mathrm{L}_\mu := G \times_T \mathbf{C} \to G_{\mathbf{C}} \times_B \mathbf{C},$$

intertwining the respective left actions of G and turning L_μ into a $G_{\mathbf{C}}$-homogeneous complex-analytic line bundle over $G_{\mathbf{C}}/B$.

(iii) *The identification in (ii) yields an isomorphism between the space:*

$$\Gamma_{\mathrm{hol}}(G_{\mathbf{C}}/B, \mathrm{L}_\mu), \quad \text{of holomorphic sections of } \mathrm{L}_\mu, \text{ defined over } G_{\mathbf{C}}/B,$$

and the space of solutions f of (4.12.5,6), in any space $\mathcal{F}$, and $\mathcal{F}'$, of functions, and distributions, respectively, on G, such that $\mathrm{M}(G) \subset \mathcal{F} \subset \mathcal{F}' \subset \mathrm{M}(G)^*$, cf. (4.12.3).

(iv) *The space $\Gamma_{\mathrm{hol}}(G_{\mathbf{C}}/B, \mathrm{L}_\mu)$ is nonzero, if and only if $\mu = -\lambda$, with λ the highest weight of an irreducible representation π of G. In this case the left action of G, and $G_{\mathbf{C}}$ on $\Gamma_{\mathrm{hol}}(G_{\mathbf{C}}/B, \mathrm{L}_\mu)$ is an irreducible representation of G, and $G_{\mathbf{C}}$, respectively, with highest weight equal to $-s_P \cdot \lambda$. Each irreducible representation of G is equivalent to precisely one of these; this implies also that every representation of G in a finite-dimensional complex vector space V has a complex-analytic extension: $G_{\mathbf{C}} \to \mathbf{GL}(V)$.*

Proof. (i) In the notation of Lemma 4.12.4 and its proof, we begin by extending the homomorphism $t \mapsto t^\mu : T \to \mathbf{C}^\times$ to a complex-analytic homomorphism: $h \mapsto h^\mu : H \to \mathbf{C}^\times$ (such a homomorphism is clearly unique), by means of the definition:

$$(t \exp X)^\mu = t^\mu \, \mathrm{e}^{\mu(X)}, \quad \text{for } t \in T, X \in \mathfrak{a} = i\mathfrak{t}.$$

Here we have extended $\mu : \mathfrak{t} \to i\mathbf{R}$ to a complex linear form $\mu : \mathfrak{h} \to \mathbf{C}$.

Next we observe that the mapping $\Phi : h \cdot n \mapsto h : B \to H$ is well-defined and real-analytic. Now N is a normal subgroup of B, because its Lie algebra $\mathfrak{n}$ is an ideal in $\mathfrak{b}$, as observed after (4.12.8), see also Proposition 1.11.5. For $h, h' \in H$ and $n, n' \in N$, we get that $(h \cdot n)(h' \cdot n') = hh' \cdot ({h'}^{-1} n h') n'$, with ${h'}^{-1} n h' \in N$; and from this we read off that Φ is a group homomorphism, and actually a homomorphism of complex Lie groups, in view of Proposition 1.6.4 and the fact that the projection $\mathfrak{b} \to \mathfrak{b}/\mathfrak{n} \xrightarrow{\sim} \mathfrak{h}$ is complex-linear. So $b \mapsto b^\mu := (\Phi(b))^\mu$ is a homomorphism of complex Lie groups: $B \to \mathbf{C}^\times$, extending $\mu \in \widehat{T}$.

The uniqueness is obtained as follows. We have $\mathfrak{n} = [\mathfrak{h}, \mathfrak{n}]$, as observed after (4.12.8); hence $\theta(\mathfrak{n}) = [\theta(\mathfrak{h}), \theta(\mathfrak{n})] = 0$, for any homomorphism of Lie algebras $\theta\colon \mathfrak{b} \to \mathbf{C}$, because $\mathbf{C}$ is Abelian. It follows that $\Theta(N) = \{1\}$, for any homomorphism of Lie groups $\Theta\colon B \to \mathbf{C}^\times$.

(ii) follows immediately from Lemma 4.12.3: if $g, g' \in G$, $c, c' \in \mathbf{C}$, $g' = gb^{-1}$, $c' = b^\mu c$, for $b \in B$, then $b = {g'}^{-1} g \in G \cap B = T$. Hence (g', c') belongs to the same T-orbit in $G \times \mathbf{C}$ as (g, c); and this proves the injectivity of the mapping: $G \times_T \mathbf{C} \to G_\mathbf{C} \times_B \mathbf{C}$. Similarly the surjectivity follows from the surjectivity of $G/T \to G_\mathbf{C}/B$.

(iii) From (i), (ii) it follows that the restriction: $f \mapsto f|_G$ defines a linear isomorphism from the space $\mathrm{C}^1(G_\mathbf{C}/B, \mathrm{L}_\mu)$ of C^1 sections of the line bundle L_μ over $G_\mathbf{C}/B$ onto the corresponding space $\mathrm{C}^1(G/T, \mathrm{L}_\mu)$. To say that $f \in \mathrm{C}^1(G_\mathbf{C}/B, \mathrm{L}_\mu)$ is complex-differentiable at $1B$ means that, regarding f as a C^1 function on $G_\mathbf{C}$, we have:

$$\mathrm{T}_1 f(iX) = i\,\mathrm{T}_1 f(X), \quad \text{for all } X \in \mathfrak{g}_\mathbf{C}.$$

Now the fact that $f(b) = b^{-\mu}$, for $b \in B$, implies that $f|_B$ is complex-analytic; hence $\mathrm{T}_1 f$ is complex-linear: $\mathfrak{b} \to \mathbf{C}$. Therefore we need only to investigate what it means that $\mathrm{T}_1 f$ is complex-linear on $\overline{\mathfrak{n}}$. Write θ for the complex-analytic extension: $\mathfrak{g}_\mathbf{C} \to \mathbf{C}$ of $T_1 f|_\mathfrak{g}$. If $Z \in \overline{\mathfrak{n}}$, then $Z = (Z + \overline{Z}) - \overline{Z}$, with $Z + \overline{Z} \in \mathfrak{g}$, $\overline{Z} \in \mathfrak{n}$, so:

$$T_1 f(Z) = \theta(Z + \overline{Z}) = \theta(Z) + \theta(\overline{Z}),$$

which is equal to $\theta(Z)$ if and only if $\theta(\overline{Z}) = 0$. So f is complex-differentiable at $1B$ if and only if $f|_G$ satisfies (4.12.5) at 1. In view of the left invariance under G, it follows that f is complex-analytic if and only if $f|_G$ satisfies (4.12.5) everywhere.

(iv) now follows from Proposition 4.12.1, together with the remarks preceding (4.12.5,6). □

(E) *Remark.* Using Cauchy's integral formula in order to get locally uniform estimates for derivatives of holomorphic sections in terms of the uniform norm, one may apply Ascoli's theorem to prove in another way that the space of holomorphic sections of L_μ over the compact manifold G/T is finite-dimensional. In our proof, such functional analysis entered via the Peter-Weyl theorem 4.6.1, whose proof also made use of Ascoli's theorem.

(F) *Remark.* The tangent mapping of $(\overline{n}, h, n) \mapsto \overline{n}hn\colon \overline{N} \times H \times N \to G_\mathbf{C}$, at $(1,1,1)$, is equal to $(Y, X, Z) \mapsto Y + X + Z\colon \overline{\mathfrak{n}} \times \mathfrak{h} \times \mathfrak{n} \to \mathfrak{g}_\mathbf{C}$, which is bijective. It follows that $\overline{N}HN$ is a neighborhood of 1 in $G_\mathbf{C}$. If $f \in \Gamma_{\mathrm{hol}}(G_\mathbf{C}/B, \mathrm{L}_{-\lambda})$ is a lowest weight vector for L (that is, a highest weight vector if the choice P of positive roots is replaced by $-P$), then f is fixed under $\mathrm{L}(\overline{N})$; so $f(\overline{n}hn) = f(1h) = h^\lambda f(1)$, for $\overline{n} \in \overline{N}$, $h \in H$, $n \in N$. In particular, if $f(1) = 0$, then $f = 0$ on $\overline{N}HN$; and because f is complex-analytic and $G_\mathbf{C}$ is connected, it follows that $f = 0$. In other words, the mapping $f \mapsto f(1)$ is injective, from the space of lowest weight vectors in $\Gamma_{\mathrm{hol}}(G_\mathbf{C}/B, \mathrm{L}_{-\lambda})$ to $\mathbf{C}$, which implies that this space is one-dimensional. In turn this yields an alternative proof that $\mathrm{L}_\lambda^{\mathrm{R}^*\mathfrak{n}}$ is irreducible.

It also follows that any complex-analytic function on $\overline{N}$ extends to a holomorphic solution of (4.12.5,6) on $\overline{N}HN$. On the power series level, and in terms of Harish-Chandra's construction discussed after Proposition 4.12.2, this corresponds to $\mathrm{U}(\mathfrak{g}_{\mathbf{C}})/\mathrm{L}(\lambda) \cong \mathrm{U}(\overline{\mathfrak{n}})$, which is infinite-dimensional if $\mathfrak{g}_{\mathbf{C}}$ is not Abelian. Using the additional elements in Γ^0, mentioned in Proposition 4.12.2.(ii), one can verify that $Y \mapsto f(\exp Y)$ is a **polynomial** on $\overline{\mathfrak{n}}$, with an explicit bound on the degree, given in terms of λ and the root structure of $\mathfrak{g}_{\mathbf{C}}$. This can be used to reduce the determination of the formal power series s in the kernel of all $u \in \Gamma^0$, as discussed in Remark (A), to an explicit problem of finite-dimensional linear algebra.

Using the Bruhat decomposition, cf. Harish-Chandra [1956], one may check that if f is a holomorphic solution of (4.12.5,6) on $\overline{N}HN$, then the condition that $\mathrm{R}^*\, Y^{\lambda(\alpha^\vee)+1} f = 0$, for all $Y \in \mathfrak{g}_{-\alpha}$, $\alpha \in S$, is equivalent to the condition that f extends as a holomorphic function over the Bruhat cells of codimension equal to 1; in view of Hartog's lemma this is equivalent to the condition that f extends to a holomorphic function on $G_{\mathbf{C}}$. This leads to a geometric proof of the fact that $\Gamma^0 = \mathrm{M}(\lambda)$ is generated by $\mathrm{L}(\lambda)$ and the $Y^{\lambda(\alpha^\vee)+1}$, for $Y \in \mathfrak{g}_{-\alpha}$, $\alpha \in S$.

(G) *Remark.* If we identify $G_{\mathbf{C}}$ with the complex affine algebraic group $G_{\mathbf{C},\mathrm{alg}}$ mentioned in the remark preceding (4.12.8), then, because the solutions of (4.12.5,6) belong to $\mathrm{M}(G)$, the space $\Gamma_{\mathrm{hol}}(G_{\mathbf{C}}/B, \mathrm{L}_\mu)$ is equal to the space $\Gamma_{\mathrm{alg}}(G_{\mathbf{C},\mathrm{alg}}/B, \mathrm{L}_\mu)$ of **algebraic sections** of L_μ over $G_{\mathbf{C},\mathrm{alg}}/B$, that is, complex-analytic **polynomials** f on $G_{\mathbf{C},\mathrm{alg}}$ that satisfy $f(xb) = b^\mu f(x)$, for all $x \in G_{\mathbf{C},\mathrm{alg}}$, $b \in B$. (Also, B is an affine algebraic subgroup of $G_{\mathbf{C},\mathrm{alg}}$, and $b \mapsto b^\mu$ a polynomial representation.) In this fashion the "explicit realization" of the irreducible representations, given such an analytic flavor in (4.12.5,6), and formulated in a (complex-analytic) differential geometric framework in the Borel-Weil theorem 4.12.5, once more is shown to be of a purely algebraic nature. Also, every finite-dimensional representation of G extends to a complex algebraic (= polynomial) one of $G_{\mathbf{C}} = G_{\mathbf{C},\mathrm{alg}}$, which is an even stronger statement than the last one in 4.12.5.(iv).

We now turn to the orbit $\{\, \pi(x)V_\lambda \mid x \in G_{\mathbf{C}} \,\}$ of the one-dimensional highest weight space V_λ, in the complex projective space $\mathbf{CP}(V)$ of complex one-dimensional linear subspaces of V. Here π is an irreducible representation of G in V; according to the Borel-Weil theorem 4.12.5, π extends to a holomorphic representation of $G_{\mathbf{C}}$ in V. In the description of the orbit, an important role will be played by:

$$(4.12.12) \qquad G_\lambda := \{\, x \in G \mid (\mathrm{Ad}\, x^{-1})^*(\lambda) = \lambda \,\},$$

the stabilizer of $\lambda \in \mathfrak{g}_{\mathbf{C}}^*$ for the coadjoint action of G. Using a W-invariant inner product on $\mathfrak{t}$, we may identify the element $i\lambda$ with some element $X \in \mathfrak{c}^{\mathrm{cl}}$, the closure of the positive Weyl chamber in $\mathfrak{t}$, cf. (3.5.11). Moreover, $G_\lambda = G_X$, a compact Lie subgroup of G, with Lie algebra equal to:

(4.12.13)
$$\mathfrak{g}_\lambda = \mathfrak{t} + \sum_{\{\alpha \in R \mid \lambda(\alpha^\vee)=0\}} (\mathfrak{g}_\alpha + \mathfrak{g}_{-\alpha}) \cap \mathfrak{g},$$

cf. (3.7.1). Recall that $G_\lambda = G_X$ is connected, cf. Theorem 3.3.1.(ii), and that it contains T.

The complex Lie subalgebra:

(4.12.14)
$$\mathfrak{p}_\lambda := \mathfrak{h} + \sum_{\{\alpha \in R \mid \lambda(\alpha^\vee) \geq 0\}} \mathfrak{g}_\alpha$$

of $\mathfrak{g}_{\mathbf{C}}$, will play a role similar to the one played by $\mathfrak{b}$ in the description of G/T in Lemma 4.12.3; these are the so-called "parabolic subalgebras" of $\mathfrak{g}_{\mathbf{C}}$. The corresponding *parabolic subgroup* P_λ is the closed complex Lie subgroup of $G_{\mathbf{C}}$ defined by:

(4.12.15)
$$P_\lambda := \mathrm{N}_{G_{\mathbf{C}}}(\mathfrak{p}_\lambda) = \{ x \in G_{\mathbf{C}} \mid \operatorname{Ad} x(\mathfrak{p}_\lambda) = \mathfrak{p}_\lambda \}.$$

(4.12.6) Lemma. *Write $(G_{\mathbf{C}})_{V_\lambda}$ for the stabilizer of V_λ in $G_{\mathbf{C}}$ for the action on $\mathbf{CP}(V)$. Then $(G_{\mathbf{C}})_{V_\lambda} = P_\lambda =$ the connected Lie subgroup of $G_{\mathbf{C}}$ with Lie algebra equal to $\mathfrak{p}_\lambda$. Furthermore, $G_\lambda \subset P_\lambda$, and this inclusion induces a diffeomorphism: $G/G_\lambda \xrightarrow{\sim} G_{\mathbf{C}}/P_\lambda$.*

Proof. From the remark on α-ladders preceding (4.11.7), we see that if $X \in \mathfrak{g}_{-\alpha}\backslash\{0\}$, $v \in V_\lambda\backslash\{0\}$, then $\pi(X)(v) = 0$, if and only if $\lambda(\alpha^\vee) = 0$. Writing an arbitrary $X \in \mathfrak{g}_{\mathbf{C}}$ as $X = Y + H + \sum_{\alpha \in P} X_{-\alpha}$, for $Y \in \mathfrak{n}$, $H \in \mathfrak{h}$, $X_{-\alpha} \in \mathfrak{g}_{-\alpha}$, and using that the spaces $V_{\lambda-\alpha}$, for $\alpha \in P$, are linearly independent, we see that $\pi(X)V_\lambda \subset V_\lambda$ if and only if $X_{-\alpha} = 0$, whenever $\lambda(\alpha^\vee) > 0$. In other words, $\mathrm{T}_1(G_{\mathbf{C}})_{V_\lambda} = \mathfrak{p}_\lambda$. Because $\mathfrak{g}_\lambda \subset \mathfrak{p}_\lambda$, and G_λ is connected, this implies that $G_\lambda \subset ((G_{\mathbf{C}})_{V_\lambda})^\circ$. Since $(G_{\mathbf{C}})_{V_\lambda}$ normalizes its own Lie algebra, it is contained in P_λ.

Because $\mathfrak{p}_\lambda \supset \mathfrak{b}$, and already $\mathfrak{g} + \mathfrak{b} = \mathfrak{g}_{\mathbf{C}}$, see the beginning of the proof of Lemma 4.12.3, it follows that the G-orbit of $1(G_{\mathbf{C}})_{V_\lambda}$ in $Q := G_{\mathbf{C}}/(G_{\mathbf{C}})_{V_\lambda}$ is open. Because the G-orbit is compact in the connected space Q, it is equal to Q. That is, the mapping $G/T \to G_{\mathbf{C}}/(G_{\mathbf{C}})_{V_\lambda}$ is surjective.

Now suppose that $x \in G$ and $\operatorname{Ad} x(\mathfrak{p}_\lambda) = \mathfrak{p}_\lambda$. Then $\operatorname{Ad} x(\mathfrak{g}_\lambda) \subset \mathfrak{p}_\lambda \cap \mathfrak{g} = \mathfrak{g}_\lambda$. Because $\mathfrak{t}$ is a maximal Abelian subspace of the Lie algebra $\mathfrak{g}_\lambda$ of the connected, compact Lie group G_λ, there exists $y \in G_\lambda$ such that $\operatorname{Ad} y$ maps the maximal Abelian subspace $\operatorname{Ad} x(\mathfrak{t})$ of $\mathfrak{g}_\lambda$ onto $\mathfrak{t}$, cf. Theorem 3.7.1.(iv). Write $z = yx$. Because the normalizer of T in G_λ acts transitively on the Weyl chambers in $\mathfrak{t}$, with respect to the root system of $\mathfrak{g}_\lambda$, we can arrange that $\operatorname{Ad} z \cdot \alpha \in P$, for all $\alpha \in P$ such that $\lambda(\alpha^\vee) = 0$. On the other hand, $\operatorname{Ad} z(\mathfrak{p}_\lambda) = \mathfrak{p}_\lambda$, because $z \in G_\lambda P_\lambda = P_\lambda$; and this implies that $\operatorname{Ad} z$ permutes $\{ \alpha \in R \mid \lambda(\alpha^\vee) \geq 0 \}$. Hence $\operatorname{Ad} z \cdot \alpha \in P$, for all $\alpha \in P$ such that $\lambda(\alpha^\vee) > 0$.

Because $\lambda(\alpha^\vee) \geq 0$, for all $\alpha \in P$, we are in the situation that $\mathrm{Ad}\, z \cdot P = P$, or $z \in T$. But this implies that $x \in G_\lambda T = G_\lambda$.

In the previous paragraph we have proved that the mapping $G/G_\lambda \to G_{\mathbf{C}}/P_\lambda$ is injective. Because it is equal to the composition of the surjective mappings:

$$G/G_\lambda \to G_{\mathbf{C}}/((G_{\mathbf{C}})_{V_\lambda})^\circ \to G_{\mathbf{C}}/(G_{\mathbf{C}})_{V_\lambda} \to G_{\mathbf{C}}/P_\lambda,$$

all mappings must be bijective; and this implies also that $((G_{\mathbf{C}})_{V_\lambda})^\circ = (G_{\mathbf{C}})_{V_\lambda} = P_\lambda$. The injectivity of the tangent mapping at $1T$ of $G/G_\lambda \to G_{\mathbf{C}}/P_\lambda$ finally yields that $\mathrm{T}_1\, P_\lambda \cap \mathfrak{g} = \mathfrak{g}_\lambda$; this implies $\mathrm{T}_1\, P_\lambda = \mathfrak{p}_\lambda$. □

In the next theorem we will make use of the *hyperplane bundle* H, which is the complex line bundle over $\mathbf{CP}(V)$, defined as $H = V \setminus \{0\} \times_{\mathbf{C}} \mathbf{C}^\times$, where $\theta \in \mathbf{C}^\times$ acts on $V \setminus \{0\} \times \mathbf{C}$ by $(v, c) \mapsto (\theta v, \theta c)$. The name comes from the identification of the fiber over $L \in \mathbf{CP}(V)$ with the space of hyperplanes in V^* that are parallel to the orthogonal complement of L in V^*, induced by the mapping:

$$(v, c) \mapsto \{\, \mu \in V^* \mid \mu(v) = c \,\},$$

if $L = \mathbf{C} \cdot v$. See Griffiths and Harris [1978], pp.145, 164. The hyperplane bundle is dual to the "universal" or "tautological" bundle, for which the fiber over L is the line L itself.

(4.12.7) [Tits, Borel-Weil theorem]. *The mapping $x \mapsto \pi(x)V_\lambda$ induces a real-analytic, and a complex-analytic, embedding from G/G_λ, and $G_{\mathbf{C}}/P_\lambda$, respectively, onto a smooth complex algebraic subvariety of $\mathbf{CP}(V)$. The complex line bundle $\mathrm{L}_{-\lambda}$ over $G_{\mathbf{C}}/B$ is obtained by pulling back a homogeneous complex line bundle $\mathrm{L}_{-\lambda}$ over $G_{\mathbf{C}}/P_\lambda$, by means of the projection $p\colon G_{\mathbf{C}}/B \to G_{\mathbf{C}}/P_\lambda$; in turn the latter one is obtained by pulling back the hyperplane bundle H over $\mathbf{CP}(V)$, by means of the embedding $G_{\mathbf{C}}/P_\lambda \to \mathbf{CP}(V)$. Finally, p^* induces an isomorphism:*

$$\Gamma_{\mathrm{hol}}(G_{\mathbf{C}}/P_\lambda, \mathrm{L}_{-\lambda}) \to \Gamma_{\mathrm{hol}}(G_{\mathbf{C}}/B, \mathrm{L}_{-\lambda}).$$

Proof. Lemma 4.12.6 implies that the $\pi(G_{\mathbf{C}})$-orbit of V_λ in $\mathbf{CP}(V)$ is a compact, smooth complex-analytic submanifold of $\mathbf{CP}(V)$. Chow's theorem, cf. Griffiths and Harris [1978], pp.167, yields that it is algebraic.

For every $p \in P_\lambda$, we have $\pi(p)V_\lambda = V_\lambda$; so there is a uniquely determined $p^\lambda \in \mathbf{C}^\times$, such that $\pi(p)(v) = p^\lambda v$, for all $v \in V_\lambda$. Clearly, $p \mapsto p^\lambda$ is a homomorphism $P_\lambda \to \mathbf{C}^\times$, which extends $b \mapsto b^\lambda\colon B \to \mathbf{C}^\times$. Both pull back statements follow by writing out the definitions.

Replacing V by M_π, we get that the solutions f of the equations (4.12.5,6) automatically also satisfy $\mathrm{R}^*(X)(f) = 0$, if $X \in \mathfrak{g}_{-\alpha}$, and $\alpha \in P$ is such that $\lambda(\alpha^\vee) = 0$. This proves the last assertion. □

Example. If $G = \mathbf{SU}(n) \subset \mathbf{GL}(\mathbf{C}^n)$, then $G_{\mathbf{C}} = \mathbf{SL}(n, \mathbf{C})$. With the choice of $\mathfrak{t}$ and of positive roots $\alpha_{p,q}$, for $p < q$, as in Section 3.12, the space $\mathfrak{n}$, $\mathfrak{h}$, and $\overline{\mathfrak{n}}$, becomes equal to the space of strictly upper triangular, diagonal, and strictly lower triangular matrices with trace equal to zero, respectively. (This choice is just opposite to the one in Lemma 4.11.3, keeping one awake.) So B consists of the upper triangular unipotent matrices, identifying $G_{\mathbf{C}}/B$ with the *complex flag variety* as discussed at the end of Section 3.12.

The G/P_λ are the *partial complex flag varieties*. Among these, the *Grassmann varieties* occur precisely when $\lambda(\alpha^\vee) = 0$, for all but one of the simple roots α. That is, when λ is an integral multiple of a fundamental weight, cf. (4.10.11). These are the highest weights of the irreducible representations of G which can be realized as the space of holomorphic sections of a homogeneous complex line bundle over a Grassmann variety. The fundamental representations themselves, as described in the example preceding (4.10.12), correspond to the complex line bundle which assigns to each k-dimensional linear subspace U of $\mathbf{C}^n$, the complex one-dimensional linear subspace of $\bigwedge^k \mathbf{C}^n$ consisting of all $u_1 \wedge \ldots \wedge u_k$, such that $u_1, \ldots, u_k \in U$.

For general compact, connected Lie groups, the $G/G_\lambda = G_{\mathbf{C}}/P_\lambda$ are called the *generalized complex flag varieties*, or just *complex flag varieties* if there is no danger of confusion. Note that actually the G_λ run over all G_X, for $X \in \mathfrak{g}$. Also, every parabolic subgroup P of $G_{\mathbf{C}}$ is conjugate, by means of an element of $G_{\mathbf{C}}$, to P_λ, for some dominant weight λ.

For $n = 2$, we have $G/T = G_{\mathbf{C}}/B$ is equal to the complex projective line $\mathbf{CP}(\mathbf{C}^2)$; the identification is obtained by mapping the matrix x to the line through its first column. The mapping:

$$(a, c) \mapsto \begin{pmatrix} a & 0 \\ c & a^{-1} \end{pmatrix} : (\mathbf{C} \setminus \{0\} \times \mathbf{C}) \to G_{\mathbf{C}}$$

has an open image in $G_{\mathbf{C}}/N$. So a complex-analytic function on $G_{\mathbf{C}}/N$ is determined by its pull back to $\mathbf{C} \setminus \{0\} \times \mathbf{C}$. Now the sections of the line bundle $\mathrm{L}_{-\lambda}$ over $G_{\mathbf{C}}/B$, defined by the character:

$$\begin{pmatrix} p & q \\ 0 & p^{-1} \end{pmatrix} \mapsto p^{-\lambda} : B \to \mathbf{C}^\times,$$

correspond to the complex-analytic functions f on $\mathbf{C} \setminus \{0\} \times \mathbf{C}$, such that $f(ap, cp) = p^\lambda f(a, c)$, for $a \in \mathbf{C} \setminus \{0\}$, $c \in \mathbf{C}$, $p \in \mathbf{C}^\times$, that is, the f which are homogeneous of degree λ. On the other hand, we know that $\mathrm{R}^*(X)^{\lambda+1} = 0$ on M_π, if $X \in \mathfrak{g}_{-\alpha}$, $\alpha \in P$, because the ladders of the irreducible representation with highest weight equal to λ are never longer than λ. This means that $(\frac{d}{dc})^{\lambda+1} f(1, c) = 0$; or $c \mapsto f(1, c)$ is a polynomial of degree $\leq \lambda$. Inserting $f(a, c) = a^\lambda f(1, \frac{c}{a})$, we obtain that f is a homogeneous polynomial on $\mathbf{C}^2$ of degree λ. Because the space of these polynomials has dimension equal to $\lambda + 1 =$ the dimension of the irreducible representation with highest weight equal to λ, the conclusion is that **the irreducible representations of $\mathbf{SU}(2)$**

can be identified with the spaces of homogeneous complex polynomials on $\mathbf{C}^2$ of degree λ, for $\lambda \in \mathbf{Z}_{\geq 0}$, on which $\mathbf{SU}(2) \subset \mathbf{GL}(\mathbf{C}^2)$ acts in the natural way. As we have seen in Fig.4.10.1, this representation is of real type, and of quaternionic type, if λ is even, and odd, respectively.

We finally mention that, inspired by the Borel-Weil theorem, one can give a proof of Weyl's character formula (4.9.14) in terms of Lie algebra cohomology, cf. Kostant [1961]. This is based on a computation by Bott and Kostant of the cohomology groups of the sheaf of holomorphic sections of the holomorphic line bundles $\mathrm{L}_{-\lambda}$ over $G_{\mathbf{C}}/B$, for arbitrary $\lambda \in \widehat{T}$; a remarkable feat, because in general the explicit computation of such cohomology groups is a difficult problem in algebraic geometry.

4.13 The Nonconnected Case

In this section, G is a compact Lie group that is allowed to be nonconnected. Let π be a representation of G in a finite-dimensional vector space V. The first step which comes to mind is the decomposition:

(4.13.1)
$$V = \sum_{[\tau] \in \widehat{G^\circ}} V_\tau .$$

Here, for any $[\tau] \in \widehat{G^\circ}$, the space V_τ is the τ-isotypical subspace of V as defined preceding Lemma 4.4.1, see also Proposition 4.4.2. We have:

(4.13.2)
$$\dim V_\tau = m_\tau \, \mathrm{d}_\tau ,$$

where $m_\tau := [\pi|_{G^\circ} : \tau] \in \mathbf{Z}_{\geq 0}$, the *multiplicity* of τ in $\pi|_{G^\circ}$. We say that $[\tau] \in \widehat{G^\circ}$ *occurs* in π if $m_\tau \neq 0$; because

$$\sum_{[\tau] \in \widehat{G^\circ}} m_\tau \, \mathrm{d}_\tau = \dim V$$

is finite, only finitely many $[\tau] \in \widehat{G^\circ}$ can occur in π.

If $x \in G$, then $\mathbf{Ad}\, x\colon y \mapsto x y x^{-1}$ is an automorphism of G°. For any $[\tau] \in \widehat{G}$, the operator $\pi(x)$ intertwines $\pi|_{V_\tau}$ with the representation $\pi \circ \mathbf{Ad}\, x$ of G° on $\pi(x)(V_\tau)$. In other words, the definition:

(4.13.3)
$$x \cdot \tau := \tau \circ \mathbf{Ad}\, x^{-1}, \quad \text{for } x \in G, [\tau] \in \widehat{G^\circ},$$

induces an action $(xG^\circ, [\tau]) \mapsto [x \cdot \tau]$ of the component group G/G° on $\widehat{G^\circ}$, such that:

(4.13.4)
$$\pi(x)(V_\tau) = V_{x \cdot \tau}, \quad \text{for all } x \in G, [\tau] \in \widehat{G^\circ}.$$

Clearly, $m_{x\cdot\tau} = m_\tau$, if $x \in G$, $[\tau] \in \widehat{G^\circ}$.

Also, **if π is irreducible, then only one G/G°-orbit in $\widehat{G^\circ}$ occurs in π.** In particular, in this case there is only one multiplicity $m \in \mathbf{Z}_{>0}$, such that $m_\tau = m$, for all $[\tau] \in \widehat{G^\circ}$ that occur in π.

The $[\tau] \in \widehat{G^\circ}$ are considered to be "known" from Sections 4.9–12. In order to formulate some of the obvious conclusions for π, we fix from now on a choice of a maximal Abelian subspace $\mathfrak{t}$ of the Lie algebra $\mathfrak{g}$ of G. Also a choice P of positive roots, with corresponding Weyl chamber $\mathfrak{c}$ in $\mathfrak{t}$, as in (3.5.11,12). If $T = \exp \mathfrak{t}$ is the corresponding maximal torus of G°, then the decomposition of $\pi|_T$ into isotypical subspaces takes the form:

(4.13.5)
$$V = \bigoplus_{\mu \in \widehat{T}} V_\mu,$$

where for any $\mu \in \widehat{T}$ we have written:

(4.13.6)
$$V_\mu := \{\, v \in V \mid \pi(t)(v) = t^\mu v, \text{ for all } t \in T \,\}.$$

As in the text preceding (4.9.1), we have identified $\mu \in \widehat{T}$ here with an element of $i\mathfrak{t}^* \subset \mathfrak{h}^*$. It is called a *weight* of π if:

(4.13.7)
$$m_\mu := \dim_{\mathbf{C}} V_\mu,$$

the multiplicity of μ in $\pi|_T$, is nonzero. Again the set of weights of π will be denoted by $\Omega(\pi)$. For the character we get correspondingly:

(4.13.8)
$$\chi_\pi(t) = \sum_{\mu \in \widehat{T}} m_\mu \, t^\mu, \quad \text{for } t \in T.$$

On the other hand,

(4.13.9)
$$\chi_\pi(t) = \sum_{[\tau] \in \widehat{G^\circ}} m_\tau \chi_\tau(t), \quad \text{for } t \in T,$$

where the characters χ_τ of the irreducible representations τ of the connected, compact Lie group G° are given on T by the Weyl character formula 4.9.1. This is an explicit formula in terms of the highest weight λ of τ, which originally was defined in Section 4.9 as the weight of τ, for which no $\lambda + \Sigma(Q)$, with Q a nonvoid subset of P, is a weight of τ. Here we have used the notation $\Sigma(Q) := \sum_{\alpha \in Q} \alpha$. As a consequence, we obtained for instance that the set of weights of τ is equal to the intersection of $\lambda + \Lambda_R$ with the convex hull of the Weyl group orbit of λ, cf. Proposition 4.10.1.

If we write:

(4.13.10)
$$m_{\tau,\mu} := [\tau|_T : \mu], \quad \text{for } [\mu] \in \widehat{T},$$

then, if π is irreducible,

$$m_\mu = m \sum_{[\tau]\in\widehat{G}} m_{\tau,\mu}, \tag{4.13.11}$$

all sums in this section being finite. It follows for instance that the set $\Omega(\pi)$ of weights of π is equal to the union of the $(\lambda + \Lambda_R) \cap \operatorname{conv}(W \cdot \lambda)$, with λ equal to the highest weight of an irreducible representation τ in $\widehat{G^\circ}$ such that $m_\tau \neq 0$. This already gives some information about the character of π, at least about its restriction to T.

In analogy with Section 4.9, let us call $\lambda \in \widehat{T}$ a *highest weight* of π, if no $\lambda + \Sigma(Q)$, with Q a nonvoid subset of P, is a weight of π. The set of highest weights of π will be denoted by $\Omega(\pi)^{\mathfrak{n}}$. It is a subset of the set of highest weights of the $[\tau] \in \widehat{G^\circ}$ such that $m_\tau \neq 0$; and nonvoid by the same argument that we used in the proof of Weyl's character formula 4.9.1 in order to prove the existence of highest weights. We shall also write:

$$V^{\mathfrak{n}} := \sum_{\lambda\in\Omega(\pi)^{\mathfrak{n}}} V_\lambda, \tag{4.13.12}$$

for the corresponding *highest weight space*. In (4.13.30) we shall see that this terminology agrees with the one of Section 4.11.

In order to obtain also some more information about χ_π on the other connected components of G, we will perform a reduction to certain irreducible representations of the subgroup:

$$U := \mathrm{N}_G(\mathfrak{c}) := \{\, u \in G \mid \operatorname{Ad} u(\mathfrak{c}) = \mathfrak{c} \,\}, \tag{4.13.13}$$

the normalizer in G of the Weyl chamber $\mathfrak{c}$. This is a Lie subgroup of G with Lie algebra equal to $\mathfrak{t}$, so $U^\circ = T := \exp \mathfrak{t}$, a maximal torus of G°. Furthermore, Proposition 3.15.1.(ii) says that the inclusions $U \subset G$, $T \subset G^\circ$ induce an isomorphism: $U/T \to G/G^\circ$ of the component groups, which shows that U is "as nonconnected as G". One should be warned that, although the Lie algebra of U is Abelian, U need not to be Abelian; and even the dimension of $\mathfrak{s} := \mathfrak{t}_x$ can vary if x is taken in different connected components of U.

The condition $\operatorname{Ad} u(\mathfrak{c}) = \mathfrak{c}$ implies that $u \cdot \alpha := \operatorname{Ad} u \cdot \alpha \in P$, for all $\alpha \in P$. On the other hand, (4.13.4), with G, and G°, replaced by U, and T, respectively, reads:

$$\pi(u)(V_\mu) = V_{u\cdot\mu}, \quad \text{for } u \in U, \mu \in \widehat{T}. \tag{4.13.14}$$

Because $u \in U$ acts on $\widehat{T}$ by means of linear transformations, $u \cdot P = P$ implies that $u \cdot \Omega(\pi)^{\mathfrak{n}} = \Omega(\pi)^{\mathfrak{n}}$, so (4.13.14) yields:

$$\pi(u)(V^{\mathfrak{n}}) = V^{\mathfrak{n}}, \quad \text{for } u \in U. \tag{4.13.15}$$

This defines a representation of U in $V^{\mathfrak{n}}$, which will be denoted by:

(4.13.16) $$\theta := (\pi|_U)|_{V^{\mathfrak{n}}}.$$

From Proposition 3.15.2.(iii) we recall that we can write $U = G_X$ for a suitable $X \in \mathfrak{c}$. From Lemma 3.15.3 it then follows that, for each $u \in U$, there exist $z \in U_x$, of principal orbit type for the action of G on itself by conjugation, and arbitrarily close to u. This allows us to choose an element z of principal orbit type in each connected component of U, as in the discussion following Lemma 3.15.4. One major point in that discussion is that every element of the connected component of u in G is conjugate, already by an element of G°, with an element of $zS \subset zT \subset U$, here $S = (G_z)^{\circ}$. This implies that **the characters of the representations of G are determined by their restrictions to U.**

As in Weyl's character formula 4.9.1, we will obtain an explicit formula for $\chi_\pi|_U$, following the same argument. Only this time keeping track of the various contributions is a bit lengthy, leading to a small orgy of formulae.

Provide V with a basis which consists of bases of the V_μ, with $\mu \in \Omega(\pi)$. If $u \in U$, then (4.13.14) yields that

$$\chi_\pi(u) = \sum_{\mu \in \Omega(\pi)^{uT}} \chi_{\pi,\mu}(u),$$

if we write:

(4.13.17) $$\Omega(\pi)^{uT} := \{\, \mu \in \Omega(\pi) \mid u \cdot \mu = \mu \,\},$$

and

(4.13.18) $$\chi_{\pi,\mu}(u) := \mathrm{tr}\big(\pi(u)|_{V_\mu}\big), \quad \text{for } u \in U, \mu \in \Omega(\pi)^{uT}.$$

Replacing u by ut, for $t \in T$, we get:

(4.13.19) $$\chi_\pi(ut) = \sum_{\mu \in \Omega(\pi)^u} \chi_{\pi,\mu}(u)\, t^\mu.$$

Here we think of u as a fixed choice of an element of U. If t runs over T, then ut runs over the connected component uT of U. (Note that $ut \cdot \mu = \mu$, if $u \cdot \mu = \mu$ and $t \in T$; this explains the notation in (4.13.17). This set of weights only depends on the choice of an element of the, finite, component group U/T.) In the same fashion the character of the representation $\theta = (\pi|_U)|_{V^{\mathfrak{n}}}$ of U is given by:

(4.13.20) $$\chi_\theta(ut) = \sum_{\mu \in \Omega(\pi)^{uT,\mathfrak{n}}} \chi_{\pi,\mu}(u)\, t^\mu, \quad \text{for } t \in T,$$

where we have written $\Omega(\pi)^{uT,\mathfrak{n}} := \Omega(\pi)^{uT} \cap \Omega(\pi)^{\mathfrak{n}}$.

The Weyl integration formula of Proposition 3.15.6, which will be applied to $f = \chi_\pi \overline{\chi_\pi}$, contains "half the Jacobian factor":

$$\delta(ut) = \prod_{\sigma\in P(u)} (1 - c_\sigma(u)^{-1} t^{-\Sigma(\sigma)}) = \sum_{Q\subset P(u)} \prod_{\sigma\in Q} -c_\sigma(u)^{-1} t^{-\Sigma(Q)}.$$

Here $P(u)$ denotes the collection of $\mathrm{Ad}\, u$-cycles in P, cf. (3.15.6); and $c_\sigma(u)$ is the eigenvalue of $\mathrm{Ad}\, x^{\#(\sigma)}$ on $\mathfrak{g}_\alpha$, with $\alpha\in\sigma$. Finally:

$$\Sigma(\sigma) = \sum_{\alpha\in\sigma} \alpha, \quad \Sigma(Q) = \sum_{\sigma\in Q} \Sigma(\sigma) = \sum_{\alpha\in U\cdot Q} \alpha.$$

Compare with (3.15.11). Note that $|c_\sigma(u)| = 1$. Combining this with (4.13.19), we get:

(4.13.21) $$(\delta\chi_\pi)(ut) = \sum_{\phi\in\widehat{T}} d_\phi(u)\, t^\phi,$$

where

(4.13.22) $$d_\phi(u) := \sum_{Q\subset P(u)} \chi_{\pi,\phi+\Sigma(Q)}(u) \prod_{\sigma\in Q} -c_\sigma(u)^{-1}.$$

The sum in (4.13.21) only ranges over the $\phi\in\widehat{T}$ such that $\phi+\Sigma(Q)\in\Omega(\pi)^{uT}$. Furthermore, if $\phi\in\Omega(\pi)^{\mathfrak{n}}$, then, since the convex cone generated by P is proper, this can only happen if Q is void. That is,

(4.13.23) $$d_\lambda(u) = \chi_{\pi,\lambda}(u) = \chi_{\theta,\lambda}(u), \quad \text{if } \lambda\in\Omega(\pi)^{uT,\mathfrak{n}};$$

and this expresses that $t\mapsto(\delta\chi_\pi)(ut)$ contains $t\mapsto\chi_\theta(ut)$ as a partial sum in its Fourier decomposition, cf. (4.13.20).

After these preliminaries, we will investigate the consequences of the fact that the integral of $\chi_\pi\overline{\chi_\pi}$ over G is equal to 1, cf. Theorem 4.3.4.(iv). Applying Proposition 3.15.6 with $f=\chi_\pi\overline{\chi_\pi}$, we get:

(4.13.24) $$1 = \#(U/U^\circ)^{-1} \sum_{uT\in U/T} \#(W(u))^{-1} \sum_{\phi\in\widehat{T}} |d_\phi(u)|^2.$$

Note that averaging over U gives each connected component uT of U total measure equal to $\#(U/T)^{-1}$. That is, if dt denotes averaging over T, then averaging over U is equal to $d(ut) = \#(U/T)^{-1} dt$ on each connected component uT of U. Also note that:

$$\int_T \Big(\sum_\phi d_\phi(u)\, t^\phi\Big)\overline{\Big(\sum_\phi d_\phi(u)\, t^\phi\Big)}\, dt = \sum_{\phi\in\widehat{T}} |d_\phi(u)|^2.$$

Recall the subgroup $W(uT)$ of the Weyl group W, consisting of the $s\in W$ which commute with $\mathrm{Ad}\, u|_{\mathfrak{t}}$, cf. (3.15.12). If $g\in G^\circ\cap \mathrm{N}(\mathfrak{t})$ represents $s\in W(uT)$, in the sense that $s=\mathrm{Ad}\, g|_{\mathfrak{t}}$, then, as we have seen in the paragraph preceding (3.15.12),

$$t_g := u^{-1} g^{-1} u g \in T. \tag{4.13.25}$$

Note that $g^{-1}(ut)g = ut_g s^{-1}(t)$, for $t \in T$.

The invariance of χ_π under conjugacy by elements of G yields:

$$\begin{aligned}\sum_\phi d_\phi(u)\, t_g^\phi\, t^{s\cdot\phi} &= (\delta\chi_\pi)(ut_g s^{-1}(t)) = (\delta\chi_\pi)(g^{-1}utg) = \delta(g^{-1}utg)\, \chi_\pi(ut) \\ &= \frac{\delta(g^{-1}utg)}{\delta(ut)} \sum_\phi d_\phi(u)\, t^\phi.\end{aligned}$$

On the other hand, writing $ut = y$, we get:

$$\begin{aligned}\delta(g^{-1}yg) &= \det(\mathrm{I} - \mathrm{Ad}(g^{-1}yg)^{-1})|_{\mathfrak{n}} \\ &= \det \mathrm{Ad}\, g^{-1} \circ (\mathrm{I} - \mathrm{Ad}\, y^{-1}) \circ \mathrm{Ad}\, g|_{\mathfrak{n}} = \det(\mathrm{I} - \mathrm{Ad}\, y^{-1})|_{\mathrm{Ad}\, g(n)} \\ &= \prod_{\sigma \in P(u)} (1 - c_{s\cdot\sigma}(y)^{-1}).\end{aligned}$$

In view of:

$$\sum_{\alpha \in -\sigma} \mathfrak{g}_\alpha = \overline{\sum_{\alpha \in \sigma} \mathfrak{g}_\alpha},$$

we get:

$$c_{-\sigma}(y) = \overline{c_\sigma(y)} = c_\sigma(y)^{-1},$$

because $\mathrm{Ad}\, y$ is a real transformation, with eigenvalues on the unit circle. So

$$\delta(g^{-1}yg) = \prod_{\{\,\tau \in P(u) | s^{-1}\cdot\tau \subset P\,\}} (1 - c_\tau(y)^{-1}) \prod_{\{\,\tau \in P(u) | s^{-1}\cdot\tau \subset -P\,\}} (1 - c_\tau(y)).$$

Because $(1-c)/(1-c^{-1}) = -c$, it follows that:

$$\begin{aligned}\frac{\delta(g^{-1}yg)}{\delta(y)} &= \prod_{\{\,\tau \in P(u) | s^{-1}\cdot\tau \subset -P\,\}} -c_\tau(y) = \prod_{\{\,\tau \in P(u) | s^{-1}\cdot\tau \subset -P\,\}} -c_\tau(u)\, t^{\Sigma(\tau)} \\ &= b_s(u)\, t^{\rho - s\cdot\rho},\end{aligned}$$

where we have written:

$$b_s(u) := \prod_{\{\,\tau \in P(u) | s^{-1}\cdot\tau \subset -P\,\}} -c_\tau(u), \tag{4.13.26}$$

and we have used that:

$$\sum_{\{\,\tau \in P(u) | s^{-1}\cdot\tau \subset -P\,\}} \Sigma(\tau) = \sum_{\alpha \in P \setminus s\cdot P} \alpha = \rho - s\cdot\rho,$$

cf. (4.9.12). Inserting this in the previous relation which expressed the conjugacy invariance of χ_π, we get:

$$\sum_{\phi} d_\phi(u)\, t_g^\phi\, t^{s\cdot\phi} = \sum_{\phi} b_s(u)\, d_\phi(u)\, t^{\phi+\rho-s\cdot\rho}.$$

Or, comparing coefficients,

$$d_{s\cdot\phi+s\cdot\rho-\rho}(u) = a_{s,\phi}(u)d_\phi(u), \quad \text{for } s \in W(uT), \tag{4.13.27}$$

where

$$a_{s,\phi}(u) := t_g^\phi\, b_s(u)^{-1} \tag{4.13.28}$$

has absolute value equal to 1.

For the proper interpretation of the character formula (4.13.29) which we will obtain below, we note here that the number $t_g^\phi = (u^{-1}g^{-1}ug)^\phi$ does not change if g is replaced by gt, for $t \in T$, if $\phi \in \Omega(\pi)^{uT}$. Indeed, in this case:

$$u^{-1}t^{-1}g^{-1}ugt = (u^{-1}g^{-1}ug)(g^{-1}u^{-1}gt^{-1}g^{-1}ug)t = (u^{-1}g^{-1}ug)(u^{-1}t^{-1}u)t,$$

and

$$(u^{-1}t^{-1}u)^\phi = (t^{-1})^{u\cdot\phi} = (t^{-1})^\phi.$$

As in the argument preceding Weyl's character formula 4.9.1, the shifted Weyl group action $(s,\mu) \mapsto s\cdot\mu + s\cdot\rho - \rho$ is free on the highest weights $\lambda \in \Omega(\pi)^{\mathfrak{n}}$. It follows that each $W(u)$-orbit has exactly $\#(W(u))$ many elements. In view of (4.13.27,28), the corresponding coefficients all have the same absolute value, equal to $|\chi_{\pi,\mu}(u)|$, if we use (4.13.23). Thus, (4.13.24) leads to $1 = A + B$, where:

$$\begin{aligned} A &= \#(U/T)^{-1} \sum_{uT\in U/T}\ \sum_{\mu\in\Omega(\pi)^{uT,\mathfrak{n}}} |\chi_{\pi,\mu}(u)|^2 \\ &= \sum_{uT\in U/T} \#(U/T)^{-1} \int_T |\chi_\theta(ut)|^2\, dt = \int_U |\chi_\theta(u)|^2\, du, \end{aligned}$$

and B is the sum in (4.13.24) over all ϕ which are not contained in a shifted $W(x)$-orbit of some $\mu \in \Omega(\pi)^{uT,\mathfrak{n}}$. Now we know from Corollary 4.3.5.(ii), that A is a positive integer, and because $B \geq 0$ implies that $A \leq 1$, it follows that $A = 1$ and $B = 0$. The first equality implies that θ is an irreducible representation of U, cf. Corollary 4.3.5.(iv), and the second one that $d_\phi(u) = 0$, whenever ϕ does not belong to a shifted $W(uT)$-orbit of some $\mu \in \Omega(\pi)^{uT,\mathfrak{n}} = \Omega(\theta)^{uT}$. Using (4.13.27) and (4.13.23), we have proved the following generalization of Weyl's character formula 4.9.1:

(4.13.1) Proposition. *Let π be an irreducible representation of G in V, let $V^{\mathfrak{n}} = \sum_{\lambda \in \Omega(\pi)^{\mathfrak{n}}} V_\lambda$, where V_μ is the weight space of $\mu \in \widehat{T}$ as defined in (4.13.6), and $\Omega(\pi)^{\mathfrak{n}}$ denotes the set of weights λ of π such that $\lambda + \sigma(Q)$ is not a weight of π for any nonvoid subset Q of P. Let U be the normalizer in G of the Weyl chamber $\mathfrak{c}$, as in (4.13.13).*

Then $\theta := (\pi|_U)|_{V^{\mathfrak{n}}}$ is an irreducible representation of U in $V^{\mathfrak{n}}$. Furthermore, the character of π is given, in terms of the character of θ, for $u \in U$, by:

(4.13.29) $\delta(u)\chi_\pi(u) =$

$$\sum_{s \in W(uT)} \prod_{\{\tau \in P(u) | s^{-1} \cdot \tau \subset -P\}} -c_\tau(u)^{-1} \sum_{\lambda \in \Omega(\theta)^{uT}} (u^{-1}g^{-1}ug)^\lambda \, \chi_{\theta,\lambda}(u).$$

Here $g \in G^\circ \cap \mathrm{N}(\mathfrak{t})$ is chosen such that $\mathrm{Ad}\, g|_{\mathfrak{t}} = s$, and $s \in W(uT)$ means that $s \in W$ commutes with $\mathrm{Ad}\, u|_{\mathfrak{t}}$. Finally,

$$\Omega(\theta)^{uT} = \{ \lambda \in \Omega(\theta) = \Omega(\pi)^{\mathfrak{n}} \mid u \cdot \lambda = \lambda \},$$
$$\chi_{\theta,\lambda}(u) = \mathrm{tr}\big(\theta(u)|_{V_\lambda}\big) = \mathrm{tr}\big(\pi(u)|_{V_\lambda}\big), \quad \text{if } \lambda \in \Omega(\theta)^{uT}, u \in U.$$

(4.13.2) Corollary. *$\Omega(\pi)^{\mathfrak{n}}$ is equal to the set of the highest weights of the $[\tau] \in \widehat{G^\circ}$ which occur in $\pi|_{G^\circ}$. These $[\tau] \in \widehat{G^\circ}$ occur in $\pi|_{G^\circ}$ with a common multiplicity equal to m, and have a common dimension:*

$$d = \mathrm{d}_\tau = \prod_{\alpha \in P} \frac{(\lambda + \rho)(\alpha^\vee)}{\rho(\alpha^\vee)}, \quad \text{for } \lambda \in \Omega(\pi)^{\mathfrak{n}}.$$

If n denotes the number of $[\tau] \in \widehat{G^\circ}$ which occur in $\pi|_{G^\circ}$, then $\mathrm{d}_\pi = nmd$. Finally,

(4.13.30) $$V^{\mathfrak{n}} = \{ v \in V \mid \pi(X)(v) = 0, \text{ for all } X \in \mathfrak{n} \},$$

the space of highest weight vectors in V as defined in (4.11.7).

Proof. We have already observed, after (4.13.4), that all $[\tau] \in \widehat{G^\circ}$ that occur in $\pi|_{G^\circ}$, do so with the same multiplicity m. Because $\dim V_\tau = m\, \mathrm{d}_\tau$, (4.13.4) also implies that they have the same dimensions.

Applying (4.13.29) to $u = t \in T$, we get:

$$\prod_{\alpha \in P} (1 - t^{-\alpha})\chi_\pi(t) = \sum_{s \in W} \det s \; t^{s \cdot \rho - \rho} \sum_{\lambda \in \Omega(\pi)^{\mathfrak{n}}} m t^{s \cdot \lambda}.$$

In view of Weyl's character formula (4.9.14), this yields $\chi_\pi = m \sum_{\lambda \in \Omega(\pi)^{\mathfrak{n}}} \chi_\lambda$. Comparing this with (4.13.9), we obtain the first statement, and the formula for d_π is an immediate consequence. Also (4.13.30) now follows, by applying Cartan's highest weight theorem 4.11.4 to the irreducible components of $\pi|_{G^\circ}$.

□

Another immediate application of (4.13.29) is obtained by substituting $\chi_\pi = 1$, the character of the trivial representation of G. This yields the following generalization of (4.9.15):

(4.13.31)

$$\delta(u) := \det(\mathrm{I} - \mathrm{Ad}\, u^{-1})_{\mathfrak{n}} = \sum_{s \in W(uT)} \prod_{\{\tau \in P(u) \mid s^{-1} \cdot \tau \subset -P\}} -c_\tau(u)^{-1}.$$

Compare with (3.15.11), and recall that $c_\tau(u)$ is equal to the eigenvalue of $\mathrm{Ad}\, u^{\#(\tau)}$ on $\sum_{\alpha \in \tau} \mathfrak{g}_\alpha$, where τ denotes an $\mathrm{Ad}\, u$-cycle in P.

Using a matrix representation for $\mathrm{Ad}\, u|_{\mathfrak{g}_\tau}$ as in the paragraph preceding (3.15.7), we see that $c_\tau(u) = (-1)^{m-1} \det \mathrm{Ad}\, u|_{\mathfrak{g}_\tau}$. Hence, using that $(-1)^{\#(P \setminus s \cdot P)} = \det s$, we get:

$$\prod_{\{\tau \in P(u) \mid s^{-1} \cdot \tau \subset -P\}} -c_\tau(u) = \det s \, \det \mathrm{Ad}\, u|_{\mathfrak{g}_{P \setminus s \cdot P}},$$

if we write $\mathfrak{g}_Q = \sum_{\alpha \in Q} \mathfrak{g}_\alpha$, for any subset Q of roots, as in (3.15.7). With these notations, we may rewrite (4.13.31) as:

$$\delta(u) = \sum_{s \in W(uT)} \det s \, \det \mathrm{Ad}\, u^{-1}|_{\mathfrak{g}_{P \setminus s \cdot P}}. \qquad (4.13.32)$$

The term $(u^{-1} g^{-1} u g)^\lambda \, \chi_{\theta, \lambda}(u)$ in (4.13.29) is actually equal to the trace of $\pi(u) \circ \pi(u^{-1} g^{-1} u g) = \pi(g^{-1}) \circ \pi(u) \circ \pi(g)$ on V_λ, which in turn is equal to the trace of $\pi(u)$ on $\pi(g)(V_\lambda) = V_{s \cdot \lambda}$, similarly to (4.13.4,14). Here $\lambda \in \Omega(\pi)^{uT}$, which is equivalent to $s \cdot \lambda \in \Omega(\pi)^{uT}$, because $\mathrm{Ad}\, u|_{\mathfrak{t}}$ commutes with s. Taking the sum over all $\lambda \in \Omega(\pi)^{uT, \mathfrak{n}}$ therefore amounts to taking the trace of $\pi(u)$ over:

$$\begin{aligned} \pi(g)(V^{\mathfrak{n}}) &= \{ w \in V \mid \pi\big(\mathrm{Ad}\, g(X)\big)(w) = 0, \text{ for all } X \in \mathfrak{n} \} \\ &= \{ w \in V \mid \pi(Y)(w) = 0, \text{ for all } Y \in \mathfrak{g}_{s \cdot P} \}. \end{aligned}$$

So, if we write, with a slight abuse of notation,

$$V^{s \cdot \mathfrak{n}} = \{ v \in V \mid \pi(X)(v) = 0, \text{ for all } X \in \mathfrak{g}_{s \cdot P} \}, \qquad (4.13.33)$$

then, combined with the observations preceding (4.13.32), we can rewrite (4.13.29) as

$$\delta(u) \chi_\pi(u) = \sum_{s \in W(uT)} \det s \, \det \mathrm{Ad}\, u^{-1}|_{\mathfrak{g}_{P \setminus s \cdot P}} \, \mathrm{tr}\big(\pi(u)|_{V^{s \cdot \mathfrak{n}}}\big). \qquad (4.13.34)$$

This neat formula has the slight disadvantage that it is not in terms of the character of the representation $\theta = (\pi|_U)|_{V^{\mathfrak{n}}}$ only, but that it also involves the characters of the representations $\theta^s := (\pi|_{U^s})|_{V^{s \cdot \mathfrak{n}}}$ of the subgroups $U^s = \{ u \in U \mid \mathrm{Ad}\, u|_{\mathfrak{t}} \text{ commutes with } s \}$, for $s \in W$, of U.

Because the $\lambda \in \Omega(\pi)^{\mathfrak{n}}$ are the highest weights of the $[\tau] \in \widehat{G^\circ}$ which occur in $\pi|_{G^\circ}$, these are dominant weights of $\widehat{T}$, cf. Proposition 4.9.4.(ii). That is, all weights of θ are dominant.

If θ is an arbitrary irreducible representation of U in a finite-dimensional vector space W, then we may apply the introductory remarks of this section with G replaced by U, and conclude that the weights of θ form a single U/T-orbit in $\widehat{T}$, with some common multiplicity. The action of U on the weights leaves the set of dominant weights invariant, because each $u \in U$ maps P to P and therefore also each $\alpha^\vee$, for $\alpha \in P$, to some $\beta^\vee$, with $\beta \in P$, cf. Lemma 5.2.2 below. So, if any of the weights of θ is dominant, then they all are dominant, and in this case θ will be called a *dominant irreducible representation* of U. We can now formulate the following generalization of Theorem 4.9.5.

(4.13.3) Proposition. *The mapping $\pi \mapsto (\pi|_U)|_{V^{\mathfrak{n}}}$ induces a bijection between $\widehat{G}$ and the set of equivalence classes of dominant irreducible representations of U.*

Proof. The injectivity follows because (4.13.29) determines the character of π in terms of the character of θ, first on $U \cap G^{\mathrm{reg}}$, then on G^{reg} in view of the conjugacy invariance, and finally on G, owing to the continuity of the character and the density of G^{reg} in G. Of course we also use that $\chi_\pi = \chi_{\pi'}$ if and only if $\pi \simeq \pi'$, cf. Corollary 4.3.5.(iii). (One may also prove the injectivity as in the proof of Proposition 4.11.6, starting with the graph of the equivalence between θ and θ', instead of the highest weight vector (v, v').)

For the surjectivity, we copy the argument in the proof of Theorem 4.9.5. For this, we only have to observe that for each dominant irreducible representation θ of U, the quotient of the right hand side in (4.13.29) by δ defines a real-analytic function f_θ on $U \cap G^{\mathrm{reg}}$, which is invariant under conjugation by elements of U. Reading its construction backwards, we see that, on uT, it is also invariant under conjugation by $g \in G^\circ \cap \mathrm{N}(\mathfrak{t})$ such that $gT \in W(uT)$. In view of $G = G^\circ U$, the reasoning preceding (3.15.12) shows that f_θ extends to a real-analytic function on G^{reg} which is invariant under conjugation by arbitrary elements of G. Applying Proposition 3.15.6, one verifies that the f_θ form an L^2-orthonormal system if the $[\theta]$ range over the equivalence classes of dominant irreducible representations of U. As in the proof of Theorem 4.9.5, the essential point is that the L^2-norm is equal to 1, because the shifted Weyl group action is free on dominant weights. If θ is not of the form $(\pi|_U)|_{V^{\mathfrak{n}}}$, for some $[\pi] \in \widehat{G}$, then f_θ would be orthogonal to all χ_π, for $[\pi] \in \widehat{G}$, and therefore equal to 0, a contradiction. □

The weights of θ form the intersection with the dominant chamber of an orbit of the coadjoint action of the nonconnected group G on $i\mathfrak{g}^*$, in the same way as in the Remark after Theorem 4.9.5. Consequently again $\widehat{G}$ is parametrized by a collection of coadjoint orbits.

Cartan's highest weight theorem can be generalized to the nonconnected case, as follows: V is equal to $V_0 :=$ the smallest $\pi(\overline{\mathfrak{n}})$-invariant subspace generated by $V^{\mathfrak{n}}$. Indeed, because $u \cdot P = P$, we also have $\mathrm{Ad}\, u(\overline{\mathfrak{n}}) = \overline{\mathfrak{n}}$, for all $u \in U$; and it follows that V_0 is $\pi(u)$-invariant. Because it is $\pi(\mathfrak{g}_{\mathbf{C}})$-invariant as usual, it is $\pi(G^\circ)$-invariant, and because $G = G^\circ U$, it is $\pi(G)$-invariant; hence equal to V in view of the irreducibility of π. Furthermore, for any representation σ of G in a vector space W, we have $[\sigma : \pi] = [(\sigma|_U)|_{W^{\mathfrak{n}}} : \theta]$; and this generalizes Corollary 4.11.5.

Applying the results above to the right regular representation, one obtains that the θ-isotypical subspace of $\mathrm{M}(G)^{\mathrm{R}^*\mathfrak{n}}$ is equal to $\mathrm{M}_\pi \cap \mathrm{M}(G)^{\mathrm{R}^*\mathfrak{n}}$, and

$$\mathrm{d}_\pi = [\mathrm{R}^* : \pi] = [(\mathrm{R}^*|_U)|_{\mathrm{M}(G)^{\mathrm{R}^*\mathfrak{n}}} : \theta] = \dim \mathrm{I}(\theta, (\mathrm{R}^*|U)|_{\mathrm{M}(G)^{\mathrm{R}^*\mathfrak{n}}}),$$

cf. Proposition 4.4.4.(ii). If θ is a dominant irreducible representation of U in the vector space W, then $f \in \mathrm{I}(\theta, (\mathrm{R}^*|_U)|_{\mathrm{M}(G)^{\mathrm{R}^*\mathfrak{n}}})$ will be identified with the real-analytic function $x \mapsto (w \mapsto (f(w))(x)) : G \to W^*$, also denoted by f, and which satisfies:

$$\mathrm{R}^*(X)f = 0, \quad \text{for all } X \in \mathfrak{n}, \tag{4.13.35}$$

and

$$\theta(u)^*(f(x)) = f(xu), \quad \text{for all } x \in G, u \in U. \tag{4.13.36}$$

The last condition means that f is a section of the homogeneous vector bundle $G \times_U W^*$ over G/U, defined by the representation $\theta^\vee$ of U in W^*. The space Γ of solutions f of (4.13.35,36) is $\mathrm{L}(G)$-invariant. By the dimension count, the left regular representation defines an irreducible representation of G in Γ, which is equivalent to $\pi^\vee$, thus generalizing Proposition 4.12.1.

The Borel-Weil theorem 4.12.5 can be generalized by defining B as the normalizer of $\mathfrak{b}$ in $G_{\mathbf{C}}$. In the same way as in Lemma 4.12.3, we get a diffeomorphism: $G/U \to G_{\mathbf{C}}/B$. Furthermore, Lemma 4.12.4 now says that $(u, X, Y) \mapsto u \exp X \exp Y$ is a diffeomorphism from $U \times \mathfrak{a} \times \mathfrak{n}$ onto B. This leads to a unique extension of $\theta^\vee$ to a holomorphic representation of B in W^*; and the solutions of (4.13.35,36) are just the holomorphic sections of the corresponding $G_{\mathbf{C}}$-homogeneous, holomorphic vector bundle over $G_{\mathbf{C}}/B$.

For the computation of the character using Lie algebra cohomology, cf. Kostant [1961].

We conclude this section with some remarks on the computation of the character of an irreducible representation θ, in a finite-dimensional vector space W, of a compact Lie group U with Abelian Lie algebra $\mathfrak{t}$, so that $U^\circ = T = \exp \mathfrak{t}$ is a torus. We will consider a reduction to irreducible representations of finite groups as the end of our task of applying Lie group techniques to the representation theory of compact groups. This despite the fact that the remaining representation theory of finite groups is a highly nontrivial matter, cf. the remark at the end of Section 4.5.

Applying the observations which we made in the beginning of this section, with G, π, and G°, replaced by U, θ, and T, respectively, we obtain that here $\Omega(\theta)$ is a single U/T-orbit in $\widehat{T}$, each $\lambda \in \Omega(\theta)$ occurring with the same multiplicity m. For $\lambda \in \widehat{T}$, define the subgroup:

$$U_\lambda := \{ u \in U \mid u \cdot \lambda = \lambda \} \tag{4.13.37}$$

of U; it has the same Lie algebra $\mathfrak{t}$ as U. If $\lambda \in \Omega(\theta)$, then the mapping $u \mapsto u \cdot \lambda$ induces a bijection from U/U_λ onto $\Omega(\theta)$. By (4.13.4), the weight space W_λ is $\theta(U_\lambda)$-invariant.

(4.13.4) Lemma.

(i) For each $\lambda \in \Omega(\theta)$, we have that $\sigma(\lambda, \theta) := (\theta|_{U_\lambda})|_{W_\lambda}$ is an irreducible representation of U_λ in W_λ, and:

$$\chi_\theta(u) = \sum_{\{ \lambda \in \Omega(\theta) \mid u \in U_\lambda \}} \chi_{\sigma(\lambda,\theta)}(u), \quad \textit{for } u \in U. \tag{4.13.38}$$

(ii) For each $\lambda \in \widehat{T}$ and each irreducible representation σ of U_λ in a vector space V, with λ as its (only) weight, $\theta := \operatorname{Ind} \sigma_{U_\lambda}^{U}$ is an irreducible representation of U, and $\sigma \sim \sigma(\lambda, \theta)$.

(iii) The mapping $\theta \mapsto \sigma(\lambda, \theta)$ induces a bijection:

$$\{ [\theta] \in \widehat{U} \mid \lambda \in \Omega(\theta) \} \xrightarrow{\sim} \{ [\sigma] \in \widehat{U_\lambda} \mid \lambda \in \Omega(\sigma) \}.$$

Proof. (i) Let W_λ^0 be a nonzero $\theta(U_\lambda)$-invariant linear subspace of W_λ. Let A denote a section of $U \to U/U_\lambda$, write $W^0 = \sum_{a \in A} \theta(a)(W_\lambda^0)$. If $u \in U$, $a \in A$, then $ua = bv$, for some $b \in A$, $v \in U_\lambda$; hence:

$$\theta(u) \circ \theta(a)(W_\lambda^0) = \theta(b) \circ \theta(v)(W_\lambda^0) = \theta(b)(W_\lambda^0) \subset W^0.$$

This shows that W^0 is $\theta(U)$-invariant. Because W is irreducible, $W^0 = W$. On the other hand, (4.13.4) yields that $\theta(a)(W_\lambda^0) \subset W_{a \cdot \lambda}$; so the sum is direct, and in particular $W_\lambda = W^0 \cap W_\lambda = W_\lambda^0$, proving that $\sigma(\lambda, \theta)$ is irreducible. The character formula (4.13.38) is just (4.13.19).

(ii) Recall the definition (4.7.2) of the induced representation θ; it is the left action on the space of $f\colon U \to V$, such that $f(uh) = \sigma(h)^{-1}(f(u))$, for $u \in U$, $h \in U_\lambda$. We have $\mathrm{d}_\theta = \#(U/U_\lambda)\, \mathrm{d}_\sigma$. Furthermore, if $t \in T$, $u \in U$, then:

$$f(t^{-1}u) = f(u(u^{-1}t^{-1}u)) = (u^{-1}tu)^\lambda f(u) = t^{u \cdot \lambda} f(u);$$

and this shows that $\Omega(\theta) = \{ u \cdot \lambda \mid u \in U \}$. The Frobenius reciprocity theorem 4.7.1.(ii) shows that if $[\pi] \in \widehat{U}$ occurs with positive multiplicity in θ, then σ occurs with positive multiplicity in $\pi|_{U_\lambda}$; hence $\mathrm{d}_\theta \geq \mathrm{d}_\pi \geq \#(U/U_\lambda)\, \mathrm{d}_\sigma = \mathrm{d}_\theta$. So equality holds, θ is irreducible, and $\sigma \sim (\theta|_{U_\lambda})|_{W_\lambda}$.

(iii) The injectivity follows because (4.13.38) shows that χ_π is uniquely determined by $\chi_{\sigma(\lambda,\theta)}$, and representations are equivalent if and only if they have the same character, cf. Corollary 4.3.5.(iii). The surjectivity follows from (ii). □

Lemma 4.13.4 allows us to restrict our attention from now on to the case that θ is an irreducible representation of U with **only one weight** λ. If $\lambda = 0$, then $\theta|_T$ is trivial, and $\theta = \sigma \circ \pi$ for some irreducible representation σ of the finite group U/T; here π denotes the canonical projection.

So we assume now that $\lambda \neq 0$. Note that $u \cdot \lambda = \lambda$ implies that $\ker \lambda := \{\, t \in T \mid t^\lambda = 1 \,\}$ is a normal subgroup of U. Clearly it is contained in $\ker \theta$, so dividing away $\ker \lambda$, we may assume that $t \mapsto t^\lambda$ is an isomorphism: $T \to C := \{\, z \in \mathbf{C} \bigm| |z| = 1 \,\}$. We shall simplify the notation by assuming that $T = C$ and that $\theta(t)$ is equal to multiplication by t, if $t \in C$. Also note that $ucu^{-1} = c$, for all $u \in U$, $c \in C$, because each element of u fixes the weight. That is, U is a so-called *"central extension of the circle by a finite group"*. Again using the Frobenius reciprocity theorem, one sees that every such group U always has at least one irreducible representation θ_0 such that $\theta_0(c)$ is equal to multiplication by c, for all $c \in C$.

The group $F = \{\, u \in U \mid \det \theta_0(u) = 1 \,\}$ is a finite normal subgroup of U. We have $U = F \cdot C$, with $F \cap C = C(d) := \{\, z \in \mathbf{C} \mid z^d = 1 \,\}$, if d equals the dimension of θ_0; and actually the homomorphism: $(f, c) \mapsto fc \colon F \times C \to U$ induces an isomorphism of groups:

$$F \times_{C(d)} C \xrightarrow{\sim} U. \tag{4.13.39}$$

The mapping $\theta \mapsto \theta|_F$ now is a bijection from the set of the irreducible representations θ of U such that $\theta(c)$ equals multiplication by c, for all $c \in C$, onto the set of the irreducible representations τ of F such that $\tau(c)$ equals multiplication by c, for all $c \in C(d) \subset F$. Up to equivalence, these τ are precisely the irreducible components of $\operatorname{Ind} i_{C(d)}^F$, where i denotes the representation $c \mapsto (z \mapsto cz)$ of $C(d)$ in $\mathbf{C}$. The formula $\chi_\theta(fc) = c\chi_\tau(f)$, for $f \in F$, $c \in C$, shows the relation between the characters.

4.14 Exercises

(4.1) Exercise. Let G be a compact Lie group (i) Suppose G acts through affine transformations in a real affine space W and let $C \subset W$ be a nonempty convex G-invariant subset. Prove that C contains a fixed point for the action of G, and in particular, that any compact group of affine transformations of W has a fixed point. (ii) Let $\pi \colon G \to \mathbf{GL}(V)$ be a finite-dimensional representation in a real vector space V. We denote by $P(V)$ the algebra

of polynomial functions on V and define a representation of G in $P(V)$ by $(\pi(g)f)(x) = f(\pi(g)^{-1}x)$. Prove that the orbits of G in V are separated by elements in $P(V)^G$, that is, for every two distinct orbits O_1 and O_2 in V there exists $f \in P(V)^G$ with $f(x) > 0$ for $x \in O_1$ and $f(x) < 0$ for $x \in O_2$.
Hint: First separate the orbits with a continuous function, next apply the Weierstrass approximation theorem to separate them with a polynomial function, and finally apply (i).
(iii) Deduce from (ii) that the orbits of G in a real vector space V are algebraic varieties in V.

(4.2) Exercise. Let G be a compact Lie group and (π, V) an irreducible unitary representation of G in a finite-dimensional Hilbert space V. Show that $\mathrm{d}_\pi \int_G \chi_\pi(gxg^{-1}y)\,dg = \chi_\pi(x)\chi_\pi(y)$, for all x, $y \in G$. Conversely, prove that $\int_G f(gxg^{-1}y)\,dg = f(x)f(y)$, for all x, $y \in G$, implies that f is of the form ${\mathrm{d}_\pi}^{-1}\chi_\pi$.

(4.3) Exercise. Let G be a compact Lie group. (i) Show that an irreducible unitary representation (π, U) of $G \times G$ in a finite-dimensional Hilbert space U is equivalent to the exterior tensor product of two irreducible unitary representations of G.
Hint: Let $\overline{\pi}(g) = \pi(g, 1)$. Prove that the representation $(\overline{\pi}, U)$ of G is a direct sum of, say n, copies of an irreducible unitary representation (ρ, V) of G. Set $W = \mathrm{Hom}_G(V, U)$ and prove $\dim W = n$. Define a representation (σ, W) of G by $\sigma(g)A = \pi(1, g) \circ A$. Finally define $\phi\colon V \otimes W \to U$ by $\phi(v \otimes A) = Av$ and verify that $\phi \in \mathrm{Hom}_{G\times G}(\rho \otimes^{\mathrm{e}} \sigma, \pi)$.
(ii) Let (τ, H) be an irreducible unitary representation of G in a finite-dimensional Hilbert space H. Show that the linear span of the $\tau(g)$ in $\mathbf{GL}(H)$, for $g \in G$, is equal to $\mathbf{L}(H, H)$.

(4.4) Exercise. Denote by P_n the linear space of homogeneous polynomials with complex coefficients of degree n in two complex variables z_1 and z_2. Define a representation π_n of $G = \mathbf{SU}(2)$ in P_n by:

$$(\pi_n(g)p)(z) = p(zg), \qquad \text{for } g \in G,\ p \in P_n,\ z = (z_1, z_2).$$

(i) Verify that (π_n, P_n) is a unitary representation of G if we provide P_n with the Hermitian inner product:

$$< p, q > = \frac{2^n}{(2\pi)^2} \int_{\mathbf{C}^2} \mathrm{e}^{-\frac{1}{2}zz^*}\, p(z)q^*(z)\,dz = (\partial(p)\overline{q})(0) = \sum_{0\le k\le n} a_k \overline{b_k} k!(n-k)!,$$

where $z^* = \begin{pmatrix}\overline{z_1}\\ \overline{z_2}\end{pmatrix}$, $q^*(z) = \overline{q(z)}$, $\partial(p) = \sum_k a_k \frac{\partial^n}{\partial^k z_1 \partial^{n-k} z_2}$ if $p(z) = \sum_k a_k z_1^k z_2^{n-k}$, $\overline{q}(z) = \sum_k \overline{b_k} z_1^k z_2^{n-k}$.
Hint: Note that $z \mapsto \mathrm{e}^{-\frac{1}{2}zz^*}\, p(z)q^*(z)$ modulo numerical factors is equal to the Fourier transform of $z \mapsto \partial(pq^*)\,\mathrm{e}^{-\frac{1}{2}zz^*}$.

(ii) Let $H \in \mathfrak{su}(2)$ be as in Exercise 2.1 and define a basis $\{ p_k \mid 0 \le k \le n \}$ for P_n by $p_k(z) = z_1^k z_2^{n-k}$. Prove $\pi_n(\exp itH)p_k = \mathrm{e}^{i(2k-n)t}\, p_k$ and conclude:

$$\chi_{\pi_n}(\exp itH) = \sum_{0 \le k \le n} \mathrm{e}^{i(2k-n)t} = \frac{\sin(n+1)t}{\sin t}.$$

Deduce from Exercise 3.15 that (π_n, P_n) is an irreducible representation of G. (iii) Use Abelian Fourier analysis to show that the χ_{π_n}, for $n \ge 0$ generate a linear subspace that is uniformly dense in $\mathrm{C}(G)^{\mathrm{Ad}}$. Prove that the character χ of an irreducible representation of G different from all the π_n satisfies $< \chi, \chi_{\pi_n} >= 0$, and deduce that $\chi = 0$. In other words, every irreducible unitary representation of G is equivalent to some (π_n, P_n). (iv) Verify $(\pi_n(g)p_i)(z) = z_2^n\, (az - \bar{b})^i (bz + \bar{a})^{n-i}\big|_{z=\frac{z_1}{z_2}}$, for $g = \begin{pmatrix} a & b \\ -\bar{b} & \bar{a} \end{pmatrix} \in G$. Let $c_{ij}^n(g)$ be the coefficient of $z_2^n (\frac{z_1}{z_2})^j$ in the polynomial $\pi_n(g)p_i$. Prove:

$$c_{ij}^n(g) = \frac{1}{j!}\frac{d^j}{dz^j}\bigg|_{z=0} (bz + \bar{a})^{n-i}(az - \bar{b})^i.$$

Introduce $abz + |a|^2$ as a new variable and show:

$$c_{ij}^n(g) = \frac{(-1)^i}{j!} a^{i+j-n} b^{j-i} P_{ij}^n(|a|^2), \quad \text{where} \quad P_{ij}^n(x) = \frac{d^j}{dx^j}(x^{n-i}(1-x)^i)$$

is a Jacobi polynomial. (v) Prove that the matrix coefficient $< \pi_n(g)p_i, p_j >$ is given by $j!\,(n-j)!\,c_{ij}^n(g)$, and deduce that the functions on G:

$$\sqrt{n+1}\sqrt{\frac{j!\,(n-j)!}{i!(n-i)!}}\; c_{ij}^n, \qquad \text{for } 0 \le n,\ 0 \le i, j \le n$$

form a complete orthonormal system in $\mathrm{L}^2(G)$. (vi) Use Cauchy's integral formula to prove that this orthonormal system also is given by:

$$\begin{pmatrix} a & b \\ -\bar{b} & \bar{a} \end{pmatrix} \mapsto \frac{\sqrt{n+1}}{2\pi\sqrt{-1}}\sqrt{\frac{j!\,(n-j)!}{i!(n-i)!}} \oint_\gamma (az - \bar{b})^i (bz + \bar{a})^{n-i} z^{-j-1}\, dz$$

where $\gamma \subset \mathbf{C}$ is the unit circle traversed once around counterclockwise.

(4.5) Exercise. Let G be a connected, compact Lie group. Show that G possesses irreducible unitary representations of arbitrary high dimension if it is noncommutative. Prove that G has only finitely many distinct equivalence classes of representations of a given dimension if it is semisimple.

(4.6) Exercise. Let G be a simply connected, compact Lie group, T a maximal torus in G with Weyl group W, P a choice of positive roots of $\mathfrak{t}$ and $\rho = \frac{1}{2}\sum_{\alpha \in P} \alpha$. (i) Verify that $\rho \in \widehat{T}$, the lattice of weights of T, and that therefore

$\Delta(t) = t^\rho \delta_P(t) = \sum_{s\in W} \det s \; t^{s\cdot\rho}$ is well-defined, for $t \in T$. (ii) For $u \in \mathrm{C}^\infty(G)^{\mathrm{Ad}\,G}$ and $f \in \mathrm{C}^\infty(G)$, show:

$$\int_G (uf)(x)\,dx = \#(W)^{-1} \int_T u(t) |\,\Delta(t)\,|^2 \int_{G/T} f(xtx^{-1})\,d(xT)\,dt.$$

Define $F_f \in \mathrm{C}^\infty(T)$ by:

$$F_f(t) = \#(W)^{-1} \Delta(t) \int_{G/T} f(xtx^{-1})\,d(xT).$$

Note that $F_f(t) = 0$, for $t \in T \setminus G^{\mathrm{reg}}$. Use $\overline{\Delta} = (-1)^{\#(P)}\Delta$ to prove:

$$\int_G (uf)(x)\,dx = (-1)^{\#(P)} \int_T u(t)\Delta(t)F_f(t)\,dt.$$

(iii) For $\alpha \in P$ select $\alpha^\vee \in [\mathfrak{g}_\alpha, \mathfrak{g}_{-\alpha}]$ such that $\alpha(\alpha^\vee) = 2$. Define the differential operator $\omega \in \mathrm{U}(\mathfrak{g}_{\mathbf{C}})$ on G by $\omega = \prod_{\alpha\in P} \alpha^\vee$. Now verify:

$$(\omega F_f)(1) = c\,f(1), \qquad c = \#(W)^{-1}(\omega\Delta)(1) = \prod_{\alpha\in P} \rho(\alpha^\vee).$$

Hint: Every factor of Δ vanishes at 1, and so:

$$(\omega F_f)(1) = \#(W)^{-1}(\omega\Delta)(1) \int_{G/T} f(1)\,d(xT) = \#(W)^{-1}(\omega\Delta)(1)f(1).$$

Furthermore $\alpha^\vee(t^\mu) = \mu(\alpha^\vee)t^\mu$ and $\det s = (-1)^{\#(P\setminus s\cdot P)}$, and so $(\omega\Delta)(t) = \prod_{\alpha\in P}\rho(\alpha^\vee)\sum_{s\in W} t^{s\cdot\rho}$.
(iv) Because F_f belongs to $\mathrm{C}^\infty(T)$ it equals the sum of its Fourier series on T; in particular:

$$(\omega F_f)(1) = \sum_{\lambda\in\widehat{T}} \int_T (\omega F_f)(t)\,t^\lambda\,dt.$$

Use integration by parts to find:

$$\int_T (\omega F_f)(t)\,t^\lambda\,dt = (-1)^{\#(P)} \prod_{\alpha\in P} \lambda(\alpha^\vee) \int_T F_f(t)\,t^\lambda\,dt.$$

Set $\Theta_\lambda(f) = (-1)^{\#(P)} \int_T F_f(t)\,t^\lambda\,dt$ and verify that $f \mapsto \Theta_\lambda(f)$ is a distribution on G. Obtain using (iii):

$$f(1) = \sum_{\lambda\in\widehat{T}} \Theta_\lambda(f) \prod_{\alpha\in P} \frac{\lambda(\alpha^\vee)}{\rho(\alpha^\vee)}.$$

The summation actually is over the set of regular elements $\lambda \in \widehat{T}$, that is, those λ satisfying $\prod_{\alpha\in P}\lambda(\alpha^\vee) \neq 0$. Furthermore $\Theta_{s\cdot\lambda}(f) = \det s\;\Theta_\lambda(f)$ and so $\lambda \mapsto \prod_{\alpha\in P}\lambda(\alpha^\vee)\Theta_\lambda(f)$ is W-invariant. Conclude:

$$f(1) = \#(W) \sum_{\lambda \in \widehat{T} \text{ regular dominant}} d_{\lambda-\rho} \Theta_\lambda(f);$$

here $d_{\lambda-\rho}$ is the dimension of a representative of the class of irreducible representations of G with highest weight $\lambda-\rho$. (v) On the other hand, deduce from the Peter-Weyl theorem and Weyl's character formula:

$$\begin{aligned} f(1) &= \sum_{\lambda \in \widehat{T} \text{ dominant}} d_\lambda \int_G f(x)\chi_\lambda(x)\, dx = \\ &= \sum_{\lambda \in \widehat{T} \text{ dominant}} d_\lambda (-1)^{\#(P)} \int_T \chi_\lambda(t) \Delta(t) F_f(t)\, dt = \\ &= \sum_{\lambda \in \widehat{T} \text{ dominant}} d_\lambda (-1)^{\#(P)} \int_T F_f(t) \sum_{s \in W} \det s\, t^{s\cdot(\lambda+\rho)}\, dt = \\ &= \sum_{\lambda \in \widehat{T} \text{ regular dominant}} d_{\lambda-\rho} \sum_{s \in W} \det s\, (-1)^{\#(P)} \int_T F_f(t) t^{s\cdot\lambda}\, dt = \\ &= \#(W) \sum_{\lambda \in \widehat{T} \text{ regular dominant}} d_{\lambda-\rho} \Theta_\lambda(f). \end{aligned}$$

4.15 Notes

For finite groups, almost the entire theory of Sections 4.1–4.5, 4.7, 4.8 has been developed by Frobenius [1896-1899]. This includes, for instance, the orthogonality relations, but not Schur's lemma 4.1.1 or the complete reducibility statement of Corollary 4.2.2.

The point of departure for Frobenius (see Hawkins [1978] for more historical details) was a question of Dedekind about the so-called *group determinant*. In our notation (4.4.3), this is the homogeneous polynomial function $D\colon f \mapsto \det \mathrm{R}^*(f)$ on $\mathbf{C}^G$, the space of complex-valued functions on the finite group G. For **Abelian** G, Dedekind had observed that D is a product of linear factors $f \mapsto \sum_{x \in G} f(x)\chi(x)$ (modulo constant factors). Here the χ are the characters of G, defined by Dedekind in 1879 as the homomorphisms: $G \to \mathbf{C}^\times$. See Lemma 4.4.1.(iii) and Corollary 4.1.2. Working out some examples of noncommutative finite groups G, Dedekind conjectured that the number of linear factors in D is equal to the order of G/G', see the end of Section 4.5 for the definition of G'.

Frobenius attacked the problem by concentrating first on the subspace $C = (\mathbf{C}^G)^{\mathbf{Ad}\, G}$ of conjugacy-invariant functions, or "class functions" in his terminology. The $R^*(f)$, for $f \in C$, form a commuting family of linear transformations; and Frobenius defines the characters of G, modulo constant factors, as the $\chi \in C$ which are common eigenvectors of the $R^*(f)$, for $f \in C$. He proves that the characters form a basis of C ("Peter-Weyl for finite groups"), and then decomposes $\mathbf{C}^G$ into the images of the $R^*(\chi)$. We recognize these as the M_π, for $[\pi] \in \widehat{G}$, the common eigenspaces in $\mathbf{C}^G$ of the $R^*(f)$, with $f \in C$, see Lemma 4.4.1.(iii). It follows that D factorizes as the product of the d_π-th powers of the irreducible polynomials $\Phi_\pi : f \mapsto \det(m_\pi^{-1}(f_\pi))$, where $f_\pi = d_\pi\, \chi_\pi * f$ equals the M_π-component of f. Frobenius also observes that the restriction of Φ_π to C is equal to the d_π-th power of the linear form $f \mapsto < f, \chi_{\pi^\vee} >$. The identification of the characters with the traces of the irreducible representations of G can be found in [1897], p.1104 = p.92, but this fact is not very important in Frobenius' theory.

Molien [1897] arrived at similar conclusions from another direction. In [1893] he had more or less proved the famous structure theorems of Wedderburn [1908] about "systems of hypercomplex numbers", that is, finite-dimensional associative algebras over **C**. By applying these, in [1897], to the *group algebra* $\mathbf{C}^G$, provided with the convolution product $*$, cf. (4.4.1), he obtained the decomposition of $\mathbf{C}^G$ into the irreducible subalgebras, our M_π.

The complete reducibility of finite-dimensional representations of finite groups is due to Maschke [1899], who used an invariant Hermitian inner product. The existence of the latter was observed a little earlier, independently, by several mathematicians, see for instance Moore [1898].

Simplified expositions of Frobenius' theory were given by Burnside [1904], stressing the use of invariant inner products in the proof of the orthogonality relations, and by Schur [1905], who introduced his lemma (Lemma 4.1.1) as a basic tool.

The induced characters, cf. (4.7.4), together with the reciprocity relation 4.7.1.(ii), were discovered by Frobenius [1898]. The corresponding induced representations were introduced only much later, in Speiser [1927], p.196.

Proposition 4.8.7 is due to Frobenius and Schur [1906]. It was generalized to compact Lie groups by Schur [1924, III]. The interest of Frobenius in the three types of representations is related to his previous proof, in [1878], §14, that the only finite-dimensional, associative division algebras over **R** are **R**, **C**, and **H**.

One can argue that the history of representation theory of **infinite** compact groups G, in particular of compact Lie groups of positive dimension, starts with the theory of Fourier series, that is, expansions of the functions on the circle group $C = \mathbf{R}/\mathbf{Z}$ in terms of the functions $x \mapsto e^{2\pi i n x}$, with $n \in \mathbf{Z}$. For this, one has to view representation theory as a theory about natural decompositions of the space of functions on G. Also, one has to interpret the harmonic oscillations as characters of C, rather than as the eigenfunctions of the differential operator $\frac{d}{dx}$. Of course these properties are closely related; Remark (D) after Proposition 4.4.2 is a more sophisticated version of this observation.

A next step in the development of the theory for compact Lie groups was the formulation by Hurwitz [1897] of the averaging principle 4.2.1. He carried it out for the rotation group, for which he had an explicit description of the Haar measure. However, as Schur [1924, I] points out, he formulated the principle in such a way, that it applied to any compact group for which one has a Haar measure, whose existence is easy to prove for compact Lie groups. Schur then uses the averaging principle in order to generalize Frobenius' theory to compact Lie groups. In [1924, II] he determines the characters of $\mathbf{SO}(n, \mathbf{R})$, the special orthogonal group. Earlier, in his thesis [1901], Schur had determined the characters of the (complex) polynomial representations of $\mathbf{GL}(n, \mathbf{C})$. As remarked after Proposition 4.10.3, this coincides with the determination of all continuous finite-dimensional representations of $\mathbf{SU}(n)$.

Inspired by the results above, Weyl [1925,26] then derives his character formula 4.9.1 and dimension formula 4.9.2, for arbitrary compact, connected, semisimple Lie groups. In the proof, Weyl combines Schur's extension to compact Lie groups of the Frobenius' character theory with the theory of Cartan [1913] of representations of semisimple Lie algebra's, and his own structure theory of compact, connected Lie groups, including his integration formula. Our proof of the character formula in Section 4.9 differs from his only in that we do not use Cartan's highest weight theorem (Theorem 4.11.2).

The completeness of the characters, and matrix coefficients, respectively was established by Peter and Weyl [1927]. They stated the result for compact Lie groups, because only for these the invariant measure was available at the time. As observed in the introduction of Section 4.6, the Peter-Weyl theorem is a generalization of the completeness of Fourier series, to compact groups which need neither be Abelian, nor finite. In this way, the Peter-Weyl theorem can be viewed as the common generalization of Fourier expansion and Frobenius' theory of characters for finite groups.

The Peter-Weyl theorem gives a direct proof of the fact that every dominant weight of T is the highest weight of some irreducible representation, cf. Theorem 4.9.5. The corresponding statement for semisimple Lie algebras had been proved before by Cartan [1913], by first establishing the existence of irreducible representations with highest weight equal to the fundamental weights, cf. (4.10.11); it then follows for all dominant weights by taking successive Cartan products, cf. Definition 4.10.4. Especially the first step involves extensive calculations, going through the classification of simple Lie algebras.

Haar [1933] proves the existence of left and right invariant measures, respectively, on metrizable, locally compact topological groups. He explicitly mentions the generalization of the Hilbert-Schmidt theory and the Peter-Weyl theorem to compact metrizable groups as a major application. The uniqueness of invariant measures is due to von Neumann [1936], who also removed the assumption of metrizability and, in [1934], extended the concept of averaging to a maximal category of locally compact topological groups, including certain noncompact ones. In the book of Weil [1940], the integration theory of locally compact groups, with applications to their representation theory, got its first complete exposition. This has been an important stimulus for the subsequent development of the representation theory of noncompact groups.

Now we turn to the "weight exercises" in Section 4.10. For the proof of the convexity proposition 4.10.1, we consulted Bourbaki [1975], Chap.8, §7, Exercice 1, and No.2, Cor.2 de Prop.3. We learned Proposition 4.10.8 from Bröcker and tom Dieck [1985], Ch.VI, Section 4, who in turn refer to Iwahori [1959]. For the general theory of real, complex, and quaternionic types, respectively, in Section 4.8, the Appendix II in Bourbaki [1982] has been helpful. Finally, formula (4.10.16) is due to Kostant [1959], and (4.10.17) to Steinberg [1961].

Although we did not use it in the proof of Weyl's character formula, we consider Cartan's highest weight theorem as a key in the understanding of the representation itself, instead of only the character. For this reason it plays a central role in our exposition of the Borel-Weil theorem in Section 4.12. Apart from Theorems 4.12.5 and 4.12.7, the exposition of Serre [1954] contains several other results of Borel and Weil about the topology and geometry of the "generalized complex flag varieties" $G/G_\lambda = G_{\mathbf{C}}/P_\lambda$. We have added the name of Tits to Theorem 4.12.7, because in [1955], pp.112–113, he independently obtained the relation between projective embeddings and irreducible representations.

Proposition 4.13.1 and 4.13.3 are due to Kostant [1961], Section 7. Only, he took (4.13.30) as the definition of the highest weight space, and obtained the character formula in the form (4.13.32,34). Furthermore, his proof is based on computations of the cohomology of the Lie algebra $\mathfrak{n}$, with values in an irreducible representation space of G. It is therefore completely different from ours, which instead uses the orthogonality relations for characters and a generalization of Weyl's integration formula to the nonconnected case. T. A. Springer encouraged us to write this down.

References for Chapter Four

Van den Ban, E.P. (1997) Induced representations and the Langlands classification. in: Bailey, T.N., Knapp, A.W. (eds.) Proc. Sympos. Pure Math. **61** (1997) 123-155. Amer. Math. Soc., Providence

Bourbaki, N. (1965) Éléments de Mathématique. Topologie Générale, Chap. 1 et 2. Hermann, Paris

Bourbaki, N. (1966) Éléments de Mathématique. Espaces Vectoriels Topologiques, Chap. 1 et 2. Hermann, Paris

Bourbaki, N. (1964) Éléments de Mathématique. Espaces Vectoriels Topologiques, Chap. III-V. Hermann, Paris

Bourbaki, N. (1975) Éléments de Mathématique. Groupes et Algèbres de Lie, Chap. 7 et 8. Hermann, Paris

Bourbaki, N. (1982) Éléments de Mathématique. Groupes et Algèbres de Lie, Chap. 9. Masson, Paris

Bröcker, T., tom Dieck, T. (1985) Repesentations of Compact Lie Groups. Springer-Verlag, New York Berlin Heidelberg Tokyo

Burnside, W. (1904) On the reduction of a group of transformations of homogeneous linear substitutions of finite order. Acta Math. **28** (1904) 369-387

Burnside, W. (1904) On the representation of a group of finite order as an irreducible group of linear substitutions and the direct establishment of the relations between the group characters. Proc. London Math. Soc. **1** (1904) 117-123

Cartan, É. (1913) Les groupes projectifs qui ne laissent invariante aucune multiplicité plane. Bull. Soc. Math. France **41** (1913) 53-96 = Œuvres Complètes, Partie I, pp. 355-398. Gauthier-Villars, Paris 1952

Curtis, C.W., Reiner, I. (1981/1987) Methods of Representation Theory, Volumes I and II. John Wiley & Sons, New York Chichester Brisbane Toronto

Demazure, M. (1974) Une nouvelle formule des caractères. Bull. Sc. Math. (2) **98** (1974) 163-172

Freudenthal, H., de Vries, H. (1969) Linear Lie Groups. Academic Press, New York London

Frobenius, F.G. (1878) Über lineare Substitutionen und bilineare Formen. J. reine angew. Math. **84** (1878) 1-63 = Gesammelte Abh., Bd. II, pp. 343-405. Springer-Verlag, Berlin Heidelberg New York 1968

Frobenius, F.G. A series of articles in: Sitzungsber. Königl. Preuss. Akad. Wiss., Phys.-Math. Kl. = Gesammelte Abh., Bd. III. Springer-Verlag, Berlin Heidelberg New York 1968:
  (a) Über Gruppencharaktere. **1896**, pp. 985-1021 = pp. 1-37;
  (b) Über die Primfactoren der Gruppendeterminante. **1896**, pp. 1343-1382 = pp. 38-77;
  (c) Über die Darstellung der endlichen Gruppen durch lineare Substitutionen. **1897**, pp. 994-1015 = pp. 82-103;
  (d) Über Relationen zwischen den Charakteren einer Gruppe und denen ihrer Untergruppen. **1898**, pp. 501-515 = pp. 104-118;
  (e) Über die Composition der Charaktere einer Gruppe. **1899**, pp. 330-339 = pp. 119-147

Frobenius, F.G., Schur, I. (1906) Über die reellen Darstellungen der endlichen Gruppen. Sitzungsber. Königl. Preuss. Akad. Wiss., Phys.-Math. Kl. **1906**, pp. 186-208 = Gesammelte Abh., Band. III, pp. 355-377. Springer-Verlag, Berlin Heidelberg New York 1968

Grauert, H. (1958) On Levi's problem and the imbedding of real-analytic manifolds. Ann. of Math. (2) **68** (1958) 460-472

Griffiths, P., Harris, J. (1978) Principles of Algebraic Geometry. J.Wiley and Sons, New York Chichester Brisbane Toronto

Haar, A. (1933) Der Massbegriff in der Theorie der kontinuierlichen Gruppen. Ann. of Math. (II) **34** (1933) 147-169

Harish-Chandra (1951) On some applications of the universal enveloping algebra of a semi-simple Lie algebra. Trans. Amer. Math. Soc. **70** (1951) 28-95 = Collected Papers, Vol. I, pp. 292-360. Springer-Verlag, New York Berlin Heidelberg Tokyo 1984

Harish-Chandra (1956) On a lemma of F. Bruhat. J. Math. Pures Appl. (9) **35** (1956) 203-210 = Collected Papers, Vol. II, pp. 223-230. Springer-Verlag, New York Berlin Heidelberg Tokyo 1984

Harish-Chandra (1967) Characters of semi-simple Lie groups. Symposia on Theoretical Physics **4** (1967) 137-142 = Collected Papers, Vol. III, pp. 655-660. Springer-Verlag, New York Berlin Heidelberg Tokyo 1984

Hawkins, T. (1978) The creation of the theory of group characters. In: History of Analysis. Rice Univ. Studies **64** (1978) 57-71

Helgason, S. (1962) Differential Geometry and Symmetric Spaces. Academic Press, New York London

Hirsch, M.W., Smale, S. (1974) Differential Equations, Dynamical Systems, and Linear Algebra. Academic Press, New York San Francisco London

Hörmander, L. (1983) The Analysis of Linear Partial Differential Operators, Vol. I. Springer-Verlag, Berlin Heidelberg New York Tokyo

Hurwitz, A. (1897) Über die Erzeugung der Invarianten durch Integration. Göttinger Nachrichten **1897**, pp. 71-90 = Mathematische Werke, Band II, pp. 546-564. Birkhäuser, Basel 1933

Iwahori, N. (1959) On real irreducible representations of Lie algebras. Nagoya Math. J. **14** (1959) 59-83

Kostant, B (1959) A formula for the multiplicity of a weight. Trans. Amer. Math. Soc. **93** (1959) 53-73

Kostant, B. (1961) Lie algebra cohomology and the generalized Borel-Weil theorem. Ann. of Math. **74** (1961) 329-387

Maschke, H. (1899) Beweis des Satzes, dass diejenigen endlichen linearen Substitutionsgruppen, in welchen einige durchgehend verschwindende Coefficienten auftreten, intransitiv sind. Math. Ann. **52** (1899) 363-368

Molien, T. (1893) Über Systeme höheren complexer Zahlen. Math. Ann. **41** (1893) 83-156; Berichtigung, ibidem **42** (1893) 308-312

Molien, T. (1897) Eine Bemerkung zur Theorie der homogenen Substitutionsgruppen. Sitzungsber. d. Naturforscher-Ges. Dorpat **18** (1897) p. 259. See also: Über die Invarianten der linearen Substitutionsgruppen. Sitzungsber. Königl. Preuss. Akad. Wiss., Phys.-Math. Kl. **1897**, pp. 1152-1156.

Moore, E.H. (1898) An universal invariant for finite groups of linear substitutions: with applications in the theory of the canonical form of a linear substitution of finite period. Math. Ann. **50** (1898) 213-219

von Neumann, J. (1936) The uniqueness of Haar's measure. Mat. Sbornik **1** (43) (1936) 721-734 = Collected Works, Vol. IV, pp. 91-104. Pergamon Press, Oxford 1962

von Neumann, J. (1934) Almost periodic functions in a group I. Trans. Amer. Math. Soc. **36** (1934) 445-492 = Collected Works, Vol. II, pp. 454-501. Pergamon Press, Oxford 1961

Peter, F., Weyl, H. (1927) Die Vollständigkeit der primitiven Darstellungen einer geschlossenen kontinuierlichen Gruppe. Math. Ann. **97** (1927) 737-755 = Gesammelte Abh., Bd. III, pp. 58-75. Springer-Verlag, Berlin Heidelberg New York 1968

Schmid, W. (1985) Recent developments in representation theory. In: Hirzebruch, F., Schwermer, J., Suter, S. (eds.) Arbeitstagung Bonn 1984. Lecture Notes in Math. **1111**, pp. 135-153. Springer-Verlag, Berlin Heidelberg New York Tokyo

Schur, I. (1901) Über eine Klasse von Matrizen, die sich einer gegebenen Matrix zuordnen lassen. Diss., Berlin 1901 = Gesammelte Abh., Bd. I, pp. 1-72. Springer-Verlag, Berlin Heidelberg New York 1973

Schur, I. (1905) Neue Begründung der Theorie der Gruppencharakteren. Sitzungsber. Königl. Preuss. Akad. Wiss., Phys.-Math. Kl. **1905**, pp. 406-432 = Gesammelte Abh., Bd. I, pp. 143-169. Springer-Verlag, Berlin Heidelberg New York 1973

Schur, I. (1924) Neue Anwendung der Integralrechnung auf Probleme der Invariantentheorie. Sitzungsber. Preuss. Akad. Wiss., Phys.- Math. Kl. **1924** I. Mitteilung, pp. 189-208; II. Über die Darstellung der Drehungsgruppe durch lineare homogene Substitution. pp. 297-321; III. Vereinfachung des Integralkalküls. Realitätsfragen. pp. 346-355 = Gesammelte Abh., Bd. II, pp. 440-494. Springer-Verlag, Berlin Heidelberg New York 1973

Serre, J-P. (1954) Représentations linéaires et espaces homogènes Kählériens des groupes de Lie compacts. Séminaire Bourbaki, Exp. No. **100** (1954) 1-8

Speiser, A. (1927) Theorie der Gruppen von endlicher Ordnung. 2er Auflage. Julius Springer, Berlin

Steinberg, R. (1961) A general Clebsch-Gordan theorem. Bull. Amer. Math. Soc. **67** (1961) 406-407

Tits, J. (1955) Sur certaines classes d'espaces homogènes de groupes de Lie. Acad. Roy. Belg. Cl. Sci. Mém. Coll. **29**, No. 3

Treves, F. (1967) Topological Vector Spaces, Distributions and Kernels Academic Press, New York London

Vogan Jr., D.A. (1997-I) Cohomology and group representations. in: Bailey, T.N., Knapp, A.W. (eds.) Proc. Sympos. Pure Math. **61** (1997) 219-243. Amer. Math. Soc., Providence

Vogan Jr., D.A. (1997-II) The orbit method and unitary representations for reductive Lie groups. In: Ørsted, B., Schlichtkrull, H. (eds.) Algebraic and Analytic Methods in Representation Theory. 1997, pp. 243-339. Academic Press, San Diego London Boston New York Sydney Tokyo Toronto

van der Waerden, B.L. (1966) Algebra I. Springer-Verlag, Berlin Heidelberg New York

Wedderburn, J.H.M. (1908) On hypercomplex numbers. Proc. London Math. Soc. (2) **6** (1908) 77-118

Weil, A. (1940) l'Intégration dans les Groupes Topologiques. Hermann, Paris

Weyl, H. (1925/26) Theorie der Darstellung kontinuierlicher halb-einfacher Gruppen durch lineare Transformationen, I, II, III, Nachtrag. Math. Z. **23** (1925) 271-309; **24** (1926) 328-376, 377-395, 789-791 = Gesammelte Abh., Band II, pp. 543-647. Springer-Verlag, Berlin Heidelberg New York 1968

Walter, W. (1986) Gewöhnliche Differentialgleichungen. 3er Aufl. Springer-Verlag, Berlin Heidelberg New York

Chapter A

Appendix

A: Some Notions from Differential Geometry

Here we collect some basic definitions of a differential geometrical nature. For the notion of a manifold of class C^k, with $1 \leq k \leq \omega$ (here ω means real-analytic), we refer to standard textbooks, see for instance, Dieudonné [1972], Kobayashi and Nomizu [1963] or Spivak [1979-I]. Many different definitions of the notion of a fibration are extant in the literature, and therefore the terminology might cause confusion. So, for the reader's convenience, we now give definitions of a fibration or fiber bundle, of a covering, and of a principal fiber bundle.

(A.1) Definition. *A C^k fibration or a C^k fiber bundle is a surjective mapping $\pi\colon M \to B$ of manifolds which satisfies the following definition of local triviality. For each $b \in B$ there exist an open neighborhood U of b in B, a manifold F and a diffeomorphism $\tau\colon \pi^{-1}(U) \to U \times F$ such that $\tau(x) = \big(\pi(x), \phi(x)\big)$, for all $x \in \pi^{-1}(U)$. Here all manifolds and mappings are supposed to be of class C^k.*

(A.2) *Remark.* We call M the *total space*, B the *base space*, π the *projection*, and τ a *local trivialization.*

Notice that the closed submanifold $M_b := \pi^{-1}(\{b\})$ of M, the *fiber over* $b \in B$, is diffeomorphic to F, and hence diffeomorphic to $M_{b'}$ if b' is sufficiently close to b. All fibers are diffeomorphic to the same manifold F if, for instance, B is connected.

Suppose U and V are as above with corresponding local trivializations $\tau_U = (\pi, \phi_U)$ and $\tau_V = (\pi, \phi_V)$. If $x \in \pi^{-1}(U \cap V)$, then $\tau_U(x) = (b, f)$ and $\tau_V(x) = (b, f')$; so that we have the *retrivialization*:

$$\tau_V \circ \tau_U^{-1}\colon (b, f) \mapsto (b, f')\colon (U \cap V) \times F_U \to (U \cap V) \times F_V \tag{A.1}$$

with $f' = \phi_{VU}(b, f)$, where:

$$f \mapsto \phi_{VU}(b, f) = \phi_{VU}(b)(f)\colon F_U \to F_V$$

is a diffeomorphism depending smoothly on $b \in U \cap V$.

(A.3) Definition. *A* C^k *fibration* $\pi\colon M \to B$ *is said to be a* C^k covering *if its fibers are discrete. This definition is equivalent to having a surjection* $\pi\colon M \to B$ *with the following property. For each* $b \in B$ *there exists a connected open neighborhood* U *of* b *in* B *such that the connected components of* $\pi^{-1}(U)$ *are of the form* M_f *and the restriction* $\pi_f\colon M_f \to U$ *of* π *to* M_f *is a diffeomorphism of* M_f *onto* U*, for all* $f \in F$*. If the number of points in every fiber is equal to* $n \in \mathrm{N}$*, we speak of a* n-fold *or* n-sheeted covering.

(A.4) Definition. *A* C^k principal fiber bundle *with* structure group G *is a* C^k *fibration* $\pi\colon M \to B$ *together with a* C^k *action of a Lie group* G *on* M*, having the following properties.*

(i) *The action of* G *on* M *is free, and is written as a right action* $(g, x) \mapsto x \cdot g$*, for* $g \in G$*,* $x \in M$ *(see Section 1.11).*

(ii) *The base space* B *is equal to the orbit space* M/G *and* $\pi\colon M \to B$ *is the corresponding natural projection.*

(iii) *Every local trivialization* $\tau\colon x \mapsto \big(\pi(x),\, \phi(x)\big)\colon \pi^{-1}(U) \to U \times G$ *satisfies:*

$$\phi(x \cdot g) = \phi(x)\, g \qquad (x \in \pi^{-1}(U),\, g \in G). \tag{A.2}$$

Notice that we have used in (iii) above that the fiber M_b equals $\{\, x \cdot g \mid g \in G \,\}$ if $\pi(x) = b$, with $b \in B$, $x \in M$. Every fiber is diffeomorphic to the structure group G.

(A.5) Definition. *Suppose* U *and* V *are as in Definition A.4 with corresponding local trivializations* $\tau_U = (\pi,\, \phi_U)$ *and* $\tau_V = (\pi,\, \phi_V)$*. If* $x \in \pi^{-1}(U \cap V)$*, then we obtain from (A.2):*

$$\phi_V(x \cdot g)\phi_U(x \cdot g)^{-1} = \phi_V(x)\phi_U(x)^{-1} \qquad (g \in G), \tag{A.3}$$

which shows that $\phi_V(x)\phi_U(x)^{-1}$ *depends only on* $\pi(x)$ *and not on* x *itself. Hence we can define the* transition map*:*

$$\phi_{VU}\colon U \cap V \to G \quad \text{by} \quad \phi_{VU}(\pi(x)) = \phi_V(x)\phi_U(x)^{-1}. \tag{A.4}$$

With this notation we have: if $\tau_U(x) = (b,\, g)$*, then* $\tau_V(x) = (b,\, \phi_{VU}(b)g)$*. It follows that in this case the retrivialization* $\tau_V \circ \tau_U^{-1}$ *of Formula (A.2) takes the form:*

$$\tau_V \circ \tau_U^{-1}\colon (b,\, g) \mapsto (b,\, \phi_{VU}(b)g)\colon (U \cap V) \times G \to (U \cap V) \times G; \tag{A.5}$$

notice that it is realized by a left multiplication of the second factor.

(A.6) *Remarks.* If W is also a neighborhood in B as above, it is easy to verify that the following *cocycle condition* (cf. Hirzebruch [1966], p.38) is satisfied:

(A.6) $$\phi_{WV}(b)\phi_{VU}(b) = \phi_{WU}(b) \qquad (b \in U \cap V \cap W).$$

Conversely, given an open cover $\{U\}$ of B and C^k maps ϕ_{VU} that satisfy the cocycle condition, a principal fiber bundle $\pi\colon M \to B$ with structure group G can be obtained by *glueing together* the sets $U \times G$ by means of the ϕ_{VU}. That is, as a point set M must be $\bigcup(U \times G)$ with points $(b, g) \in U \times G$ and $(b', g') \in V \times G$ identified if $b' = b \in U \cap V$ and $g' = \phi_{VU}(b)\, g$, and with the manifold structure on M induced by the inclusions $U \times G \to M$. The resulting bundle then has the functions ϕ_{VU} as transition functions, see Kobayashi and Nomizu [1963], Proposition I.5.2 for full details.

If the action of G on M is from the left instead of from the right, one of course has to interchange systematically left and right in the results above.

B: Ordinary Differential Equations

Because they play such a central role in our treatment, we recall here the basic facts about the existence, uniqueness, and smooth dependence on initial values and parameters, of solutions of ordinary differential equations on manifolds. We also give some ideas of the proofs, but refer to standard textbooks for more details.

We shall assume throughout this appendix that M is a C^{k+1} manifold with $1 \leq k \leq \omega$ (here ω means real-analytic), so that the tangent bundle $\mathrm{T}\,M$ of M is of class C^k. A C^k *vector field* on M is a C^k mapping $v\colon M \mapsto \mathrm{T}\,M$ such that:

(B.1) $$v(x) \in \mathrm{T}_x\, M \qquad (x \in M).$$

A *solution curve* of the vector field v is a differentiable mapping $\gamma\colon I \to M$, where I is an open interval in $\mathbf{R}$, that satisfies the *differential equation*:

(B.2) $$\frac{d\gamma}{dt}(t) = v\big(\gamma(t)\big) \qquad (t \in I).$$

Note that the vector field v does not depend explicitly on t; therefore the differential equation is said to be *autonomous*.

For the exposition it will be convenient to start with the following **local uniqueness** result, which will be proved later on.

(B.1) Proposition. *Let v be a vector field on M of class C^1. If $\gamma\colon I \to M$ and $\gamma'\colon I' \to M$ are solution curves of v, if $t_0 \in I \cap I'$ and $\gamma(t_0) = \gamma'(t_0)$, then there is an open interval I_0 around t_0 in $I \cap I'$ such that $\gamma(t) = \gamma'(t)$, for all $t \in I_0$.*

If $\gamma\colon I \to M$ and $\gamma'\colon I' \to M$ are arbitrary solution curves of v, the proposition implies that $J = \{\, t \in I \cap I' \mid \gamma(t) = \gamma'(t) \,\}$ is an open subset of

$I \cap I'$. The set J is also closed in $I \cap I'$ because of the continuity of γ and γ'. Now $I \cap I'$ is connected, hence $J = \emptyset$ or $J = I \cap I'$. That is, Proposition B.1 implies the following **global uniqueness**.

(B.2) Corollary. *If $\gamma\colon I \to M$ and $\gamma'\colon I' \to M$ are arbitrary solution curves of $v \in \mathrm{C}^1$ and if $\gamma(s) = \gamma'(s)$, for some $s \in I \cap I'$, then $\gamma(t) = \gamma'(t)$, for all $t \in I \cap I'$.*

For given $s \in \mathbf{R}$, $x \in M$, denote by $I(s, x, v)$ the union of all the domains of definition I of solution curves $\gamma\colon I \to M$ of v satisfying $s \in I$, $\gamma(s) = x$. If $t \in I(s, x, v)$, then there exists a $\gamma\colon I \to M$ as before such that $t \in I$; and we write:

$$\Phi_v^{t,s}(x) := \gamma(t). \tag{B.3}$$

This definition does not depend on the choice of the solution curve $\gamma\colon I \to M$ of v satisfying s, $t \in I$, $\gamma(s) = x$, in view of Corollary B.2. The mapping $t \mapsto \Phi_v^{t,s}(x)$ is a solution curve of v with:

$$\Phi_v^{s,s}(x) = x; \tag{B.4}$$

and clearly it is the *maximal solution curve with initial value x for $t = s$*. If we write:

$$M_v^{t,s} = \{\, x \in M \mid t \in I(s, x, v) \,\}, \tag{B.5}$$

then $\Phi_v^{t,s}\colon x \mapsto \Phi_v^{t,s}(x)\colon M_v^{t,s} \to M$ is said to be the *flow of v from time s to time t*.

If $x \in M_v^{s,u}$, then $t \mapsto \Phi_v^{t,s}(\Phi_v^{s,u}(x))$ is a solution curve of v which at time $t = s$ has the value $\Phi_v^{s,u}(x)$. Since $t \mapsto \Phi_v^{t,u}(x)$ has the same properties, we read off from Corollary B.2 that:

$$\begin{aligned} &\text{if} \quad x \in M_v^{s,u}, \quad \Phi_v^{s,u}(x) \in M_v^{t,s}, \quad \text{then} \quad x \in M_v^{t,u} \\ &\qquad\qquad \text{and} \quad \Phi_v^{t,u}(x) = \Phi_v^{t,s}\bigl(\Phi_v^{s,u}(x)\bigr). \end{aligned} \tag{B.6}$$

This is called the *group property* of the flows. Exploiting the fact that (B.2) is autonomous, we also see that if $x \in M_v^{t,s}$ and $\tau \in \mathbf{R}$, then $x \in M_v^{t-\tau,s-\tau}$ and $\Phi_v^{t,s}(x) = \Phi_v^{t-\tau,s-\tau}(x)$. The reason is that, as a function of t, both sides are solution curves for v, with the same value x at $t = s$. Reading this for $\tau = s$, we see that it is sufficient to know:

$$M_v^t := M_v^{t,0}, \qquad \Phi_v^t := \Phi_v^{t,0}\colon M_v^t \to M. \tag{B.7}$$

The mapping Φ_v^t is said to be the *time t flow of v*; and the $\Phi_v^{t,s}$ are found back from these by means of the formula:

$$\Phi_v^{t,s} = \Phi_v^{t-s}, \tag{B.8}$$

while (B.6) gets translated into:

$$\text{(B.9)}\qquad \begin{aligned} &\text{if}\quad x \in M_v^p,\quad \Phi_v^p(x) \in M_v^q,\quad \text{then}\quad x \in M_v^{p+q} \\ &\qquad\qquad \text{and}\quad \Phi_v^{p+q}(x) = \Phi_v^q\big(\Phi_v^p(x)\big). \end{aligned}$$

So apart from the complications with the domains of definition, this just says that $t \mapsto \Phi_v^t$ is a homeomorphism from $(\mathbf{R}, +)$ to the group of transformations in M.

Denote by $\mathcal{X}^k(M)$ the space of C^k vector fields on M, provided with the C^k topology of locally uniform convergence of all derivatives up to the order k, in local coordinate patches. For $k = \omega$, this has to be replaced by uniform convergence of the complex-analytic extensions of v to some fixed neighborhood in $\mathbf{C}^n$ of the coordinate patches in $\mathbf{R}^n$. We are now ready to formulate the main theorem of this appendix.

(B.3) Theorem. *We have the following results.*

(i) $\Omega^1 = \{\, (t, x, v) \in \mathbf{R} \times M \times \mathcal{X}^1(M) \mid x \in M_v^t \,\}$ *is an open subset of* $\mathbf{R} \times M \times \mathcal{X}^1(M)$*, which contains* $\{0\} \times M \times \mathcal{X}^1(M)$.

(ii) $(t, x, v) \mapsto \Phi_v^t(x)$ *is a* C^k *mapping from the open subset* $\Omega^k = \Omega^1 \cap (\mathbf{R} \times M \times \mathcal{X}^k(M))$ *in* $\mathbf{R} \times M \times \mathcal{X}^k(M)$ *to* M.

(iii) If $\epsilon \mapsto v_\epsilon$ *is differentiable:* $J \to \mathcal{X}^1(M)$*, with* J *an interval in* $\mathbf{R}$*, then we have the* variational formula*:*

$$\text{(B.10)}\qquad \frac{\partial}{\partial \epsilon}\Phi_{v_\epsilon}^t(x) = \int_0^t \mathrm{T}_{\Phi_{v_\epsilon}^s(x)}\big(\Phi_{v_\epsilon}^{t-s}\big)\,\frac{\partial v_\epsilon}{\partial \epsilon}\big(\Phi_{v_\epsilon}^s(x)\big)\, ds \in \mathrm{T}_{\Phi_{v_\epsilon}^t(x)}\, M.$$

(iv) If T *is a finite boundary point of* $I(s, x, v)$*, then there exists, for each compact subset* K *of* M*, a neighborhood* I *of* T *in* $\mathbf{R}$ *such that* $\Phi_v^t(x) \notin K$ *for all* $t \in I \cap I(s, x, v)$*. In particular, if* M *is compact, then* $\Omega^1 = \mathbf{R} \times M \times \mathcal{X}^1(M)$.

Proposition B.1 and Theorem B.3 are based on the following local version, in a local coordinate patch U, where U is identified with an open subset of $\mathbf{R}^n$, and $\mathrm{T}U = U \times \mathbf{R}^n$, $v(x) = \big(x, f(x)\big)$ for some $f \in \mathrm{C}^k(U, \mathbf{R}^n)$. An important feature is that we keep explicit control over the size of the domains of definition.

(B.4) Lemma. *Fix* $x_0 \in \mathbf{R}^n$, $r > 0$, $B = \{x \in \mathbf{R}^n \mid \|x - x_0\| \le r\}$ *and* $f_0 \in \mathrm{C}^1(B, \mathbf{R}^n)$*. For arbitrary* $f \in \mathrm{C}^1(B, \mathbf{R}^n)$*, write:*

$$M_0(f) = \sup\{\, \|f(x)\| \mid x \in B \,\}, \qquad M_1(f) = \sup\{\, \|\mathrm{D} f(x)\| \mid x \in B \,\}.$$

Choose $\delta > 0$ *such that* $\delta\, M_0(f_0) < r$ *and* $\delta\, M_1(f_0) < 1$*. Then, for each* $x \in \mathbf{R}^n$ *and* $f \in \mathrm{C}^1(B, \mathbf{R}^n)$ *such that* $\|x - x_0\| + \delta\, M_0(f) < r$ *and* $\delta\, M_1(f) < 1$*, there is a unique differentiable solution* $\gamma\colon\ ]-\delta, \delta\,[\to \mathbf{R}^n$ *of:*

$$\text{(B.11)} \qquad \frac{d\gamma}{dt}(t) = f(\gamma(t)) \qquad (|t| < \delta),$$

with the initial value:

$$\text{(B.12)} \qquad \gamma(0) = x.$$

If we restrict f further to $\mathrm{C}^k(B, \mathbf{R}^n)$, then the mapping assigning to these x and f the solution γ is of class C^k. Finally, if f depends differentiably on the parameter ϵ, then with obvious notations:

$$\text{(B.13)} \qquad \frac{\partial}{\partial \epsilon}\gamma(t, x, \epsilon) = \int_0^t \frac{\partial \gamma}{\partial x}(t - s, \gamma(s, x, \epsilon), \epsilon)\frac{\partial f}{\partial \epsilon}(\gamma(s, x, \epsilon), \epsilon)\, ds.$$

Proof. Integrating (B.11) and (B.12), for differentiable $\gamma\colon\,]-\delta, \delta[\, \to \mathbf{R}^n$, from 0 to t we obtain the integral equation:

$$\text{(B.14)} \qquad \gamma(t) = x + \int_0^t f(\gamma(s))\, ds =: (F(\gamma, x, f))(t) \qquad (|t| < \delta).$$

Conversely, if $\gamma\colon\,]-\delta, \delta[\, \to \mathbf{R}^n$ is continuous and satisfies (B.14), then it is differentiable (because the integral is a differentiable function of t), and differentiation yields (B.11). Also, reading (B.14) for $t = 0$ we get (B.12).

Let Γ be the Banach space of mappings $\gamma\colon\,]-\delta, \delta[\, \to \mathbf{R}^n$ that are bounded and continuous, provided with the supremum norm. Denote by $\underline{x}_0$ the mapping on $]-\delta, \delta[$ that is constant equal to x_0. Then:

$$\mathcal{B} := \{\, \gamma \in \Gamma \mid \|\gamma - \underline{x}_0\| \leq r \,\}$$

is a convex closed subset of Γ, while $F(\gamma, x, f)$ in (B.14) is defined for $\gamma \in \mathcal{B}$, $x \in \mathbf{R}^n$ and $f \in \mathrm{C}^1(B, \mathbf{R}^n)$. We have the straightforward estimates:

$$\text{(B.15)} \qquad \|F(\gamma, x, f) - \underline{x}_0\| \leq \|x - x_0\| + \delta\, M_0(f);$$

$$\text{(B.16)} \qquad \|F(\gamma, x, f) - F(\gamma', x, f)\| \leq \delta\, M_1(f)\|\gamma - \gamma'\| \qquad (\gamma, \gamma' \in \mathcal{B}).$$

We see that the assumptions for x, f imply that $\gamma \mapsto F(\gamma, x, f)$ maps $\mathcal{B}$ into $\mathcal{B}$. Furthermore, $\gamma \mapsto F(\gamma, x, f)$ is a contraction in the complete metric space $\mathcal{B}$. It follows that there is a unique $\gamma \in \mathcal{B}$ such that $\gamma = F(\gamma, x, f)$. Translating this back we have shown the existence of the unique solution γ of (B.11) and (B.12).

From Definition (B.14) of F we immediately read off that F is a C^k mapping from $\mathcal{B} \times \mathbf{R}^n \times \mathrm{C}^k(B, \mathbf{R}^n)$ to Γ. From (B.16) (or by a direct computation) it follows that the operator norm of the total derivative $\frac{\partial F}{\partial \gamma}(\gamma, x, f)$ of $\gamma \mapsto F(\gamma, x, f)\colon \mathcal{B} \to \Gamma$ is dominated by $\delta\, M_1(f)$; and hence $\frac{\partial}{\partial \gamma}(\gamma - F(\gamma, x, f)) = \mathrm{I} - \frac{\partial F}{\partial \gamma}(\gamma, x, f)$ is invertible when $\delta\, M_1(f) < 1$. The usual smoothness conclusion in the implicit function theorem in Banach spaces (see Dieudonné [1969], Theorem 10.2.1) now gives that the solution γ

of $\gamma - F(\gamma, x, f) = 0$ depends in a C^k fashion on the parameters (x, f). For $k = \omega$, apply the implicit function theorem in the complex-analytic setting, to the complex extension of the vector field.

Replacing γ in (B.14) by $t \mapsto \gamma(t, x, \epsilon)$ and differentiating with respect to ϵ, we get:

$$\frac{\partial \gamma}{\partial \epsilon}(t, x, \epsilon) = \int_0^t \Big(\frac{\partial f}{\partial x}\big(\gamma(s, x, \epsilon), \epsilon\big) \frac{\partial \gamma}{\partial \epsilon}(s, x, \epsilon) + \frac{\partial f}{\partial \epsilon}\big(\gamma(s, x, \epsilon), \epsilon\big) \Big)\, ds.$$

Differentiating this equality with respect to t, we see that $\frac{\partial \gamma}{\partial \epsilon}$ satisfies the inhomogeneous linear differential equation:

(B.17)

$$\frac{d}{dt}\frac{\partial \gamma}{\partial \epsilon}(t) = \frac{\partial f}{\partial x}\big(\gamma(t, x, \epsilon), \epsilon\big) \frac{\partial \gamma}{\partial \epsilon}(t, x, \epsilon) + \frac{\partial f}{\partial \epsilon}\big(\gamma(t, x, \epsilon), \epsilon\big), \qquad \frac{\partial \gamma}{\partial \epsilon}(0, x, \epsilon) = 0.$$

Differentiating (B.11) with respect to x, we obtain:

$$\frac{d}{dt}\frac{\partial \gamma}{\partial x}(t, x, \epsilon) = \frac{\partial f}{\partial x}\big(\gamma(t, x, \epsilon), \epsilon\big) \frac{\partial \gamma}{\partial x}(t, x, \epsilon);$$

and therefore $t \mapsto \frac{\partial \gamma}{\partial x}(t, x, \epsilon)$ is a solution of the homogeneous equation associated with (B.17). The classical variation-of-constants formula of Lagrange yields (B.13) as the solution of (B.17). □

Proof of Proposition B.1. The assertion now follows from the uniqueness part of Lemma B.4. We may assume that $t_0 = 0$ (cf. (B.8)), $\gamma(0) = x_0 = \gamma'(0)$. By shrinking I_0 sufficiently we get $I_0 = \,]-\delta, \delta[$ with δ as in Lemma B.4; note that the assumptions remain valid if $\delta > 0$ is taken smaller. By taking I_0 sufficiently small, we also get that $\gamma(I_0) \subset B$, $\gamma'(I_0) \subset B$; so that $\gamma|_{I_0}$ and $\gamma'|_{I_0}$ are solutions as in Lemma B.4. □

Proof of Theorem B.3. The existence part of Lemma B.4 shows that the set Ω^1 in Theorem B.3 is a neighborhood of $\{0\} \times M \times \mathcal{X}^1(M)$ in $\mathbf{R} \times M \times \mathcal{X}^1(M)$. Combining this with (B.9), one verifies in a straightforward fashion that Ω^1 is an open subset of $\mathbf{R} \times M \times \mathcal{X}^1(M)$. In the same way, combining (B.9) with the chain rule for C^k mappings, one gets from the C^k dependence of the local solutions on (x, v), that $\Phi_v^t(x)$ depends in a C^k fashion on (x, v), with values in the space of continuous functions of t, for all $(t, x, v) \in \Omega^k$. The differentiability with respect to t now follows too, if we consider the differential equation:

$$\frac{d}{dt}\Phi_v^t(x) = v\big(\Phi_v^t(x)\big). \tag{B.18}$$

From this we read off that if $(t, x, v) \mapsto \Phi_v^t(x)$ is C^l, for $l \leq k$, then $(t, x, v) \mapsto \frac{d}{dt}\Phi_v^t(x)$ is C^l. This yields even one more degree of differentiability in t than stated in the theorem. In the case of $k = \omega$ one has to show that $t \mapsto \Phi_v^t(x)$

has a complex-analytic extension to complex time. This can be demonstrated by replacing t by $z\,t$, for $z \in \mathbf{C}$, $t \in \mathbf{R}$, which amounts to replacing v by $z\,v$, and treating z as a complex parameter on which the solution will depend in a complex-analytic way.

The variational formula (B.10) is just (B.13) if $|t|$ is sufficiently small. The extension to all $(t, x, v) \in \Omega^1$ again follows by "continuous induction on t", if we use (B.9) and the chain rule. Indeed, suppose (B.10) holds. Applying (B.10) with x, and t, replaced by $\Phi_v^t(x)$, and τ with $|\tau| < \delta$, respectively, we get:

$$\begin{aligned}
\frac{d}{d\epsilon}\Big|_{\epsilon=0} \Phi_{v+\epsilon\,w}^{t+\tau}(x) &= \frac{d}{d\epsilon}\Big|_{\epsilon=0} \Phi_{v+\epsilon\,w}^{\tau}\big(\Phi_{v+\epsilon\,w}^{t}(x)\big) \\
&= \frac{d}{d\epsilon}\Big|_{\epsilon=0} \Phi_{v+\epsilon\,w}^{\tau}\big(\Phi_v^t(x)\big) + \mathrm{T}_{\Phi_v^t(x)}\big(\Phi_v^\tau\big)\,\frac{d}{d\epsilon}\Big|_{\epsilon=0} \Phi_{v+\epsilon\,w}^{t}(x) \\
&= \int_0^\tau \mathrm{T}_{\Phi_v^\sigma(\Phi_v^t(x))}\big(\Phi_v^{\tau-\sigma}\big)\,w\big(\Phi_v^\sigma(\Phi_v^t(x))\big)\,d\sigma \\
&\quad + \mathrm{T}_{\Phi_v^t(x)}\big(\Phi_v^\tau\big)\int_0^t \mathrm{T}_{\Phi_v^s(x)}\big(\Phi_v^{t-s}\big)\,w\big(\Phi_v^s(x)\big)\,ds \\
&= \int_0^\tau \mathrm{T}_{\Phi_v^{t+\sigma}(x)}\big(\Phi_v^{t+\tau-(t+\sigma)}\big)\,w\big(\Phi_v^{t+\sigma}(x)\big)\,d\sigma \\
&\quad + \int_0^t \mathrm{T}_{\Phi_v^s(x)}\big(\Phi_v^{t+\tau-s}\big)\,w\big(\Phi_v^s(x)\big)\,ds \\
&= \int_0^{t+\tau} \mathrm{T}_{\Phi_v^s(x)}\big(\Phi_v^{t+\tau-s}\big)\,w\big(\Phi_v^s(x)\big)\,ds.
\end{aligned}$$

Statement (iv) in the theorem expresses that solutions are not globally defined if and only if they run off to infinity in a finite time. Examples, such as $\frac{dx}{dt}(t) = x(t)^2$ with $x(t) \in \mathbf{R}$, show that in noncompact manifolds this easily can occur. (Therefore it is a nontrivial statement that in a Lie group the solution curves of the left, and right, invariant vector fields exist globally.)

For the proof, we may assume that $s = 0$. Suppose that there are a sequence $t_i \in I(0, x, v)$ and a compact subset K of M such that $\lim_{i\to\infty} t_i = T$ and $\Phi_v^{t_i}(x) \in K$ for all i. Let U be a compact neighborhood of K in M. We first show that there is a neighborhood I of T such that $\Phi_v^t(x) \in U$, for all $t \in I \cap I(0, x, v)$. Otherwise one can find t, $s \in I(0, x, v)$, arbitrarily close to T, such that $\Phi_v^s(x) \in K$, $\Phi_v^t(x) \in \partial U$, the boundary of U in M, while $\Phi_v^u(x) \in U$, for all u between s and t. The boundedness of v on U yields a uniform bound for $\frac{d}{du}\Phi_v^u(x)$; and because s and t get arbitrarily close to each other, $\Phi_v^s(x)$ and $\Phi_v^t(x)$ approach each other as s, $t \to T$. But this is a contradiction with $\Phi_v^s(x) \in K$, $\Phi_v^t(x) \notin U^{\mathrm{int}} \supset K$. Next the fact that $\Phi_v^t(x) \in U$, for all $t \in I \cap I(0, x, v)$, combined with the boundedness of v on U, yields that $t \mapsto \Phi_v^t(x)$ is a Cauchy sequence as $t \to T$; so it converges to some $y \in U$. Applying the local existence theorem at y, we see that the solution $t \mapsto \Phi_v^t(x)$ can be extended beyond $t = T$, in contradiction with the assumption on T. $\square$

References for Appendix

Dieudonné, J. (1969) Foundations of Modern Analysis, Second Ed. Academic Press, New York London

Dieudonné, J. (1972) Treatise on Analysis, Volume III. Academic Press, New York San Francisco London

Hirzebruch, F. (1966) Topological Methods in Algebraic Geometry, Third Enlarged Edition. Springer-Verlag, New York

Kobayashi, S., Nomizu, K. (1963) Foundations of Differential Geometry, Vol. I. Interscience Publishers, New York London

Spivak, M. (1979-I) A Comprehensive Introduction to Differential Geometry, Vol. 1, 2nd Ed. Publish or Perish, Berkeley

Subject Index

Made in the USA
Monee, IL
28 April 2026

49137072R00197